Optical Bistability

Optical Bistability

Edited by

Charles M. Bowden

US Army Missile Laboratory
Redstone Arsenal, Alabama

and

Mikael Ciftan

and

Hermann R. Robl

US Army Research Office
Research Triangle Park, North Carolina

Plenum Press · New York and London

Library of Congress Cataloging in Publication Data

International Conference on Optical Bistability (1980: Asheville, N.C.)
  Optical bistability.

  "Invited papers presented at the International Conference on Optical Bistability,
held June 3-5, 1980, in Asheville, North Carolina" — Verso of t.p.
  Bibliography: p
  Includes index.
  1 Optical bistability — Congresses. 2. Nonlinear optics—Congresses. I. Bowden,
Charles M. II. Ciftan, M. III. Robl, Hermann R. IV. Title.
QC446.15 I6 1980   535'.2   81-2559
ISBN 0-306-40722-1   AACR2

Invited papers presented at the International Conference on Optical Bistability,
held June 3—5, 1980, in Asheville, North Carolina

© 1981 Plenum Press, New York
A Division of Plenum Publishing Corporation
233 Spring Street, New York, N.Y. 10013

Printed in the United States of America

This volume is dedicated to the memory of

MARIO GRONCHI

Born in Como, Italy, on December 2, 1942; received his degree
in Physics in 1968 at the Univeristy of Pisa. Served at the
University of Milan as Professor of Quantum Electronics; made
distinct contributions in the field of Quantum Optics, especially to the subjects of fluctuations in laser systems, superfluorescence, and optical bistability. Died in Milan on June 5, 1980.

PREFACE

During the past few decades we have witnessed at least two major innovations in science which have had substantial impact on technology as well as science itself, pervasive enough to modify many facets of our daily lives. We refer, of course, to the transistor and the laser. It is striking that now with the advent of optical bistability we may have opened the door to another such field, which combines these two aspects (transistor and laser) and has the possibility for important device applications as well as providing a unique window into the as yet not thoroughly explored frontiers of nonequilibrium statistical physics.

This has prompted us to organize an international conference on the subject of optical bistability to provide an adequate means for assessing the current state of the art of this important field and to stimulate further significant developments by means of intense technical exchange and interaction among the leading scientists in this subject area.

This volume is a collection of the invited papers presented at the International Conference on Optical Bistability in Asheville, North Carolina, June 3-5, 1980. The conference was sponsored jointly by the U.S. Army Research Office in Durham, North Carolina, and by the U.S. Army Missile Command, Redstone Arsenal, Alabama. The conference was attended by more than forty active leading scientists in the important and rapidly-expanding field of optical bistability from various countries and various parts of the United States. The conference was organized and administrated by a committee consisting of the editors of this volume.

So persuasive are the arguments presented at this conference in support of the combined scientific and technological potential of the rapidly-developing field of optical bistability that it is not difficult to perceive that many useful devices may emerge which indeed combine certain unique and beneficial features of both the transistor and the laser for numerous innovative applications. This conclusion may also be gleaned from the panel discussion at the conference, which brings forth the critical aspects of this subject yet to be fully investigated and understood. We have made every effort to retain the flavor and workshop nature of this international conference by including the panel discussion in detail, the "panel" being all the participants who were present that last day. The organizers believe that inclusion of the discussion will serve as a valuable guide to future research.

The organizers wish to thank all the participants for a most stimulating and fruitful conference. We are also grateful to each contributor to this volume, which we believe is a significant contribution to the future development of the field.

C.M. Bowden
M. Ciftan
H.R. Robl

15 September 1980

CONTENTS

CONTENTS

CONDITIONS AND LIMITATIONS IN INTRINSIC OPTICAL BISTABILITY

S. L. McCall and H. M. Gibbs[*]

Bell Laboratories
Murray Hill, N.J. 07974

First, we shall review the conditions for achieving optical
bistability in plane Fabry-Perot structures, and then mention ex-
pectations for cases when the optical cavity is more general. Using
present experimental results, estimates of the minimum switching
energies, times, etc., are made. Hybrid devices are not considered.

In the absorptive case, the state equation is

$$E_I = E_T + \Gamma v \tag{1}$$

where E_I and E_T are the input and output fields, suitably normalized
(e.g., Ref.1). The out-of-phase polarization v, which is a current, depends on the cavity-enclosed material. We normalize v so
that $v/E_T \to 1$ as $E_T \to 0$, and then $\Gamma = \alpha L/T$, where α is the absorption
constant of the material, L the length, and T the mirror transmis-
sion.

If, for example

$$v = E_T/(1 + E_T^2) \;, \tag{2}$$

a simple absorber result, then $\Gamma > 8$ allows bistability[2]. In general

$$\Gamma^{-1} < |dv/dE_T| \tag{3}$$

is required[1], where dv/dE_T is assumed to be negative in some region,
and the most negative value is used in Eq. 3.

---

[*]Present Address: Optical Sciences Center, University of Arizona,
Tucson, AZ. 85721.

In the purely dispersive case, a result similar in appearance applies. The state equation[1] is

$$E_I = E_T + i\Gamma u + i\beta E_T \tag{4}$$

where u is the in-phase-polarization, and β a detuning parameter. It then follows that

$$\frac{T}{2L} < P_T \left| \frac{d\Delta n}{dP_T} \right| \tag{5}$$

is required for bistability, where u is written in terms of a non-linear refractive index.

For the mixed case, the state equation[1] is

$$E_I = E_T + i\beta E_T + \Gamma(v - iu) . \tag{6}$$

Writing $u = \sigma_R E_T$, $u = \sigma_I E_T$, it has been shown[3] that

$$\Gamma^{-1} < \mathrm{Sup}\left[- \frac{d}{dP_T}(\sigma_R P_T) + P_T \sqrt{\left(\frac{d\sigma_R}{dP_T}\right)^2 + \left(\frac{d\sigma_I}{dP_T}\right)^2} \right] , \tag{7}$$

is the bistability condition, where $P_T = (E_T)^2$.

The above results all seem to neglect standing waves, and are all in the limit $T \to 0$, $\alpha L \to 0$, $\alpha L/T$ finite. To include standing waves, we define[4] σ_0 , $\sigma_{\pm 2k}$, $\sigma_{\pm 4k}$ by

$$u + iv = (\sigma_0 + \sigma_{2k} e^{2ikz} + \ldots)(E_F e^{ikz} + E_B e^{-ikz}) , \tag{8}$$

where $u + iv$ here only varies rapidly on a wavelength scale, and E_F and E_B represent intracavity fields. Then define[4]

$$\sigma_R + i\sigma_I = \sigma_0 + \sigma_{2k} . \tag{9}$$

In the αL, $T \to 0$ limit, then everything works out as before.[3]

In particular, if we choose a model wherein

$$v - iu = \frac{(1 - i\Delta\omega T_2)E_T}{1 + \Delta\omega^2 T_2^2 + (E_T)^2} \tag{10}$$

then inequality (7) yields

$$\Gamma > 4\left(1 + \sqrt{1 + \Delta\omega^2 T_2^2}\right) .$$ (11)

The standing wave result is more complicated, as is shown with
(--) in Fig. 1. Essentially, standing wave effects require larger
Γ by 20-25%. Switch-up and switch-down powers are not changed
much at the input, but are lower by roughly a factor of two at the
output, as shown in Fig. 2.

In more complicated cavities, there are two effects. One is
that either the transverse or longitudinal characteristics of the
cavity mode may change. The other is that suitable averaging must
be done. Assume the mode is forced to be Gaussian. For a simple
absorber with no standing wave effects, the differential conductivity[4] is

$$-\frac{1}{2}\left[E_T^2 \ln(1 + E_T^2) - 2/(1 + E_T^2)\right]$$

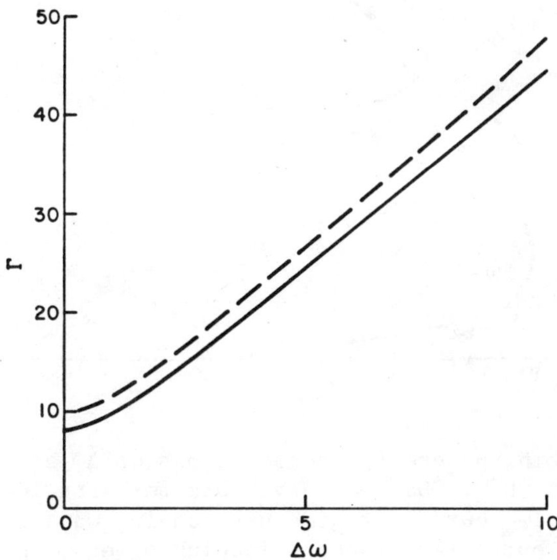

Fig. 1.   Required values of $\Gamma = \alpha L/T$ as a function of $\Delta\omega$. The
          lower solid curve is for the no-standing wave case,
          while the dashed curve is for the standing wave model.

with a maximum negative conductivity of 0.0602 for $E_T \approx 3.65$, in-stead of the more familiar 0.125 at $E_T = \sqrt{3}$. For mixed or disper-sive bistability, one should take into account[5] the transverse changes of the cavity mode.

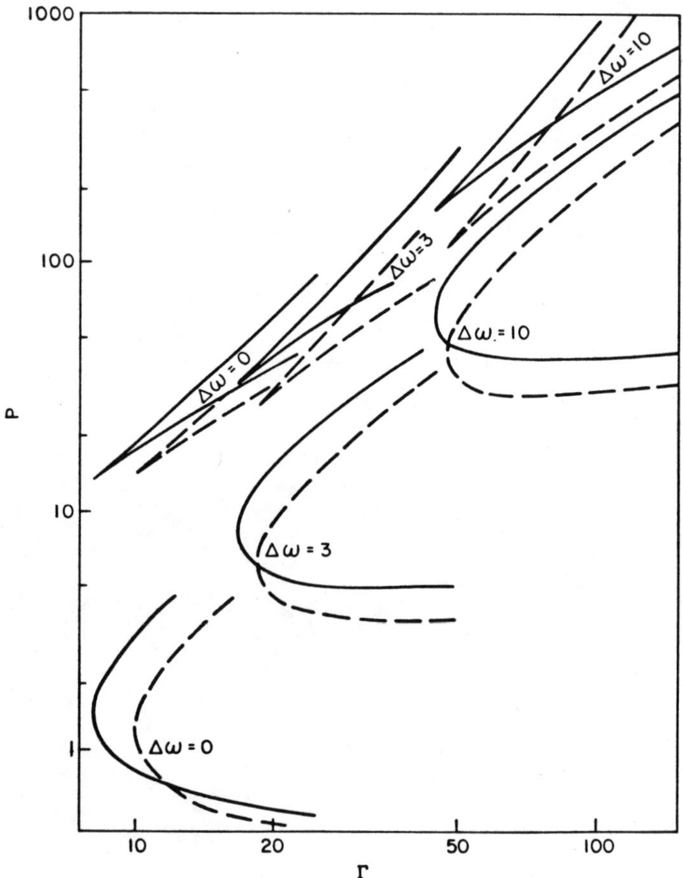

Fig. 2.   Switching powers for three values of $\Delta\omega$ as a function of $\Gamma = \alpha L/T$. Dashed curves are for standing wave models and solid curves are for no-standing wave models. The upper curves are input switching powers, and the lower curves are output switching powers. Recalling that powers are defined by $P_I = (E_I)^2/T$ and $P_T = (E_T)^2/T$, the variation of $\Gamma = \alpha L/T$ should be regarded as due to αL changing and T constant.

One procedure for understanding transverse mode-changing effects, required probably in many cases, is computer analysis of the problem including diffraction, transverse, etc., effects. This approach will be discussed later in the conference by Mattar.

In a true sense, the mixed and dispersive case with standing waves in a plane Fabry-Perot cavity includes cavity-mode changes. When one speaks of an optical cavity about a half-wave in dimension, both transverse and longitudinal cavity changes should be included.

Although standing-wave effects hurt bistability when working near an absorbing saturable resonance, it is not clear that other mode-changing effects will be detrimental. Certainly in purely dispersive bistability, when the medium is a Kerr medium, mode-changing effects help bistability, but so do standing-wave effects.[5] As an aside, the non-linear interface bistability effect[6], though potentially important, is not considered here.

So far all the calculations spoken of above refer to steady-state. It is true that negative slope regions are unstable. It is not true that positive slope regions are always stable. This was first shown for the short cavity case and demonstrated using a hybrid device.[7] Under suitable conditions, the regenerative oscillations had a surprisingly constant frequency. For the simple absorber short cavity case, positive slope regions are stable, working either on resonance or in mixed mode. If the cavity length is allowed to be a length roughly corresponding to a Rabi frequency, regenerative oscillations can occur in the ring-cavity plane-wave case.[8]

If a positive slope region is unstable, the device can either oscillate or prematurely switch. This can happen when, for example, thermal effects are mixed with electronic effects. Regenerative oscillations can be used to convert D.C. light to pulsed light.[7]

Next, the limit of small, fast, low-power bistable optical devices will be examined. First, two theoretical limits are established, and then extrapolations from present results are made. Then comparison is made with existing and futuristic semiconductor and Josephson technology. It is clear today that for transmitting information, light is better. A 5-psec light pulse has been transmitted undistorted for a kilometer through a fiber.[9] Try that with an electric pulse.

The first theoretical limit is based on semiclassical equations. For bistability one needs $\alpha L/T > 8$. An optical cavity can have a waist about λ^2 in area. Practically, choose $T = .05$, so $\alpha L \sim 1$ is needed. Use a single atom whose absorption line is only broadened by lifetime effects. The absorption cross section is about λ^2. Then semiclassically, this device should be bistable.

The second theoretical limit is based on the first. The first will "glitch" too often. Consequently, one needs about 1000 two-level atoms, or the statistical equivalent. This amounts to a switching energy (using 1.5 eV photons) of 2.4×10^{-16} joules, about one-fourth femtojoule.

Can this be achieved? We have results on GaAs[10] and InSb[11] devices available. The n_2 in GaAs (0.4 cm^2/KW) is several times the n_2 in InSb (0.1 cm^2/KW). Also, more information is available to us about GaAs, so the extrapolation will be based on GaAs results.

The measured switching energy in the GaAs device was 600,000 fJ over $(50\ \mu)^2$. The wavelength of light in the material is 2300 Å, so that an optical cavity could concentrate the light into an active diameter of about $\frac{1}{4}\ \mu$. The switching energy would then be 15 fJ.

The active length of the GaAs device was $4.5\ \mu$. Peak absorptivities of $7\ \mu^{-1}$ have been observed in good GaAs. The GaAs device did not have this high absorptivity. With such absorptivity, one should need only about $2/7\ \mu$ length of active material. The switching energy should then reduce to 0.95 fJ. That is about four times the theoretical limit given above.

By using lower-frequency light, the switching energy might be reduced. The light photon energy must be several factors of ten times the operating temperature, however, In fact, the limit of 1000 quanta-switch evidently applies to semiconductor and Josephson devices also, yielding limits of about $\frac{1}{4}$ fJ and about 10^{-18} joules, respectively.

Can room-temperature operation be achieved? Bistability was achieved up to 120°K. Furthermore, with use of superlattices, the exciton feature can be moved further[12] from the band absorption edge. Lifetimes can be shortened by introducing impurities, hopefully in a way that does not significantly broaden the exciton feature. The only other limitation is the cavity-build-up time. A cavity $1\ \mu$ long with $T = 0.1$ will take about 0.12 psec to build up.

How does this compare with futuristic semiconductor technology? Gates with transit times of 10 psec seem to be the present state of art, and 1 psec may be a limit governed by electron velocity and a required thickness of about 1000 Å. These speeds must be

typically multiplied by three factors of 3. One 3 is for fanout.
Another is for charging lines. Another is for additional capacity
unavoidably associated with the gate in use. Thus an effective
gate delay of about 27 psec seems to be a limit. Other semiconductors may reduce this time by a factor perhaps 3.

Josephson devices have been demonstrated which switch in 15
psec, and probably 1 psec is achievable. For bistable optical
devices, these factors either are not there or are already included.
If switching is done not through input mirrors, bistable optical
devices already have a fan-out capability.

We do not know that optical devices can out-perform semiconductor devices. We do not know that they cannot.

REFERENCES

1.   H. M. Gibbs, S. L. McCall, and T. N. C. Venkatesan, Phys.
     Rev. Lett. $\underline{36}$, 1135 (1976).
2.   A. Szöke, V. Daneu, J. Goldhar, and N. A. Kurnit, Appl. Phys.
     Lett. $\underline{15}$, 376 (1969).
3.   S. L. McCall and H. M. Gibbs, Opt. Commun., to be published.
4.   S. L. McCall, Phys. Rev. A$\underline{9}$, 1515 (1974).
5.   J. H. Marburger and F. S. Felber, Phys. Rev. A$\underline{17}$, 335 (1978).
6.   P. W. Smith, J. P. Hermann, W. J. Tomlinson, and P. J.
     Maloney, Appl. Phys. Lett. $\underline{35}$, 846 (1979).
7.   S. L. McCall, Appl. Phys. Lett. $\underline{32}$, 284 (1978).
8.   R. Bonifacio and L. A. Lugiato, Lett. Nuovo Cimento $\underline{21}$, 510
     (1978).
9.   D. M. Bloom, L. F. Mollenauer, C. Lin, D. W. Taylor, and
     A. M. DelGaudio, Opt. Lett. $\underline{4}$, 297 (1979).
10. H. M. Gibbs, S. L. McCall, T. N. C. Venkatesan, A. C. Gossard,
     A. Passner, and W. Wiegmann, in Digest of 1979 IEEE/OSA
     Conference on Laser Engineering and Applications (Institute
     of Electrical and Electronics Engineers, New York, 1979);
     in Laser Spectroscopy IV, H. Walther and K. W. Rothe, eds.
     (Springer-Verlag, Berlin, 1979); Appl. Phys. Lett. $\underline{35}$,
     451 (1979).
11. H. M. Gibbs, S. L. McCall, and T. N. C. Venkatesan, U. S.
     Patent 4,012,699 (1977) and U. S. Patent 4,121,167 (1978);
     D.A.B. Miller, S.D. Smith, and A. Johnston, Appl. Phys.
     Lett. $\underline{35}$, 658 (1979); D.A.B. Miller and S.D. Smith, Opt.
     Commun. $\underline{31}$, 101 (1979).
12. R. Dingle, Festkörperprobleme XV (Advances in Solid State
     Physics) (Braunschweig, Pergamon-Vieweg) 21, 1975.

SEMICLASSICAL AND QUANTUM STATISTICAL DRESSED MODE DESCRIPTION OF

OPTICAL BISTABILITY

V. Benza and L. A. Lugiato

Istituto di Fisica dell'Università
Via Celoria 16
20133 Milano, Italy

Abstract: Since the self-pulsing is a many-mode problem, it
cannot be described by the mean field theory of OB, even for $\alpha L \ll 1$,
$T \ll 1$. Hence it seems at first hopeless to obtain a simple description of this phenomenon. In fact, the direct numerical solution
of the Maxwell-Bloch equations (MBE) amounts to little more than a
crude registration of data, without any predictive power. In order
to get an insight into the behavior of these self-pulsing instabilities it is necessary to give an analytical or quasi-analytical
description. We have achieved this goal by elaborating a formalism
that we have called "dressed mode theory of OB." This formalism
is based on Haken's formulation of generalized Ginzburg-Landau
equations for Phase Transitions in systems far from thermal equilibrium. This method translates the MBE into an infinite set of
equations in time only for the mode variables. We call them
"dressed modes" because, even if each mode has a dominant field or
atomic character, they incorporate in part the atom-field interaction. By selecting the dressed modes that play the dominant role
and using the adiabatic elimination principle, we reduce the problem to a pair of coupled equations in time only, that describe
self-pulsing, fully including both nonlinearity and propagation.
The results obtained from these two equations agree very satis-
factorily with the data obtained from the numerical solutions of
the MBE. The picture of self-pulsing that arises from this bidimensional phase space is quite appealing and leads to new predictions. In particular, we find new types of hysteresis cycles,that
involve both cw and pulsing solutions. In this picture the system
appears as multistable rather than bistable, in the sense that some
of the stable states are cw, others are pulsing. This dressed-
mode formalism has also been extended to the quantum statistical
theory. Here the starting point is the many-mode master equation

for absorptive and dispersive OB that has been recently derived by one of us (LAL). This equation holds for a ring cavity provided $\alpha L \ll 1$, $T \ll 1$. Using the quantum statistical dressed mode treatment, we have studied the spectrum of transmitted light by considering *all* the modes of the cavity. Thanks to the flexibility of the formalism, we can obtain a quite general expression of the spectrum without adiabatically eliminating either the atomic or the field variables. We find that when the self-pulsing instability is approached the spectrum develops sidebands in correspondence to the modes that are going unstable. Thus once again the rise of an instability is heralded by the fluctuations of the system; crossing the instability threshold, the self-pulsing becomes manifest at a macroscopic level.

I.   INTRODUCTION

In treating a complicated many-body problem, one deals with a very large number of variables. The crucial point is the selection of the restricted set of variables that really control the system. This selection allows obtaining a reasonably simple description of the phenomenon, in terms of few variables only. In the case of optical bistability (OB) this goal is achieved via the so-called mean field limit, i.e., the double limit[1] $\alpha L \to 0$, $T \to 0$. In this limit one reduces to the mean field treatment (MFT), that is a *one-mode* theory[3]. In the MFT one deals with *very few equations in time only*, so that the problem can be treated analytically or quasi-analytically, in the sense that the numerical work is reduced to a minimum.

However there are situations, as in self-pulsing[2], in which *all the modes* of the cavity become relevant even in the mean field limit.   Therefore, in order to describe the self-pulsing one must give up the simplicity of the mean field model and rely directly on the Maxwell-Bloch equations. In this paper we illustrate an approach (the "dressed-mode" theory) that allows selection of the right variables even in the difficult case of self-pulsing, thereby reducing the problem to a set of very few equations, which are again functions of time only. This theory has produced an analytical or quasi-analytical treatment of self-pulsing, giving physical insight and leading to new predictions.

Recently, we have generalized the dressed-mode formalism to the quantum statistical case. The starting point for this generalization is provided by a many-mode master equation for OB (both absorptive and dispersive) in a ring cavity derived by one of us[5]. Using the quantum-statistical dressed-mode theory, we have treated the problem of describing the spectrum of transmitted light taking into account *all* the modes of the cavity, in order to find the behavior of this spectrum when one approaches the self-pulsing instability.

In Sec. 2 we describe the dressed mode formalism by reformulating the Maxwell-Bloch equations as a set of time evolution equations for the dressed-mode variables. Using the adiabatic elimination principle, in Sec. 3 we reduce the problem to a couple of differential equations in time only. In Sec. 4 we describe the results obtained from the analysis of these equations. The quantum-statistical dressed-mode theory, sketched in Sec. 5, is used in Sec. 6 to calculate the spectrum of transmitted light including all the modes of the cavity. Finally, in Sec. 7, we summarize the main results of the paper.

II.   THE DRESSED MODE FORMALISM

We consider a system of N two-level atoms embedded in a sample of length L and volume V contained in a ring cavity (see Fig. 1 in Ref. 2). A coherent field E_I of frequency ω_o is injected into the cavity and gets partially transmitted (E_T) and partially reflected (E_R). For simplicity, we assume that the atomic system is homogeneously broadened and that the transition frequency of the atoms coincides with the frequency ω_o of the incident field. Furthermore, we assume that the length of the cavity is equal to an integer number of wavelengths of the incident field. These assumptions imply that we consider purely absorptive OB. The dynamics of the coupled system, atoms plus radiation field, is described by the Maxwell-Bloch equations (MBE):

$$\frac{\partial E}{\partial t} + c \, \frac{\partial E}{\partial z} = -g \, S \, , \tag{2.1a}$$

$$\frac{\partial D}{\partial t} = - \, \frac{\mu}{\hbar} \, ES - \gamma_{\parallel} \left(D - \frac{N}{2} \right) \, , \tag{2.1b}$$

$$\frac{\partial S}{\partial t} = \frac{\mu}{\hbar} \, ED - \gamma_{\perp} S \, , \tag{2.1c}$$

where:
   $E(z,t)$ is the slowly varying envelope of the electric field;
   $D(z,t)$ is one half the difference between the population of
          the lower and the population of the upper level;
   $S(z,t)$ is the macroscopic polarization field;
   μ is the modulus of the dipole moment of the N atoms
   g is a coupling constant given by

$$g = \frac{4\pi\omega_o}{V} \, \mu \, , \tag{2.2}$$

$\gamma_\parallel = T_1^{-1}$ and $\gamma_\perp = T_2^{-1}$ are the longitudinal and transverse atomic relaxation rates.

The boundary condition for the electric field is

$$E(o,t) = \sqrt{T}\, E_I + RE(L,t - \Delta t) \quad . \tag{2.3a}$$

Furthermore, E_R, E_T, E_I and E are related by:

$$E(L,t) = E_T(t)/\sqrt{T} \, , \tag{2.3b}$$

$$E_R(t) = \sqrt{RT}\, E(L,t-\Delta t) - \sqrt{R}\, E_I = \sqrt{R}\, [E_T(t-\Delta t) - E_I], \tag{2.3c}$$

where $T = 1-R$ is the transmission coefficient of the upper mirrors and Δt is the time the light takes to travel from the exit mirror to the entrance mirror (see Fig. 1 of Ref. 2),

$$\Delta t = (L + 2\ell)/c \quad . \tag{2.4}$$

We take the fields E, D, S real; in fact, in absorptive OB the inclusion of the phases of E and S does not lead to new instabilities. The stationary solutions of (2.1) can be explicitly obtained[1]; in particular, in the double "mean field" limit:

$$\alpha L \ll 1, \quad T \ll 1 \quad , \quad \frac{\alpha L}{2T} \equiv C = \text{constant} \, , \tag{2.5}$$

where α is the linear absorption coefficient

$$\alpha = \frac{\mu g N}{2\hbar c \gamma_\perp} \, , \tag{2.6}$$

the solution reduces to

$$y = x + \frac{2Cx}{1 + x^2} \tag{2.7}$$

where y and x are the normalized amplitudes

$$y = \frac{\mu E_I}{\hbar\sqrt{\gamma_\perp \gamma_\parallel T}} \quad , \quad x = \frac{\mu E_T}{\hbar\sqrt{\gamma_\perp \gamma_\parallel T}} \quad .$$

Equation (2.7) shows a bistable behavior when $C > 4$ (Fig. 1). The dressed-mode formalism is based on Haken's theory of generalized Ginzburg-Landau equations for phase transition-like phenomena in systems far from thermal equilibrium. After its general formulation[6], this theory has been applied to the second threshold of the laser[7] and to chemical instabilities[8].

We[4] have recently simplified this method and extended it to the case of steady-state without spacial uniformity, as one

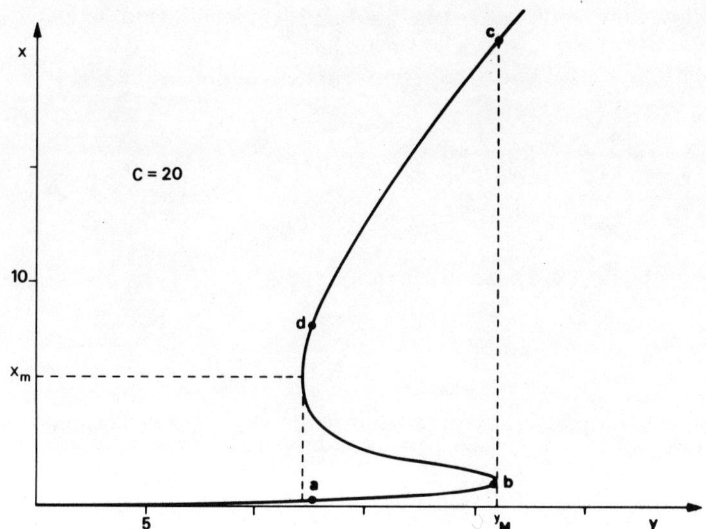

Fig. 1.　Hysteresis cycle of the normalized transmitted field x as a function of the normalized incident field y. The points a), b), c), d) are for future reference (see Fig. 9).

encounters in OB. First, we consider a stationary solution of the MBE $E_{st}(z)$, $D_{st}(z)$, $S_{st}(z)$ and introduce the deviations from the steady-state values:

$$e(z,t) = E(z,t) - E_{st}(z),$$

$$d(z,t) = D(z,t) - D_{st}(z), \qquad (2.8)$$

$$s(z,t) = S(z,t) - S_{st}(z).$$

Note that these deviations are not assumed small in general. By introducing Eq. (2.8) into (2.1), we reformulate the MBE as follows:

$$\frac{\partial e}{\partial t} = - c\, \frac{\partial e}{\partial z} - g s \;, \qquad (2.9a)$$

$$\frac{\partial d}{\partial t} = - \frac{\mu}{\hbar}\, (E_{st} s + S_{st} e) - \gamma_\parallel d - \frac{\mu}{\hbar}\, es \;, \qquad (2.9b)$$

$$\frac{\partial s}{\partial t} = \frac{\mu}{\hbar}\, (E_{st} d + D_{st} e) - \gamma_\perp s + \frac{\mu}{\hbar}\, ed \;. \qquad (2.9c)$$

If one drops the last term in Eqs. (2.9b) and (2.9c), one obtains the equations of the linear stability analysis[2]. However, in

order to describe self-pulsing we must go beyond the linearized
theory and therefore we must keep the nonlinear term in Eqs. (2.9b)
and (2.9c). Now, introducing the three-component vector

$$\underset{\sim}{q}(z,t) = \begin{pmatrix} e(z,t) \\ d(z,t) \\ s(z,t) \end{pmatrix} , \qquad (2.10)$$

we can write Eqs. (2.9) in a compact way:

$$\frac{\partial \underset{\sim}{q}}{\partial t} = \hat{L}\underset{\sim}{q} + \underset{\sim}{\psi}_{NL} , \qquad (2.11)$$

where $\hat{L}$ is the linear operator that appears in the linear part of
Eqs. (2.9):

$$\hat{L} = \begin{pmatrix} -c\,\dfrac{\partial}{\partial z} & 0 & -g \\[2ex] -\dfrac{\mu}{\hbar}\,S_{st} & -\gamma_{\parallel} & -\dfrac{\mu}{\hbar}\,E_{st} \\[2ex] \dfrac{\mu}{\hbar}\,D_{st} & \dfrac{\mu}{\hbar}\,E_{st} & -\gamma_{\perp} \end{pmatrix} \qquad (2.12)$$

while $\underset{\sim}{\psi}_{NL}$ contains the nonlinear terms:

$$\underset{\sim}{\psi}_{NL} = \begin{pmatrix} 0 \\[2ex] -\dfrac{\mu}{\hbar}\,es \\[2ex] \dfrac{\mu}{\hbar}\,ed \end{pmatrix} . \qquad (2.13)$$

Note that $\hat{L}$ is not only a matrix but is also a differential operator in the space variable z.

The linear stability analysis[2] amounts to the search of the
eigenvalues and of the eigenstates of $\hat{L}$. One finds that these
eigenstates are labeled by two indices $n = 0,\pm1,\pm2,\ldots$ and $j = 1,$
2,3. The index n labels the frequencies of the cavity, in particular n=0 corresponds to the resonant frequency.

$$\hat{L}\underset{\sim}{Q}_{nj} = \lambda_{nj}\,\underset{\sim}{Q}_{nj} . \qquad (2.14)$$

Let us consider the eigenvalues λ_{nj} in the limit $\alpha L \ll 1$, $T \ll 1$.
One has[2]

$$\lambda_{n1} = -i2\pi n \frac{c}{L} + \mathcal{O}(T) \ ,$$

$$\lambda_{n\genfrac{}{}{0pt}{}{2}{3}} = - \frac{1}{2} \left\{ \gamma_\perp + \gamma_\| \pm \sqrt{(\gamma_\perp - \gamma_\|)^2 - 4\gamma_\perp \gamma_\| x^2} \right\} \qquad (2.15)$$

$$+ \mathcal{O}(T) \ ,$$

where $L = 2(L + \ell)$ is the total length of the ring cavity. Hence
for j=1 the dominant part of the eigenvalues is equal to the difference between the nth cavity frequency and the resonant frequency.
For j=2,3, λ depends only on saturated atomic decay rates. Therefore the eigenvalues with j=2,3, have a dominant atomic character,
whereas the eigenvalues with j=1 have a dominant field character.

At this point, we say that the indices n and j label the
modes of our system: according to the previous discussion, we shall
call the modes with j=1 "field modes" and those with j=2,3 "atomic
modes." Despite this field or atomic dominant character, these
modes incorporate in part the atom-field interaction. Precisely,
it is the part of this interaction that is contained in the linearized equations. As we have seen, these modes arise naturally from
the linear stability analysis. Now we can expand the vector $\underset{\sim}{q}$ in
Eqs. (2.11) in terms of the eigenmodes $\underset{\sim}{O}_{nj}$ (see (2.14)):

$$\underset{\sim}{q}(z,t) = \sum_{nj} \xi_{nj}(t) \, \underset{\sim}{O}_{nj}(z) \ , \qquad (2.16)$$

the coefficients ξ of this expansion are the dressed mode variables,
that depend only on time. By inserting this expansion into Eqs.
(2.11) one easily obtains the time evolution equations for the mode
variables ξ_{nj}, i.e., the equations that govern dressed mode dynamics:

$$\dot{\xi}_{nj} = \lambda_{nj}\xi_{nj} + \sum_{n'n''} \sum_{j'j''} \Gamma(nj,n'j',n''j'')\xi_{n'j'}\xi_{n''j''} \ . \qquad (2.17)$$

The coefficients Γ are complicated functions of the parameters indicated and are explicitly given in Ref. 4. In Eq. (2.17) we can
distinguish a linear term and several quadratic terms. If we drop
the nonlinear terms, we obtain the exponential time evolution of
the linear stability analysis, which gives simple damping for the
stable modes and amplification for the unstable modes. In order
to describe self-pulsing we must keep the nonlinear terms that describe how the stable modes eventually stabilize the unstable ones
thereby producing the undamped spiking behavior.

In conclusion, the dressed mode formalism translates the
Maxwell-Bloch equations into a set of time evolution equations for
the dressed mode variables. Contrary to the MBE, these equations
contain only the derivative with respect to time.

III. ADIABATIC ELIMINATION

Until now, we have discussed only some mathematics. We have simply reformulated the original MBE (2.1) in an equivalent form (Eq. (2.17)). The advantage of this new formulation lies in the fact that now we can drastically reduce the number of variables involved. The first reduction is obtained by suitably selecting the dressed mode variables. Of course this selection is guided by physical intuition. For simplicity, let us consider the following situation: only the two field dressed modes $\lambda_{11}, \lambda_{-11}$ that are adjacent to the resonance frequency are unstable. In this case, as indicated by the numerical results the only variables that play an important role are the three dressed modes λ_{oj} corresponding to the resonant frequency, plus the six dressed modes $\lambda_{\pm 1j}$ corresponding to the adjacent frequencies.

This reduction strongly simplifies the problem, but is clearly not enough, because we have still to deal with nine differential equations in time. The second crucial step to simplify the problem is the adiabatic elimination of the rapidly varying variables. Haken[6] has given a general iterative procedure to perform the adiabatic elimination in an approximate way.

In collaboration with Meystre, we[9] have applied this perturbative method to our problem. It turns out that in our case this method works when the amplitude of the oscillations is sensibly smaller than that of the stationary field. In fact, only in this case do we find agreement with the numerical data obtained by directly solving the MBE. Furthermore, this method fails to give a description of the precipitation to the lower branch[2]. Therefore this procedure is limited and does not allow one to get a complete insight into the behavior of these instabilities.

On the other hand, we have realized that for small mirror transmittivity the adiabatic elimination can be performed practically in an exact way, without relying on any perturbative scheme. In such a way, we give an example of a nontrivial bifurcation problem that is treated in an exact way. Therefore let us consider directly this exact adiabatic elimination.

First, one sees that for $\alpha L \ll 1$, $T \ll 1$, the atomic modes vary on a time scale much shorter than the field ones. Hence one can adiabatically eliminate the atomic modes, and the field modes play the role of order parameters. This elimination can be performed exactly thanks to a fortunate circumstance. In fact, let us indicate by ξ_F and ξ_A the generic field and atomic mode variable, respectively. The time evolution equation for ξ_A has the following structure:

$$\dot{\xi}_A = \lambda_A \xi_A + \Gamma_{AF}\xi_A\xi_F + \Gamma_{FF}\xi_F\xi_F + \Gamma_{AA}\xi_A\xi_A \quad . \tag{3.1}$$

One finds that for $\alpha L \ll 1$, $T \ll 1$

$$\lambda_A, \ \Gamma_{AF}, \ \Gamma_{FF} = \mathcal{O}(T^o),$$
$$\Gamma_{AA} = \mathcal{O}(T) \quad . \tag{3.2}$$

Thus, the term with Γ_{AA} in (3.1) can be dropped. Hence, by setting $\dot{\xi}_A = 0$ (adiabatic elimination of the atomic modes) Eq. (3.1) becomes

$$0 = \lambda_A \xi_A + \Gamma_{AF}\xi_A\xi_F + \Gamma_{FF}\xi_F\xi_F \quad . \tag{3.3}$$

This system of equations is linear with respect to ξ_A, so that one easily obtains the explicit expression of the atomic modes ξ_A in terms of the field modes ξ_F. By substituting these expressions into the equations for the field modes, we obtain closed time evolution equations for the field modes alone. Precisely, these modes are ξ_{11}, ξ_{01} and ξ_{-11}. Since $\xi_{-11} = (\xi_{11})^*$ one can reduce the problem to two variables only, namely $u = |\xi_{11}|$ and $v = \xi_{01}$. These variables have a direct physical meaning. In fact, u is the half-amplitude of the oscillations of the transmitted field in the self-pulsing regime and v is the difference between the mean value of the oscillations and the (unstable) steady state value of the transmitted field. Thus, we have reduced the problem from the level of MBE to a set of two coupled equations in time only:

$$\dot{u} = f(u,v) \ ,$$
$$\dot{v} = g(u,v) \ , \tag{3.4}$$

where f and g are rational functions whose explicit expressions are of no concern here. Equations (3.4), which include both nonlinearity and propagation, describe the competition between the resonant mode v and the unstable mode u.

IV.   DESCRIPTION OF THE RESULTS

At this point, we can reason in terms of a simple bidimensional phase space (see Eqs. (3.4), which gives an appealing picture of the behavior of the self-pulsing instabilities. In particular, the stable stationary solutions u_{st}, v_{st} of Eqs. (3.4) correspond to cw states of the system when $u_{st} = 0$, to self-pulsing states when $u_{st} \neq 0$. Hence, the behavior of the system when we vary some external parameter is immediately understood by looking at the displacements of the stationary solutions of Eqs. (3.4) in the phase plane.

In order to illustrate the results that we have obtained from
this treatment, let us consider in detail the instability region.
In fact, as shown in[2] the self-pulsing instability can arise only
in the high transmission branch (see Fig. 2). Precisely, the adjacent mode is unstable provided the difference $2\pi c/L$ between its
frequency and the resonance frequency lies within a suitable range.
Figure 3 shows this range in correspondence to each point in the
high transmission branch for $\alpha L \ll 1$, $T \ll 1$, $C = \alpha L/2T = 20$, $\gamma_\perp =
\gamma_\parallel = \gamma$. Let us analyze the behavior of the system when we vary the
external parameters thereby exploring the plane of Fig. 3. Let us
consider first the upper part of the instability region, by moving
along the vertical line a) in Fig. 3. This means that we vary
$\alpha_1/\gamma = 2\pi c/L\gamma$, which can be obtained by varying the length L of
the cavity. Figure 4 shows the half-amplitude of the oscillations
vs. α_1/γ and compares the theoretical curve obtained from Eqs. (3.4)
with the numerical curve obtained by directly solving the MBE (the
latter curve has been obtained by Meystre). The agreement between
the two curves is excellent; the 10% discrepancy can be ascribed
either to numerical errors or to the neglect of higher frequencies
in the derivation of Eqs. (3.4). Note that our theory predicts
the correct behavior also when the amplitude of the oscillations
is comparable with the stationary field. The most striking feature
which emerges from Fig. 4 is that one can find self-pulsing even
out of the instability range. This means that the system can exhibit regular spiking behavior even when the stationary state in

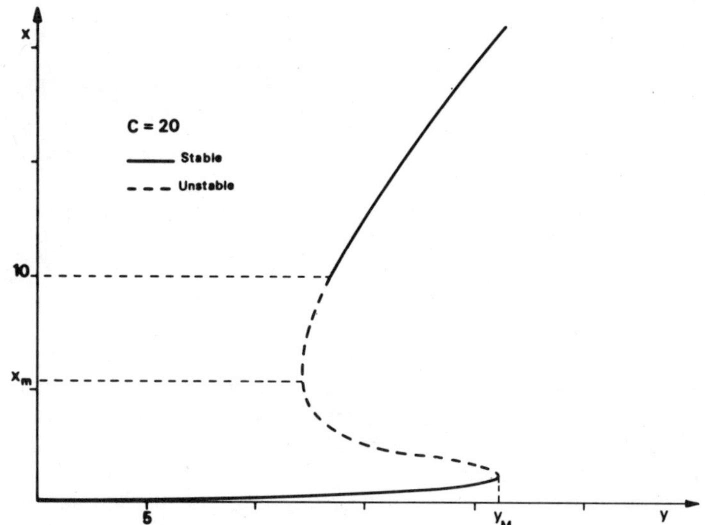

Fig. 2.   Stable and unstable steady states in the hysteresis
           cycle of transmitted vs. incident field. Mean field
           limit, $C = 20$, $\gamma_\perp = \gamma_\parallel = \gamma$.

the only stable stationary solution of Eqs. (3.4) is that corresponding to the cw low transmission state.

We stress that the behavior of the system that we have found
using Eqs. (3.4) agrees in all cases and in all features with the
results given by the direct numerical solution of the MBE[16]. However, the latter procedure amounts to little more than a crude
registration of data, whereas in the dressed mode theory this behavior is simply obtained from the analysis of the displacements
of the steady state solutions of Eqs. (3.4) in the phase plane
(u,v). The dressed mode treatment, together with the results given
in this section. leads us to look at optical bistability with new
eyes. In fact, now our bistable system does no longer appear simply as a device capable of two different stationary states, but
rather as a *multistable* system in which some of the stable steady
states are cw, some are pulsing. One can reach all the (cw and
pulsing) branches of the system by suitably maneuvering the external
parameters. A striking example of multistable situation is given
in Fig. 5. In fact, on the left of the instability range we have
three possible stable states, of which one is pulsing and two are
cw (both the low and the high transmission steady states are stable
in this situation).

Thus phenomenology of OB gets deeply enriched. From a formal
viewpoint, this enrichment is already evident from Eqs. (3.4) that
treat cw and pulsing solutions on the same footing. Using the
language of laser physics, we can say that the self-pulsing regime
is a many-mode operation of the system, whereas the cw regime is
a one-mode operation.

V. MANY-MODE QUANTUM STATISTICAL THEORY

We have generalized the dressed mode formalism to the quantum
statistical theory. Here the starting point is a many-mode master
equation for OB that holds for a ring cavity and in the limit
$\alpha L \ll 1$, $T \ll 1$.[5] This equation covers both the absorptive and the
dispersive cases. Its ingredients are on the one hand the creation
and annihilation operators A_n, A_n^+ of the nth cavity frequency, and
on the other hand the macroscopic dipole and inversion operators
R_n^+, R_n^-, R_{3n} introduced in Ref. 10. If one neglects all the frequencies of the cavity except the resonant one, thereby keeping
only the operators A_0, A_0^+, R_0^+, R_0^-, R_{30}, this master equation reduces to the usual quantum statistical mean field model[11]. Using
the procedure of Ref. 12, we have translated the many-mode master
equation into a classical-looking c-number differential equation
for a suitable quasi-probability distribution for the macroscopic
variables involved. In the limit of a very large number of atoms,
this last equation reduces to a Fokker-Planck equation (FPE). This
FPE is then linearized around a stable stationary state. The

eigenstates and the eigenvalues of the drift matrix of the linearized FPE introduce the dressed modes into this quantum statistical framework. Finally, with a straightforward change of variables one obtains a (linearized) FPE for the dressed mode variables ξ_{nj}. This last FPE has a diagonal drift matrix.

VI. SPECTRUM OF TRANSMITTED LIGHT

We have used the quantum statistical dressed mode theory to describe the spectrum of transmitted light taking into account *all* the modes of the cavity. Our aim is to find the behavior of this spectrum when we approach the self-pulsing instability. The spectrum of transmitted light is given by

$$S(\omega) = \mathrm{Re} \int_{o}^{\infty} \frac{dt}{\pi} \exp(-i\omega t) <E^{(-)}(L,t)E^{(+)}(L,0)>_{st} \qquad (6.1)$$

where $E^{(+)}$ and $E^{(-)}$ are the positive and negative frequency parts of the electric field, and the correlation function is calculated on a stable steady state. By expressing $E^{(\pm)}$ in terms of the creation and annihilation operators of the cavity frequencies, we obtain

$$\begin{aligned} &<E^{(-)}(L,t)E^{(+)}(L,o)>_{st} \\ &\qquad \propto \sum_{n,n'} \exp[-i(\kappa_{n'} - \kappa_{n})L]<A_{n'}^{+}(t)A_{n}(o)>_{st} \quad , \end{aligned} \qquad (6.2)$$

where κ_n is the wave vector of the nth frequency. Let us now subdivide A_n into stationary and fluctuating parts:

$$A_n(t) = A_{n,st} \exp(-i\omega_o t) + \delta A_n \quad , \qquad (6.3)$$

where, indicating by N_s the saturation photon number, one has[5]

$$A_{n,st} = \sqrt{N_s} \; x \; \delta_{n,o} \quad . \qquad (6.4)$$

Note that only the resonant mode n=0 has a nonvanishing stationary value for $\alpha L \ll 1$, $T \ll 1$. Hence

$$<A_{n'}^{+}(t)A_{n}(o)>_{st} = N_s x^2 \exp(i\omega_o t)\delta_{n,o}\delta_{n',o} + <\delta A_{n'}^{+}(t)\delta A_n(o)>_{st}. \qquad (6.5)$$

From Eqs. (6.1), (6.2) and (6.5) we see that the spectrum is composed of a coherent and an incoherent part:

$$S_{coh}(\omega) \propto N_s x^2 \delta(\omega-\omega_o) \quad , \qquad (6.6)$$

$$S_{inc}(\omega) \propto \mathrm{Re} \int_{o}^{\infty} \frac{dt}{\pi} \exp(-i\omega t) \sum_{n,n'} \exp[-i(\kappa_{n'}-\kappa_n)L]$$
$$\cdot \ \langle \delta A_{n'}^{+}(t)\delta A_n(o)\rangle_{st} \quad . \tag{6.7}$$

The incoherent part arises from the fluctuations of the system around the steady state. Since the deviation δA_n can be expressed as a linear combination of dressed mode variables ξ_{nj}, the correlation function $\langle \delta A_{n'}^{+}(t)\delta A_n(o)\rangle_{st}$ is in turn a linear combination of $\langle \xi_{n'j'}(t)\xi_{nj}(o)\rangle_{st}$. The latter correlation functions are easily calculated using the linearized FPE for the ξ variables mentioned in the previous section. One finds the following structure:

$$\langle \delta A_{n'}^{+}(t)\delta A_n(o)\rangle_{st} = \delta_{n,n'} \sum_{j,j'} C_{j,j'}^{n} \exp[(i\omega_o + \lambda_{nj})t] \ , \tag{6.8}$$

where the coefficients $C_{j,j'}^{n}$, are functions of the eigenstates $\underset{\sim nj}{O}$ introduced in Eq. (2.14). By substituting Eq. (6.8) into Eq. (6.7), one obtains the expression of the incoherent part of the spectrum. Thanks to the power of the dressed mode formalism, we have avoided having to perform any adiabatic elimination. Hence the expression of the spectrum given by Eqs.(6.7) and (6.8) is more general than all the ones derived up to now[13-15], even if one drops the contribution of the nonresonant modes.

Let us now give the explicit expression for the spectrum in the "good cavity" limit, i.e., when the empty cavity width

$$\kappa = cT/L$$

is much smaller than the atomic linewidth $\gamma_\perp$ and $\gamma_\parallel$. This situation is most interesting to us, because in absorptive OB the self-pulsing instability can arise only for $\kappa \ll \gamma_\perp, \gamma_\parallel$. One has that the incoherent part of the spectrum is a sum of Lorentzians centered on the cavity frequencies ω_n:

$$S_{inc}(\omega) \propto \sum_n \left\{ v_n \frac{\bar{\lambda}_n/\pi}{(\omega-\omega_n)^2 + \bar{\lambda}_n^2} + w_n \frac{\bar{\lambda}_{\phi n}/\pi}{(\omega-\omega_n)^2 + \lambda_{\phi n}^2} \right\} \quad . \tag{6.9}$$

The widths $\bar{\lambda}_n$ and $\bar{\lambda}_{\phi n}$ are given by

$$\bar{\lambda}_n = \kappa \left\{ 1 + \frac{2C\gamma_\perp}{1+x^2} \ \mathrm{Re} \ \frac{\gamma_\parallel(1-x^2) - i\alpha_n}{G(-i\alpha_n,x)} \right\} \quad ,$$

$$\lambda_{\phi n} = \kappa \left\{ 1 + \frac{2C\gamma_\perp^2}{1+x^2} \ \frac{1}{\gamma_\perp^2 + \alpha_n^2} \right\} \quad , \tag{6.10}$$

where

$$\alpha_n = \frac{2\pi nc}{L} = \omega_n - \omega_0 \, , \qquad n = 0, \pm 1, \pm 2, \ldots \, , \tag{6.11}$$

$$G(-i\alpha_n, x) = (\gamma_\perp - i\alpha_n)(\gamma_\parallel - i\alpha_n) + \gamma_\perp \gamma_\parallel x^2 \, . \tag{6.12}$$

The weights v_n, w_n in Eq. (6.9) (which are not always positive) are given by

$$v_n = \frac{x^2}{1+x^2} \frac{2x^2 + \left(1 - \dfrac{\gamma_\parallel}{\gamma_\perp}\right)\left[1 + \left(\dfrac{\alpha_n}{\gamma_\parallel}\right)\right]}{\left|\dfrac{G(-i\alpha_n, x)}{\gamma_\perp \gamma_\parallel}\right|^2} \, , \tag{6.13}$$

$$w_n = \frac{x^2}{1+x^2} \frac{1}{1 + \left(\dfrac{\alpha_n}{\gamma_\perp}\right)^2} \, .$$

If in the sum (6.9) one drops all the terms except the one for n=0, one recovers the expression given in Ref. 15 for the good cavity case.

Figure 9 illustrates the behavior of the incoherent part of the spectrum for $\kappa \ll \gamma_\perp, \gamma_\parallel$, $\gamma_\perp = \gamma_\parallel = \gamma$, $C = 20$, $\alpha_1/\gamma = 5$, $\alpha_1/\kappa = 200$, as described by Eqs. (6.9–6.13). Precisely, Figs. 9a,b,c,d show the lineshape in correspondence to the four states on the hysteresis cycle indicated by the same letters in Fig. 1. As we see from Fig. 9a, in the cooperative branch the Lorentzian peaks corresponding to the adjacent cavity frequencies are higher and narrower than the resonant contribution which is very flat (width $\simeq C\kappa$). Approaching the upper bistability threshold, the resonant peak shrinks due to the usual line narrowing[15], until it completely dominates the other contributions (Fig. 9b). When the system jumps to the high transmission branch, the resonant peak remains much higher than the two adjacent peaks (Fig. 9c). Here all the peaks have a width of a few κ. Finally, Fig. 9d shows what happens when we approach the self-pulsing instability by decreasing x. As shown by Fig. 3, for $\alpha_1/\gamma = 5$ the instability range begins at x=7.68. Approaching this value from above the adjacent mode peaks exhibit line narrowing and dominate the resonant peak. This is due to the fact that the two adjacent modes are going unstable, whereas the resonant mode remains stable here. This is exactly the converse of what happens when approaching the upper bistability threshold (Fig. 9b). Hence, once again, the rise of an instability is announced at the level of fluctuations by a line narrowing in the spectrum. After entering into the instability interval, the instability becomes manifest at the macroscopic level and one has self-pulsing.

The quantum statistical dressed mode formalism allows in principle to describe the fluctuations in the self-pulsing regime. The
analysis of this problem is left for a future publication.

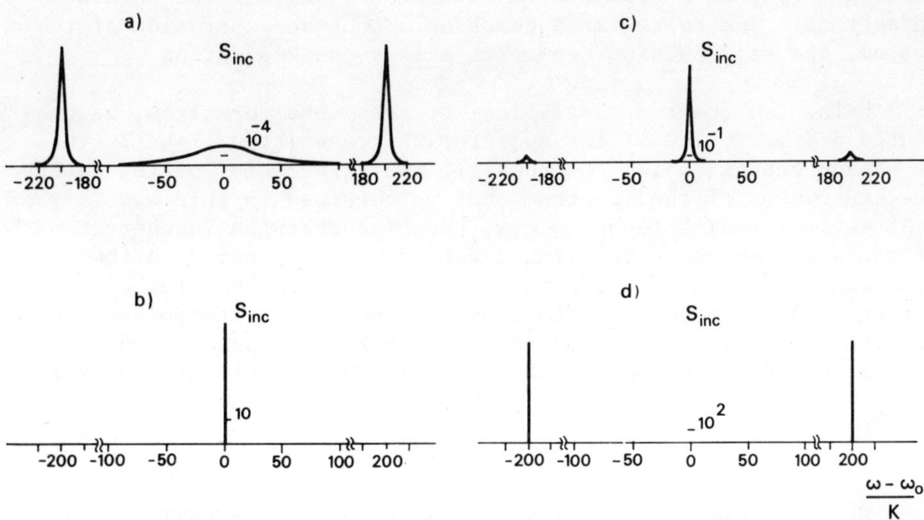

Fig. 9.   Behavior of the incoherent part of the spectrum of trans-
           mitted light for $\gamma_\perp = \gamma_\parallel = \gamma$, $C = 20$, $\alpha_1/\gamma = 5$, $\alpha_1/\kappa = 200$
           (see Eqs. (6.9–6.13)). The figures a)–d) correspond
           to the four points indicated in Fig. 1. a) $y = 14.2$,
           b) $y = 21.025$, c) $y = 22$, d) $y = 12.8$.

VII.   CONCLUSION

Using the semiclassical dressed mode formalism, we reduce the
problem of self-pulsing from the level of Maxwell-Bloch equations
to a bidimensional phase space. In this space, the cw states and
the self-pulsing states of the system are treated on the same footing. In fact, all of them correspond to stationary points in the
phase plane. We describe the behavior of the system under a variation of the external parameters, by simply looking at the displacements of the stable stationary states in the phase plane. We show
that the self-pulsing behavior can appear and disappear both as a
first and as a second order phase transition. We find hysteresis
cycles of a new type, that involve both cw and pulsing solutions.
The structure of these cycles suggests the most suitable procedure
to obtain the highest pulses. The agreement with the numerical
data, obtained by directly solving the Maxwell-Bloch equations is

excellent both in the case of self-pulsing and in the case of precipitation to the low transmission branch.

Thus, our dressed mode treatment leads to a new enriched picture of optical bistability in which the system appears as multistable rather than bistable, in the sense that some of the stable
states are cw, others are pulsing. One can reach all the (cw and
pulsing) branches of the system by suitably varying the external
parameters. The cw regime appears as a one-mode operation of the
system, the self-pulsing regime as a many-mode operation.

Using the quantum statistical dressed mode formalism, we presented a description of the spectrum of transmitted light in the
cw steady states, which includes all the frequencies of the cavity.
The expression of the spectrum that we obtained in this way is the
most general available up to now. We find that the incoherent part
of the spectrum has Lorentzian sidebands corresponding to the cavity
frequencies adjacent to the resonant frequency. The width and the
height of all the peaks changes as we cover the hysteresis cycle.
In particular, the sidebands dominate over the resonant peak as
soon as we approach the unstable (self-pulsing) part of the cycle.

ACKNOWLEDGMENTS

We are grateful to P. Meystre for his precious collaboration
and to R. Bonifacio for several stimulating discussions.

REFERENCES

1.   R. Bonifacio and L. A. Lugiato, Lett. Nuovo Cimento $\underline{21}$, 505
      (1978).
2.   R. Bonifacio, M. Gronchi and L. A. Lugiato, paper in this
      volume.
3.   See the introduction in Ref. 2.
4.   V. Benza and L. A. Lugiato, Zeits. f. Phys. B$\underline{35}$, 383 (1979).
5.   L. A. Lugiato, Zeits. f. Physik, in press.
6.   H. Haken, Zeits. f. Phys. B$\underline{21}$, 105 (1975); B$\underline{22}$, 69 (1975).
7.   H. Haken and H. Ohno, Opt. Comm. $\underline{16}$, 205 (1976) and Phys.
      Lett. $\underline{59}$A, 261 (1976).
8.   H. Haken, "Synergetics - An Introduction," Springer-Verlag,
      Berlin 1977.
9.   V. Benza, L. A. Lugiato and P. Meystre, Opt. Comm. $\underline{33}$, 113
      (1980).
10.   R. Bonifacio and L. A. Lugiato, Phys. Rev. A$\underline{11}$, 1507 (1975).
11.   R. Bonifacio and L. A. Lugiato, Phys. Rev. A$\underline{18}$, 1129 (1978).
12.   M. Gronchi and L. A. Lugiato, Lett. Nuovo Cimento $\underline{21}$, 593
      (1979).

13. R. Bonifacio and L. A. Lugiato, Phys. Rev. Lett. $\underline{40}$, 1023, 1538 (1978).
14. G. S. Agarwal, L. M. Narducci, R. Gilmore and Da Hsuan Feng, Opt. Lett. $\underline{2}$, 88 (1978) and Phys. Rev. A$\underline{18}$, 620 (1978).
15. L. A. Lugiato, Nuovo Cimento B$\underline{50}$, 89 (1979).
16. M. Gronchi, V. Benza, L. A. Lugiato, P. Meystre and M. Sargent III, submitted for publication.

INSTABILITIES IN OPTICAL BISTABILITY:TRANSFORM FROM CW TO PULSED

R. Bonifacio, M. Gronchi, and L. A. Lugiato

Istituto di Fisica dell'Università
Via Celoria, 16
20133 Milano, Italy

Abstract: This paper reviews the results obtained by our group
on the stationary solution for OB in a ring cavity and on the instabilities that can arise in these solutions. First, we solve exactly and analytically the Maxwell-Bloch equations with the proper
boundary conditons both in the absorptive and in dispersive case,
both for homogeneous broadening and for Lorentzian inhomogeneous
broadening. In the limit $\alpha L \to 0$, $T \to 0$, with $\alpha L/T$ arbitrary, the exact solution reduces to the previously calculated mean field state
equation. In this case, we give explicit analytic bistability conditions which show in particular that the purely absorptive case
is the one in which one finds the largest hysteresis cycle. On the
other hand, the exact solution shows that the mean field limit case
is the optimal situation to observe bistability. In fact, an increase of T causes a decrease of the size of the cycle, until for
T large enough the bistable behavior disappears. We show also that
under suitable conditions a part of the curve of transmitted vs.
incident light *with positive slope* can become unstable. In the dispersive case, this situation can occur also in the absence of bistability, whereas in the purely absorptive case this instability
can arise only in bistable situations, and precisely in the high
transmission branch of the hysteresis cycle. When there is instability the system either precipitates to the low transmission
branch, thereby producing a net reduction of the cycle, or evolves
towards an undamped spiking situation (self-pulsing). In the latter case, the system works as an all-optical device which transforms CW light into pulsed light. We give analytic instability conditions in the case $\alpha L \ll 1$, $T \ll 1$ (mean field limit, again!), in
which the dynamics of the system is governed by the *modes* of the
cavity. The self-pulsing instability occurs when the nonlinear
dynamics of the system transfers part of the incident energy from

the resonant mode to other cavity modes, thereby producing the
spiking behavior. Hence, with respect to these off-resonance modes,
the system behaves as a type of laser that works without population
inversion. Finally, we have analyzed the instabilities in the case
of a Fabry-Perot. It turns out that self-pulsing is much more difficult to be achieved than in the case of the ring cavity.

I. INTRODUCTION

Absorptive optical bistability (OB) has been theoretically
predicted by Szöke et al.[1] Some years later[2] McCall proved that
under suitable conditions the same system can show differential gain
with transistor action, and on the other hand treated absorptive OB
in a Fabry-Perot cavity by a numerical analysis of the Maxwell-
Bloch equations. This work suggested the experiments of Gibbs,
McCall and Venkatesan in Na, in which both transistor operation and
bistability were observed[3]. The analysis of the data showed that
the observed bistability was of dispersive type, with few excep-
tions. The mechanism which produces dispersive OB was explained
with the help of a simple phenomenological "cubic" model.

These results stimulated theoretical and experimental activity;
see especially Felber and Marburger[4], Bonifacio and Lugiato[5], Smith
and Turner[6]. Successively, other experiments on OB have been performed by Venkatesan and McCall[7], Bishofberger and Shen[8],
Grischkowski[8], Garmire et al.[9], Gibbs et al.[10], Miller et al.[11],
Sandle and Gallagher[12], Arimondo et al.[13].

Crucial for the development of the theory of OB has been the
so-called mean field theory (MFT) initiated in Refs.5 and 14, in
which the problem of OB is reduced to three equations in time only,
under the assumption that the field inside the cavity remains basically uniform. Hence the MFT has allowed an analytical approach to
OB, producing the following main results[5,14]:

1)   It gives a theory of OB that includes also the dynamics of the
     system, whereas the previous treatments considered almost ex-
     clusively the steady-state behaviour. The linear stability
     analysis has led to new predictions concerning the transient.
     In particular, we showed that also the approach to the steady
     state exhibits a hysteresis cycle. When the value of the in-
     cident field gets near to the upper bistability threshold from
     below or to the lower bistability threshold from above one
     finds a lengthening of the time of approach to the steady
     state (critical slowing down). This phenomenon has been ex-
     perimentally observed by Garmire et al.[9] Furthermore, we in-
     troduced the basic distinction between the "good cavity" and
     the "bad cavity" cases. In the first case, the approach to
     the stationary situation is always monotonic, whereas in the

second, one finds Rabi oscillations in the approach to the high intensity transmission branch.

2)   Via the regression hypothesis, the results on the transient imply as many predictions for the spectrum of transmitted light. Hence one finds a spectral hysteresis cycle that exhibits line narrowing in correspondence to the bistability thresholds, as a consequence of the critical slowing down. In the bad cavity case, when crossing the upper bistability threshold one finds the abrupt appearance of a triplet with well separated sidebands (discontinuous Dynamical Stark Effect).

3)   The MFT incorporates OB in the general framework of cooperative phenomena, by pointing out the role of atomic cooperation in this phenomenon. In particular, it shows that OB is the prototype of a first order phase transition (far from thermal equilibrium) in quantum optics.

4)   The MFT allows a simple quantum statistical formulation, based on a master equation that holds both for the good and for the bad cavity case.

Reference 5 stimulated a considerable series of theoretical papers, the first of which were Agarwal et al.[15], Carmichael and Walls[16], Willis[17], Arecchi and Politi[18]. At the quantum statistical level, the mean field model was used to give a treatment of the spectrum of the transmitted light, that made rigorous and completed the results obtained via the regression hypothesis[19-23]. On the other hand, the quantum statistical mean field model produced a description of the photon statistics of the transmitted light in the good cavity case. In Ref. 19 and 24 we treated extensively this problem, by discussing the bimodal character of the distribution function in the bistability region and describing the behaviour of the mean value and of the fluctuations of the transmitted light. This behaviour completes the analogy between OB and first-order phase transitions, showing, on the other hand, the nonthermodynamic character of the transition which stems from the fact that the diffusion term of the Fokker-Planck equation is intensity-dependent. These results have been further developed in Refs. 25-31.

At the semiclassical level, the mean field model has been used to investigate various aspects of the transient behaviour, in particular with regard to switching characteristics[32-35].

The mean field model for a ring cavity has been generalized, both at the semiclassical and quantum statistical level, to the case of mixed absorptive and dispersive bistability[36]. The general bistability conditions have been worked out in Refs. 37-39. In particular, in Ref. 39 we analyze in detail the behaviour of the hysteresis cycle when the parameters in play are varied, and show

that in the purely absorptive case one finds the largest cycle.
The limits of validity of the "cubic" model of Ref. 3 are discussed
in the case of two-level atoms. Further analyses of dispersive bistability can be found in Refs. 40-43.

In 1978 we gave the first *exact analytical* treatment of OB
fully including propagation effects, by solving the Maxwell-Bloch
equations at steady state with boundary conditions for a ring cavity,
in the purely absorptive case[44]. We proved that in the double limit
$\alpha L \to 0$, $T \to 0$ (αL = unsaturated absorption coefficient, T = mirror
transmittivity), with $C = \alpha L/2T$ constant, one recovers exactly the
MFT solution. This result pointed out the limit of validity of the
MFT. The "mean field" limit $\alpha L \to 0$, $T \to 0$ had been already implicitly
used by Szöke et al., but its crucial role in the framework of an
exact solution for OB was first shown in Ref. 44. Contemporarily,
in Ref. 36 we extended the mean field limit to the general absorp-
tive-dispersive case: $\alpha L \to 0$, $T \to 0$ and $\phi = (\omega_c - \omega_o)(L/c) \to 0$, where ω_o
is the frequency of the incident field and ω_c is the frequency of
the cavity that is nearest to resonance. Using this triple limit,
in which $C = \alpha L/2T$ and $\theta = \phi/T$ are taken constant and arbitrary, the
general mean field model for OB in a ring cavity of Ref. 36 follows
exactly from the MBE, as explicitly shown in Ref. 39. The exact
analytical solution of Ref. 44 (ring cavity) has been extended to
the mixed absorptive-dispersive case independently in Refs. 45, 46,
47. The generalization to the case of Lorentzian inhomogeneous
broadening is given in Ref. 48.

Correspondingly, the analysis of Ref. 44 has been generalized
to the case of Fabry-Perot cavity. Meystre[49] first analyzed the
mean field limit in this framework and the deviations from the MFT
that one finds when T is not small enough. To treat these problems
he used some equations obtained in Ref. 14 by suitably truncating
the infinite hierarchy of equations derived in Ref. 50. The analysis of Ref. 49 has been extended in Refs. 51 and 52. On the other
hand, Carmichael and Hermann[53,54] solved, both numerically[53] and
analytically[54], the steady-state Maxwell-Bloch equations derived
by McCall[2]. These equations include the standing-wave effects
more completely than the time-dependent equations of Refs. 14 and
49. They showed that there are deviations of the order of 15% with
respect to the results obtained from the truncated hierarchy. This
analysis is extended to the mixed absorptive-dispersive case in
Ref. 55, where it is shown that in the mean field limit one obtains
for a Fabry-Perot a state equation that differs from that valid for
a ring cavity[36] (again, quantitatively the differences are of the
order of 15%). This state equation for a Fabry-Perot coincides,
apart from a change of sign to convert the absorber into an amplifier, with that previously derived by Spencer and Lamb for a laser
with injected signal[56]. Very recently, McCall and Gibbs derived
from this state equation the general bistability conditions for a

Fabry-Perot in both absorptive and dispersive OB[57].

Again in 1978[58], we performed the stability analysis of our exact analytical solution for absorptive OB in a ring cavity and showed that under suitable conditions a part of the high transmission branch can become unstable. In these conditions, the system works as a transformer of cw into pulsed light[58,59]. An analogous behaviour has been independently suggested and concretely realized by McCall for a hybrid electro-optical device[60]. The working principles of this device are completely different from those of our self-pulsing system, and have a more limited interest from a theoretical viewpoint.

The self-pulsing instabilities are the main subject of this paper.

We apologize for the incompleteness of our list of references, especially with regard to the development of the device aspects, that has led in particular to the construction of miniaturized all-optical bistable devices[10]. For an adequate discussion, we refer to Ref. 61 and to the papers of Gibbs, McCall and P. Smith, Miller and S. D. Smith in this volume.

II. SEMICLASSICAL TREATMENT OF STEADY STATE

In order to describe theoretically the phenomenon of OB, it is easier to consider a ring cavity (Fig. 1) than a Fabry-Perot, because in the ring cavity one has to deal with propagation only in one direction, thus avoiding standing wave difficulties. For simplicity we assume that mirrors 3 and 4 have 100% reflectivity. We call R and T (with $R+T=1$) the reflectivity and transmittivity of mirrors 1 and 2. We describe the dynamics of the coupled system, atoms plus radiation field, by the well known one-sided Maxwell-Bloch equations, which are here written in general for an inhomogeneously broadened system of N two-level atoms.

$$\frac{\partial S_\omega}{\partial t} = \frac{\mu}{\hbar} ED_\omega - [\gamma_\perp + i(\omega-\omega_0)]S_\omega \tag{1.1}$$

$$\frac{\partial D_\omega}{\partial t} = -\frac{\mu}{2\hbar}(ES_\omega^* + E^*S_\omega) - \gamma_\parallel\left[D_\omega - \frac{N(\omega)}{2}\right], \tag{1.2}$$

$$\frac{\partial E}{\partial t} = c\frac{\partial E}{\partial z} = -g\int d\omega\, S_\omega . \tag{1.3}$$

In Eqs. (1) $N(\omega)d\omega$ is the number of atoms with frequency between ω and $\omega + d\omega$. One has of course $\int d\omega N(\omega) = N$. We call S_ω and D_ω

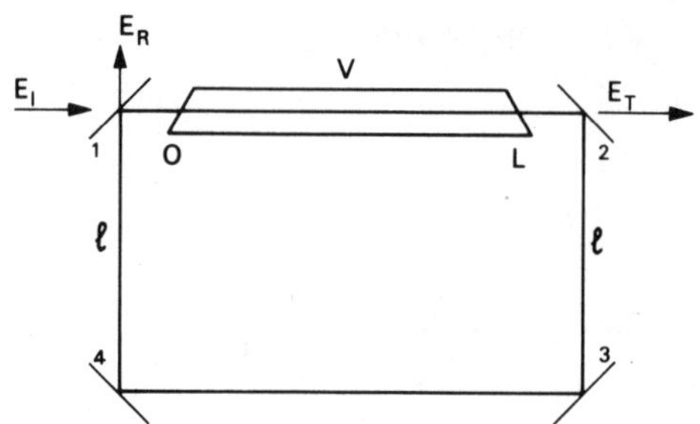

Fig. 1. Ring Cavity. $\underline{E}_I$, $\underline{E}_T$ and $\underline{E}_R$ are the incident, transmitted and reflected field respectively.

the polarization and the population difference (precisely, one half the difference between the populations of the lower and of the upper level) associated with the isochromat of frequency ω. E is the slowly varying envelope of the electric field. μ is the modulus of the dipole moment of the atoms and g is a coupling constant given by

$$g = 4\pi\omega_o/V \qquad (2)$$

where ω_o is the frequency of the incident field and V the volume of the atomic sample. $\gamma_\parallel$ and $\gamma_\perp$ are the inverse of the atomic relaxation times T_1 and T_2 respectively. Of course the field equation is supplemented by some proper boundary conditions (see Fig. 1):

$$E_T(t) = \sqrt{T} \; E \; (L,t) \qquad (3.a)$$

$$E(o,t) = \sqrt{T} \; E_I + RE(L,t-\Delta t)\exp(-i\theta T) \quad . \qquad (3.b)$$

In Eq. (3) E_I and E_T are the incident and transmitted field amplitudes respectively, $\Delta t = (2\ell+L)/c$ is the time the light takes to travel from mirror 2 to mirror 1 and θ is the cavity detuning parameter

$$\theta = (\omega_c - \omega_o)/\kappa \qquad , \qquad \kappa = cT/L \quad , \qquad (4)$$

where ω_c is the frequency of the cavity that is nearest to resonance and $L = 2(L+\ell)$ is the total length of the cavity. κ is the empty cavity width. The second contribution on the r.h.s. of (3.b) describes a feedback mechanism due to the mirrors, which is essential to give rise to bistability.

Let us first consider the steady state ($\partial E/\partial t = \partial S_\omega/\partial t = \partial D_\omega/\partial t = 0$). It is suitable to introduce the dimensionless amplitudes (y is taken real for definiteness)

$$F(z) = \frac{\mu E(z)}{\hbar\sqrt{\gamma_\perp \gamma_\parallel}} \quad , \quad y = \frac{\mu E_I}{\hbar\sqrt{\gamma_\perp \gamma_\parallel}\, T} \quad . \tag{5}$$

One easily derives from (1) the following equation for the stationary field:

$$\frac{\partial F}{\partial z} = -\alpha F \chi(|F|^2) \quad , \tag{6}$$

where α is the linear absorption coefficient

$$\alpha = \mu\, g\, N/2\hbar c\gamma_\perp$$

and χ is the complex nonlinear susceptibility

$$\chi(|F|^2) = \int d\omega\, \frac{N(\omega)}{N}\left(1 - i\,\frac{\omega-\omega_o}{\gamma_\perp}\right)\left[1 + |F|^2 + \frac{(\omega-\omega_o)^2}{\gamma_\perp^2}\right]^{-1} \quad . \tag{7}$$

Equation (6) contains only 50% of the physics of optical bistability. The other 50% is contained in the boundary conditions (3). In fact the bistability arises from the very fact that the field at z=0 is in general quite different from the incident field because there is the feedback contribution. To describe the response of the system we must derive the relation which expresses the transmitted intensity $X = |F(L)|^2$ as a function of the incident intensity $Y=y^2$. By combining the solution of Eq. (6) with the boundary condition (3.b) we find the parametric prepresentation for this function given in Ref. 45:

$$X = \frac{2}{\rho^2-1}\,[\alpha L - (1 + \Delta^2)\,\ell n\,\rho] \quad , \tag{8}$$

$$Y = \frac{2}{T^2}\,\frac{\alpha L - (1+\Delta^2)\,\ell n\,\rho}{\rho^2-1}\left[\rho^2 + R^2 - 2R\rho\,\cos(\Delta\,\ell n\,\rho-\theta T)\right]$$

where

$$\rho = |F(0)|/|F(L)| \quad , \quad \Delta = (\omega_A - \omega_o)/\gamma_\perp \quad . \tag{9}$$

Δ is the atomic detuning parameter and ω_A is the central frequency of the atomic line. On resonance ($\Delta = \theta = 0$) Eq. (8) reduces to the exact solution for purely absorptive OB[44]:

$$\ell n\left[1 + T(\frac{y}{x} - 1)\right] + \frac{x^2}{2}\left\{\left[1 + T(\frac{y}{x} - 1)\right]^2 - 1\right\} = \alpha L \quad , \tag{10}$$

where $x = \sqrt{X}$. Equation (8) holds for a homogeneously broadened system and includes all propagation effects. Also in the case of an inhomogeneously broadened system we have obtained an analytical solution assuming a lorentzian atomic line:

$$N(\omega) = (N/\pi T_2^*)\,[\,(\omega - \omega_A)^2 + (T_2^*)^{-2}\,]^{-1} \quad . \tag{11}$$

The latter solution is more complicated and is reported in Ref. 48.

Before analyzing the exact solution, let us state a theorem which shows that the solution drastically simplifies in the limit of small mirror transmissivity. Precisely, let us consider the following triple limit (mean field limit):

$$T \to 0 \ , \quad \alpha L \to 0, \quad (\omega_c - \omega_o)L/c \to 0 \quad , \tag{12}$$

with $C \equiv \alpha L/2T$ and $\theta = (\omega_c - \omega_o)L/cT$ constant. In this limit the exact solution reduces to the state equation of the Mean Field Theory (MFT) of OB in a ring cavity[36]:

$$Y = X\{[1 + \chi_1(X)]^2 + [\theta - \chi_2(X)]^2\} \ , \tag{13}$$

$$\chi_1(X) = 2C\mathrm{Re}\chi(X) = \frac{\sigma + \sqrt{1+X}}{\sqrt{1+X}}\,\frac{2C}{\Delta^2 + (\sigma + \sqrt{1+X})^2} \tag{14.a}$$

$$\chi_2(X) = 2C\mathrm{Im}\chi(X) = \frac{2C\Delta}{\Delta^2 + (\sigma + \sqrt{1+X})^2} \ , \tag{14.b}$$

where $\sigma = (\gamma_\perp T_2^*)^{-1}$. In particular Eqs. (13) and (14) for $\sigma = 0$ follow from (8) performing the mean field limit (12). The mean field theorem (12), (13) shows that the MFT works very well when T is small enough. Note that for αL small the field inside the sample becomes more and more uniform. Furthermore the conditions

$$(\omega_c - \omega_o)L/c \ll 1, \qquad (\omega_c - \omega_o)L/cT \qquad \text{finite}$$

mean that the cavity detuning must be much smaller than the free spectral range, but of the same order of magnitude as the cavity width. A relation of type (13) between incident and transmitted field was first given in Refs. 1 and 3 on the basis of phenomenolical arguments. Our approach derives this formula from first principles as an analytical solution of the Maxwell-Bloch equations with boundary conditions, pointing out its limit of validity and giving explicit expressions for $\chi_{1,2}$ in the case of the ring cavity and inhomogeneously broadened lorentzian lineshape. The homogeneous case is trivially obtained by putting $\sigma=0$ in (13). On the other hand (8) contains propagation effects, which must be measurable far from the mean field limit (12), when (13) does not hold. These propagation effects, as we shall see, induce an anomalous transient behaviour (self-pulsing) which is unpredictable in the framework of the mean field theory. Figure 2 shows the shape of

the curves of transmitted vs. incident light when there is bistability. The part of the curves with negative slope are unstable,
as we shall see later, so that one finds a hysteresis cycle. Curve
e) in Fig. 2.1 is the mean field result for C=50, $\Delta=\theta=\sigma=0$ (purely
absorptive case); curve e) in Fig. 2.2 is the mean field result for
C=50, $\Delta=10$, $\theta=2.25$, $\sigma=0$ (dispersive case). In both Figs. 2.1, 2.2
the curves a,b,c,d show the exact solution (8) for different values
of αL and of the transmittivity, chosen in such a way that $C=\alpha L/2T$
is constant equal to 50. For high values of αL and T, as in curve
a) there is no bistability whereas the bistable behaviour increases
by decreasing αL and T. In this way one approaches the mean field
result which is already a good approximation for $\alpha L \simeq 1$. For C and
T fixed the mean field curve is a better approximation in the dispersive case (Fig. 2.2) than the absorptive one (Fig. 2.1). This
is due to the fact that absorption is reduced in the dispersive
case so that the variation of the field in space is not strong even
for αL large.

Let us now consider in detail the mean field state equation
(13). When $\Delta,\theta \ll 1$ one can neglect the term $[\theta-\chi_2]^2$ so that one
reduces to the equation for purely *absorptive* optical bistability,
in which only the absorptive part of the susceptibility is important.
On the contrary, for $\Delta \gg 1$ the absorptive part χ_1 is negligible
and one has the equation for purely *dispersive* optical bistability,
in which only the dispersive part of the polarization plays a role.
The state equation (13), in the case of homogeneous broadening
($\sigma = 0$) has been analyzed in Ref. 39. In this case one can analytically derive the conditions which ensure a bistable response. One
has two conditions:

$$2C > \Delta\theta - 1 \quad , \tag{15.a}$$

$$(2C - \Delta\theta + 1)^2(C + 4\Delta\theta - 4) > 27C(\Delta + \theta)^2 \quad . \tag{15.b}$$

Condition (15.a) guarantees that the curve Y(X) given by (13) has
an inflection point $X_{inf} > 0$; condition (15.b) ensures that the
slope of the curve at X_{inf} is negative. The analysis of the bistability conditions (15) leads to the following results. First,
in the case of purely absorptive OB one has bistability for C > 4
as it is well known. Second, for C < 4 not only absorptive but also
dispersive bistability is impossible. Finally, when C > 4 the
hysteresis cycle is largest for $\Delta=\theta=0$, i.e., in the absorptive case.
The conclusion is that in the case of a homogeneously broadened
system purely absorptive bistability is more suitable than dispersive OB, at least when the mean field approximation holds.

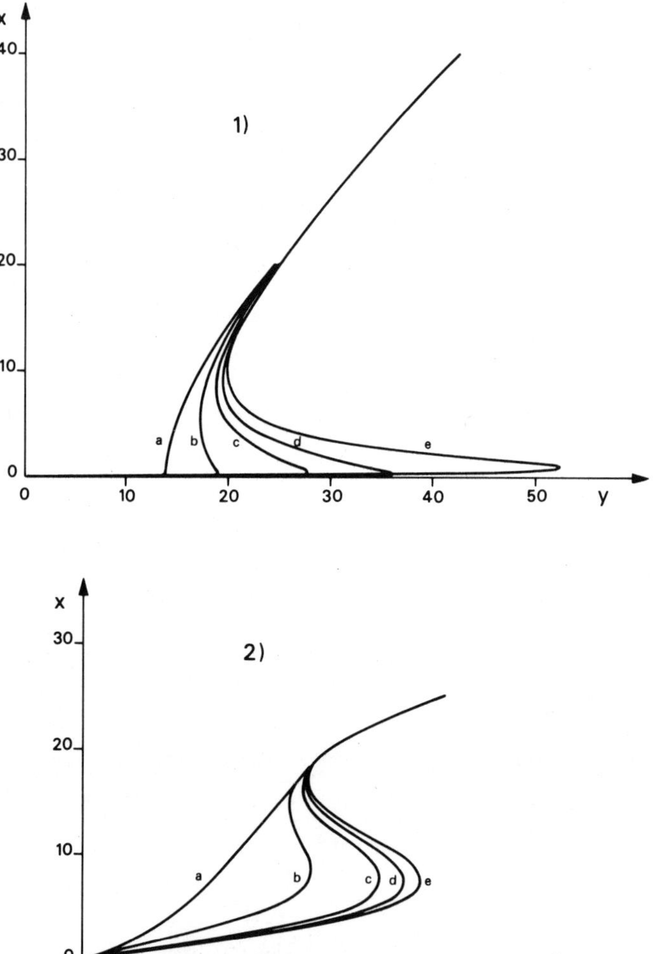

Fig. 2.   Plot of the transmitted amplitude $x = \sqrt{X}$ vs. the incident amplitude y in the homogeneously broadened case. In both Figs. 2.1 and 2.2 curves a), b), c), d) show the exact stationary solution; curve e) is the mean field result. In Fig. 2.1 C=50, $\Delta=\theta=0$; in Fig. 2.2 C=50, $\Delta=10$, $\theta=2.25$. For curves a) $\alpha L=100$, T=1; for curves b) $\alpha L=50$, T=0.5; for curves c) $\alpha L=20$, T=0.2; for curves d) $\alpha L=10$, T=0.1.

On the other hand let us consider the case of Lorentzian in-homogeneous broadening. In general, for Δ, θ and σ fixed one has bistability when C is larger than a suitable threshold value $C_{min}(\Delta, \theta, \sigma)$ which increases rapidly with σ. When $\sigma \gg 1$ one finds values of C such that absorptive bistability is impossible, where-as dispersive bistability is possible. The situation is illustrated in Fig. 3. The conclusion is that in the case of inhomogeneously broadened systems dispersive bistability can be more suitable than absorptive bistability.

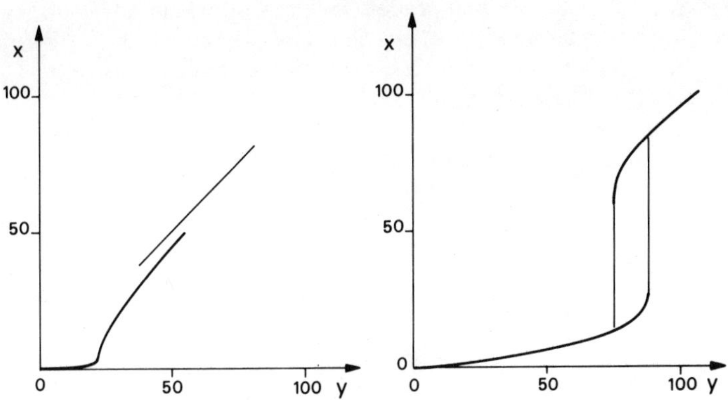

Fig. 3.   Mean field curves for the transmitted amplitude $x = \sqrt{X}$
vs. the incident amplitude y in the case of inhomogeneous
broadening (Lorentzian atomic line). For $\Delta = \theta = 0$
(absorptive case) one has no bistability whereas for
$\Delta = 30$, $\theta = 1$ one has a hysteresis cycle. In both curves
$C = 180$ and $\sigma = 15$.

III.   LINEAR STABILITY ANALYSIS

Let us now discuss the stability of the stationary solutions, and for definiteness let us consider the purely absorptive case and assume homogeneous broadening. Hence we drop the label ω in S_ω, D_ω and treat S and E as real quantities. Te check the stability of the steady state, we must consider an initial condition in which the system is slightly displaced from the stationary state and in-vestigate whether the system returns to the stationary state or not. This regression to the steady state is described by the lin-earized Maxwell-Bloch equations for the deviations of the fields from their stationary values:

$$\delta S(z,t) = S(z,t) - S_{st}(z) \quad ,$$

$$\delta D(z,t) = D(z,t) - D_{st}(z) \quad ,$$

$$\delta E(z,t) = E(z,t) - E_{st}(z) . \tag{16}$$

The time evolution of these deviations is a linear combination of exponentials, e.g.

$$\delta E(z,t) = \sum_{nj} e_{nj}(z) \exp(\lambda_{nj} t) + c.c. \quad , \tag{17}$$

where the eigenvalues λ_{nj} are the solutions of a complicated equation. The index n labels the modes of the cavity; n=0 corresponds to the mode that is resonant with the incident field. The index j runs from 1 to 3 (because the Maxwell–Bloch equations are three in number). For $T \ll 1$ one finds[58,59]

$$\lambda_{n1} = -i\alpha_n - \kappa \left\{ 1 + \frac{2C\gamma_\perp}{1+x^2} \frac{\gamma_\parallel(1 - x^2) - i\alpha_n}{(\gamma_\perp - i\alpha_n)(\gamma_\parallel - i\alpha_n) + \gamma_\perp \gamma_\parallel x^2} \right\} + O(T^2) \quad , \tag{18}$$

$$\lambda_{\substack{n2 \\ n3}} = - \frac{1}{2} \left\{ \gamma_\perp + \gamma_\parallel \pm \sqrt{(\gamma_\perp - \gamma_\parallel)^2 - 4\gamma_\perp \gamma_\parallel x^2} \right\} + O(T) \quad ,$$

where x is the stationary value of the normalized transmitted field and

$$\alpha_n = 2\pi nc/L \quad , \qquad n = 0, \pm1, \pm2, \ldots \quad . \tag{19}$$

Since x is real one has that $\lambda_{-n1} = (\lambda_{n1})^*$, $\lambda_{-n2} = (\lambda_{n3})^*$.

The stationary state is unstable if and only if at least one of the eigenvalues λ_{nj} has a positive real part. From (18), we see that only λ_{n1} can yield a positive real part for $T \ll 1$. Let us first consider the resonant mode n=0; one has from (18):

$$\lambda_{01} = - \kappa \left\{ 1 + 2C \frac{1 - x^2}{(1 + x^2)^2} \right\} = - \kappa \frac{dy}{dx} \quad . \tag{20}$$

Hence, as anticipated, all the points of the curve $x = x(y)$ with *negative* slope are unstable, because the resonant mode is unstable. Let us now consider the off-resonant modes $n \neq 0$, and for the sake of simplicity let us concentrate on the case $\gamma_\perp = \gamma_\parallel \overset{def}{=} \gamma$ (the general case is discussed in Ref. 59). The analysis of the condition $Re\lambda_{n1} > 0$ leads to the following conclusion, which is illustrated in Figs. 4 and 5. For $C > 2(1 + \sqrt{2})$ the points in the high

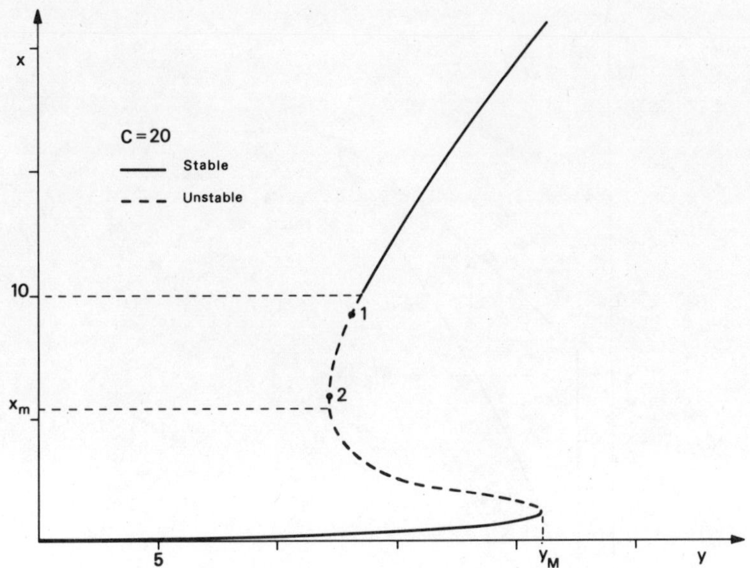

Fig. 4.   Stable and unstable steady states in the hysteresis cycle
           of transmitted vs. incident light for $T \ll 1$, $\alpha L \ll 1$,
           $C = \alpha L / 2T = 20$ and $\gamma_\perp = \gamma_\parallel = \gamma$. The points indicate the
           values of y and x in correspondence of which we show (in
           Fig. 6) the numerical solution of the Maxwell–Bloch
           equations.

transmission branch such that $x < C/2$ are unstable provided at least
one of the discrete values α_n lies in the range $\alpha_{min} < |\alpha_n| < \alpha_{max}$ where
(Fig. 5):

$$\alpha_{\substack{max \\ min}} = \gamma \left[x^2 - C - 1 \pm \sqrt{C^2 - 4x^2} \right]^{1/2} . \qquad (21)$$

Hence when these conditions hold a part of the curve $x = x(y)$ with
positive slope is unstable, because the off-resonance modes such
that $\alpha_{min} < |\alpha_n| <_{max}$ are unstable there. Note that since $2(1+\sqrt{2}) > 4$
this situation can arise only when there is a hysteresis cycle.

IV.   SELF–PULSING

One asks what happens in correspondence to the unstable points
in the high transmission branch. *A priori* one has three different
possibilities: for a fixed incident field y the system i) precipitates to the corresponding stationary state in the low transmission
branch, which is always stable, or ii) evolves to a time periodic
state (limit cycle), so that the transmitted light becomes an

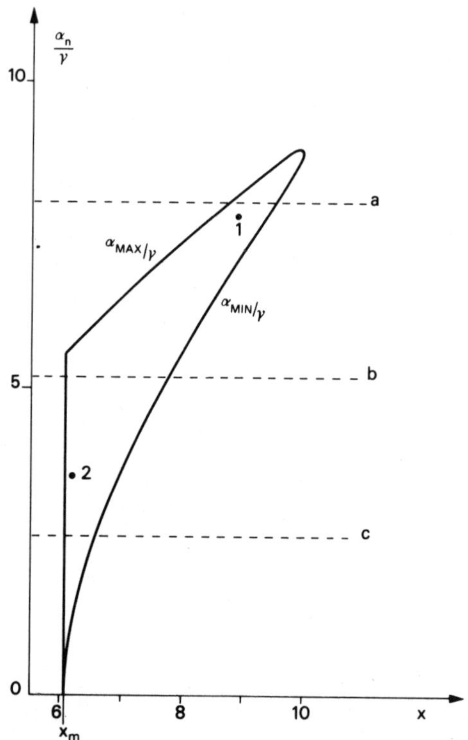

Fig. 5. Plot of the functions α_{max} and α_{min} vs. x, which shows the frequency interval for instability for $\gamma_\perp = \gamma_\parallel = \gamma$ (Eq. 21). The points indicate the values of the quantity $\alpha_1/\gamma = 2\pi c/L\gamma$ in correspondence with the two graphs in Fig. 6.

undamped regular sequence of pulses, or iii) evolves to a chaotic situation, in which it exhibits a completely irregular sequence of pulses of the type one finds in the Lorenz model[62]. In case ii) one speaks of self-pulsing because the system shows this pulsing behavior spontaneously, without any external manipulation. This phenomenon is similar to what occurs beyond the so-called second threshold of the laser[63,64]. In order to answer the question at the beginning of this section, we have numerically solved the MBE (1). At the moment we have evidence only of cases i) and ii)[65].

Let us first consider the case in which only the two adjacent modes $n = \pm 1$ are unstable, so that $\alpha_{min} < \alpha_1 < \alpha_{max}$. We have fixed $\gamma_\perp = \gamma_\parallel = \gamma$, C=20, T=0.1 or T=0.01 and $L = 5L$. Figure 6 shows an example of self-pulsing and an example of precipitation; the

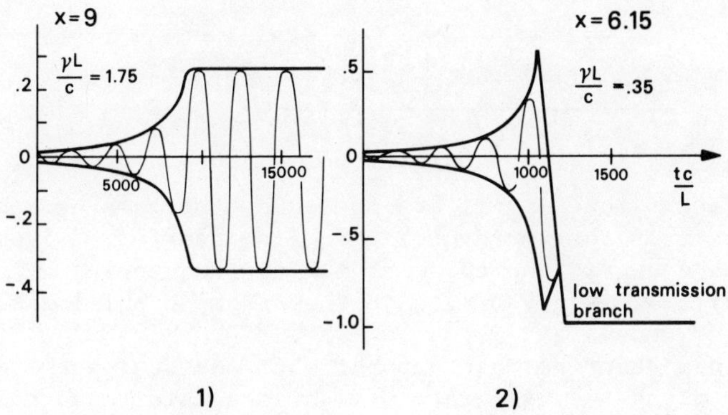

Fig. 6.   Time evolution of the transmitted field when the sta-
           tionary state in the high transmission branch is unstable.
           The time width of the pulses has been increased by a
           factor $\sim$300. The transmittivity is T = 0.1.

values of x and α_1/γ which correspond to these two diagrams are
shown in Figs. 4 and 5.   The frequency of the oscillations is
roughly equal to the detuning α_1 = $2\pi c/L$ of the unstable modes with
respect to the resonant mode, with an error of the order of T.   In
the self-pulsing case, the mean value x_m of the oscillations is
always lower than the steady state value x, showing a kind of at-
traction exerted by the stable steady state.   One has roughly

$$x(t) = x_m + A \cos\left(\frac{2\pi c}{L} + \phi_o\right) \quad , \qquad (22)$$

where A is the half-amplitude of the oscillations and the phase ϕ_o
depends on the initial condition.   Hence the electric field
$E(z,t) = E(z,t)\cos[\omega_o(t-z/x)]$ has the structure

$$E(L,t) = E_m \cos[\omega_o t + \phi_1] + A\left\{\cos\left[\left(\omega_o + \frac{2\pi c}{L}\right)t + \phi_2\right]\right.$$

$$\left. + \cos\left[\left(\omega_o - \frac{2\pi c}{L}\right)t + \phi_3\right]\right\} \quad . \qquad (23)$$

Hence in the self-pulsing situation the system acts as a transformer of cw light into pulsed light and behaves as a novel type
of laser which works without population inversion.   Note in this
connection from Eq. (18) that $\mathrm{Re}\lambda_{11}$ has a gain-minus-loss structure

$$\text{Re}\,\lambda =$$

$$= \frac{2C\kappa\gamma_\perp}{1+x^2}\; \frac{\gamma_\parallel\,(x^2-1)\,[\gamma_\perp\gamma_\parallel\,(1+x^2)-\alpha_1^2]-\alpha_1^2(\gamma_\perp+\gamma_\parallel)}{[\gamma_\perp\gamma_\parallel\,(1+x^2)-\alpha_1^2]^2+\alpha_1^2(\gamma_\perp\gamma_\parallel)^2} - \kappa. \tag{24}$$

When the adjacent modes $n = \pm 1$ are unstable, the gain exceeds the
losses and part of the incident light gets transferred from the incident frequency ω_0 to the adjacent cavity frequencies $\omega_0 \pm 2\pi c/L$
(see Eq. (23)). This gives rise to the undamped pulsing behavior.

Let us now describe what happens when one decreases the incident field y along the high intensity branch, starting from a value
of y such that x > C/2 (see Fig. 4)[66]. As y decreases, x decreases
whereas $\alpha_1 = 2\pi c/L$ remains constant. Hence the point $\{x, \alpha_1/\gamma\}$ moves
in the plane of Fig. 5 along a horizontal line as a),b),c) from
the right to the left. Let us first consider the case of the line
a). When the point $\{x, \alpha_1/\gamma\}$ enters from the right into the instability region bounded by the lines α_{max}/γ, α_{min}/γ, $x = x_m$ the self-
pulsing behavior appears *abruptly*, with oscillations of finite
amplitude. In other words, crossing the right boundary of the instability region the system shows a first-order-like phase transition from a stationary to a self pulsing behavior. When y (i.e. x)
is decreased, the amplitude of the oscillations continuously decreases until in correspondence to the left boundary of the instability region the oscillations vanish and the system is back again
to a stationary state. Hence crossing the left boundary one finds
a second-order-like phase transition from a self-pulsing to a stationary behavior. Let us now consider somewhat larger values of
L, as in the case of the line b). When we cross the left boundary
everything happens as in the case of the line a). However, before
arriving at the left boundary, the system precipitates to the low
transmission branch. Finally, in the case of line c) the system
precipitates as soon as one enters into the instability region.

Let us now consider the case of many unstable modes, which
can be obtained by increasing the length L of the ring cavity.
This is the situation in which one expects the shortest pulses.
We have only preliminary results in this case which is very complicated. We have obtained a nice periodic sequence of pulses
(Fig. 7) with a repetition period equal to the transit time in the
cavity L/c. These pulses have a complicated structure of short
sub-pulses, which reminds one of the behavior of lasers in the
mode-locked regime. This behavior has a very long duration in
time (~ 600 transit times), but it finally disappears and the system
precipitates to the low transmission branch. We do not yet know
whether this precipitation is due to numerical errors or not.

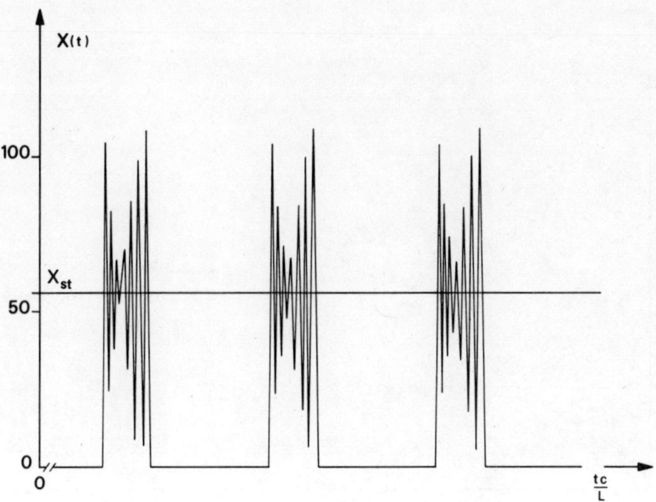

Fig. 7.   Spiking behavior in a situation with 20 unstable modes
           ($C = 20$, $X = 56.25$, $L = 55L$, $\gamma L/c = 0.5$, $T = 0.1$).

        Recently, the self-pulsing instability has been analyzed in
the case of *dispersive* OB in homogeneously broadened systems in a
ring cavity[67]. The main novelty of this dispersive case is that
one can obtain the self-pulsing instability even without a hysteresis cycle (Fig. 8). This suggests that, in order to observe
self-pulsing, the dispersive case can offer definite advantages
over the absorptive one. In fact, in the latter case the incident
field must be increased beyond the upper bistability threshold y_M
(Fig. 4) to allow the system to jump to the high transmission
branch. This may require an exceedingly intense incident field.
On the other hand, in the dispersive case one can get instability
also without bistability and therefore one obtains self-pulsing
for much lower values of the incident field. Another advantage
is that in the dispersive case one can *a priori* exclude the possibility of precipitation.

        Finally, the problem of self-pulsing instabilities for ab-
sorptive OB has been analyzed in the case of a Fabry-Perot[68,69].
One finds that the threshold for the onset of instability is much
higher in that case ($C \gtrsim 60$ instead of $C > 2 (1 + \sqrt{2})$ for a ring
cavity).

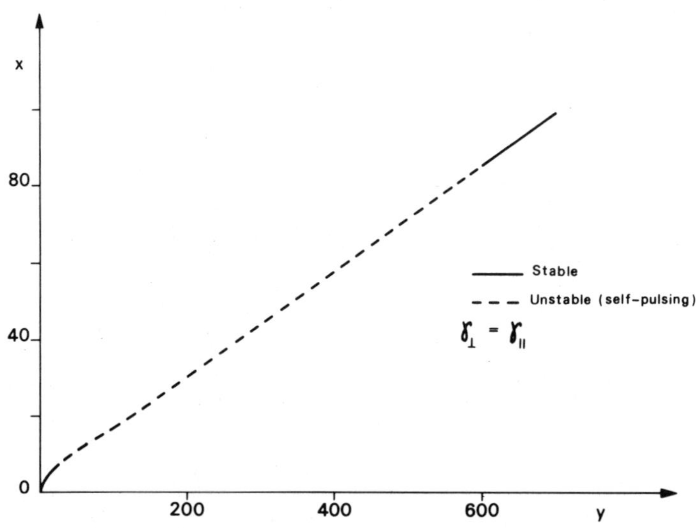

Fig. 8.   Stable and unstable stationary states in a dispersive
         case (C = 30, $\Delta = \theta = 7$).

REFERENCES

1.   A. Szöke, V. Daneu, J. Goldhar and N. A. Kurnit, Appl. Phys.
     Lett. 15, 376 (1969); see also J. W. Austin and L. G.
     Deshazer, J. Opt. Soc. Am. 61, 650 (1971); E. Spiller,
     J. Appl. Phys. 43, 1673 (1972) and H. Seidel, US Patent
     3,610,731 (1971).
2.   S. L. McCall, Phys. Rev. A9, 1515 (1974).
3.   H. M. Gibbs, S. L. McCall and T. N. C. Venkatesan, Phys. Rev.
     Lett. 36, 1135 (1976).
4.   F. S. Felber and J. H. Marburger, Appl. Phys. Lett. 28, 731
     (1976) and Phys. Rev. A17, 335 (1978).
5.   R. Bonifacio and L. A. Lugiato, Opt. Commun. 19, 172 (1976).
6.   P. W. Smith and E. H. Turner, Appl. Phys. Lett. 30, 280 (1977).
7.   T. N. C. Venkatesan and S. L. McCall, Appl. Phys. Lett. 30,
     282 (1977).
8.   T. Bishofberger and Y. R. Shen, Appl. Phys. Lett. 32, 156
     (1978) and Phys. Rev. A19, 1169 (1979); D. Grischkowsky,
     J. Opt. Soc. Am. 68, 641 (1978).
9.   E. Garmire, J. H. Marburger, S. D. Allen and H. G. Winful,
     Appl. Phys. Lett. 34, 374 (1979).
10. H. M. Gibbs, S. L. McCall, T. N. C. Venkatesan, A. C. Gossard,
     A. Passner and W. Wiegmann, Appl. Phys. Lett. 35, 451
     (1979).

11. D. A. B. Miller, S. D. Smith and A. Johnston, Appl. Phys. Lett. 35, 658 (1979).

12. W. J. Sandle and A. Gallagher, in this volume.

13. E. Arimondo, A. Gozzini, L. Lovitch and E. Pistelli, in this volume.

14. R. Bonifacio and L. A. Lugiato in "Coherence and Quantum Optics IV," Ed. by L. Mandel and E. Wolf, Plenum Publishing Co., New York, 1977 and Phys. Rev. A18, 1129 (1978).

15. G. S. Agarwal, L. M. Narducci, D. H. Feng and R. Gilmore, in "Coherence and Quantum Optics IV," Ed. by L. Mandel and E. Wolf, Plenum Publishing Co., New York, 1977.

16. H. J. Carmichael and D. F. Walls, Journ. of Phys. B10, L685 (1977).

17. C. R. Willis, Opt. Comm. 23, 151 (1977).

18. F. T. Arecchi and A. Politi, Lett. Nuovo Cimento 23, 65 (1978).

19. R. Bonifacio and L. A. Lugiato, Phys. Rev. Lett. 40, 1023, 1538 (1978).

20. G. S. Agarwal, L. M. Narducci, R. Gilmore and D. H. Feng, Opt. Lett. 2, 88 (1978); Phys. Rev. A18, 620 (1978); 20, 545 (1979).

21. L. A. Lugiato, Nuovo Cimento B50, 89 (1979).

22. G. S. Agarwal and S. P. Tewari, Phys. Rev. A, in press.

23. F. Casagrande and L. A. Lugiato, Nuovo Cimento B55, 173 (1980).

24. R. Bonifacio, M. Gronchi and L. A. Lugiato, Phys. Rev. A18, 2266 (1978).

25. C. R. Willis, Opt. Commun. 26, 62 (1978).

26. A. Schenzle and H. Brandt, Opt. Commun. 27, 85 (1978) and 31, 401 (1979).

27. F. T. Arecchi and A. Politi, Opt. Commun. 29, 361 (1979).

28. R. F. Gragg, W. C. Schieve and A. R. Bulsara, Phys. Lett. A68, 294 (1978) and Phys. Rev. A19, 2052 (1979).

29. P. D. Drummond and D. F. Walls, Journ. Phys. A13, 725 (1980); J. Chrostowski and A. Zardecki, Opt. Commun. 29, 230 (1979).

30. L. A. Lugiato, J. Farina and L. M. Narducci, Phys. Rev., A22, 253 (1980).

31. P. Hanggi, A. R. Bulsara and R. Janda, Phys. Rev. A, in press.

32. R. Bonifacio and P. Meystre, Opt. Commun. 27, 147 (1978); 29, 131 (1978).

33. P. Meystre and H. Hopf, Opt. Commun. 29, 235 (1979).

34. V. Benza and L. A. Lugiato, Lett. Nuovo Cimento 26, 405 (1979).

35. F. Hopf, P. Meystre, P. D. Drummond and D. F. Walls, Opt. Commun. 31, 245 (1979).

36. R. Bonifacio and L. A. Lugiato, Lett. Nuovo Cimento 21, 517 (1978).

37. S. S. Hassan, P. D. Drummond and D. F. Walls, Opt. Commun. 27, 480 (1978).

38. G. P. Agrawal and H. J. Carmichael, Phys. Rev. A19, 2074 (1979).

39. R. Bonifacio, M. Gronchi and L. A. Lugiato, Nuovo Cimento B53, 311 (1979).

40. P. Schwendimann, Journ. of Physics 12, A, L 39 (1979).

41. C. M. Bowden and C. C. Sung, Phys. Rev. A19, 2392 (1979); and C. M. Bowden, this volume.

42. C. R. Willis and J. Day, Opt. Commun. 28, 137 (1979).

43. S. P. Tewari, Opt. Acta 26, 145 (1979).

44. R. Bonifacio and L. A. Lugiato, Lett. Nuovo Cimento 21, 505 (1978).

45. R. Bonifacio, L. A. Lugiato and M. Gronchi in "Laser Spectroscopy IV," Ed. by H. Walther and W. K. Rothe, Springer-Verlag, 1979.

46. K. Ikeda, Opt. Commun. 30, 257 (1979).

47. R. Roy and M. S. Zubairy, in press.

48. M. Gronchi and L. A. Lugiato, Opt. Lett. 5, 108 (1980).

49. P. Meystre, Opt. Commun. 26, 277 (1978).

50. J. A. Fleck, Jr., Appl. Phys. Lett. 13, 365 (1968).

51. F. Abraham, R. K. Bullough and S. S. Hassan, Opt. Comm. 29, 109 (1979) and 32 (1980).

52. R. Roy and M. S. Zubairy, Opt. Commun. 32, 163 (1980) and Phys. Rev. A21, 274 (1980).

53. H. J. Carmichael, Opt. Acta 27, 147 (1980).

54. J. A. Hermann, Opt. Acta 27, 159 (1980).

55. H. J. Carmichael and J. A. Hermann, Zeit. f. Physik, 38, 365 (1980).

56. M. B. Spencer and W. E. Lamb Jr., Phys. Rev. A5, 884 (1972). Note however that the laser with injected signal is not bistable.

57. S. L. McCall and H. M. Gibbs, Opt. Commun. 33, 335 (1980).

58. R. Bonifacio and L. A. Lugiato, Lett. Nuovo Cimento 21, 510 (1978).

59. R. Bonifacio, M. Gronchi and L. A. Lugiato, Opt. Commun. 30, 129 (1979).

60. S. L. McCall, Appl. Phys. Lett. 32, 284 (1978).

61. H. M. Gibbs, S. L. McCall and T. N. C. Verkatesan, Optics News, Summer 1979.

62. H. Haken, Phys. Lett. 53A, 77 (1975).

63. H. Risken and K. Nummedal, Journ. Appl. Phys. 49, 4662 (1968).

64. R. Graham and H. Haken, Zeits. f. Physik 213, 420 (1968).

65. An example of chaotic behavior is given in Ref. 46 for a case of dispersive multistability. The nature of this behavior is completely different from that of our self-pulsing behavior. In fact, the instability that leads to chaos in Ref. 46 never arises in the absorptive case.

66. M. Gronchi, V. Benza, L. A. Lugiato, P. Meystre and M. Sargent III, submitted for publication.

67. L. A. Lugiato, Opt. Commun. 33, 108 (1980).

68. F. Casagrande, L. A. Lugiato and M. L. Asquini, Opt. Commun. 32, 492 (1980).

69. M. Sargent III, Sov. J. Quantum Electronics, in press.

BISTABLE SYSTEMS IN NONLINEAR OPTICS

D. F. Walls

Physics Department
University of Waikato
Hamilton, New Zealand, and
Joint Institute for Laboratory Astrophysics*
University of Colorado and National Bureau of Standards
Boulder, Colorado 80309 U.S.A.

P. D. Drummond

Department of Physics and Astronomy
University of Rochester
Rochester, New York 14627 U.S.A.

and

K. J. McNeil

Physics Department
University of Waikato
Hamilton, New Zealand

Abstract: A review of intracavity nonlinear optical systems
exhibiting bistability is presented. We consider a coherently driven
Fabry Perot interferometer enclosing an intracavity medium with a
non-linear polarizability. As an example of a system with a second-
order nonlinear susceptibility $\chi^{(2)}$ we consider sub/second harmonic
generation and for a system with a third-order nonlinear susceptibility $\chi^{(3)}$ we consider a nonlinear dispersive medium such as Kerr
liquid. The conditions under which these systems display bistability

---

*Visiting Fellow, 1979–80, Joint Institute for Laboratory Astro-
 physics.

are derived. A quantum mechanical analysis enables a calculation of the spectrum and photon statistics of the transmitted light as well as the lifetime of the metastable states.

I. INTRODUCTION

Systems driven far from equilibrium have been shown to display a rich variety of instabilities in a diverse number of fields ranging from hydrodynamics, nonlinear electronics, plasma physics, chemical reactions to nonlinear optics.[1] We shall consider systems in nonlinear optics which display characteristics typical of many systems driven far from equilibrium. In the basic configuration common to all these systems a nonlinear dissipative system is driven far from equilibrium by an external pumping source. In the particular configuration we shall study in nonlinear optics a medium with a nonlinear optical susceptibility is placed inside an optical cavity which is driven externally by a coherent driving field. Dissipation of the cavity mode is included via the cavity losses. Our description differs from other models for optical bistability which consider a microscopic model for the atoms of the intracavity medium.[2] In our case we consider the medium to be characterized by a macroscopic polarization nonlinear in the electric field amplitudes[3]

$$\underset{\sim}{P} = \chi^{(1)} \cdot \underset{\sim}{E} + \chi^{(2)} : \underset{\sim}{E} \cdot \underset{\sim}{E} + \chi^{(3)} : \underset{\sim}{E} \cdot \underset{\sim}{E} \cdot \underset{\sim}{E} \tag{1.1}$$

where $\chi^{(n)}$ is a $(n + 1)$th rank susceptibility tensor.

We shall consider how optical bistability may arise in systems characterized by $\chi^{(2)}$ (sub/second harmonic generation) and by $\chi^{(3)}$ (nonlinear dispersion). This by no means exhausts the variety of nonlinear optical systems that will exhibit bistability. Instabilities in intracavity parametric oscillators were studied by Oshman and Harris.[4] It may be shown that such systems may exhibit bistability (M. Steyn Ross, private communication). Arecchi and Politi[5] have considered an optical bistable device arising from a two photon absorber placed in a cavity. A number of instabilities occurring in intracavity nonlinear optical systems including Raman active medium have been studied by Lugovoi[6] who derived the conditions required for bistability.

In the section that follows a semiclassical analysis is adopted to yield the state equation for bistability together with the corresponding stability conditions and switching times. The spectrum and photon statistics of the transmitted light together with an estimate of the metastable lifetimes are calculated via a quantum mechanical analysis.

II. DISPERSIVE BISTABILITY – HAMILTONIAN AND SEMICLASSICAL ANALYSIS

In this section we shall consider a medium where we can neglect $\chi^{(2)}$ and concern ourselves with the nonlinear dispersive effects arising from $\chi^{(3)}$. A semiclassical analysis of a nonlinear dispersive Fabry Perot interferometer has been given by Marburger and Felber[7] and a full quantum treatment by Drummond and Walls.[8]

We consider the intracavity field to be represented by a single mode field

$$\underset{\sim}{E} = i \left(\frac{\hbar\omega}{2\varepsilon_o}\right)^{\frac{1}{2}} [a\underset{\sim}{u}(\underset{\sim}{r}) - a^{\dagger}\underset{\sim}{u}^*(\underset{\sim}{r})] \tag{2.1}$$

where a and $a^{\dagger}$ are the boson annihilation and creation operators and $\underset{\sim}{u}(\underset{\sim}{r})$ is the normal mode function. The intracavity field is driven by a coherent driving field and interacts with the nonlinear dispersive medium. This interaction is reversible and may be described by a Hamiltonian H_{rev}. In addition there is an irreversible interaction due to cavity damping which we describe by the Hamiltonian H_{irrev}. The total Hamiltonian is then

$$H = H_{rev} + H_{irrev}$$

$$H_{rev} = \hbar\omega_c \, a^{\dagger}a + \hbar\chi'' \, a^{\dagger 2}a^2 + i\hbar[a^{\dagger}\varepsilon(t) \, a^{-i\omega_p t} - a\varepsilon^*(t) \, e^{i\omega_p t}]$$

$$H_{irrev} = a^{\dagger}\Gamma_c + a\Gamma_c^{\dagger} \tag{2.2}$$

where the rotating wave approximation has been made in the derivation of the anharmonic term and the anharmonicity parameter is defined by

$$\chi'' = \left(\frac{3\hbar\omega^2}{8\varepsilon_o^2}\right)\int \chi^{(3)}(\underset{\sim}{r})|u(\underset{\sim}{r})|^4 \, d^3r \ . \tag{2.3}$$

ω_c is the fundamental cavity resonance, ω_p is the driving frequency, $\varepsilon(t)$ is the driving field amplitude, Γ_c, $\Gamma_c^{\dagger}$ are the reservoir operators for cavity damping. A closely related model to that described by the Hamiltonian above has been studied by Selloni and Schwendiman.[9]

The master equation for the density operator of the cavity field mode is obtained via standard techniques.[10] In a frame rotating at frequency ω_p we have

$$\dot{\rho} = -i\Delta\omega[a^{\dagger}a,\rho] - i\chi''[a^{\dagger 2}a^2,\rho] + [\varepsilon(t)a^{\dagger} - \varepsilon^*(t)a,\rho]$$

$$+ \kappa'(2a\rho a^{\dagger} - \rho a^{\dagger}a - a^{\dagger}a\rho) + 2\kappa' n_{th}[[a,\rho],a^{\dagger}] \qquad (2.4)$$

where κ' is the cavity damping rate, n_{th} is the thermal occupation number due to Gaussian fluctuations in the thermal reservoir Γ_c, and the detuning $\Delta\omega = \omega_c - \omega_p$.

In the semiclassical approximation (i.e., assuming that all correlation functions factorize) the equation for the mean field amplitude $\alpha = \langle a \rangle$ is

$$\frac{\partial\alpha}{\partial t} = \varepsilon(t) - \kappa\alpha + 2\chi |\alpha|^2 \alpha \qquad (2.5)$$

where we have defined the parameters $\kappa = \kappa' + i\Delta\omega$ giving the linear absorption and dispersion and $\chi = i\chi''$ giving the nonlinear dispersion. In the presence of a two-photon absorber, Eq. (2.5) has the same form with $\chi = \chi' + i\chi''$ where χ' gives the nonlinear absorption.

If we define the mean steady state photon number $n = \alpha_0\alpha_0^*$ the steady state of Eq. (2.5) yields

$$|\varepsilon|^2 = n[\kappa'^2 + (\Delta\omega + 2n\chi'')^2] . \qquad (2.6)$$

The cavity photon number exhibits bistability with respect to the driving field intensity provided the detuning exceeds a critical value $\Delta\omega^2 > 3\kappa'^2$ and the sign of the detuning is opposite to that of the anharmonicity, $\Delta\omega\chi'' < 0$. The state equation is shown in Fig. 1. A linear stability analysis reveals that the section with negative slope is unstable. In physical terms the bistability arises as follows. Initially the cavity is detuned well away from resonance and does not transmit the incident light. However, as the intensity of the incident light is increased the intensity-dependent refractive index tunes the cavity into resonance and a high transmission follows.

III. DISPERSIVE BISTABILITY - QUANTUM FLUCTUATIONS

The master Eq. (2.4) may be transformed to a c number Fokker Planck equation using quasi-probability distributions. However, the Fokker Planck equation resulting from use of the Glauber P representation[11] does not have positive definite diffusion and hence its solutions are not always defined (except as generalized functions). For this reason we prefer to use the nondiagonal P representation defined by[12]

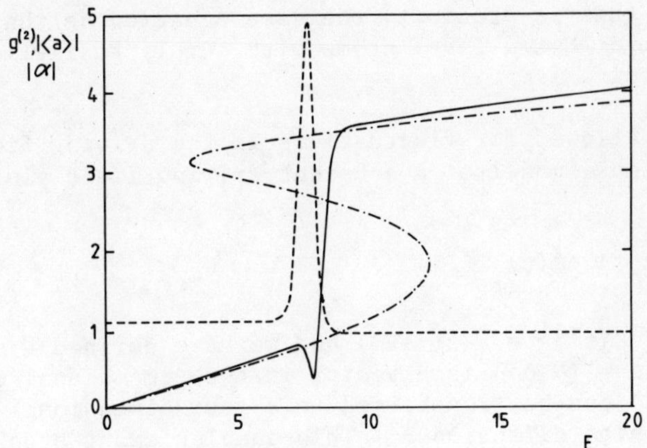

Fig. 1. Chain curve, semiclassical value of steady state field
         amplitude $|\alpha|$ as a function of driving field E; full curve,
         quantum-mechanical mean of steady state field amplitude
         $|<\alpha>|$ as a function of driving field E; broken curve, second
         order correlation function $g^{(2)}(0)$ as a function of driving
         field E. Detuning $\Delta\omega\chi'' < 0$ (parameters $\Delta\omega = -10$, $\kappa' = 1$,
         $\chi'' = 0.5$).

$$\rho = \int_{D} P(\alpha,\beta) \left[\frac{|\alpha><\beta*|}{<\beta*|\alpha>} \right] d\mu(\alpha,\beta) \tag{3.1}$$

as this always has solutions on an appropriate domain. Here $|\gamma>$ is
a coherent state, D is the integration domain and $d\mu$ is the integra-
tion measure. For later use we write $(\alpha,\beta) = (\alpha,\alpha^\dagger)$ where $(\alpha,\alpha^\dagger)$
are not complex conjugate. However there is the following correspond-
ence principle between operators and c numbers

$$\alpha \leftrightarrow a , \qquad \alpha^\dagger \leftrightarrow a^\dagger .$$

The Fokker Planck equation in the generalized P representation corresponding to Eq. (2.4) is

$$\frac{\partial}{\partial t} \underset{\sim}{P}(\alpha) = \frac{\partial}{\partial \alpha} \left(\kappa\alpha + 2\chi\alpha^2\alpha^\dagger - \varepsilon_o(t) \right) - \chi \frac{\partial^2}{\partial \alpha^2} \alpha^2$$

$$+ \frac{\partial}{\partial \alpha^\dagger} \left(\kappa*\alpha^\dagger + 2\chi*\alpha^{\dagger 2}\alpha - \varepsilon_o^*(t) \right)$$

$$\left. - \chi* \frac{\partial^2}{\partial \alpha^{\dagger 2}} \alpha^{\dagger 2} + \Gamma \frac{\partial^2}{\partial \alpha \partial \alpha^\dagger} \right] \underset{\sim}{P}(\alpha) . \tag{3.2}$$

We note that this is precisely the same equation as that obtained from a medium of N two level atoms with purely radiative damping in the dispersive limit.[13]

We have allowed for fluctuations in the driving field by considering a simple model of a coherent driving field plus thermal fluctuations

$$\varepsilon(t) = \varepsilon_o + \delta\varepsilon(t) \tag{3.3}$$

where $<\delta\varepsilon^*(t)\delta\varepsilon(t')> = \Gamma_\varepsilon \delta(t-t')$ and we have defined $\Gamma = \Gamma_\varepsilon + \kappa' n_{th}$. In Eq. (3.2) $\underset{\sim}{\alpha} = (\alpha, \alpha^\dagger)$ is a vector in two-dimensional complex space. Equation (3.2) can be transformed to a four-dimensional equation with positive definite diffusion.[12] This enables exact stochastic differential equations in the Ito calculus to be derived as follows[14]

$$\frac{\partial}{\partial t}\begin{pmatrix}\alpha \\ \alpha^\dagger\end{pmatrix} = \begin{pmatrix}\varepsilon_o(t) - \kappa\alpha - 2\chi\alpha^2\alpha^\dagger \\ \varepsilon_o^*(t) - \kappa^*\alpha^\dagger - 2\chi^*\alpha^{\dagger 2}\alpha\end{pmatrix}$$
$$+ \begin{pmatrix}-2\chi\alpha^2 & \Gamma \\ \Gamma & -2\chi^*\alpha^{\dagger 2}\end{pmatrix}^{\frac{1}{2}}\begin{pmatrix}\eta_1(t) \\ \eta_2(t)\end{pmatrix} \tag{3.4}$$

where $\eta_1(t)$, $\eta_2(t)$ are δ correlated random Gaussian functions;

$$<\eta_i(t)\eta_j(t')> = \delta_{ij}\delta(t-t') . \tag{3.5}$$

This implies that α_i, $\alpha_i^\dagger$ are not complex conjugate except in the mean.

These equations define a probability distribution in two-dimensional[7] complex phase space. If we were to regard each complex variable as a pair of real variables the underlying space would be a four-dimensional real space. The extra dimensions relative to a classical description allow nonclassical processes like photon anti-bunching[15] to occur. However any normal ordered moment or correlation function can still be obtained in a straightforward way as a complex phase space moment or correlation evaluated relative to a real positive probability distribution.

A. Spectrum of the Transmitted Light

To calculate the spectrum of the transmitted light we adopt a linearized fluctuation analysis. We let

$$\underset{\sim}{\alpha} = \underset{\sim}{\alpha_o} + \delta\underset{\sim}{\alpha}(t) \tag{3.6}$$

where α_o is the steady state solution. To first order in the fluctuations, Eqs. (3.4) reduce to

$$\frac{\partial}{\partial t} \delta\underset{\sim}{\alpha}(t) = -\underset{\approx}{A} \delta\underset{\sim}{\alpha}(t) + \underset{\approx}{D}^{\frac{1}{2}}(\alpha_o) \underset{\sim}{\eta}(t) . \tag{3.7}$$

Here $\underset{\approx}{A}$ is the generalized drift and $\underset{\approx}{D}$ is the diffusion array evaluated at $\underset{\sim}{\alpha} = \alpha_o$. These have the general form

$$\underset{\approx}{D} = \begin{bmatrix} -2\chi\alpha_o^2 & \Gamma \\ \Gamma & -2\chi^*\alpha_o^{*2} \end{bmatrix}$$

$$\underset{\approx}{A} = \begin{bmatrix} \kappa+4\chi n & 2\chi\alpha_o^2 \\ 2\chi^*\alpha_o^{*2} & \kappa^*+4\chi^*n \end{bmatrix} \quad \text{where } n = \alpha_o^*\alpha_o . \tag{3.8}$$

The spectrum may be obtained directly following the method of Chaturvedi, et al[16] which gives

$$S(\omega_c + \omega) = n\delta(\omega) + \frac{1}{2\pi} \{[\underset{\approx}{A} + Ii\omega]^{-1} D[\underset{\approx}{A}^T - Ii\omega]^{-1}\}_{21} \tag{3.9}$$

$$= n\delta(\omega) + \frac{1}{2\pi|\lambda(\omega)|^2} [8\kappa'\chi''^2 n^2 + \Gamma(|\kappa+4\chi n+i\omega|^2 + 4\chi''^2 n^2)] \tag{3.10}$$

where

$$\lambda(\omega) = (i\omega + \kappa^* + 4\chi^*n)(i\omega + \kappa + 4\chi n) - 4|\chi|^2 n^2 .$$

This spectrum contains a delta function peak corresponding to the radiation transmitted at the input frequency plus two quasi-Lorentzian peaks symmetrically located about the input frequency at frequencies

$$\omega^{\pm} = \omega_p \pm [\Delta\omega^2 + 8\Delta\omega\chi''n + 12n^2\chi''^2] . \tag{3.11}$$

These lines coalesce to give a single peaked spectrum when

$$\Delta\omega^2 + 8\Delta\omega\chi''n + 12n^2\chi''^2 \leq 0 . \tag{3.12}$$

For a linear interferometer there is an "elastic" (i.e., δ function) peak at the driving frequency ω_p and a Lorentzian peak at the interferometer tuning ω_c due to thermal fluctuations.

In the nonlinear interferometer the tuning point varies since the cavity refractive index is power dependent. In the region of two peaks one peak occurs at the "dressed" cavity frequency ω_c', the other occurs at $2\omega_p - \omega_c'$. In the absence of thermal fluctuations ($\Gamma = 0$) the spectrum arises entirely from the quantum fluctuations and is completely symmetric relative to ω_p. As the threshold point is approached from below the initially separated spectral lines coalesce, then one line diverges at the threshold point. Above the threshold a double peaked spectrum appears on the upper branch as shown in Fig. 2. On reducing the driving intensity the lines coalesce then diverge at the lower threshold. Thus the dispersive bistable high Q interferometer exhibits a new feature not found in the absorptive case.[17] Our results predict a transition from a one peak to a double peak spectrum in switching the interferometer from below to above threshold.

When thermal fluctuations are included the line closest to the effective interferometer tuning is enhanced at low driving intensity. However, above threshold a crossing over effect occurs and the line furthest away from the original linear tuning becomes enhanced. This asymmetry in the spectrum is similar to that observed in the case of detuned atomic fluorescence with a fluctuating driving field.[18-20] In this case there is a symmetric spectrum with a coherent input that becomes asymmetric when fluctuations are included in the input.

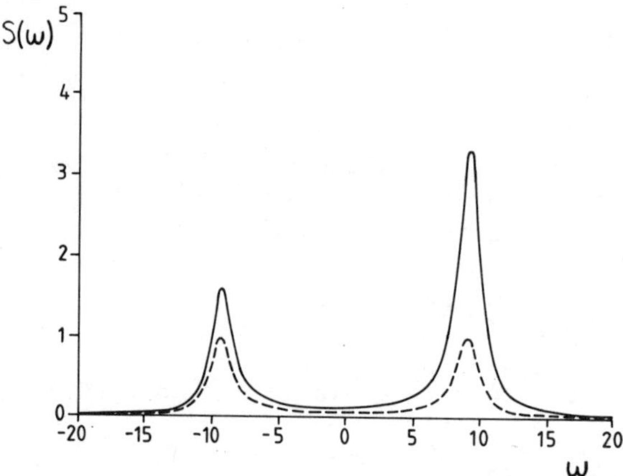

Fig. 2.   Spectrum of the transmitted light above threshold. Broken curve quantum fluctuations only, full curve quantum plus thermal fluctuations.

B. Exact Photon Statistics

In certain cases it is possible to obtain an exact steady state solution for $P(\underset{\sim}{\alpha})$. Consider the Fokker Planck equation

$$\frac{\partial}{\partial t}\, P(\underset{\sim}{\alpha}) = [\frac{\partial}{\partial\mu}\, A_\mu(\underset{\sim}{\alpha}) + \frac{1}{2}\, \partial_\mu\partial_\nu D_{\mu\nu}(\underset{\sim}{\alpha})]P(\underset{\sim}{\alpha}) \ . \tag{3.13}$$

A steady state solution exists provided the potential conditions[21]

$$\partial_\mu V_\nu = \partial_\nu V_\mu \tag{3.14}$$

are satisfied where

$$V_\rho(\underset{\sim}{\alpha}) = [D_{\rho\nu}(\alpha)]^{-1}\, [2A_\nu(\underset{\sim}{\alpha}) + \partial_\sigma D_{\nu\sigma}(\underset{\sim}{\alpha})] \ . \tag{3.15}$$

These conditions are not satisfied in general for the Fokker Planck Eq. (3.2). One may consider the thermal $\Gamma \gg |\chi n|$ or quantum $\Gamma \ll |\chi n|$ limits. For dispersive optical bistability no potential solution exists in the thermal limit. That is in contrast to absorptive optical bistability where a potential solution has been obtained in the thermal limit.[22,23]

In the quantum limit, that is for a perfectly coherent driving field and $kT \ll \hbar\omega_L$, we can neglect thermal fluctuations, that is set $\Gamma = 0$. In this case, Eq. (3.2) satisfies the potential conditions and the steady state solution is given by

$$P^{ss}(\underset{\sim}{\alpha}) = \exp[-\Phi(\underset{\sim}{\alpha})] \tag{3.16}$$

where the potential

$$\Phi(\underset{\sim}{\alpha}) = \int^{\underset{\sim}{\alpha}} V_\rho(\underset{\sim}{\alpha}')d\alpha'_\rho \ . \tag{3.17}$$

Defining the phase of the driving field so that E_0 is real we find

$$P^{ss}(\underset{\sim}{\alpha}) = \alpha^{(c-2)}\, \alpha^{\dagger(d-2)}\, \exp\left[\frac{\varepsilon_0}{\chi}\, (\frac{1}{\alpha} + \frac{1}{\alpha^\dagger}) + 2\alpha\alpha^\dagger\right] \tag{3.18}$$

where

$$c = \frac{\kappa}{\chi} \ , \qquad d = \left(\frac{\kappa}{\chi}\right)^* \ .$$

It can be seen immediately that the usual integration domain of the complex plane with $\alpha^\dagger = \alpha^*$ is not possible since the potential diverges for $\alpha\alpha^\dagger \to \infty$. This means that no Glauber P representation exists in the steady state (except as a generalized function). Instead it is necessary to use a generalized P representation given by Eq. (3.1). This implies an integration domain with $\alpha^\dagger \neq \alpha^*$ defined so that the distribution function vanishes correctly at the boundary. This means choosing new paths of integration for α, $\alpha^\dagger$, that are line integrals on the individual $(\alpha, \alpha^\dagger)$ complex planes. In other words, the new domain will be a complex manifold embedded in the space C^2. The integration is described in Ref. 8.

From the solution for $P^{ss}(\alpha)$ exact expressions for equal time correlation functions of the field may be obtained. These include the mean field amplitude $<a>$ the mean photon number $<a^\dagger a>$ and the second-order correlation function $g^{(2)}(\tau)$ of the field evaluated at $\tau = 0$

$$g^{(2)}(0) = \frac{<a^\dagger a^\dagger aa>}{<a^\dagger a>^2} . \tag{3.19}$$

The results for $<a>$ and $g^{(2)}(0)$ are

$$<a> = \frac{1}{c} \left|\frac{\varepsilon_o}{\chi}\right| F\left(1 + c, d, 2 \left|\frac{\varepsilon_o}{\chi}\right|^2\right) / F\left(c, d, 2 \left|\frac{\varepsilon_o}{\chi}\right|^2\right) \tag{3.20}$$

$$g^{(2)}(0) = \frac{cd\, F\left(c, d, 2 \left|\frac{\varepsilon_o}{\chi}\right|^2\right) F\left(c + 2, d + 2, 2 \left|\frac{\varepsilon_o}{\chi}\right|^2\right)}{(c+1)(d+1)\left[F\left(c + 1, d + 1, 2 \left|\frac{\varepsilon_o}{\chi}\right|^2\right)\right]^2} \tag{3.21}$$

where $F( ) = {}_0F_2( )$ are generalized Gauss hypergeometric series.[24] These quantities are plotted in Fig. 1 where they are compared with the semiclassical value for $<a>$.

It is evident that whereas the semiclassical equation predicts a bistability and hysteresis, the exact steady state equation which includes quantum fluctuations does not exhibit bistability or hysteresis. In the semiclassical analysis the upper and lower branches are found to be stable to small perturbations. In a actual device there will be fluctuations due to thermal noise and quantum noise which may be sufficient to drive the system from one steady state through the unstable region to the other steady state. The lifetimes of the "stable" steady states will thus be noninfinite, so these steady states are metastable.

The extent to which bistability is observed in practice will depend on the fluctuations, which in turn determine the time for random switching from one branch to the other. The driving field must be pumped in time intervals shorter than this random switching time in order for bistability to be observed.

The variance of the fluctuations as displayed by $g^{(2)}(0)$ shows an increase near the transition point due to an enhancement of the fluctuations in this region. The dip in the steady state mean at the transition point is due to out-of-phase fluctuations between the upper and lower branches. While thermal fluctuations always increase $g^{(2)}(0)$ above the input value of unity, for a coherent driving field the quantum noise term can decrease $g^{(2)}(0)$ to a value less than one. This is known as photon antibunching, a uniquely quantum mechanical feature.[15] On the upper branch a $g^{(2)}(0)$ less than one may be obtained, however, a linearized fluctuation analysis reveals that $g^{(2)}(0) = 1 - (1/3n)$ indicating that photon antibunching will only be observed at low intensities.

IV. HARMONIC GENERATION - HAMILTONIAN AND SEMICLASSICAL ANALYSIS

We consider now effects arising from the second-order term in the susceptibility. We consider the interaction of a light mode at frequency ω_1 with another mode at approximately twice the frequency $\omega_2 \approx 2\omega_1$ in a nonlinear crystal with a significant second-order susceptibility. The nonlinear crystal is placed within a Fabry Perot interferometer and both modes are driven with external coherent phase locked driving fields. We assume that the input frequencies are at precisely ω_p and $2\omega_p$ but allow for a mistuning with cavity resonances. The modes may be damped via cavity losses. A full analysis of this system has been given by Drummond et al.[25-27] The system may be described by the following Hamiltonian

$$H = H_{rev} + H_{irrev}$$

$$H_{rev} = \hbar\omega_1 a_1^\dagger a_1 + \hbar\omega_2 a_2^\dagger a_2 + i\hbar \frac{\chi}{2} (a_1^{\dagger 2} a_2 - a_1^2 a_2^\dagger)$$

$$+ i\hbar\left(\epsilon_1 a_1^\dagger e^{-i\omega_p t} - \epsilon_1^* a_1 e^{i\omega_p t}\right)$$

$$+ i\hbar\left(\epsilon_2 a_2^\dagger e^{-2i\omega_p t} - \epsilon_2^* a_2 e^{i\omega_p t}\right) \tag{4.1}$$

$$H_{irrev} = \left(a_1 \Gamma_{c_1}^\dagger + a_1^\dagger \Gamma_{c_1}\right) + \left(a_2 \Gamma_{c_2}^\dagger + a_2^\dagger \Gamma_{c_2}\right)$$

where a_1, a_2 are the boson operators for the modes of frequency ω_1 and ω_2 respectively. χ is the coupling constant for the interaction between the two modes and the spatial mode function is chosen so that χ is real. Γ_{c1}, Γ_{c2} are heat bath operators which represent cavity losses for the two modes and ε_1 and ε_2 are proportional to the coherent driving field amplitudes. Using standard techniques we obtain the following master equation for the reduced density operator of the system

$$\frac{\partial \rho}{\partial t} = \frac{1}{i\hbar}[H_{rev}, \rho] + \sum_{i=1}^{2} \kappa_i(2a_i\rho a_i^\dagger - a_i^\dagger a_i\rho - \rho a_i^\dagger a_i)$$

$$+ \sum_{i=1}^{2} 2\kappa_i n_i^{th}[[a_i,\rho], a_i^\dagger] \tag{4.2}$$

where κ_i are the damping constants of the modes and n_i^{th} are the thermal photon numbers of the heat baths.

The semiclassical equations (i.e., assuming all correlation functions factorize) are: $(\alpha_i = \langle a_i \rangle)$

$$\dot{\alpha}_1 = -i\Delta_1\alpha_1 - \kappa_1\alpha_1 + \chi\alpha_1^*\alpha_2 + \varepsilon_1$$

$$\dot{\alpha}_2 = -i\Delta_2\alpha_2 - \kappa_2\alpha_2 - (\chi/2)\alpha_1^2 + \varepsilon_2 \tag{4.3}$$

together with the complex conjugate equations. Here α_1, α_2 are defined relative to a rotating frame at frequencies ω_p and $2\omega_p$ so that the frequency mismatches are $\Delta_1 = \omega_1 - \omega_p$, $\Delta_2 = \omega_2 - 2\omega_p$. We initially consider the case of zero detuning $\Delta_1 = \Delta_2 = 0$.

We shall be interested in the steady states of these equations. The stability of the steady states may be determined by a linear stability analysis. This reveals the perturbations evolve as $e^{\lambda t}$ where the eigenvalues λ are

$$\lambda_1,\lambda_2 = -\frac{1}{2}[-|\chi\alpha_2^o| + \kappa_1 + \kappa_2] \pm \frac{1}{2}[(-|\chi\alpha_2^o| + \kappa_1 - \kappa_2)^2 - 4|\chi\alpha_1^o|^2]^{\frac{1}{2}}$$

$$\lambda_3,\lambda_4 = -\frac{1}{2}[(|\chi\alpha_2^o| + \kappa_1 + \kappa_2) \pm \frac{1}{2}[(|\chi\alpha_2^o| + \kappa_1 - \kappa_2)^2 - 4|\chi\alpha_1^o|^2]^{\frac{1}{2}}.$$

$$\tag{4.4}$$

These eigenvalues will be used to investigate the stability of the systems studied in the following sections.

A. Subharmonic Generation

First we shall consider the situation where only mode 2 is pumped so that $\varepsilon_1 = 0$, $\varepsilon_2 \neq 0$. This is the process of subharmonic generation where input photons of frequency $2\omega_1$ are converted to photons of frequency ω_1.

From Eqs. (4.3) with $\varepsilon_1 = 0$, we find the following steady state solutions for this system

$$\alpha_1^o = 0 \qquad , \qquad \alpha_2^o = \frac{\varepsilon_2}{\kappa_2} \quad , \quad |\varepsilon_2| < \varepsilon_2^c$$

$$\alpha_1^o = \pm\left[\frac{2}{\chi}\,(\varepsilon_2 - \varepsilon_2^c)\right]^{\frac{1}{2}} \qquad , \qquad \alpha_2^o = \frac{\kappa_1}{\chi} \quad , \quad \varepsilon_2 \leq \varepsilon_2^c$$

$$\alpha_1^o = \pm i\left[\frac{2}{\chi}\,(|\varepsilon_2| - \varepsilon_2^c)\right]^{\frac{1}{2}} , \qquad \alpha_2^o = \frac{-\kappa_1}{\chi} \quad , \quad \varepsilon_2 \leq -\varepsilon_2^c \qquad (4.5)$$

where $\varepsilon_2^c = (\kappa_1\kappa_2/\chi)$ is the threshold value of ε_2. Thus the system exhibits behavior similar to a second-order phase transition at $\varepsilon_2 = \pm\varepsilon_2^c$ where the below threshold solution $\alpha_1^o = 0$ becomes unstable and the system moves onto a new stable branch. This behavior has been noted in an electrical circuit context by Woo and Landauer[28] and in a quantum optical context by Graham.[29]

B. Second Harmonic Generation

If only the lower frequency mode is pumped ($\varepsilon_2 = 0$), we have second harmonic generation where input photons of frequency ω_1 are converted to photons of frequency $2\omega_1$. The steady state solution of Eq. (4.3) with $\varepsilon_2 = 0$ for $|\alpha_2^o|^2$ is a monotonically increasing single valued function of $|\varepsilon_1|^2$. When we examine the stability of this steady state a more interesting phenomenon emerges. In particular we shall examine eigenvalues λ_1 and λ_2 which may be written in the following form after α_1^o is eliminated

$$\lambda_1,\lambda_2 = -\frac{1}{2}[-|\chi\alpha_2^o| + \kappa_1 + \kappa_2] \pm [\{-|\chi\alpha_2^o| + (\kappa_1-\kappa_2)\} + 8\kappa_2|\chi\alpha_2^o|]^{\frac{1}{2}}.$$

$$(4.6)$$

When ε_1 is increased to the point where $|\chi\alpha_2^o| \to \kappa_1 + \kappa_2$ this pair of eigenvalues becomes

$$\lambda_1,\lambda_2 \to 0 \pm i\omega \qquad\qquad\qquad\qquad (4.7)$$

where $\omega = [\kappa_2(2\kappa_1 + \kappa_2)]^{\frac{1}{2}}$. That is, the initially negative real parts vanish, leaving a nonzero imaginary part. Thus the light modes in the cavity undergo a hard mode transition where the steady state becomes unstable and is replaced by a limit cycle behavior. This is illustrated in Fig. 3. The period of the oscillations is of the order of the cavity lifetime. The strength of the driving field corresponding to the onset of self pulsing is given by

$$|\varepsilon_1^c| = \frac{1}{\chi}(2\kappa_1 + \kappa_2)[2\kappa_2(\kappa_1 + \kappa_2)]^{\frac{1}{2}}. \qquad (4.8)$$

Thus the threshold driving parameter to observe self pulsing is of the order of κ^2/χ. This corresponds to a threshold power of the order of watts.

C.   Sub/Second Harmonic Generation

If both modes are coherently driven so that both ε_1 and ε_2 are nonzero both the second harmonic process and the subharmonic process can compete. We shall consider the case where ε_1 and ε_2 are in phase and taken to be real. In the steady state we find from Eq. (4.3)

$$\alpha_1^o = \frac{1}{\kappa_1}[\varepsilon_1 + \chi\alpha_1^{o*}\alpha_2^o] \; ; \quad \alpha_2^o = \frac{1}{\kappa_2}[\varepsilon_2 - (\chi/2)\alpha_1^{o2}] \; . \qquad (4.9)$$

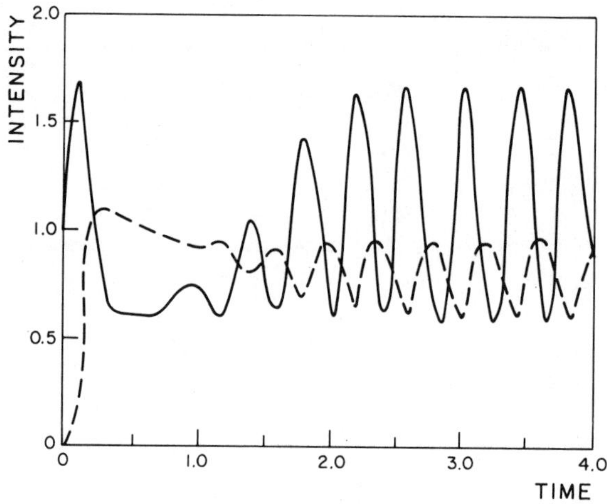

Fig. 3.   Self pulsing in second harmonic generation $|\alpha_1|^2$ (solid line) and $|\alpha_2|^2$ (dotted line) as functions of time. Numerical solution of Eqs. (4.3) with $\chi = 10$, $\kappa_1 = \kappa_2 = 3$, $\varepsilon_1 = 10$. Initial condition $\alpha_1(t=0) = 1.0 + i0.1$; $\alpha_2(t=0) = 0$.

In general α_1^o need not necessarily be real. We let $\alpha_1^o = r\,e^{i\phi}$. Equation (4.9) then gives the following equations for r and ϕ

$$r^3 - \frac{2\varepsilon_2}{\chi}\,r\cos 2\phi + \frac{2\kappa_1\kappa_2}{\chi^2}\,r - \left(\frac{2\kappa_2\kappa_1}{\chi^2}\right)\cos\phi = 0 \; ,$$

$$(4.10)$$

$$\sin 2\phi + \left(\frac{\kappa_2\varepsilon_1}{\chi\varepsilon_2 r}\right)\sin\phi = 0 \; .$$

Obviously under certain conditions solutions with $\phi \neq n\pi$ (i.e., $\sin\phi \neq 0$) can exist. We shall call these "out of phase" solutions. The "in phase" solutions correspond to $\sin\phi = 0$ in which case the amplitude r obeys the following "equation of state"

$$r^3 + (B-A)r = C \tag{4.11}$$

where

$$A = \frac{2\varepsilon_2}{\chi} \; , \qquad B = \frac{2\kappa_1\kappa_2}{\chi^2} \; , \qquad C = \frac{2\kappa_2\varepsilon_2}{\chi^2} \; .$$

The nature of the roots of this cubic equation for r is determined by the value of

$$\delta = \left(\frac{C}{2}\right)^2 + \left(\frac{B-A}{3}\right)^3 \; . \tag{4.12}$$

For $\delta < 0$ there are three real roots in which case the amplitude versus input curve is multiple valued. The bifurcation point $\delta = 0$ defines the following locus in the $(\varepsilon_1,\varepsilon_2)$ plane which defines the boundary of bistability

$$\varepsilon_2 = \varepsilon_2^c + \frac{3}{2}\,\chi^2\left(\frac{\kappa_2\varepsilon_1}{\chi^2}\right)^{2/3} \tag{4.13}$$

where $\varepsilon_2^c = \kappa_1\kappa_2/\chi$ is the critical point for subharmonic generation. From Eq. (4.13) it can be seen that for $\varepsilon_2 \leq \varepsilon_2^c$ there is no value of ε_1 which will give $\delta < 0$. Hence a bistable region exists only when $\varepsilon_2 > \varepsilon_2^c$. This is shown in Fig. 4 where the amplitude versus ε_1 curves are shown for $\varepsilon_2 < \varepsilon_2^c$, $\varepsilon_2 = \varepsilon_2^c$ and $\varepsilon_2 > \varepsilon_2^c$. The eigenvalues (4.4) show that the upper and lower branches are stable and the middle branch is unstable. The two stable branches have a relative phase of $180°$, the upper branch corresponding to $\cos\phi = 1$ and the lower branch to $\cos\phi = -1$. This is similar to the bistability which

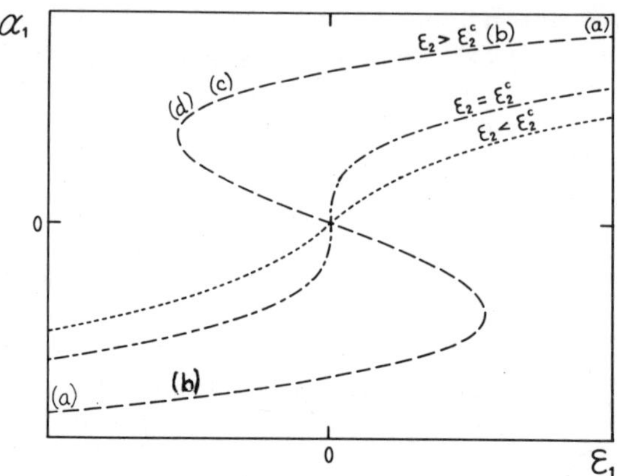

Fig. 4.   Sub/second harmonic generation steady state: α_1 as a
function of ε_1 for ε_2 below, at and above threshold.

occurs in a magnetic system except that no saturation occurs in our
system. In fact we may make the following analogy with a magnetic
system

$$\varepsilon_1 \leftrightarrow H , \quad \alpha_1 \leftrightarrow B , \quad \varepsilon_2 \leftrightarrow T^{-1} , \quad \frac{\kappa_1 \kappa_2}{\chi} \leftrightarrow T_c^{-1} . \qquad (4.14)$$

The above characteristics show that this system has application as
a switching device in optical communication systems. Since the
amplitude α_1 is an odd function of the pump ε_1 the device will function as a phase switch rather than an intensity switch. In the metastable region this system displays a memory effect since the output
retains its phase when the input phase is reversed. A small increase
in the intensity of the input will then cause the phase of the output to switch.

D.   Switching Time

An important quantity when considering the application of this
system as a switching device is the time taken to reach the new
steady state when the pumping is suddenly changed. In order to
simplify the study of the time dependent behavior we shall consider
the adiabatic case where $\gamma_2 \gg \gamma_1$. In the case where $\Delta_1 = \Delta_2 = 0$
and we choose ε_1 and ε_2 to be real and positive then for α_1 initially
real it will remain real and Eqs. (4.3) reduce to

$$\frac{\partial \alpha_1}{\partial \hat{t}} = -[\alpha_1^3 + (B-A)\alpha_1 - C] \tag{4.15}$$

where

$$\hat{t} = \frac{\chi^2}{2\kappa_2}\, t \ .$$

This equation may be solved exactly to give

$$e^{-\hat{t}} = \left(\frac{\alpha_1 - a}{\alpha_1(0) - a}\right)^u \left(\frac{\alpha_1 - b}{\alpha_1(0) - b}\right)^v \left(\frac{\alpha_1 - b^*}{\alpha_1(0) - b^*}\right)^w \tag{4.16}$$

where a and b are the real and complex roots respectively of the cubic Eq. (4.11) and

$$u = [(a-b)(a-b^*)]^{-1} , \qquad v = [(b-a)(b-b^*)]^{-1}$$

$$w = [(b^*-a)(b^*-b)]^{-1} \tag{4.17}$$

in the partial fraction expansion of $[(\alpha_1-a)(\alpha_1-b)(\alpha_1-b^*)]^{-1}$.

The time dependence of α_1 is shown in Fig. 5 where α_1 is plotted as a function of time for various initial conditions. In each case, the input ε_1 is suddenly switched from $-\varepsilon_1(0)$ to $+\varepsilon_1(0)$ at $t = 0$ and the ensuing passage of α_1 to its new steady state is shown. The figure graphically depicts the existence of two different time scales. The first is the time taken to traverse the region near the turning point on the characteristic where $\delta \to 0$. This time becomes very

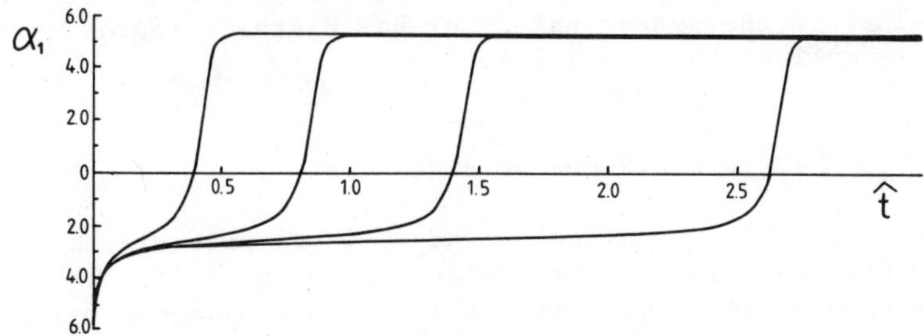

Fig. 5.   Sub/second harmonic generation time dependent behavior: α_1 as a function of $2\gamma_2 t/\kappa^2 = \hat{t}$ calculated from Eq. (4.16) with $\varepsilon_2/\kappa = 20.0$. Plots for various ε_1/κ, decreasing towards the critical value $\varepsilon_1^c/\kappa = 1.72$.

large (i.e., critical slowing down) as $\varepsilon_1(0)$ tends to the critical value given by Eq. (4.13) which corresponds to $\delta = 0$.

The second time scale is the simple relaxation away from the unstable region to the steady state. As the figure shows, this time scale is virtually independent of the input ε_1. Similar behavior has been observed in numerical[30] and analytic studies[31] of the time dependent behavior of optical bistability in atomic systems.

E. Out-of-Phase Solutions

The out-of-phase solutions are solutions of Eqs. (4.10) for which $\sin \phi \neq 0$ and which thus obey

$$\cos \phi = - \frac{C}{2Ar} \tag{4.18}$$

$$r^3 + (A+B)r = 0.$$

The only physical solution for r is

$$r = \left[\frac{2}{\chi} \left(|\varepsilon_2| - \varepsilon_2^c \right) \right]^{\frac{1}{2}} , \qquad \varepsilon_2 < -\varepsilon_2^c . \tag{4.19}$$

The eigenvalues of Eq. (4.6) corresponding to this solution show that this solution is stable providing

$$(\varepsilon_1)^2 \leq \frac{8\chi}{\kappa_2^2} |\varepsilon_2|^2 \left[|\varepsilon_2| - \varepsilon_2^c \right] . \tag{4.20}$$

Thus provided condition (4.20) is satisfied, the solution (4.18) is another stable solution in addition to the in-phase solution of Eq. (4.10). That is the system has a further bistable region whose boundary is defined by Eq. (4.20).

F. Dispersive Bistability

If we allow for detunings Δ_1 and Δ_2 it is possible to obtain bistability in the presence of a single driving field. The steady state solution to Eqs. (4.3) with a finite Δ_1, Δ_2 and a driving field at the fundamental frequency only ($\varepsilon_2 = 0$) is

$$|\varepsilon_1|^2 = (\kappa_1^2 + \Delta_1^2)I + \chi^2 \left(\frac{\kappa_1\kappa_2 - \Delta_1\Delta_2}{\kappa_2^2 + \Delta_2^2} \right) I^2 + \frac{\chi^2}{\kappa_2^2 + \Delta_2^2} I^3 \tag{4.21}$$

where $I = |\alpha_1|^2$.

This cubic equation predicts a bistable behavior in I versus $|\varepsilon_1|^2$ provided

$$\Delta_1\Delta_2 - \kappa_1\kappa_2 > \sqrt{3} \left|\kappa_1\Delta_2 + \Delta_1\kappa_2\right| . \tag{4.22}$$

In the adiabatic limit ($\gamma_2 \gg \gamma_1$) and with $\varepsilon_2 = 0$ the time dependent equation for α_1 from Eq. (4.3) reduces to

$$\dot{\alpha}_1 = \varepsilon_1 - (\kappa_1 + i\Delta_1)_1 - \frac{\chi^2}{2} \left(\frac{\kappa_2 - i\Delta_2}{\kappa_2^2 + \Delta_2^2}\right) |\alpha_1|^2 \alpha_1 . \tag{4.23}$$

This has the same form as Eq. (2.5) for dispersive optical bistability. In this case, the nonlinear susceptibility $\chi^{(3)}$ defined by

$$\chi^{(3)} = \frac{\chi^2}{2} \left(\frac{\kappa_2 - i\Delta_2}{\kappa_2^2 + \gamma_2^2}\right) \tag{4.24}$$

contains an absorptive as well as a dispersive component. The behavior of this system follows closely the analysis given in §II.

G. Phase Diagram

The various instabilities occurring in sub/second harmonic generation may be summarized in a phase diagram in the (ε_1, ε_2) plane (Fig. 6). The phase diagram can be thought of as being defined by the curves which are the boundaries separating the various steady state regimes.

A close analogy can be noted with the phase diagram of an equilibrium thermodynamic system. The point C is a critical point for subharmonic generation and terminates the line DC of first order transitions. The line BB' is a critical line, with B' having a similar behavior to C. The line CE markes the boundary of metastable behavior for the system. Finally the line ABE marks the boundary of hard mode oscillations. The diagram for negative ε_1 is the mirror image of the diagram for positive ε_1. The point B has the characteristic of a triple point, except that the line ABE (as it is a hard mode threshold) would not occur in an equilibrium thermodynamic system. Further structure could be obtained by including the relative phase of the driving fields ε_1 and ε_2 to obtain a three-dimensional phase diagram. In this case the critical line and boundary of hard mode oscillations would become critical surfaces.

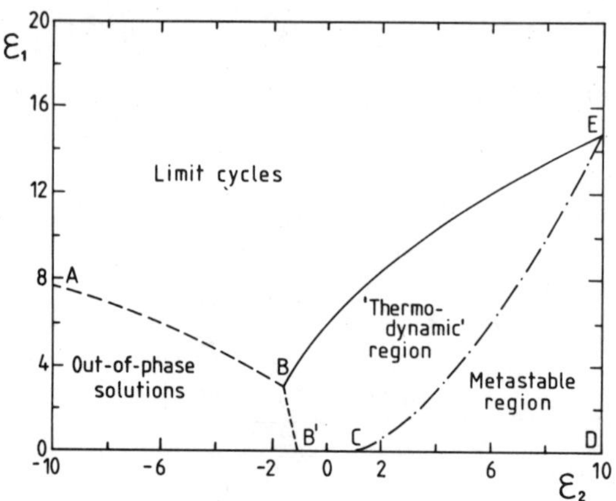

Fig. 6.   Phase diagram for $\gamma_1 = \gamma_2 = \kappa = 1.0$.

V.   HARMONIC GENERATION – QUANTUM FLUCTUATIONS

Adapting the generalized P representation Eq. (3.1) to a two mode system the master Eq. (4.2) may be transformed to the Fokker Planck equation

$$\frac{\partial}{\partial t} P(\underset{\sim}{\alpha}) = \left\{ \frac{\partial}{\partial \alpha_1} \left[(\kappa_1 + i\Delta_1)\alpha_1 - \varepsilon_1 - \chi\alpha_1^\dagger \alpha_2 \right] \right.$$

$$+ \frac{\partial}{\partial \alpha_1^\dagger} \left[(\kappa_1 - i\Delta_1)\alpha_1^\dagger - \varepsilon_1^* - \chi\alpha_1\alpha_2 \right]$$

$$+ \frac{\partial}{\partial \alpha_2} \left[(\kappa_2 + i\Delta_2)\alpha_2 - \varepsilon_2 + \frac{\chi}{2} \alpha_1^2 \right]$$

$$+ \frac{\partial}{\partial \alpha_2^\dagger} \left[(\kappa_2 - i\Delta_2)\alpha_2^\dagger - \varepsilon_2^* + \frac{\chi}{2} \alpha_1^2 \right]$$

$$+ 1/2 \left. \left[\frac{\partial^2}{\partial \alpha_1^2} (\chi\alpha_2) + \frac{\partial^2}{\partial \alpha_1^{\dagger 2}} (\chi\alpha_2^\dagger) + \Gamma_1 \frac{\partial^2}{\partial \alpha_1 \partial \alpha_1^\dagger} + \Gamma_2 \frac{\partial^2}{\partial \alpha_2 \partial \alpha_2^\dagger} \right] \right\} P(\underset{\sim}{\alpha})$$

$$(5.1)$$

where

$$\Gamma_i = 2\kappa_i n_i^{th} , \qquad \underset{\sim}{\alpha} = [\alpha_1, \alpha_1^\dagger, \alpha_2, \alpha_2^\dagger] ,$$

and we have transformed to the rotating frames of the driving fields

$$\alpha_1 \rightarrow \alpha_1 \, e^{-i\omega_p t} \quad , \qquad \alpha_2 \rightarrow \alpha_2 \, e^{-2i\omega_p t} \quad .$$

By defining a Fokker Planck equation in an eight-dimensional space with equivalent observables, and equation with positive semi-definite diffusion can be developed. Using the Ito rules, the resulting stochastic differential equations are[14]:

$$\frac{\partial}{\partial t} \begin{pmatrix} \alpha_1 \\ \alpha_1^\dagger \end{pmatrix} = \begin{pmatrix} \varepsilon_1 + \chi\alpha_1^\dagger\alpha_2 - (\kappa_1 + i\Delta_1)\alpha_1 \\ \varepsilon_1^* + \chi\alpha_1\alpha_2^\dagger - (\kappa_1 - i\Delta_1)\alpha_1^\dagger \end{pmatrix} + \begin{pmatrix} \kappa\alpha_2 & \Gamma_1 \\ \Gamma_1 & \kappa\alpha_2^\dagger \end{pmatrix}^{\frac{1}{2}} \begin{pmatrix} \eta_1(t) \\ \eta_2^\dagger(t) \end{pmatrix}$$

$$\frac{\partial}{\partial t} \begin{pmatrix} \alpha_2 \\ \alpha_2^\dagger \end{pmatrix} = \begin{pmatrix} \varepsilon_2 - \frac{\chi}{2}\alpha_1^2 - (\kappa_2 + i\Delta_2)\alpha_2 \\ \varepsilon_2^* - \frac{\chi}{2}\alpha_1^{\dagger 2} - (\kappa_2 - i\Delta_2)\alpha_2^\dagger \end{pmatrix} + \begin{pmatrix} 0 & \Gamma_2 \\ \Gamma_2 & 0 \end{pmatrix}^{\frac{1}{2}} \begin{pmatrix} \eta_2(t) \\ \eta_2^\dagger(t) \end{pmatrix} \qquad (5.2)$$

where $\eta_i(t)$, $\eta_i^\dagger(t)$ are delta correlated stochastic forces

$$\langle \eta_i(t)\eta_j(t')\rangle = \delta_{ij}\delta(t-t') \quad . \tag{5.3}$$

Thus these equations define a probability distribution in a four-dimensional complex phase space, the underlying real phase space being eight dimensional.

A. Spectrum of the Transmitted Light

We shall calculate the spectrum of the transmitted light using a linearized fluctuation analysis as defined by Eqs. (3.6) and (3.7) where

$$\underset{\sim}{A} = \begin{bmatrix} -\kappa_1 & \chi\alpha_2^o & \chi\alpha_1^{o*} & 0 \\ \chi\alpha_2^{o*} & -\kappa_1 & 0 & \chi\alpha_1^o \\ -\chi\alpha_1^o & 0 & -\kappa_2 & 0 \\ 0 & -\chi\alpha_1^{o*} & 0 & -\kappa_2 \end{bmatrix} \qquad (5.4)$$

and

$$
\underset{\sim}{D} = \begin{bmatrix} \chi\alpha_2^o & \Gamma_1 & 0 & 0 \\ \Gamma_1 & \chi\alpha_2^{o*} & 0 & 0 \\ 0 & 0 & 0 & \Gamma_2 \\ 0 & 0 & \Gamma_2 & 0 \end{bmatrix}
$$

The spectrum is then given directly by Eq. (3.9). We shall neglect
thermal fluctuations, i.e., set $\Gamma_1 = \Gamma_2 = 0$ and calculate the spectrum
due to purely quantum fluctuations. The spectrum of the fundamental
light $S_F(\omega)$ is given by

$$
S_F(\omega) = \left(\frac{\chi\alpha_2^o}{\kappa_2}\right)^2 \left[\left(\frac{\omega}{\kappa_2}\right)^2 + 1\right] \left\{\left(\frac{\chi\alpha_1^o}{\kappa_2}\right)^2 + \frac{\kappa_1}{\kappa_2}\left[\left(\frac{\omega}{\kappa_2}\right)^2 + 1\right]\right\} N^{-1}(\omega).
$$

$$(5.5)$$

The spectrum of the second harmonic light $S_{SH}(\omega)$ is given by

$$
S_{SH}(\omega) = \left(\frac{\chi\alpha_2^o}{\kappa_2}\right)^2 \left\{\left(\frac{\chi\alpha_1^o}{\kappa_2}\right)^2 + \frac{\kappa_1}{\kappa_2}\left[\left(\frac{\omega}{\kappa_2}\right)^2 + 1\right]\right\} N^{-1}(\omega) \qquad (5.6)
$$

where

$$
\frac{N(\omega)}{\pi\kappa_2} = \left[\left(\frac{\chi\alpha_1^o}{\kappa_2}\right)^2 + \frac{\kappa_1}{\kappa_2} - \left(\frac{\omega}{\kappa_2}\right)^2\right]^2 - \left(\frac{\omega}{\kappa_2}\right)^2 \left(\frac{\kappa_1}{\kappa_2} + 1\right)^2
$$

$$
\left(\frac{\chi\alpha_2^o}{\kappa_2}\right)^2 \left[1 - \left(\frac{\omega}{\kappa_1}\right)^2\right] + 2\left(\frac{\omega}{\kappa_2}\right)\left\{\left(\frac{\chi\alpha_2^o}{\kappa_2}\right)^2\right.
$$

$$
\left. - \left(\frac{\kappa_1}{\kappa_2} + 1\right)\left[\left(\frac{\chi\alpha_1^o}{\kappa_2}\right)^2 + \frac{\kappa_1}{\kappa_2} - \left(\frac{\omega}{\kappa_2}\right)^2\right]\right\} .
$$

The spectrum in the region of optical bistability is shown in Fig. 7.
The curves (a)-(b) refer to the spectrum at the steady states shown
on Fig. 4. Curve (a) is just below the threshold for the limit cycle
and we observe a sharp two-peaked spectrum. Decreasing ε_1 the peaks
move closer together and broaden (b) and finally coalesce to a single

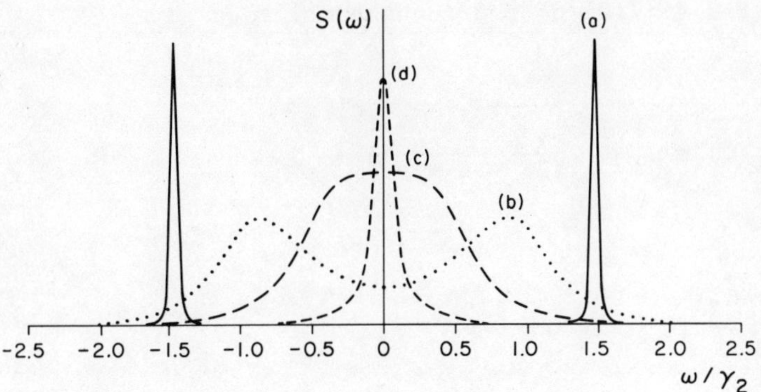

Fig. 7.   Spectrum of the fundamental mode.   Parameters $\kappa_1/\kappa_2 = 0.1$. $\chi\varepsilon_2/\kappa_2^2 = 0.5$.   (a) $\chi\varepsilon_1/\kappa_2^2 = 2.10$.   (b) $\chi\varepsilon_1/\kappa_2^2 = 0.14$. (c) $\chi\varepsilon_1/\kappa_2^2 = -0.11$.   (d) $\chi\varepsilon_1/\kappa_2^2 = -0.317$.

peak (c).   As the unstable point is approached the spectrum becomes very narrow (d), then as the transition to the new branch is made the spectrum discontinuously changes to a double-peaked spectrum (b). (Note the peak height ratio   (a):(b):(c):(d) is 850:2:7.5:350.)

B.   Photon Statistics – Linearized Analysis

We wish to investigate the photon statistics of the light modes characterized by the second order correlation function $g^{(2)}(0)$.   For simplicity we restrict our considerations to the adiabatic case where $\gamma_2 \gg \gamma_1$.   In this limit the stochastic differential equations (5.2) reduce to

$$
\frac{\partial}{\partial t}\begin{pmatrix}\alpha_1 \\ \alpha_1^\dagger\end{pmatrix} = \begin{pmatrix} \varepsilon_1 + \frac{\chi}{\kappa_2}(\varepsilon_2 - \frac{\chi}{2}\alpha_1^2)\alpha_1^\dagger - \kappa_1\alpha_1 \\[2ex] \varepsilon_1^* + \frac{\chi}{\kappa_2}(\varepsilon_2^* - \frac{\chi}{2}\alpha_1^{\dagger 2})\alpha_1 - \kappa_1\alpha_1^\dagger \end{pmatrix}
$$

$$
+ \begin{bmatrix} \frac{\chi}{\kappa_2}(\varepsilon_2 - \frac{\chi}{2}\alpha_1^2) & \Gamma_1 \\[2ex] \Gamma_1 & \frac{\chi}{\kappa_2}(\varepsilon_2^* - \frac{\chi}{2}\alpha_1^{\dagger 2}) \end{bmatrix}^{\frac{1}{2}} \begin{pmatrix} \eta_1(t) \\[2ex] \eta_1^\dagger(t) \end{pmatrix} \tag{5.7}
$$

where thermal fluctuations in the second harmonic mode have been neglected (i.e., $\Gamma_2 = 0$) and $\Delta_1 = \Delta_2 = 0$.   A linearized fluctuation analysis for $g^{(2)}(0)$ of the fundamental mode gives

$$g^{(2)}(0) = 1 + \frac{1}{n}\,\frac{\left[\dfrac{2\kappa_2\Gamma_1}{\chi^2} + A - n\right]}{(3n - A + B)} \tag{5.8}$$

where

$$A = \frac{2\varepsilon_2}{\chi}\,, \qquad B = \frac{2\kappa_1\kappa_2}{\chi^2}$$

and $r = |\alpha_1^o|^2$ is the deterministic mean where α_1^o is the solution of
Eq. (4.11). The denominator of Eq. (5.8) is always positive in the
region of stable real single valued solutions. Hence the existence
of photon bunching or antibunching is determined by the sign of the
numerator. The condition for photon antibunching ($g^{(2)}(0) < 1$) is

$$n > \frac{\varepsilon_2}{\chi} + \frac{2\kappa_2\Gamma_1}{\chi^2} \tag{5.9}$$

The antibunching may be enhanced by choosing ε_2 negative, that is,
180° out of phase with ε_1. If ε_2 is taken very large and negative
$g^{(2)}(0)$ would tend to the quantum limit of $1 - (1/n)$. However, this
limit is not physically accessible since it involves crossing the
boundary between real and complex solutions [see Eq. (4.20)]. Pro-
vided $\varepsilon_2 \approx -\varepsilon_2^c$, the stability properties are satisfied and the result
for the intensity correlation functions is (with $\Gamma_1 = 0$)

$$g^{(2)}(0) = 1 - \frac{1}{2n} \tag{5.10}$$

Normally, when a second harmonic is frequency converted to give a
field at half the frequency, there is a photon bunching effect, as
occurs here for large enough $|\varepsilon_2|$. However, when both modes are
pumped externally with inputs out of phase by 180° the additional
input field at the fundamental frequency can cause induced emission
of photons out of the fundamental mode thus enhancing the antibunching effect compared to when just one field is input. This is a
similar effect to that proposed by Stoler[32] for a transient situation.
In the present case, the effect occurs in the steady state, an advantage for experimental measurement.

C. Exact Photon Statistics

We shall demonstrate how exact solutions in certain cases may be obtained for the steady state distribution function $P(\alpha)$. Again we consider the adiabatic limit ($\kappa_2 \gg \kappa_1$). The Fokker Planck equation corresponding to Eq. (5.7) is

$$\frac{\partial}{\partial t} P(\alpha_1,\alpha_1^\dagger) = \{\frac{\partial}{\partial \alpha_1} [\kappa_1\alpha_1 - \varepsilon_1 - \frac{\chi}{2} (\varepsilon_2 - \frac{\chi}{2} \alpha_1^2)\alpha_1^\dagger]$$

$$+ \frac{\partial}{\partial \alpha_1^\dagger} [\kappa_1\alpha_1 - \varepsilon_1^* - \frac{\chi}{2} (\varepsilon_2^* - \frac{\chi}{2} \alpha_1^{\dagger 2})\alpha_1]$$

$$+ \frac{1}{2} [\frac{\partial^2}{\partial \alpha_1^2} \frac{\chi}{\kappa_2} (\varepsilon_2 - \frac{\chi}{2} \alpha_1^2) + \frac{\partial^2}{\partial \alpha_1^{\dagger 2}} \frac{\chi}{\kappa_2} (\varepsilon_2^* - \frac{\chi}{2} \alpha_1^{\dagger 2})$$

$$+ \Gamma_1 \frac{\partial^2}{\partial \alpha_1 \partial \alpha_1^\dagger}]\} P(\alpha_1,\alpha_1^\dagger) \ . \tag{5.11}$$

A general solution to this equation is difficult hence we shall consider two limiting cases.

(1) Thermal Limit

When $\Gamma_1 \gg |(\chi/\kappa_2)[\varepsilon_2 - (\chi/2)\alpha_1^2]|$ the quantum noise terms in the diffusion terms of Eq. (5.11) can be set equal to zero. The resulting equation can then be shown to satisfy the potential conditions (3.14). The steady state solution therefore has the form

$$P(\alpha_1,\alpha_1^\dagger) = N \exp[-\Phi(\alpha_1,\alpha_1^\dagger)] \tag{5.12}$$

where N is a normalization constant and

$$\Phi(\alpha_1,\alpha_1^\dagger) = \frac{2}{\Gamma_1} \{\frac{\chi^2}{4\kappa_2} (\alpha_1^\dagger\alpha_1)^2 + \kappa_1\alpha_1^\dagger\alpha_1$$

$$- \frac{\chi\varepsilon_2^*\alpha_1^2}{2\kappa_2} - \frac{\chi\varepsilon_2}{2\kappa_2} \alpha_1^{\dagger 2} - \varepsilon_1^*\alpha_1 - \varepsilon_1\alpha_1^\dagger\} \tag{5.13}$$

may be regarded as a potential function where the potential is bounded and integrable on the domain $\alpha^\dagger = \alpha^*$ so that the Glauber–Sudarshan P representation is well defined. A steady state solution for the Wigner function in this limit has been obtained by Brand et al.[33]

It is interesting to compare the probability distribution (5.13) with that for a laser with injected signal. Changing to the variables R and θ where $\alpha_1 = R\, e^{i\theta}$, the potential functions can be written:

a) <u>Laser with injected signal</u>[34-36]

$$\Phi(R,\theta) = -\frac{2}{\Gamma}\, (\chi R^4 + \kappa R^2 - 2\varepsilon R \cos\theta) \qquad (5.14)$$

where χ is the susceptibility of the laser medium, κ the cavity line width, Γ the strength of the thermal fluctuations and ε the amplitude of the injected signal.

b) <u>Sub/second harmonic generation.</u> Choosing ε_1 and ε_2 in phase and real

$$\Phi(R,\theta) = \frac{2}{\Gamma_1}\, \left(\frac{\chi^2}{4\kappa_2} R^4 + \kappa_1 R^2 - 2\varepsilon_1 R \cos\theta - \frac{\chi\varepsilon_2 R^2 \cos 2\theta}{\kappa_2}\right) . \qquad (5.15)$$

In the case of the laser, the potential has only one local minimum with respect to variation in θ so no bistability can occur. However, the potential for sub/second harmonic generation may have more than one local minimum in the θ variable giving the possibility of bistability. These two potentials are shown for comparison in Figs. 8 and 9.

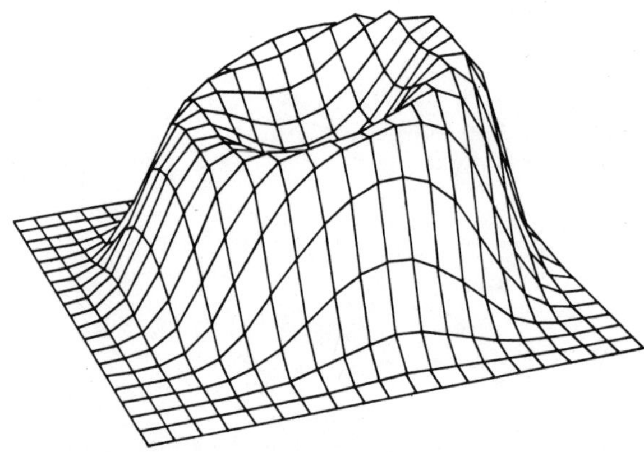

Fig. 8. Three-dimensional plot of the potential for the laser with injected signal, $2\chi/\Gamma$ = 2.0, $2\gamma/\Gamma$ = 2.0, $4\varepsilon/\Gamma$ = 0.25.

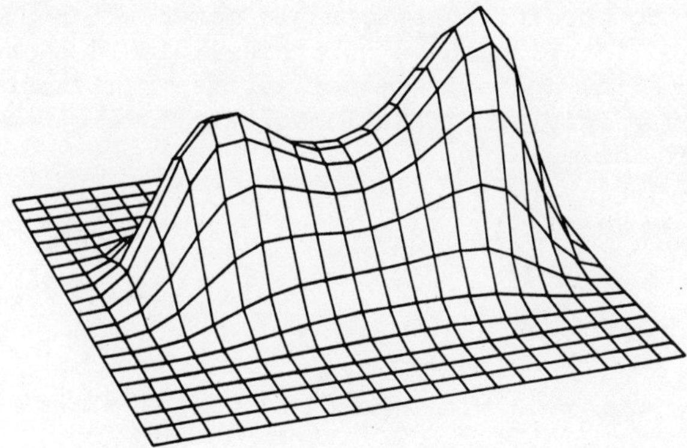

Fig. 9.   Three-dimensional plot of the potential for sub/second har-
          monic generation $\kappa^2/2\Gamma_1\gamma_2 = 2.0$, $2\gamma_1/\Gamma_1 = 1.0$, $4\epsilon_1/\Gamma_1 = 0.25$,
          $8\kappa\epsilon_2/\Gamma_1\gamma_2 = 2.0$.

(2)   "Quantum Limit"

When $\Gamma_1 << |(\chi/\kappa_2)[\epsilon_2 - (\chi/2)\alpha_1^2]|$ the thermal noise term Γ_1 may
be neglected in the Fokker Planck Eq. (5.11). Again the method of
potential solutions may be used, giving the solution

$$P(\alpha_1,\alpha_1^\dagger) = N \exp[2\alpha_1^\dagger\alpha_1 + \frac{2\bar{\kappa}_1\kappa_2}{\chi^2} \ln(c^2 - \chi^2\alpha_1^2)$$

$$+ 2 \left(\frac{\bar{\kappa}_1\kappa_2}{\chi^2}\right)^* \ln\left(c^{*2} - \chi^2\alpha_1^{\dagger 2}\right) + \frac{2\kappa_2\epsilon_1}{c\chi} \ln \left(\frac{c+\chi\alpha_1}{c-\chi\alpha_1}\right)$$

$$+ 2 \left(\frac{\kappa_2\epsilon_1}{c\chi}\right)^* \ln\left(\frac{c^*+\chi\alpha_1^\dagger}{c^*-\chi\alpha_1^\dagger}\right)]$$
$$\tag{5.16}$$

where

$$C = \sqrt{2\chi\epsilon_2} , \qquad \bar{\kappa}_1 = \kappa_1 - \frac{\chi^2}{2\kappa_2} .$$

Note that κ_1 and κ_2 need not be real in order for this solution to
exist. Thus, by taking κ_1 and κ_2 to mean $\kappa_1 + i\Delta_1$ and $\kappa_2 + i\Delta_2$
respectively, the detunings Δ_1 and Δ_2 may be included in this
analysis.

    D. F. WALLS, P. D. DRUMMOND, AND K. J. McNEIL

In order to obtain the observable moments for this case, it is necessary to integrate on a suitable manifold chosen so that the distribution (5.16) and all its derivatives vanish at the boundary of integration. Details of the integration are given in Ref. 27. The result for the moments

$$I_{nn'} = \oint\oint \alpha^{\dagger n} \alpha^{n'} P(\alpha) d\alpha^{\dagger} d\alpha \tag{5.17}$$

is

$$I_{nn'} = N \sum_{n=0}^{\infty} \frac{2^m}{m!} \left(\frac{C}{\chi}\right)^{m+n} \left(\frac{C^*}{\chi}\right)^{m+n'}$$

$$\times {}_2F_1[-(m+n),j_1,j_2,2] \, {}_2F_1[-(m+n),j_1^*,j_2^*,2] \tag{5.18}$$

where

$$j_1 = \frac{2\kappa_1\kappa_2}{\chi^2} + \frac{2\kappa_2\varepsilon_1}{C\chi} , \qquad j_2 = \frac{2\kappa_1\kappa_2}{\chi^2}$$

and the ${}_2F_1$ are hypergeometric functions.

Plots of the mean amplitude $\langle\alpha_1\rangle$, the mean intensity $\langle\alpha_1^{\dagger}\alpha_1\rangle$ and the second order correlation function $g^{(2)}(0)$ as functions of ε_1 are shown in Fig. 10 in the region of bistability ($\varepsilon_2 > \varepsilon_2^C$). the stochastic mean amplitude shows a continuous single valued transition in contrast to the semiclassical mean in Fig. 4. The second order correlation function $g^{(2)}(0)$ is less than one on both branches signifying photon antibunching. There is a peak in $g^{(2)}(0)$ at $\varepsilon_1 = 0$ corresponding to an increase in fluctuations at the transition from one stable branch to another.

D.   Metastable Lifetime

The metastable lifetime of the branches in sub/second harmonic generation may be calculated in the thermal limit using the method of Kramers[37] and Landauer.[38] We require the potential solution in the thermal limit given by Eq. (5.13). Since the turning points z_1, z_2, z_3 of Φ occur along the real axis, we write the potential (5.13) in terms of the real and imaginary parts, x and y respectively of α_1:

$$\Phi(x,y) = \frac{1}{\Gamma_1}\left[\frac{\chi}{4\kappa_2} (x^2+y^2)^2 + \kappa_1(x^2+y^2) - \frac{\chi\varepsilon_2}{\kappa_2} (x^2-y^2) - 2\varepsilon_1 x\right]. \tag{5.19}$$

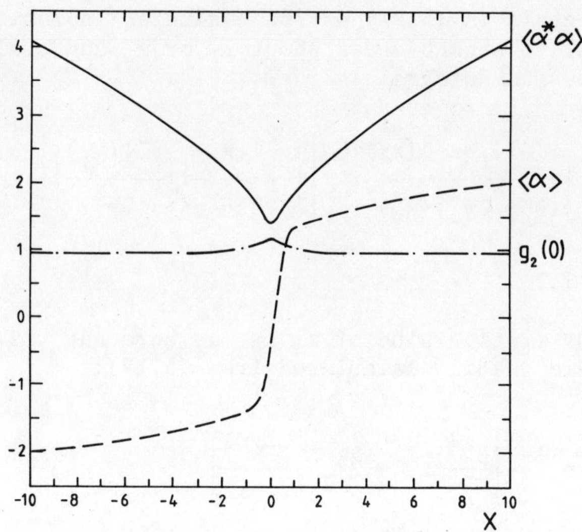

Fig. 10.   Mean amplitude, mean intensity and second order correla-
           tion function in sub/second harmonic generation as func-
           tions of ε_1, $\varepsilon_2 = 2.0$, $\gamma_1 = 1.0$, $\gamma_2 = 100$, $\kappa = 20$.

The method of Kramers[37] and Landauer[38] yields the following expression for the time constant for the transitions between the minima at
z_1 and z_3 via the saddle point at z_2:

$$\frac{1}{\tau} = \frac{1}{\tau_1} + \frac{1}{\tau_2} = \frac{\Gamma_1}{4} \left(e^{F_1 - F_2} + e^{F_3 - F_2} \right) . \tag{5.20}$$

The "free energies" F_1 and F_3 are defined in terms of local potential
integrals in the neighborhood of the minima at z_1 and z_3:

$$F_1, F_3 = -\ln \{ \int_{z_1, z_2} \exp[-\Phi(z)] d^2 z \}$$

The "effective free energy" F_2 at the saddle point z_2 is given by
expanding the potential about

$$z_2 = x_2 + iy_2$$

$$\Phi(z) \approx \Phi(z_2) - \frac{1}{2} a(x-x_2)^2 + \frac{1}{2} b(y-y_2)^2$$

$$F_2 \equiv \Phi(z_2) + \frac{1}{2} \ln(b/a) . \tag{5.22}$$

The integrals in (5.21) may be evaluated approximately by expanding $\Phi(z)$ to second order about z_1 or z_2 and evaluating the resulting gaussian integrals

$$\frac{1}{\tau_i} \approx \frac{\Gamma_1}{8\pi} \left[\frac{-(\partial^2\Phi/\partial x^2)(x_2)}{(\partial^2\Phi/\partial y^2)(x_2)}\right]^{1/2} \left[\left(\frac{\partial^2\Phi(x_i)}{\partial x^2}\right)\left(\frac{\partial^2\Phi(y_i)}{\partial y^2}\right)\right]^{1/2} e^{\Phi(x_i)-\Phi(x_2)}$$

$$i = 1,3 . \tag{5.23}$$

Here x_i is the (real) value of z_i (y_i is zero and $\partial^2\Phi(x_i)/\partial x^2$ and $\partial^2\Phi(x_i)/\partial y^2$ are readily calculated from (5.19):

$$\frac{\partial^2\Phi(x_i)}{\partial x^2} = \frac{1}{\Gamma_1} \left[\frac{3\chi^2}{\kappa_2}x_i^2 + 2\kappa_1 - \frac{2\chi\varepsilon_2}{\kappa_1}\right]$$

$$\frac{\partial^2\Phi(x_i)}{\partial y^2} = \frac{1}{\Gamma_1} \left[\frac{\chi^2}{\kappa_2}x_i^2 + 2\kappa_1 + \frac{2\chi\varepsilon_2}{\kappa_2}\right] . \tag{5.24}$$

In the thermodynamic limit the depths of the potential wells become infinite and the effect of fluctuations becomes insignificant giving two essentially stable states. A calculation of the mean first passage time using the Wigner distribution has been made by Brand et. al.[33]

VI. DISCUSSION

A survey of intracavity nonlinear optical systems that display optical bistability has been presented. As particular examples we have analyzed dispersive bistability arising from a cubic nonlinear susceptibility $\chi^{(3)}$ and sub/second harmonic generation arising from a quadratic nonlinear susceptibility $\chi^{(2)}$. The criteria to observe bistability are examined and calculations of the spectrum and photon statistics of the transmitted light together with the switching times and metastable lifetimes are presented. The spectrum of the transmitted light in dispersive bistability displays a splitting into two peaks on the upper branch in contrast to absorptive bistability where the spectrum remains single peaked.

Coherently driven intracavity second harmonic generation provides an attractive mechanism for producing a series of periodic pulses from a cw input. Sub/second harmonic generation displays bistability via a phase change of $180°$. This would provide a useful mechanism for a switching device in phase sensitive optical processes without large intensity variations.

Under certain conditions the transmitted light produced by these nonlinear optical processes will display photon antibunching, i.e., $g^{(2)}(0) < 1$. Photon antibunching has recently been observed[39,40] in resonance fluorescence from an atomic beam of two level atoms according to the predictions of Carmichael and Walls.[41] However, the atomic number fluctuations in the beam prevent a direct observation of a $g^{(2)}(0) < 1$. The experiments in nonlinear optics suggested in this article should enable a $g^{(2)}(0) < 1$ to be measured directly. An interesting feature of the nonclassical correlations obtained in this way, is that they are essentially a property of the interacting modes only. The atoms in the nonlinear medium play a passive role and achieve coupling of the field modes via virtual transitions. The type of correlation obtained is a function of the phase and amplitude of the input driving fields. An experimental verification of these correlations would provide evidence that non-classical correlation properties are inherent in the quantized radiation field, rather than due to a neoclassical mechanism in which the field is not required to be quantized.[42]

The nonlinear optical systems considered in this article offer the possibility of achieving optical bistability in solid state devices. Such systems have great potential for optical switching devices in optical communication systems. In addition, these systems are of fundamental interest in the study of far from equilibrium phase transitions. The precise degree of control achievable with highly stabilized dye lasers together with the accurate measurements of spectral characteristics and photon statistics available in the optical region make these nonlinear optical systems prime candidates for the study of quantum fluctuations in non-equilibrium systems.

ACKNOWLEDGMENTS

We wish to thank the New Zealand University Grants Committee for its continued support of this work.

REFERENCES

1. See the papers in _Progress in Theoretical Physics_, Supplement No. 64 (1978), and the monographs by H. Haken, _Synergetics_ (Springer Verlag, 1977), and G. Nicolis and I. Prigogine, _Self Organization in Nonequilibrium Systems_ (Wiley, New York 1977).

2. See the papers by S. L. McCall, L. A. Lugiato and R. Bonifacio, and W. J. Sandle, R. K. Ballagh and A. Gallagher in this volume and references therein.

3. N. Bloembergen, _Nonlinear Optics_ (Benjamin, New York, 1965).

4. M. K. Oshman and S. E. Harris, IEEE J. Quant. Electron. $\underline{4}$, 491 (1969).

5. F. T. Arecchi and A. Politi, Lett. Nuovo Cimento 21, 510 (1978).
6. V. N. Lugovoi, Optica Acta 24, 743 (1977).
7. J. H. Marburger and F. S. Felber, Phys. Rev. A17, 335 (1978).
8. P. D. Drummond and D. F. Walls, J. Phys. A13, 725 (1980).
9. A. Selloni and P. Schwendiman, Optica Acta 26, 1541 (1979).
10. W. H. Louisell, Quantum Statistical Properties of Radiation
    (Wiley, New York, 1973).
11. R. J. Glauber, Phys. Rev. 130, 2529 (1963); 131, 2766 (1963).
12. P. D. Drummond and C. W. Gardiner, J. Phys. A13, 2353 (1980).
13. P. D. Drummond and D. F. Walls (to be published).
14. L. Arnold, Stochastic Differential Equations (Wiley, New York,
    1974).
15. D. F. Walls, Nature 280, 451 (1979).
16. S. Chaturvedi, C. W. Gardiner, I. Matheson and D. F. Walls,
    J. Stat. Phys. 17, 469 (1977).
17. L. A. Lugiato, Il Nuovo Cimento 50B, 89 (1979).
18. H. J. Kimble and L. Mandel, Phys. Rev. A15, 689 (1977).
19. K. Wodkiewicz, Phys. Lett. 66A, 369 (1978).
20. P. L. Kinght, W. A. Molander and C. R. Stroud, Phys. Rev. A17,
    1547 (1978).
21. H. Haken, Rev. Mod. Phys. 47, 67 (1975).
22. R. Bonifacio and L. Lugiato, Phys. Rev. A18, 1129 (1978).
23. A. Schenzle and H. Brand, Opt. Comm. 27, 85 (1978).
24. I. S. Gradshteyn and I. M. Ryzhik, Table of Integrals, Series
    and Products, (Academic Press, New York, 1965).
25. K. J. McNeil, P. D. Drummond and D. F. Walls, Opt. Comm. 27,
    292 (1978).
26. P. D. Drummond, K. J. McNeil and D. F. Walls, Opt. Comm. 28,
    255 (1979).
27. P. D. Drummond, K. J. McNeil and D. F. Walls, Optica Acta (in
    press).
28. J. W. F. Woo and R. Landauer, IEEE J. Quant. Electron. 7, 435
    (1971).
29. R. Graham, Springer Tracts in Modern Physics, Vol. 66 (Springer
    Verlag, 1973).
30. R. Bonifacio and P. Meystre, Opt. Comm 27, 147 (1978).
31. V. Benza and L. Lugiato, Lett. Nuovo Cimento 26, 405 (1979).
32. D. Stoler, Phys. Rev. Lett. 33, 1397 (1974).
33. H. Brand, R. Graham and A. Schenzle, Opt. Comm. 32, 359 (1980).
34. W. W. Chow, M. O. Scully and E. W. van Stryland, Opt. Comm. 15,
    6 (1975).
35. L. Lugiato, Lett. Nuovo Cimento 23, 609 (1978).
36. P. D. Drummond, D. Phil. Thesis, University of Waikato, 1979
    (unpublished).
37. H. A. Kramers, Physica 7, 284 (1940).
38. R. Landauer, J. Appl. Phys. 23, 2209 (1962).
39. H. J. Kimble, M. Dagenais and L. Mandel, Phys. Rev. Lett. 39,
    691 (1977); Phys. Rev. A18, 201 (1978).
40. G. Leuchs, M. Rateike and H. Walther (to be published).

41. H. J. Carmichael and D. F. Walls, J. Phys. B9, L43, 1199 (1976).
42. D. T. Pegg, J. Phys. A13, 1389 (1980).

BISTABILITY IN IRRADIATED JOSEPHSON JUNCTIONS

G. S. Agarwal and S. R. Shenoy

School of Physics, University of Hyderabad

Hyderabad-500001, India

Abstract: In this paper we consider a suitably prepared
Josephson junction, with external resistance across it, interacting
with an external coherent radiation. We show the existence of a
non-equilibrium first order phase transition in such a system. The
non-equilibrium steady states of the system are determined by a
balance between the drive and dissipation terms. The self-consistently developed voltage shows hysteresis, the size of which
could be controlled by varying the resistance across it. Estimates
are given to show that the predicted bistability in Josephson junctions is indeed within the reach of experiments. A detailed study
of the relaxation time T_1 and the decay time T_2 is made, to show
that the system would indeed show a hysteresis behavior rather than
a simple jump behavior. The size of the critical region is also
estimated.

Recently optical bistability has attracted a good deal of attention both experimentally and theoretically[1] and various optical
systems have been suggested that could show a variety of bistable
behavior under suitable excitation conditions and choice of the
parameters of the system[2]. It may be noted that optical bistability
constitutes a very interesting example of a nonequilibrium system
which exhibits a first order phase transition[3]. From the point of
view of nonequilibrium phase transitions, it is instructive to examine if certain solid state systems would show bistable behavior.
Such a study has recently begun[4-6]. We have predicted[5] bistability
in Josephson junctions whereas a recent paper by Toyazawa[4] predicts
bistability in exciton-photon interactions. The situation we get
in the case of Josephson junctions is quite analogous to optical
bistability though there are major differences due to the different

type of physics involved. The nonequilibrium effects discussed in
this paper are completely different from the well known Fiske,
Shapiro steps[7] and also the recently discovered nonequilibrium superconductivity[8] where the drive acts on quasiparticles whereas in our
case the drive directly acts on the pairs tunnelling through the
junction.

We consider a Josephson Junction, Fig. 1, which is irradiated
by external coherent radiation and which has a resistance R across
it. We assume that there is no externally applied d.c. voltage.
The tunnelling Hamiltonian can be written in the form[9]

$$H_T = -\hbar \frac{I_j}{4e} (S^+ + S^-)$$
(1)

where I_j is the experimentally measured maximal Josephson tunnelling
current. In Eq. (1) $S^+(S^-)$ represents a pair transfer from left
(right) to right (left) and S^z counts the number of excess pairs in
the left superconductor. The commutation relations for these operators are

$$[S_z, S^\pm] = \pm S^\pm , \quad [S^+, S^-] \cong \frac{2}{m^2} S^z ,$$
(2)

where m is the number of electron states within $\pm \omega_D$ of the Fermi
surface. The difference between the pair transfer operator commutation relations and the angular momentum commutation relations
is worth noticing. The Hamiltonian giving the unperturbed energy
of the charged electrode-oxide-electrode capacitor will be

$$H_c = \frac{1}{2} \frac{(2eS^z)^2}{C} , \quad C = \ell_x \ell_y \, \varepsilon / 4\pi \ell_z ,$$
(3)

where ε is the dielectric constant of the oxide layer. It can be
shown that the interaction Hamiltonian for interaction between the
cavity mode of frequency ω_c and the wave vector $k = \pi/\ell_x$ has the
approximate form

$$H_{cav} = i\hbar T(S^- - S^+)(a + a^+) + \hbar \omega_c a^+ a ,$$
(4)

where

$$T = \frac{I_j}{2} \sqrt{\frac{\pi \ell_z}{\hbar \omega_c \ell_x \ell_y \varepsilon}} .$$
(5)

In the above it has been assumed that a static magnetic field H_o has
been applied say in the y direction, to couple the current with the
electromagnetic field. The interaction Hamiltonian for interaction
with the external field has a form similar to (4)

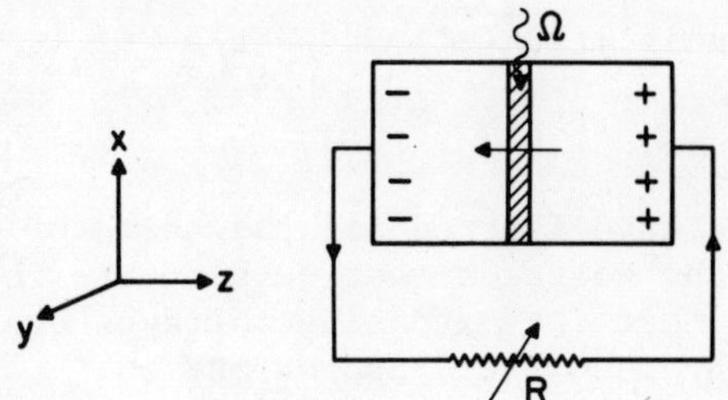

Fig. 1.   Schematic representation of our Josephson junction.

$$H_{ext} = \int i\hbar T \, (S^- - S^+)(a_{ex} + a_{ex}^+) \, W(\Omega',\Omega)d\Omega' \, ,$$

$$(6)$$

$$a_{ext} = N_{ex}^{1/2} \cos \Omega't$$

where $W(\Omega',\Omega)$ is the spectral function of the external radiation which is assumed to be centered at Ω. It will be seen later that a finite bandwidth of the coherent radiation is essential in order to describe the nonequilibrium phase transitions in Josephson junctions, for when the Josephson frequency ω jumps, there must be a component of radiation at ω available to continue to drive it. The total Hamiltonian is the sum of H_T, H_c, H_{cav}, H_{ext}. We now examine the incoherent processes; the important ones are (i) the leakage of photons in the cavity at a rate $\kappa = \omega_c/2Q$, (ii) The discharge rate of the capacitor, as the current is passing through the external resistance.

The equations of motion (quantum Langevin equations) are now found to be

$$\dot{a} = -i\omega_c a - \kappa a - T(S^+ - S^-) + f_a$$

$$\dot{S}^\pm = \pm \frac{2eV}{h} S^\pm + f^\pm, \quad V = \frac{2eS^z}{C}$$

$$(7)$$

$$\dot{S}^z = - \frac{S^z}{RC} - \frac{I_j}{4ei} (S^+ - S^-) - T(a + a^+)(S^- + S^+)$$

$$- 2N^{1/2} \cdot T \int W(\Omega',\Omega) \cos(\Omega't)d\Omega' \, (S^+ + S^-) + f^z,$$

where the f's are various random forces which account for operator characteristics and any thermal noise factors. For example, f_a has the properties (at zero temperature),

$$<f_a(t)f_a(t')> = <f_a^+(t)f_a^+(t')> = <f_a(t)> = <f_a^+(t)> = <f_a^+(t)f_a(t')>=0,$$

$$<f_a(t)f_a^+(t')> = 2\kappa\delta(t-t') \ . \tag{8}$$

In deriving (7), we have ignored all the terms of order $(1/m^2)$. We have assumed that the voltages that develop are intensive in nature, as it can be shown that the extensive voltages are likely to lead to the breakdown of the mean field approximation.

In order to simplify (7) further, we make use of the physics[9,10] of Josephson junctions, and then we obtain analytical expressions for the self-consistently developed voltage. We make use of the following assumptions: (i) ignore all the operator characteristics of $S^\pm$, S^z; (ii) ignore all thermal noise factors since the system is assumed to be at a temperature close to zero; (iii) the photon leakage rate $\kappa \gg 1/RC$ so that the photon mode is a fast mode and can be adiabatically eliminated (the later estimates show that this condition is very well satisfied); (iv) do the time averaging over time intervals of the order ω_c^{-1} and hence ignore all the rapidly oscillating terms. The last assumption also enables us to ignore the contribution $I_j \sin\theta[S^\pm = e^{\pm i\theta}]$. Under the above assumptions we have shown that the Fokker-Planck equation for the self-consistently developed voltage is given by (everything has been expressed in terms of scaled variables),

$$f = \frac{\omega}{\omega_c} \ , \quad \omega = \frac{2eV}{\hbar} = \frac{4e^2}{\hbar C} S^z \ ,$$

$$\alpha = \frac{8e^2}{\hbar} R\left(\frac{T}{\omega_c}\right)^2 \ , \quad N_c^{-1/2} = \frac{8e^2}{\hbar} R \left\{\frac{T}{\omega_c}\right\} W, \quad \mu = \left(\frac{N}{N_c}\right)^{1/2} \ , \tag{9}$$

$$\frac{\partial P}{\partial t} = \frac{1}{RC} \frac{\partial}{\partial f} \left[f - \mu + \frac{2Q\alpha}{[1+4Q^2(f-1)^2]}\right]P$$

$$+ \left(\frac{4e^2}{\hbar C\omega_c RC}\right) \frac{\partial^2}{\partial f^2} \left[\left\{\frac{Q\alpha}{(1+4Q^2(g-1)^2)} + \frac{1}{2}\right\} P\right] \ , \tag{10}$$

$$\sim \frac{1}{RC} \frac{\partial}{\partial f} \left[f - \mu + \frac{2Q\alpha}{(1+4Q^2(f-1)^2)}\right]P$$

$$+ \frac{1}{2\tau_f} \frac{\partial^2 P}{\partial f^2} \ , \quad (2\tau_f)^{-1} = \left(\frac{4e^2}{\hbar C\omega_c RC}\right) Q\alpha \left\{1 + \frac{1}{2Q\alpha}\right\} \ . \tag{11}$$

The Fokker-Planck equation (11) has the steady state solution

$$P \sim \exp\left\{-\frac{2\tau_f}{RC}\,\Phi(f)\right\} , \tag{12}$$

$$\Phi(f) = \Phi(1) + \frac{1}{2}(f^2 - 1) - \mu(f-1) + \alpha \tan^{-1} 2Q(f-1) . \tag{13}$$

The probability distribution (12) is sharply peaked at the minima of Φ provided $\frac{\tau_f}{RC} \gg 1$, which is indeed satisfied as the later estimates show. Hence the most probable values of f are the stable solutions of

$$\mu - f = \frac{2Q\alpha}{(1 + 4Q^2(f-1)^2)} . \tag{14}$$

An analysis of (14) shows that it admits three real roots provided that

$$Q^2\alpha \geq 2/3\sqrt{3} , \quad N_{C1} > N > N_{C2} \tag{15}$$

where the critical points N_{C1}, N_{C2} or equivalently μ_{C1}, μ_{C2} are given by

$$\left(\frac{\partial f}{\partial \mu}\right)^{-1} = 0 . \tag{16}$$

The bistable behavior of the junction is shown in Fig. 2 and the corresponding behavior of the potential function is shown in Fig. 3. Equation (15) also gives us a critical value of the resistance R_o below which bistable behavior will not be seen. For R close to R_o we have

$$R = R_o(1 + \delta), \quad 0 < \delta \ll 1, \quad f_o = 1 + (2\sqrt{3}\, Q)^{-1} ,$$

$$\mu_o = 1 + \sqrt{3}\,(1 + \frac{2\delta}{3})/2Q; \quad f_{C1,2} = f_o \pm \frac{1}{3}\,\delta^{1/2}/Q , \tag{17}$$

$$\mu_{C1,2} = \mu_o \pm \frac{1}{9}\,\delta^{3/2}/Q .$$

It can be shown that the bistability discussed above is accessible to experiments, for if we choose typical values of the parameters $R = 136$ ohms, $Q = 5$, $\ell_x = \ell_y = 1$ cm, $\ell_z = 20$ Å, $\varepsilon = 5$, $I_j = 30$ μA, $\lambda = 3.7 \times 10^{-6}$ cm, $\omega_c = 68$ G Hz, $H_o = .14$ Gauss, $\omega \sim 1/15$, then $\alpha = .02$, $N_c = 4.3 \times 10^4$ photons. This value of ω_c corresponds to external photons $N_c/(2\lambda + \ell_z)\ell_x\ell_y \sim 6 \times 10^{11}$ cm^{-3} which requires an external power of order .1 watt/cm^2. The photon leakage times and capacitor discharge times are of the order 10^{-10} secs, 3×10^{-6} secs

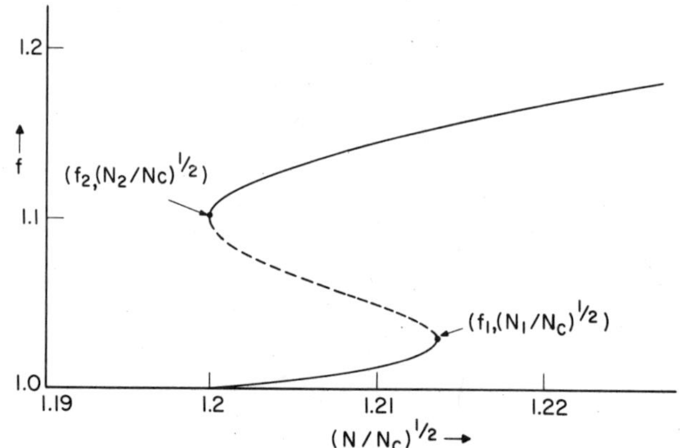

Fig. 2.   Plot of the scaled d.c. voltage versus the scaled square
root of externally supplied photon number μ. The hysteresis region is bounded by the critical points $f_{C1,2}$ $\mu_{C1,2}$.
Solid lines denote the stable solutions whereas the
dashed line gives the unstable solution.

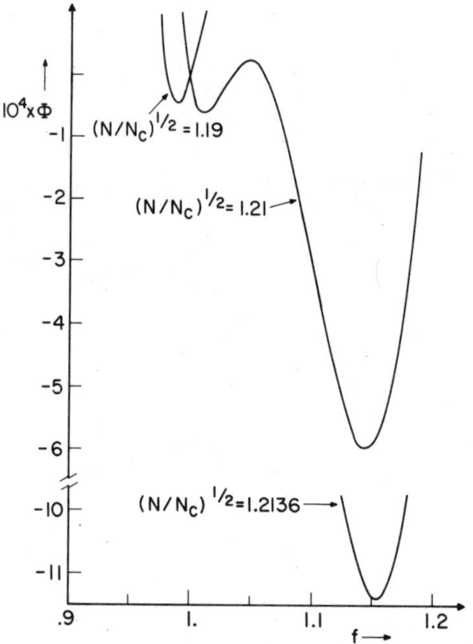

Fig. 3.   Plot of the potential Φ as a function of f for various
values of the parameter $\mu = (N/N_c)^{\frac{1}{2}}$

respectively, which justifies our original fast mode assumption for photons. The typical value of the developed voltage $\sim$ .02 mV. The power requirements and bistability conditions turn out to be sensitive to the junction parameters S, Q, I_j and R and this could explain why the predicted phenomena have not been seen earlier. The above analysis also shows that the size of the hysteresis region could be tuned by varying the resistance across the junction.

We conclude this paper by a brief discussion of the characteristic time scales of the system. The relaxation time T_1 of f(t) to a steady state f is given by

$$\dot{f} = - \frac{1}{T_1} (f(t) - f) , \tag{18}$$

where

$$T_1 = RC/\Phi'' \ (f) \to \infty \text{ as } f \to f_{C1}, \ f_{C2} . \tag{19}$$

In the critical region $1/T_1$ has the typical mean field exponent:

$$\frac{1}{T_1} = \left(\frac{1}{RC}\right) \sqrt{2} \left(\frac{\partial^2 \mu}{\partial f_{C1}^2}\right)^{-1/2} \Phi''' (f_{C1}) |\mu - \mu_{C1}|^{1/2} . \tag{20}$$

It can be further shown that the fluctuations in the steady state, in the Gaussian approximation[11], are given by

$$<(\delta f)^2> = \frac{T_1}{2\tau_f} \tag{21}$$

which, of course, diverge at the critical points (break down of Gaussian approximation). The characteristic time T_2 for making the transition from the state f_1 to state f_2 is found to be[12], with f_3 denoting the maxima of the potential Φ

$$T_2 = \frac{\pi RC \ \exp\{\frac{2\tau_f}{RC} (\Phi(f_3) - \Phi(f_1))\}}{\{\Phi''(f_1) |\Phi''(f_3)|^{1/2}\}} , \tag{22}$$

whereas in the critical region T_2 is found to have the structure

$$\frac{T_2}{RC} = \sqrt{2}(.83) \ 6^{1/3} \left(\frac{2\tau_f}{RC}\right)^{2/3} \left(\frac{\partial^2 \mu}{\partial f_{C1}^2}\right)^{-1/2} (\Phi''' (f_{C1}))^{-1/3} |\mu - \mu_{C1}|^{1/2} . \tag{23}$$

The size of the critical region turns out to be

$$\Delta\mu \ll \left[\frac{3}{8\sqrt{2}} \frac{(RC/2\tau_f)}{|\Phi'''(f_{Cl})|}\right]^{2/3} \left(\frac{\partial^2\mu}{\partial f_{Cl}^2}\right) . \tag{24}$$

For the values of the parameters given earlier $T_1 \sim 10^{-6}$ secs, $T_2 \sim 10^{84}$ secs, $\Delta\mu \sim 10^{-4}$. With these parameters the hysteresis conditions are very well met.

In conclusion the irradiated Josephson junction is worth investigating not only as an interesting new bistable system in its own right, but also as a means to deepen our understanding of non-equilibrium phase transitions in general.

REFERENCES

1. H. M. Gibbs, S. L. McCall and T. N. C. Venkatesan, Phys. Rev. Letts. <u>36</u>, 1135 (1976); R. Bonifacio and L. A. Lugiato, Phys. Rev. A<u>18</u>, 1129 (1978); G. S. Agarwal, L. M. Narducci, R. Gilmore and D. H. Feng, Phys. Rev. A<u>18</u>, 620 (1978).
2. Cf. papers presented at this conference.
3. Cf. H. Haken, "Synergetics," Springer-Verlag (1977).
4. Y. Toyzawa, Sol State Commun., <u>28</u>, 533 (1978); I. Sh. Averbukh, V. A. Kovarsky and N. F. Perelman, Phys. Letters <u>74A</u>, 36 (1979); G. S. Agarwal, to be published; H. M. Gibbs, S. L. McCall, T. N. C. Venkatesan, A. C. Gossard, A. Passner, W. Wiegmann, Appl. Phys. Letters <u>35</u>, 451 (1979); H. M. Gibbs, S. L. McCall and T. N. C. Venkatesan, this volume.
5. S. R. Shenoy and G. S. Agarwal, Phys. Rev. Letters. <u>44</u>, 1525 (1980); R <u>45</u>, 401 (1980).
6. S. R. Shenoy and G. S. Agarwal, Phys. Rev. <u>B</u> (in press).
7. Cf. L. Solymar, "Superconductive Tunnelling and Applications," Chapman and Hall, 1972, Chap. 9, 10; N. R. Werthamer and S. Shapiro, Phys. Rev. <u>164</u>, 523 (1967); J. Chen, R. S. Todd and Y. W. Kim, Phys. Rev. B<u>5</u>, 1843 (1972).
8. I. M. Eliashberg, JETP Lett. <u>11</u>, 114 (1970); T. M. Kopwijk, J. N. Van den Bergh and J. E. Mooij, J. Low. Temp. Phys. <u>26</u>, 385 (1977); A. Schmid, Phys. Rev. Letts. <u>38</u>, 922 (1977).
9. P. A. Lee and M. O. Scully, Phys. Rev. B<u>3</u>, 769 (1971).
10. See also M. J. Stephen, Phys. Rev. <u>182</u>, 531 (1969).
11. See, for example, for the technique, Agarwal et al. in Ref.(1).
12. H. A. Kramers, Phisica <u>7</u>, 284 (1940); R. Gilmore, Phys. Rev. A<u>20</u>, 2510 (1979); G. S. Agarwal and S. R. Shenoy, Phys. Rev. <u>A</u> (in press).

OPTICAL BISTABILITY EXPERIMENTS AND MEAN FIELD THEORIES

W. J. Sandle and R. J. Ballagh

Physics Department
University of Otago
Dunedin, New Zealand

A. Gallagher*

Joint Institute for Laboratory Astrophysics
National Bureau of Standards, and
University of Colorado
Boulder, Colorado 80309 U.S.A.

Abstract: We have performed experiments in absorptive and dispersive optical bistability based upon saturation of the D_1 line of atomic sodium in a tight-focussed spherical-mirror Fabry-Perot cavity. The conditions of large homogeneous broadening (due to high optical power and buffer gas broadening) allow us to interpret the experiments in terms of a two-state model for the D_1 transition. The experimental results, taken under conditions appropriate to the mean-field limit, do not confirm two-state mean-field theories of optical bistability in plane wave cavities. Inclusion of the transverse (Gaussian) mode structure in a two-state mean-field model leads to a significant reduction in the predicted rate of onset of saturation in the absorptive (as compared to the dispersive) regime. This prediction is supported by experiment. There remains the problem that both absorptive and dispersive threshold powers are, in absolute terms, an order of magnitude larger than expected.

I.   INTRODUCTION

Theories of optical bistability [1-11] have largely been concerned with idealized two-state absorbers in optical cavities, but

---

*Staff Member, Quantum Physics Division, National Bureau of Standards.

experiments [14-16] must contend with the properties of real atoms.
The main purpose of this paper is to present experimental evidence
for optical bistability (OB) taken under conditions where real atomic
behaviour can be closely represented by the two-state model, so that
tests of mean field theories of OB are possible. An important feature
of our work is the use of high optical intensities obtained with a
near-concentric focussing cavity. We shall find (both experimentally
and theoretically) that the Gaussian intensity profile obtained in
such a cavity gives rise to important modification of OB behaviour
compared to a plane-wave cavity case. In brief, absorptive bistability threshold powers are significantly raised relative to dispersive bistability thresholds. Physically, this can be understood in
terms of an averaging effect on the relationship between transmitted
and incident power arising from the distribution of cavity intensities
present. For given atom density the onset of saturation is thus
"softened". Absorptive OB then occurs only at higher values of
cooperativity ($C > 10$) and correspondingly requires higher critical
incident laser power.

II.   ATOMS AND EQUIVALENT TWO-STATE BEHAVIOUR

The dynamics of an atom irradiated by quasi-monochromatic radiation are influenced primarily by the two resonant (rotationally
degenerate) levels. In general there are a number of magnetic substates present in each level, so that the atom behaviour may still be
extremely complex. Under certain circumstances, however, the macroscopic polarization produced by real atoms in a gas cell obeys the
same form of saturation behaviour as for simplistic two-state atoms,
although the equivalent Rabi frequency and transverse relaxation
rate must now be carefully identified.

We shall concentrate on the specific case of a $J = \frac{1}{2}$ to $J = \frac{1}{2}$
dipole transition, excited by linearly (π) polarized light, since
this obviously avoids the unnecessary complications arising from
angular momentum conservation, such as optical pumping, alignment
and orientation. We note, however, that more complicated two-level
transitions can also be shown to mimic simple two-state behaviour
[12].

For a degenerate two-level atom (upper level labelled u, lower
level ℓ) with resonant frequency ω_a and irradiated by a monochromatic
laser of frequency ω_L, we make use of the spherical symmetry of the
relaxation processes and express the density matrix in terms of
irreducible tensorial components $\rho^k_q(u)$, $\rho^k_q(\ell)$, $\rho^k_q(\ell u)$. The relevant equations of motion for a $J = \frac{1}{2}$ to $J = \frac{1}{2}$ system, in which homogeneous broadening dominates inhomogeneous broadening, can be written
[13]

$$\dot{\rho}^0_0(u) = -\Gamma_u(o)\rho^0_0(u) + i\,\frac{P\varepsilon}{\sqrt{6}}\,[\rho^1_0(\ell u) - \rho^1_0(\ell u)^*] \qquad (1a)$$

$$\dot{\rho}^0_0(\ell) = -\dot{\rho}^0_0(u) \qquad (1b)$$

$$\dot{\rho}^1_0(\ell u) = -[\Gamma_{\ell u}(1) - i\,\Delta\omega]\,\rho^1_0(\ell u) + \frac{iP\varepsilon}{\sqrt{6}}\,[\rho^0_0(u) - \rho^0_0(\ell)] \quad (1c)$$

Here the detuning $\Delta\omega = \omega_a - \omega_L$, the (classical) laser field is given
by $2\,\varepsilon\,\cos\omega_L t$, $\Gamma_u(0)$ is the total rate of transfer of population from
the upper to lower level (including spontaneous radiation and in-
elastic collisions), $\Gamma_{\ell u}(1)$ is the total (radiative and collisional)
rate of destruction of optical coherence and P is the reduced dipole
matrix element.

Solving for the steady state expression for the optical coherence
gives

$$\rho^1_0(\ell u) = -\sqrt{\frac{\Gamma_u(0)}{\Gamma_{\ell u}(1)}}\;\frac{X(i + \Delta)}{2(1 + \Delta^2 + X^2)} \qquad (2)$$

where we have defined the dimensionless quantities

$$\Delta = \frac{\Delta\omega}{\Gamma_{\ell u}(1)} \qquad \text{and} \qquad X = \sqrt{\frac{2}{3}}\;\frac{P\varepsilon}{\sqrt{\Gamma_u(0)\,\Gamma_{\ell u}(1)}} \qquad (3)$$

The formal similarity of Eq. (2) with the standard result for a two-
state atom [e.g. see Ref. [7] Eq. (2.5)] is apparent, and physically,
is expected because under π radiation the $m_\ell = -1/2$ to $m_u = -1/2$
transition and the $m_\ell = 1/2$ to $m_u = 1/2$ transition proceed independently. Alghough spontaneous radiation will couple these transitions,
it is an incoherent process, and preserves equality of population
between the $m = 1/2$ and $m = -1/2$ substates. Equation (2) leads to
the standard result for the two-state saturable susceptibility, as
used in Eq. (4) below.

In many cases the atom also possesses nuclear spin which couples
to the electronic angular momentum, producing hyperfine structure in
both the lower and upper levels. This complication prevents an exact
description in terms of two levels, but if experimental conditions are
arranged to homogeneously broaden the levels to substantially more
than the hyperfine splittings, then the effects of nuclear spin can
be ignored. Collisions, for example, if occurring frequently enough
(i.e. large pressure width) cause the nuclear and electronic momenta
to remain essentially uncoupled. Also, saturation by a high powered

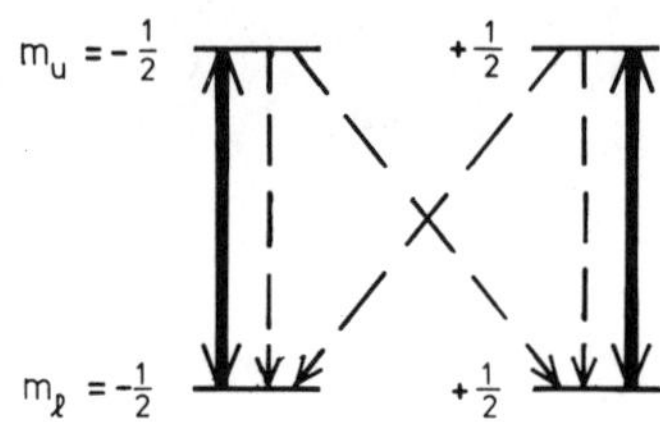

Fig. 1. $J = \frac{1}{2} \leftrightarrow J = \frac{1}{2}$ system.   Solid lines indicate laser π
transitions and dashed lines show spontaneous transitions.

laser may cause sufficient power broadening to allow neglect of the
detuning differences between hyperfine components. The behaviour
of the atom then becomes that of the underlying two electronic
angular momentum levels, a point which is discussed further elsewhere [12].

III.   THE EXPERIMENT

Bearing the above in mind, we choose as our system the
$3^2S_{\frac{1}{2}}$ - $3^2P_{\frac{1}{2}}$ D_1 line of atomic sodium placed at the waist region of
a focussing (near concentric) cavity. We have added up to 80 torr
of argon buffer gas to homogeneously broaden the transition [18].
In addition, at the high intensities used (> 1kW cm^{-2} peak at the
80μm waist), the transition is power broadened to several times the
inhomogeneous width. The conditions discussed in §2 are then met,
and to a good approximation, the D_1 line behaves effectively as a
two-state system.

Although our experiment resembles the pioneering one of Gibbs,
et al. [15] in utilizing atomic sodium in an optical cavity, the
mechanisms in the two cases are quite different. In their case the
non-linearity arose from hyperfine pumping in the ground $3^2S_{\frac{1}{2}}$ state
whereas in our case the mechanism is saturation of the D_1 transition. The experimental regimes are therefore quite different. Gibbs,
et al. employed low laser intensities, whereas the present experiment uses high laser intensities.

Briefly, the details of the experiment are as follows [16]; the
cavity is constructed from mirrors of 30 cm radius of curvature
(250 MHz free spectral range); the cavity finesse, observed far off
the sodium resonance, is 55; the sodium vapour is confined to the
cavity waist region by argon buffer gas.

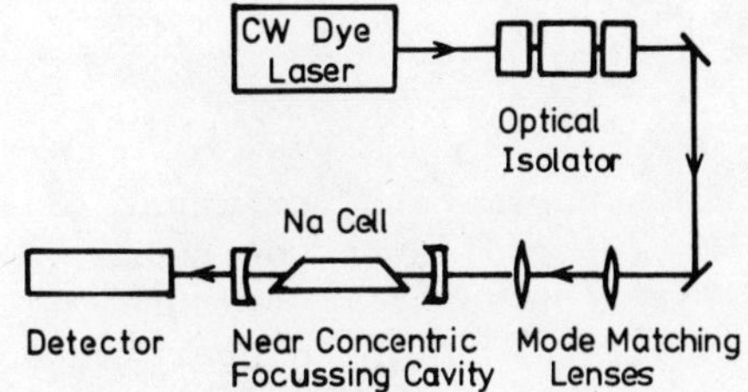

Fig. 2.   Experimental Arrangement.

IV.   EXPERIMENTAL RESULTS

A.   CONFIRMATION OF THE PHENOMENOLOGY OF OB

Although theoretical analyses of the bistability state equation
have concentrated on the case where the laser frequency is fixed and
the input power is varied, one can equally choose to fix the laser
power and vary the frequency. In our case, experimental considerations make the latter procedure more convenient. The laser frequency
is varied through an optical cavity transmission profile and the
output power is monitored for fixed input power. Typical experimental profiles are shown in Figs. 3(a), 4(a) and 4(c). In the usual
manner, we distinguish the dispersive regime ($|\Delta| \gg 1$) as in Fig.
3(a), and the absorptive regime ($\Delta \simeq 0$) as in Fig. 4(a) and Fig. 4(c).
The characteristic appearance of dispersive and absorptive profiles
is seen to differ significantly.

The difference can also be demonstrated theoretically from the
equation of (for example) a plane wave ring cavity model. Near a
cavity resonance, this can be written [4,6,7]

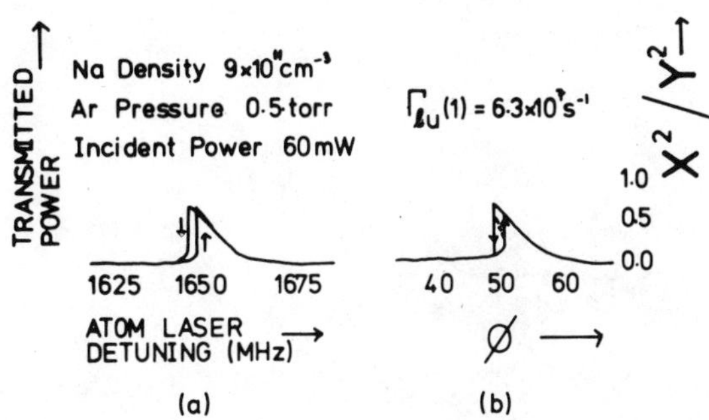

Fig. 3.   Dispersive bistability for fixed incident power and varying
          frequency.   a) Experiment   b) Theory.

$$Y^2 = X^2 \left[\left(1 + \frac{2C}{1 + \Delta^2 X^2}\right)^2 + \left(\phi - \frac{2C\Delta}{1 + \Delta^2 + X^2}\right)^2 \right] \tag{4}$$

where the dimensionless parameters for input (Y) and output (X) fields, cooperativity (C) and cavity detuning (ϕ) have the standard meaning [e.g. Ref. 7]. Curves of X^2 versus ϕ (Y fixed) are drawn in Figs. 3(b) and 4(b), (d) for the dispersive and absorptive regimes respectively. Parameter values have been used for which the curves correspond to the experimental data.

In principle, one expects to be able to calculate all the parameters from known experimental conditions. However, the intensity Y^2 of a plane wave model cannot be unambiguously related to the power of a Gaussian field mode; therefore, we have treated Y^2 as a "free" parameter, chosen to give best fit.

In the dispersive case we have used the expected values of $C = 4.5 \times 10^3$ and $\Delta = 165$ and a fitted value Y = 65 (compared to an expected [17] value of order Y ~ 270). In fitting the absorptive data to Eq. (4) (Figs. 4(b), (d)), we have found it necessary to allow both Y and C to be fitting parameters, and have used the values $\Delta = 0$, C = 5, with Y = 5.134 (shows switching) and Y = 6.132 (no switching). We note that the values expected from experimental conditions are C ~ 110, Y ~ 90.

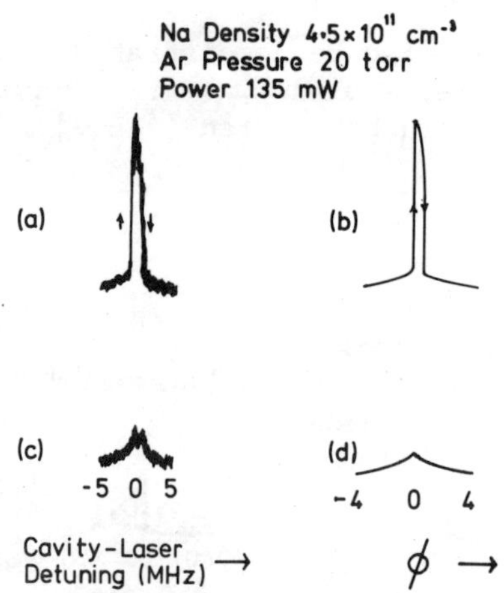

Fig. 4. Absorptive bistability for a fixed incident power and varying frequency. In (a) [experimental] and (b) [theoretical] the power is marginally above threshold, while in (c) [experimental] and (d) [theoretical] the power is marginally below threshold.

Both the dispersive and absorptive data confirm the phenomenology of OB: dispersive hysteresis is clearly evident in Fig. 3(a);
in the absorptive case switching and hysteresis are evident above
the critical threshold (Fig. 4(a)) but absent below (Fig. 4(c)).

Quantitative agreement is, however, poor in respect of the substantial difference between expected and fitted values of parameters
Y and C. The significance of this disagreement is not immediately
clear, since the applicability of the two-state model is doubtful
at the low cavity powers present before switching. (If collision
broadening had been sufficiently increased to ensure the applicability
of the two-state model, the laser power requirement for switching
would have been greater than the ~200 mW available). However, rather
than tackle the daunting task of making a complete theory including
the effects of hyperfine structure, doppler motion, etc., we follow
a different path. Predictions of mean field theories will be experimentally tested in a high power, large collisional width regime,
where the two-state model is valid.

B.   PEAK PROFILE HEIGHTS

For maximum power broadening we clearly must concentrate attention on the data at the peaks (where cavity power is highest). In
addition, the power broadening (proportional to $\sqrt{\Gamma_{\ell u}(1)}\,\varepsilon$) is enhanced
by raising the collisional width. Of course, the total width is the
(quadrature) sum of the collisional width and the power broadening,
and eventually (for fixed laser power) the collisional width will
dominate. However, this corresponds to values of X < 1, which are
not physically interesting. We have chosen 80 torr of argon as
optimum for our available laser power.

We also find it necessary to use an optimum value of the cooperativity C, determined by the requirements that the cavity be opaque
at low laser power, but transmit at the highest available power.
These conditions amount to operating at the critical threshold, for
which no significant hysteresis or switching is observed. The latter
has the additional advantage that there is no ambiguity in measured
peak heights. Figure 5 illustrates the procedure and confirms the
degree of homogeneous broadening present.

The laser frequency is here being swept by 8 GHz across the D_1
absorption line with constant incident laser power of 80 mW, and the
power transmitted by the cavity is monitored. Individual optical
cavity profiles show up as "spikes" (because of the high cavity
finesse) separated by ~ 250 MHz. Frequency scans such as above were
repeated with the laser power held constant at a range of values
between 25 mW and 130 mW. Experimental results are shown in Fig. 6
where, for clarity, the individual profiles have been suppressed and
the peak heights recorded as dots.

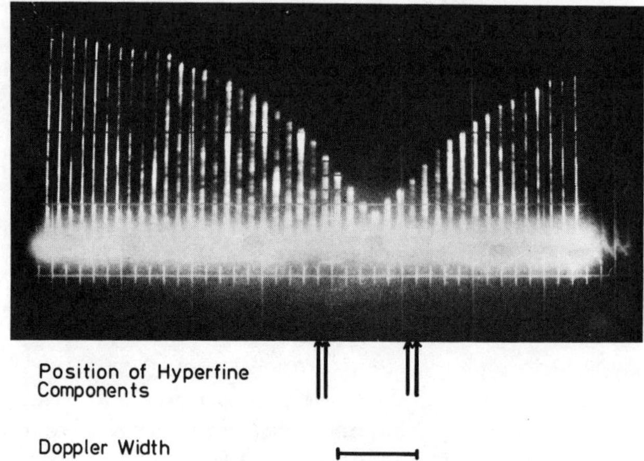

Fig. 5.　Oscilloscope trace of transmitted power versus laser frequency for fixed incident laser power (80 mW). Buffer gas pressure is 80 torr.

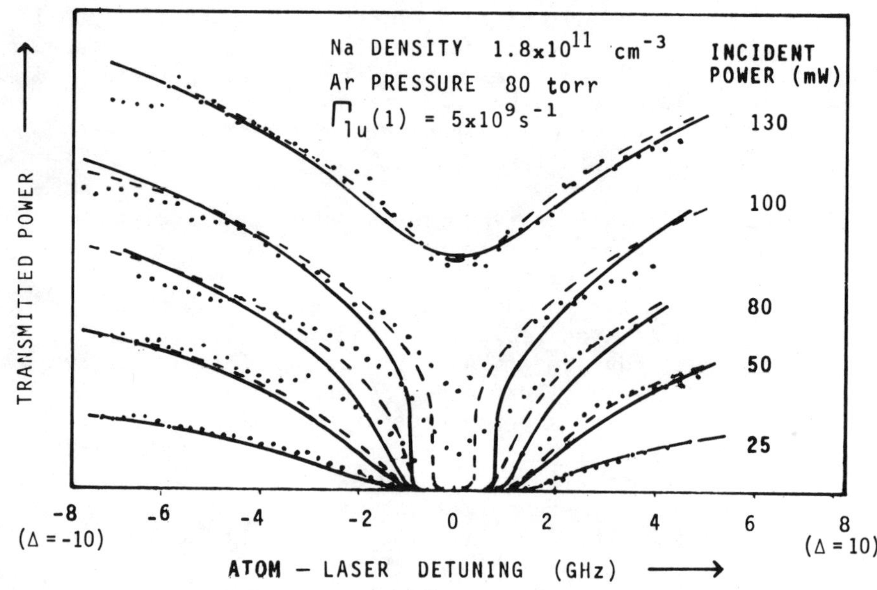

Fig. 6.　Fits of theoretical envelope equations (solid line, ring cavity, dashed line, Fabry-Perot) to experimental profile peak heights (dots).

V. COMPARISON WITH THEORETICAL PEAK PROFILE HEIGHTS

Envelope equations for the profile peaks can readily be obtained from the bistability state equations. To illustrate, consider the simplest case of a plane wave ring cavity in the mean field limit. In the cubic approximation (i.e. near a cavity resonance) the state equation is given by Eq. (4). The peak height is obtained by setting the dispersive contribution to zero;

$$\phi - \frac{2C\Delta}{1 + \Delta^2 + x^2} = 0 \tag{5}$$

giving the envelope equation for peaks

$$Y^2 = x_p^2 \left(1 + \frac{2C}{1 + \Delta^2 + x_p^2}\right)^2 \tag{6}$$

(where the subscript p refers to peak profile height).

In the case of a plane standing wave (Fabry-Perot) cavity in the mean-field limit, the equation for a single cavity profile is [8]

$$Y^2 = x^2 \left\{1 + \frac{2C}{x^2} \left[1 - \left(\frac{1 + \Delta^2}{1 + \Delta^2 + 2x^2}\right)^{\frac{1}{2}}\right]\right\}^2$$

$$+ \left\{\phi - \frac{2C\Delta}{x^2} \left[1 - \left(\frac{1 + \Delta^2}{1 + \Delta^2 + 2x^2}\right)^{\frac{1}{2}}\right]\right\}^2 \tag{7}$$

Again, setting the dispersive contribution to zero leads to the envelope equation

$$Y^2 = x_p^2 \left\{1 + \frac{2C}{x_p^2} \left[1 - \left(\frac{1 + \Delta^2}{1 + \Delta^2 + 2x_p^2}\right)^{\frac{1}{2}}\right]\right\}^2 \tag{8}$$

In Fig. 6, theoretical predictions from Eqs. (6) and (8) are compared with experimental results. As discussed in §4, for plane wave theories the conversion constant relating Y^2 to input laser power can not be established unambiguously. This leads us to treat the conversion constant for our experiment as a freely varying fitting parameter, and furthermore we allow the value to differ between the ring and standing wave cavity models. The expected value of the cooperativity is C = 10, but in an attempt to improve the fit of theory, we have treated also C as a fitting parameter, resulting in best fit values of C = 9.5 (ring cavity case) and C = 9 (standing wave cavity case). The theory has been fitted to give best agreement with the high power data, and it can be seen that the agreement for other laser powers is poor. (Fits to other than the high power data do not improve the overall quality of fit). The overall disagreement in form of the plane wave theories implies quantitative failure of the plane wave models. Furthermore, the fitted conversion factors (0.64 mW^{-1} for the ring cavity model; 0.55 mW^{-1} for the standing wave model) differ substantially from the value of 15 mW^{-1} which is our best estimate [17] of the expected conversion constant.

These problems have lead us to develop a mean-field theory for OB which accounts for the transverse Gaussian form of the cavity modes [19]. In §6 we present this theory and finally in §7, its predictions are compared with experiment.

VI.   THEORETICAL TREATMENT OF THE GAUSSIAN CAVITY MODE

The system we consider consists of a cell of two-state atoms (length 2ℓ) placed in the waist region of a spherical mirror cavity (mirror reflectance R, separation $2d$) with an incident laser beam mode matched to excite only the fundamental TEM$_{00}$ mode. The Bloch equations for stationary two-state atoms at position $\underline{r}$, exposed to intensity $I(\underline{r})$, give in steady state the standard expression for susceptibility (see e.g. [7])

$$\chi(\underline{r}) = \frac{\alpha c}{\omega_L} \frac{\Delta + i}{1 + \Delta^2 + I(\underline{r})} \tag{9}$$

where α is the resonant absorption coefficient for the two-state homogeneously broadened transition with half width $\Gamma_{\ell u}(1)$. The intensity $I(\underline{r})$ is measured in units of the saturation intensity.

The forward and backward travelling waves in a spherical mirror cavity are most conveniently described in terms of the well known orthonormal Gaussian modes [20],

$$E^f_{mn}(x,y,z) = \psi_{mn}(x,y,z)\exp\left\{i\left[\frac{\omega_L}{c}z + \frac{\omega_L}{c}\frac{\rho^2}{R(z)} - (m+n+1)\phi(z)\right]\right\} \tag{10a}$$

$$E^b_{mn}(x,y,z) = \psi_{mn}(x,y,z)\exp\left\{-i\left[\frac{\omega_L}{c}z + \frac{\omega_L}{c}\frac{\rho^2}{R(z)} + (m+n+1)\phi(z)\right]\right\} \tag{10b}$$

where

$$\psi_{mn}(x,y,z) = \frac{1}{\sqrt{2^{m+n}m!n!}} \frac{\sqrt{2/\pi}}{W(z)} H_m\left(\sqrt{2}\,\frac{x}{W}\right) H_n\left(\sqrt{2}\,\frac{y}{W}\right) \exp\left(-\frac{\rho^2}{W^2(z)}\right), \tag{11}$$

$$W^2(z) = W_o^2[1 + (\frac{z}{z_o})^2], \quad R(z) = z[1 + (\frac{z_o}{z})^2],$$

$$\phi(z) = \arctan(\frac{z}{z_o}), \quad \rho^2 = x^2 + y^2, \quad z_o = \frac{W_o^2 \omega_L}{2c}. \tag{12}$$

The quantity W_o is the minimum spot size and z_o is the Rayleigh length.

The symmetry of the incident laser field ensures that only the $m = n = 0$ mode is driven. We thus write the positive frequency component of the field as a sum of forward and backward travelling waves in the form

$$E^+(\rho,z) = E_f(\rho,z)\, \mathcal{E}_{00}^f + E_b(\rho,z)\, \mathcal{E}_{00}^b \tag{13}$$

(Complex fields in Eq. (13) are expressed in units of the saturation field). The field functions $E_f(\rho,z)$ and $E_b(\rho,z)$ allow for modification from the TEM_{00} mode caused by the (nonlinear) susceptibility of the atoms.

Within the cell, Maxwell's equations can be simplified by the usual slowly varying envelope approximation, and by ignoring the transverse gradients of $E_f(\rho,z)$ and $E_b(\rho,z)$ (which requires $\alpha\ell \ll 1, \ell < z_o$). Solutions are easily obtained in the mean-field limit (MFL) ($\alpha\ell \ll 1, R \rightarrow 1$), giving

$$E_f(\rho,z) = E_f(\rho,-\ell)\, \exp\{- \frac{A(\rho)}{2}(z + \ell)\}$$

$$\tag{14}$$

$$E_b(\rho,z) = E_b(\rho,+\ell)\, \exp\{+ \frac{A(\rho)}{2}(z - \ell)\}$$

$$(-\ell \leq z \leq \ell)$$

where the complex absorption coefficient A can be found in the homogeneous broadening case to be [7]

$$A(\rho) = \frac{\alpha(1-i\Delta)}{I(\rho)}\left[1 - \left(\frac{1 + \Delta^2}{1 + \Delta^2 + 2I(\rho)}\right)^{\frac{1}{2}}\right]. \tag{15}$$

Eqs. (14) describe the propagation of the fields within the cell, and the boundary conditions imposed by the cavity on the ends of the cell complete the specification of the problem. These boundary conditions are most easily expressed in terms of the pure Gaussian modes E^f_{mn} and E^b_{mn}, as the latter have particularly simple reflection properties at the mirrors. Thus, expanding the fields in the region $\ell \leq |z| \leq d$ as

$$E_f(\rho,z) \; E^f_{00} = \sum_{m,n} \mu^f_{mn} \; E^f_{mn} \; , \tag{16}$$

$$E_b(\rho,z) \; E^b_{00} = \sum_{m,n} \mu^b_{mn} \; E^b_{mn} \; . \tag{17}$$

we find that the amplitudes μ^f_{mn} and μ^b_{mn} must satisfy the boundary conditions

$$\mu^b_{mn}(\ell) = \mu^f_{mn}(\ell) \; \sqrt{R} \; \exp\{i\,[\frac{\omega_L}{c} \, 2d + \eta - (m+n+1)2\phi(d)]\} \tag{18}$$

$$\mu^f_{mn}(-\ell) = \mu^b_{mn}(-\ell) \; \sqrt{R} \; \exp\{i\,[\frac{\omega_L}{c} \, 2d + \eta + (m+n+1)2\phi(-d)]\}$$

$$+ \; \delta_{mo}\delta_{no} \; \sqrt{P_i} \; \sqrt{1-R} \; e^{i\eta'} \; . \tag{19}$$

We have assumed reflection from a mirror produces a phase change η, and Eq. (19) (for the leading edge of the cell) also includes a term for the transmitted laser field (phase shifted by η' at the input mirror). This contributes only to the $m = n = 0$ mode since we have assumed a mode matched input field with form $\sqrt{P_i} \; E^f_{00}$ outside the mirror, having power P_i.

The nonlinear response of the atoms couples the various modes but since this coupling is only of order $\alpha\ell$ the higher order mode amplitudes generated in a single pass are small. The resultant $m \neq 0$, $n \neq 0$ amplitudes from many passes are suppressed because these modes are not degenerate with the $m = n = 0$ mode (see Eqs. (10)) and are not resonant at the laser frequency. Consequently, in the empty regions of the cavity, we need consider only the $m = n = 0$ amplitudes. Thus clearly, $E_f(\rho,-\ell) = \mu^f_{00}(-\ell)$, and using Eq. (14) to calculate $E_f(\rho,\ell)$, we finally obtain the amplitude of the $m = n = 0$ mode at $z = \ell$ by using the orthonormality of the modes:

$$\mu^f_{00}(\ell) = \mu^f_{00}(-\ell) \int_0^\infty 2\pi\rho d\rho \; \psi_{00}^* \, e^{-A(\rho)\ell} \psi_{00} \tag{20}$$

where ψ_{00} is evaluated at $z = \ell$. With $A(\rho)$ given in Eq. (15), and using the MFL to expand $\exp[-A(\rho)\ell]$ to first power in $\alpha\ell$ and to replace $I(\rho)$ in Eq. (15) by

$$I(\rho) \cong \frac{4|\mu_{00}^{f}(\ell)|^2}{\pi W_o^2} \exp(-2\rho^2/W_o^2) \equiv I_o \exp(-2\rho^2/W_o^2) \ , \tag{21}$$

then

$$\mu_{00}^{f}(\ell) = \frac{\kappa \sqrt{1-R} \sqrt{P_i} \ e^{i\eta'}}{1-\kappa^2 R \exp\{i[4 \frac{\omega_L}{c} d - 4\phi(d) + 2\eta]\}} \ , \tag{22}$$

where

$$\kappa = 1 - (1-i\Delta) \frac{2\alpha\ell}{I_o} \ln \frac{1}{2} \left[\left(\frac{1 + \Delta^2 + 2I_o}{1 + \Delta^2}\right)^{\frac{1}{2}} + 1 \right] \ . \tag{23}$$

The output field is generated from the forward mode, and takes the form $\sqrt{P_t} \ E_{00}^{f}$ outside the cavity, where

$$P_t = (1 - R)|\mu_{00}^{f}(\ell)|^2 \tag{24}$$

Finally, under the "small detuning limit",

$$\delta = 2m\pi - (4 \frac{\omega_L}{c} d - 4\phi(d) + 2\eta) \to 0 \tag{25}$$

and, in the usual way, setting

$$\frac{R\delta}{1-R} \to \phi \ , \qquad \frac{\alpha\ell}{1-R} \to C \ . \tag{26}$$

we obtain the following cavity profile equation for the Gaussian mode Fabry-Perot:

$$Y^2 = X^2 \left\{ [1 + \frac{4C}{X^2} \ln \xi]^2 + [\frac{4C\Delta}{X^2} \ln \xi - \phi]^2 \right\} \ . \tag{27}$$

Here

$$X^2 = I_o = \frac{4P_t}{(1-R)\pi W_o^2} \ , \tag{28}$$

is the (on axis) average intensity in the atom vapour and is proportional to the transmitted power.

$$Y^2 = \frac{4P_i}{(1-R)\pi W_o^2}$$ (29)

is proportional to the input power, and

$$\xi = \tfrac{1}{2}\left\{\left[\frac{1 + \Delta^2 + 2X^2}{1 + \Delta^2}\right]^{\frac{1}{2}} + 1\right\}.$$ (30)

We note that it is a simple matter to adapt the above formalism to a ring cavity situation, and the result is presented in [19].

VII. COMPARISON OF EXPERIMENT WITH THE GAUSSIAN MODE THEORY

An envelope equation for peak profile heights in the Gaussian Fabry-Perot case is readily obtained from Eq. (27), giving

$$Y^2 = X_p^2\left\{1 + \frac{4C}{X_p^2}\ln\tfrac{1}{2}\left[\left(\frac{1 + \Delta^2 + 2X_p^2}{1 + \Delta^2}\right)^{\frac{1}{2}} + 1\right]\right\}^2$$ (31)

In Fig. 7 we have plotted the same experimental peak heights as in Fig. 6, and taking the expected value of $C = 10$, have used fitting procedures as in §5 to draw the solid lines corresponding to Eq. (31). The resulting conversion factor between Y^2 and P_i is 3.39 $(mW)^{-1}$.

Agreement between the Gaussian model and the experimental data for peak heights is seen to be much better than was the case for the plane wave models (Fig. 6). The dominant feature of the Gaussian model, compared to plane wave models, is the much less abrupt onset of saturation at line centre with increasing laser intensity. This implies that absorptive bistability occurs at larger values of the cooperativity C than for the plane wave case and consequently the threshold laser intensities are also raised. This feature is confirmed by numerical examination of Eq. (27) which leads to the result that the critical threshold for absorptive bistability in the Gaussian Fabry-Perot case is $Y = 16.21$ occurring at $C = 10.04$. By comparison, for a plane wave Fabry-Perot critical threshold occurs at $Y = 5.30$ with $C = 4.97$. We note that in the dispersive regime $(\Delta \gg 1,\ C > 2\Delta,\ \Delta^2 \gg X^2)$ the Gaussian Fabry-Perot threshold value of $Y = 1.01\ \sqrt{\Delta^3/C}$ is raised only a small amount in relation to the plane wave Fabry-Perot value of $Y = 0.72\ \sqrt{\Delta^3/C}$.

In summary, the experiments we have performed on atomic sodium in a focussing cavity confirm the qualitative behaviour of OB predicted by mean field theories involving two-state atoms in a plane

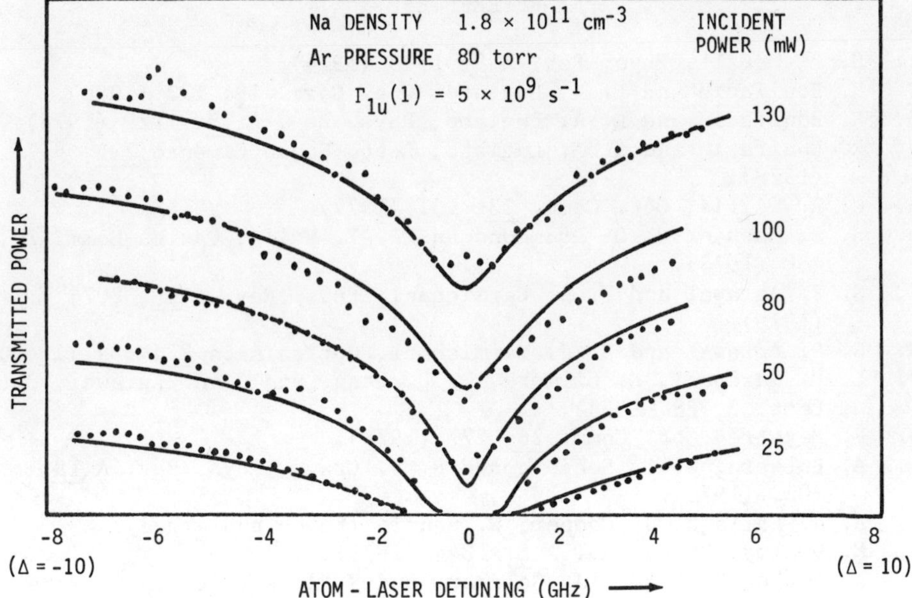

Fig. 7:   Comparison of Gaussian Fabry-Perot envelope equations
         [———], with experimental peak profile heights [•].

wave cavity. However, data taken in a regime where the two-state
behaviour and the MFL should both be applicable, fails to confirm
quantitative details of the plane wave theories. In particular, the
observed threshold laser intensity for absorptive OB occurs at significantly higher values than predicted. The use of a mean field
theory which accounts for the Gaussian profile of the focussing
cavity substantially improves agreement with the experiment. In
particular, the Gaussian mode theory accounts precisely for the
dependence (see Fig. 7) of peak heights as a function of both power
and detuning. The only remaining discrepancy involves a single
scaling factor between predicted and actual cavity powers, which may
be accounted for by experimental uncertainties in waist size and
cavity finesse or, more probably, by laser frequency fluctuations.

ACKNOWLEDGEMENTS

     It is a pleasure to thank Drs. D. Warrington and J. Cooper for
valuable contributions to this work.

REFERENCES

1. S. L. McCall, Phys. Rev. A $\underline{9}$, 1515 (1974).
2. R. Bonifacio and L. A. Lugiato, Opt. Comm. $\underline{19}$, 172 (1976).
3. R. Bonifacio and L. A. Lugiato, Phys. Rev. A $\underline{18}$, 1129 (1978).
4. R. Bonifacio and L. A. Lugiato, Lett. Nuovo Cimento $\underline{21B}$, 517 (1978).
5. C. R. Willis, Opt. Comm. $\underline{23}$, 151 (1977).
6. S. S. Hassan, P. D. Drummond and D. F. Walls, Optics Comm. $\underline{27}$, 480 (1978).
7. G. P. Agrawal and H. J. Carmichael, Phys. Rev. A $\underline{19}$, 2074 (1979).
8. G. P. Agrawal and H. J. Carmichael, Optica Acta $\underline{27}$, 651 (1980).
9. L. M. Narducci, R. Gilmore, D. H. Feng, and G. S. Agrawal, Opt. Lett. $\underline{2}$, 88 (1978).
10. P. Meystre, Opt. Comm. $\underline{26}$, 277 (1978).
11. A. Bulsara, W. C. Schieve and R. F. Cragg, Phys. Rev. A $\underline{19}$, 2052 (1979).
12. R. J. Ballagh, J. Cooper, W. Sandle (to be published).
13. M. Ducloy, Phys. Rev. A $\underline{8}$, 1844 (1973).
14. A. Szöke, V. Daneu, J. Goldhar, and N. A. Kurnit, Appl. Phys. Lett. $\underline{15}$, B76 (1969).
15. H. M. Gibbs, S. L. McCall and T. N. C. Venkatesan, Phys. Rev. Lett. $\underline{36}$, 1135 (1976); S. L. McCall, H. M. Gibbs, G. G. Churchill and T. N. C. Venkatesan, Bull. Am. Phys. Soc. $\underline{20}$, 636 (1975).
16. W. J. Sandle and A. Gallagher, submitted to Phys. Rev.
17. W. J. Sandle in Laser Physics, edited by D. F. Walls and J. D. Harvey (Academic Press, Sydney) 255 (1980).
18. D. G. McCartan and J. M. Farr, J. Phys. B $\underline{9}$, 985 (1976); R. H. Chatam, A. Gallagher and E. L. Lewis, J. Phys. B $\underline{13}$, 67 (1980).
19. R. J. Ballagh, J. Cooper, M. W. Hamilton, W. J. Sandle and D. M. Warrington, submitted for publication.
20. A. Yariv, Introduction to Optical Electronics, 2nd ed. p. 40, (Holt, Rhinehart, Winston, New York 1976).

OPTICAL BISTABILITY IN A GaAs ETALON

H. M. Gibbs[*], S. L. McCall, T. N. C. Venkatesan,
A. Passner, A. C. Gossard, and W. Wiegmann

Bell Laboratories
Murray Hill, N. J. 07974

Abstract: The first observation of optical bistability in a
passive semiconductor etalon is reviewed. The bistable etalon
consists of a GaAlAs-GaAs-GaAlAs molecular-beam epitaxially-grown
sandwich with 90% reflectivity coatings. The bistability is primarily dispersive with the nonlinear refractive index arising from
light-induced changes in exciton absorption. Using light of frequency just below the exciton peak, we observed bistability from
5 to 120°K. The holding intensity was about 1 mW/μm^2, and switching times of < 1 ns turn-on and 40 ns turn-off have been achieved.
This device illustrates the use of a material resonance and optical cavity to reduce the holding intensity and switching energy.
Further reductions are anticipated by further miniaturization and
device development. The possibility of utilizing bistable devices
in high-speed optical processing and computing motivates the development of these semiconductor etalons.

I.   INTRODUCTION

We first reported[1] optical bistability in GaAs over a year
ago, so that the details of the experiments are available elsewhere[2-5]. Consequently, here we will only summarize the important
conclusions.

Optical bistability (OB) experiments have been reviewed in
two recent articles[6]. The use of semiconductors such as GaAs and
InSb for OB was anticipated in the patents on dispersive OB[7].

[*]Present Address: Optical Sciences Center, University of Arizona,
Tucson, AZ. 85721.

Search and Conditions for Optical Bistability in GaAs

Because no information was found on the nonlinear index of the GaAs free exciton resonance, it seemed reasonable to search for absorptive OB first. The free exciton resonance in GaAs saturates as

$$\alpha_{EX} = \alpha_B + \frac{\alpha_o}{1 + I/I_s} \ , \tag{1}$$

i.e., as a homogeneously broadened two-level system except for the unsaturable absorption α_B.[3] This background absorption probably arises from spilling over of the band-to-band absorption onto the wavelength of peak exciton absorption; if so, it saturates at intensities about $10^3 I_s$.[8] Using an antireflection coated heterostructure of 0.42 μm of GaAs between 2.38 and 3.33 μm windows of $Al_{0.24}Ga_{0.76}As$ we found[3]

$$\alpha_{EX}L = 0.26 + \frac{1.71}{1 + I/600W/cm^2} \ .$$

From this value of I_s one can deduce $I_s = 150W/cm^2$ for the uniform-plane-wave saturation intensity for an optically thin GaAs exciton transition. The ratio α_o/α_B never exceeded 7 whereas a value of at least 8 is needed to observe purely absorptive optical bistability. This unsaturable background absorption then explains the failure to see absorptive optical bistability in a number of samples. One must in general have[1]

$$\frac{\alpha_o L}{T + \alpha_B L} > 4\left[1 + \left(1 + \frac{\Delta\lambda^2}{\delta\lambda^2}\right)^{\frac{1}{2}}\right] \ ; \tag{2}$$

standing wave effects increase the right-hand side by amounts less than 25%. In Eq. (2), T is the coating transmission, $\Delta\lambda$ is the laser-resonance wavelength separation, and $\delta\lambda$ is the resonance half-width at half-maximum. Thus detuning to introduce dispersive effects makes OB even more difficult to achieve. This conclusion assumes α_B is frequency independent. In fact in GaAs α_B decreases with detuning below the free exciton resonance. This realization led to the idea of using thicker samples (≈ 4 μm) and operating considerably off resonance. Monitoring the etalon transmission with a weak below-bandgap cw beam and saturating the free exciton resonance by production of free carriers by 200ps, 5145 Å pulses, it was demonstrated that index changes large enough for observing OB were available (provided they could be obtained with exciton-wavelength light)[4,5a]. (Those two-beam experiments also are interesting examples of control of one light beam by another. Such etalons can also be self-swept to reshape pulses[5b].) OB was observed at $\Delta\lambda = 15$ Å below the exciton resonance in a sample with $\alpha_o L = 10$, $\delta\lambda = 4$ Å, and a measured finesse of 8; the effective $T + \alpha_B L$ was then 0.3 and Eq. (2) was satisfied.

II. APPARATUS AND DATA

The GaAs sample used in these studies was grown by molecular beam epitaxy with 4.1 μm of GaAs between 0.21-μm $Al_{0.42}Ga_{0.58}As$ layers supported by a 150-μm GaAs substrate containing a 1-2 mm-diam etched hole for optical access. Reflective coatings with R = 0.9 were added. The input laser beam was provided by a Coherent Radiation 490 dye laser containing oxazine 750 perchlorate dye pumped by the 647- and 676-nm 7.5-W output of a Spectra Physics 171 Kr laser. The 0.6-W dye laser output, about 1 Å wide, easily tunable from 770 to 870 nm, was converted into 1-μs triangular pulses by an acousto-optic modulator. The laser beam was focused to a 10-μm diameter on the sample. A silicon photodiode with a 0.6-ns full-width half-maximum pulse response and a fast oscilloscope were used to observe the input I_I and output I_T.

Using this configuration, OB has been observed with laser wavelengths 10-25 Å longer than the wavelength of the free-exciton peak. See Fig. 1 for an example of OB in the GaAs etalon and Fig.2 for the corresponding nonlinear index and absorption changes. A holding power (between switch-up and switch-down powers) of about 200 mW was required for the 5 μm-thick sample over a diameter of 10 μm, corresponding to a holding intensity of 1 mW/μm^2. A switching energy of 8 nJ is the product of the 200-mW power and 40-ns switching time [Fig. 1(b)], but only a fraction of this energy was actually absorbed. By working at the holding power and using a 200-ps 5900 Å pulse of 0.6-nJ energy, the device was turned on in a detector-limited time less than 1 ns. The limiting turn-on time should be less than 10^{-12}s for 1-μm-thick devices, based on measurements of Shank et.al.[9] The 5900=Å pulse illuminated a circle about 50 μm in diameter. A better measure of the switching energy is then 24 pJ for a 10-μm spot, or 0.24 pJ/μm^2. This energy/unit area corresponds to a carrier density of about 1.6×10^{17}/cm^3, or one per (185 Å)^3; the exciton Bohr radius is 140 Å.

III. DISCUSSION AND FUTURE DIRECTIONS

Cavity lifetime limitations are of no concern in this short device. The ring-down time of a 90% reflectivity, 5-μm-long cavity is shorter than 1.2 ps.

Bistability was observed only up to 120°K, where kT is about twice the exciton binding energy. Systems such as superlattices with larger exciton binding energies may allow room-temperature operation. Above 150°K, bistability was observed, but was slow

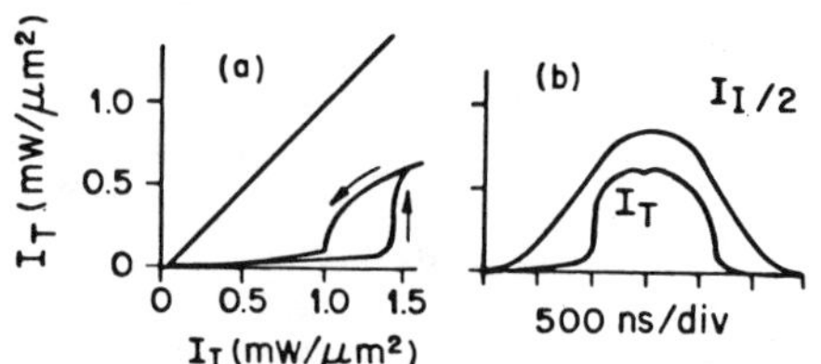

Fig. 1.   Excitonic optical bistability in GaAs at 15°K and 819.9-nm
           laser wavelength. The laser-etalon detuning depends upon
           position on the sample and was determined only crudely;
           see Fig. 2(c). (a) Bistability as seen in x-y display.
           The 45° line shows etalon transmission at the next etalon
           peak (≈830 nm). (b) Time display under the same condi-
           tions as (a). The switch-on and switch-off times are
           both about 40 ns.

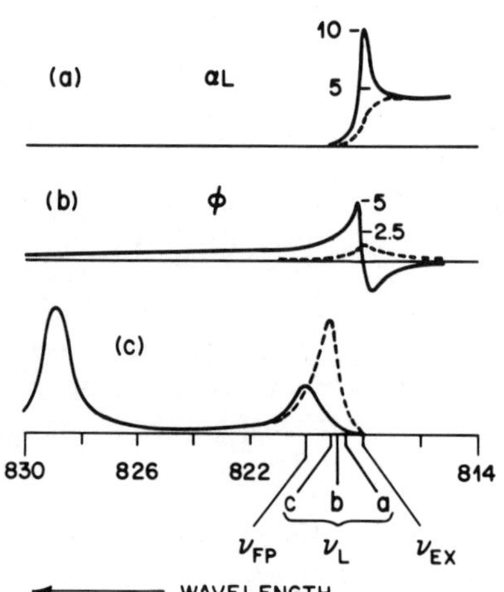

Fig. 2.   Approximate GaAs absorption αL (a), roundtrip phase shift
           ϕ (b), and Fabry-Perot transmission under conditions of
           optical bistability (c). The solid curves are for zero in-
           put intensity where the exciton absorption is fully ef-
           fective. The dashed curves are for intensities which are
           high enough to saturate the exciton feature, but low enough
           to leave the band-to-band contributions to α and ϕ unaf-
           fected. The peak absorption $\alpha_{ex}L$ at the exciton resonance
           was 10 in our sample. Relative frequency positions of lase
           ν_L, Fabry-Perot ν_{FP} and free-exciton ν_{ex} for observed
           optical bistability are shown in (c).

due to thermal contributions to the refractive index in the present sample, which was not heat sunk. Thermal and excitonic contributions oppose one another.

In other GaAs samples, we have observed peak-exciton to nearby band-to-band absorptivity ratios larger than 6. Compatible with the wavelength of about 0.23 μm in the material, and our present practical requirement $\alpha L \approx 1$, the active volume is $(0.2\ \mu m)^3$. This should allow absorbed switching energies less than 10^{-15}J, approaching a statistical limit of 10^3 carriers for reliable switching. Optical switching energies may be reduced by adding an accelerating field.

Turn-off times may be reduced below the 40 ns observed if carrier lifetimes can be shortened without destroying the exciton feature (e.g., by adding impurities or defects or by a geometry with a doped semiconductor alloy or metal near the light-exciton interaction region).

OB has also been seen by Miller, Smith and Johnston[10] in an uncoated 560-μm InSb etalon with only 15 μW/$(\mu m)^2$ holding intensity, but switching times and energies have not been reported. They report a nonlinear refractive index n_2 of 10^{-2} to 10^{-1} cm^2/kW for InSb. If one calculates the change Δn in index n for the intensity change ΔI used in the GaAs optical modulation and bistability experiments, one finds $\Delta n/\Delta I \approx 10^{-4}$cm^2/kW.[6b] This is not n_2 (because of off-resonance and optical thickness effects). From the nonlinear absorption data and a two-level model, an n_2 of 0.4 cm^2/kW is deduced[6b].

This first observation of optical bistability in a semiconductor should provide a significant step toward active integrated optics as well as provide impetus for a number of interesting areas for further research.

REFERENCES

1.   H. M. Gibbs, S. L. McCall, T. N. C. Venkatesan, A. C. Gossard, A. Passner, and W. Wiegmann, CLEA 1979 (IEEE J. Quantum Electron. QE-15, 108D (1979)) and Appl. Phys. Lett. 35, 451 (1979).
2.   H. M. Gibbs, S. L. McCall, A. C. Gossard, A. Passner, W. Wiegmann, and T. N. C. Venkatesan in "Laser Spectroscopy IV," H. Walther and K. W. Rothe, eds., Springer-Verlag, Berlin (1979), p. 441.
3.   H. M. Gibbs, A. C. Gossard, S. L. McCall, A. Passner, W. Wiegmann and T. N. C. Venkatesan, Solid State Commun. 30, 271 (1979).

4.   H. M. Gibbs, T. N. C. Venkatesan, S. L. McCall, A. Passner,
     A. C. Gossard, and W. Wiegmann, Appl. Phys. Lett. 34, 511
     (1979).

5.   T. N. C. Venkatesan, H. M. Gibbs, S. L. McCall, A. Passner,
     A. C. Gossard, and W. Wiegmann, (a) IEEE J. Quantum
     Electron QE-15, 8D (1979) and (b) Opt. Commun. 31, 228
     (1979).

6.   H. M. Gibbs, S. L. McCall, and T. N. C. Venkatesan, (a) Optics
     News 5, 6 (1979) and (b) Optical Engineering, to be pub-
     lished.

7.   H. M. Gibbs, S. M. McCall and T. N. C. Venkatesan, U. S.
     Patents 4,012,699 and 4,121,167.

8.   J. Shah, R. F. Leheny, and C. Lin, Solid State Commun. 18,
     1035 (1976).

9.   C. V. Shank, R. L. Fork, R. F. Leheny, and J. Shah, Phys. Rev.
     Lett. 42, 112 (1979).

10. D. A. B. Miller, S. D. Smith, and A. Johnston, Appl. Phys.
     Lett. 35, 658 (1979); D. A. B. Miller and S. D. Smith,
     Opt. Commun. 31, 101 (1979).

OPTICAL BISTABILITY AND MULTI-STABILITY IN THE SEMICONDUCTOR InSb

D. A. B. Miller, S. D. Smith, and C. T. Seaton

Department of Physics
Heriot-Watt University
Edinburgh, UK

Abstract: The discovery of third-order refractive nonlinearity
with $\chi^3 \sim 10^{-2}$ esu has led to the observation of optical bistability
in thin parallel crystals of InSb in various orders. The effect is
so sensitive that it is observable at power densities ~ 10 W/cm^2
(mW incident powers) using 100 µm dimension devices. Transphasor
action, that is, two-beam differential signal gain analogous to
transistor action, has been observed with gains of up to 10. The
future prospects of these devices are discussed in terms of macroscopic and microscopic processes, for both of which new theory has
been developed, and the possibility of fast, low-power switching is
discussed.

INTRODUCTION

The first observations of optical bistability in a nonlinear
Fabry-Perot interferometer fashioned from a parallel sided crystal
of InSb were made in 1979[1] independently and almost coincident in
time with observations in epitaxial films of GaAs[2]. The work on
InSb was rapidly extended, using the differential gain mode to
two beams clearly showing the "optical transistor" effect in which
one optical beam was amplified by the second with gains of up to 10
being observed[3]. The large size of the bandgap-resonant nonlinearity
has enabled optical thickness changes as large as $5\lambda/2$ to be induced
allowing the observation of bistability in fourth and fifth orders
in both transmission and reflection. The early work was conducted
at 5K; we here report extension to 77K and time-resolved operation.

II. CHARACTERISTICS OF SEMICONDUCTOR OPTICALLY BISTABLE DEVICES

The particular characteristic of semiconductor devices lies in the fact that very large dispersive nonlinearities can be obtained which facilitate the creation of intrinsic one-element devices of small size and power requirement. The required optical feedback is provided very simply by using either the natural reflection of the crystal surface (0.36 for InSb) or two layer reflection coatings. The devices therefore have relatively modest finesse which is consistent with a fast intra-cavity field build up time. A second advantage of the large nonlinear refraction is that intensity tuning is large and the interferometer can therefore be operated at relatively low power and the thickness can be small. In the first InSb devices, 500 μm thickness was used; the second generation interferometers are $\sim$ 130 μm thick and future thicknesses are likely to be $\sim$ 50 μm or less. Thus cavity field build up times can be restricted to be of the order of picoseconds. Power densities in the first observations at 5K were of the order of 10 μW/μm^2 and results reported here now reduce this by a further order of magnitude. It is relatively easy to construct a device $\sim$ 1 cm^2 in area so that the potential two-dimensional capability for all-optical data storage and beam amplification is quite considerable with limiting element dimension of a few μm. Ultimate speeds will depend only upon the speed of the nonlinear response: theoretical mechanisms for the giant nonlinear effect exploited in this work are discussed in a later section.

III. EXPERIMENTAL

The experimental realisation of a series of new effects observed in InSb is achieved by exploiting the coincidence of the 60 or 70 CO laser lines from 1930 cm^{-1} to 1660 cm^{-1} with the bandgap of InSb at 5K ($\sim$ 1900 cm^{-1}) and at 77K ($\sim$ 1840 cm^{-1}). The development of a continuously variable attenuator which, combined with a spatial filter, enables the beam intensity to be controlled over at least four orders of magnitude whilst retaining a near-perfect Gaussian profile[4] has been a key technique in uncovering a new series of low power nonlinear optical effects.[5] The first of these effects to be discovered[5] was beam broadening at intensities greater than 10 W/cm^2 ($\sim$ 10 mW incident). A macroscopic analysis of the near- and far-field beam patterns[6] enabled us to conclude that this effect was due to self-defocusing with a nonlinear refractive index n_2 defined from $n = n_1 + n_2 I$, with a value of $n_2 \sim -6 \times 10^{-5}$ cm^2/W at 1886 cm^{-1} and 5K. The *sign* of this effect also indicated that the origin was unlikely to be thermal as did a series of time-resolved experiments.

The output beam from the grating tunable CO laser (Edinburgh Instruments PL3) is passed through the attenuator, through an electro optic modulator for time-resolved measurements and then through the sample held in a cryostat with anti-reflected ZnSe windows. A detector then detects the transmission of the interferometer as the intensity is changed either by the attenuator (that is, for a steady state measurement) or the modulator (for a time-resolved measurement). The output from the interferometer is illustrated in Fig. 1 for 1895 cm^{-1} and with the sample held at 5K. It shows five distinct steps corresponding to the successive changes of optical thickness of $\lambda/2$. It is interesting to note that the increment of power between the steps increases as the intensity increases indicating a saturation in the refractive nonlinearity. The *entire* Gaussian beam power was observed at the detector and despite the inevitable incident Gaussian intensity variation across the specimen, clear optical bistability is observed in fifth order. Figure 2 shows further observations in both transmission and reflection showing bistability in both fourth and fifth orders, i.e. multi-stability. The results also show that the background absorption in the interferometer is low, i.e. ∿ 50% transmission or reflection is observed.

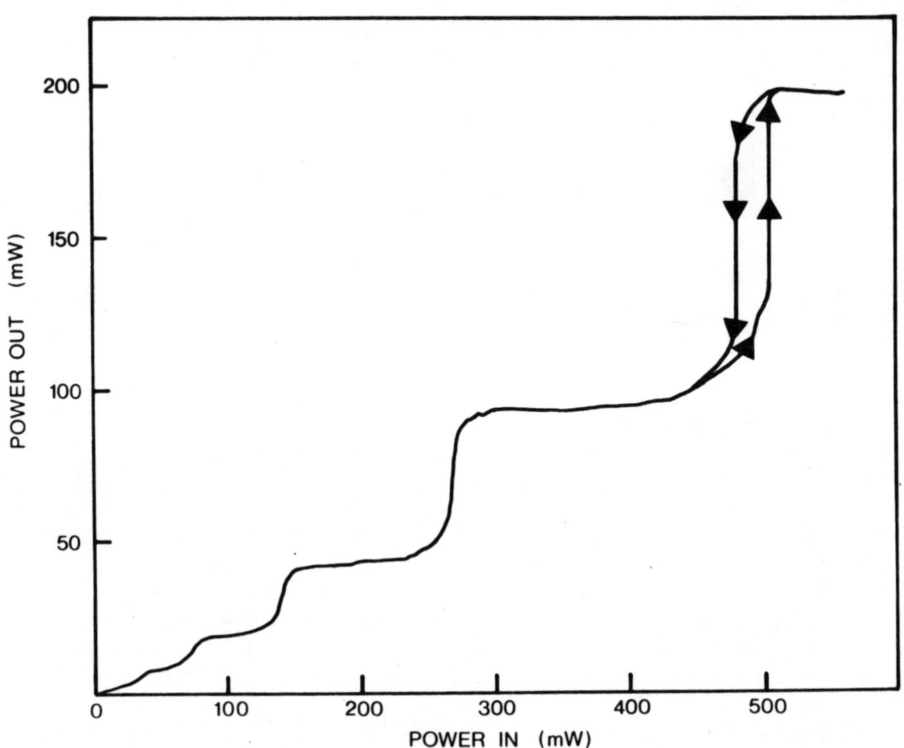

Fig. 1.   Optical bistability in InSb at 5K. Crystal thickness 560 μm with natural reflectivity (36%) faces. Incident cw CO laser wave number 1895 cm^{-1}, spot size 180 μm.

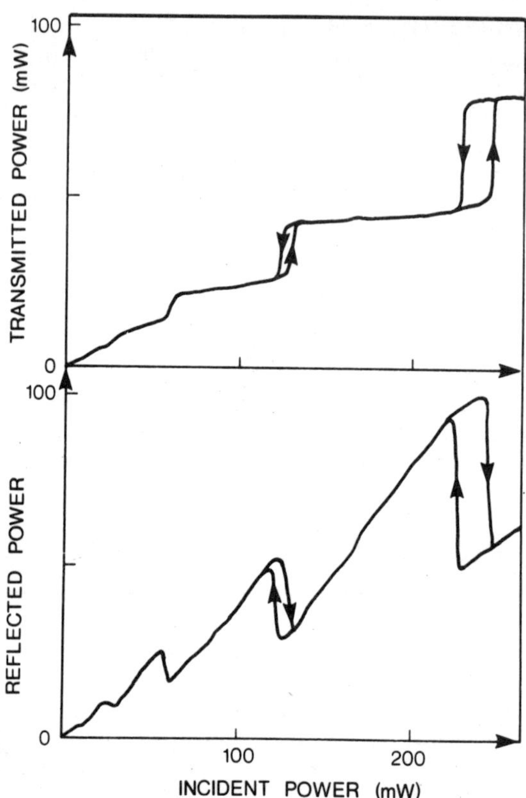

Fig. 2.   Multistability in transmission and reflection.   Sample and
laser line as for Fig. 1 with different cavity tuning (by
lateral displacement of sample).

The conclusion that the basic effect is a fundamental electronic
process led us to predict that the intensity change can be provided
by two beams rather than one.   Thus, the steps in Fig. 1 which show
differential gain (that is, the output changes more than the input)
can be induced with a second beam.   The results of this experiment
are demonstrated in Fig. 3.   At each step the second beam modulates
the transmission of the first beam and signal gains of up to 10 are
demonstrated.   This is a clear demonstration of 'optical transistor'
action and since it is caused by the change of phase thickness
induced by one beam, we term it, by analogy with the transistor,
"transphasor" action.   It should be noted that the physics of such
a two-beam device is, however, not identical to differential gain
with a single beam.   When a second beam is coincident with the
standing wave field inside the Fabry-Perot cavity the resulting
conditions are exactly those required for degenerate four-wave
mixing (DFWM) (In the case of a single beam this DFWM term vanishes

into the nonlinear refraction). Indeed we have observed DFWM in
InSb at power levels well below the levels used in the "transphasor"
experiments[7]. In the "transphasor", therefore, new beams are generated by DFWM which must affect the detailed operation of the device,
and it would not be justifiable to claim 'optical transistor' action
simply on the basis of single beam differential gain measurements.

IV.   MECHANISM OF NONLINEARITY

Since the large nonlinear effect is very effective in a device
sense we are concerned to understand the microscopic origin of the
nonlinearity. This requires considerable explanation since the value
of $\chi^{(3)}$ $(\omega:\omega,-\omega,\omega)$ which corresponds to the intensity dependent

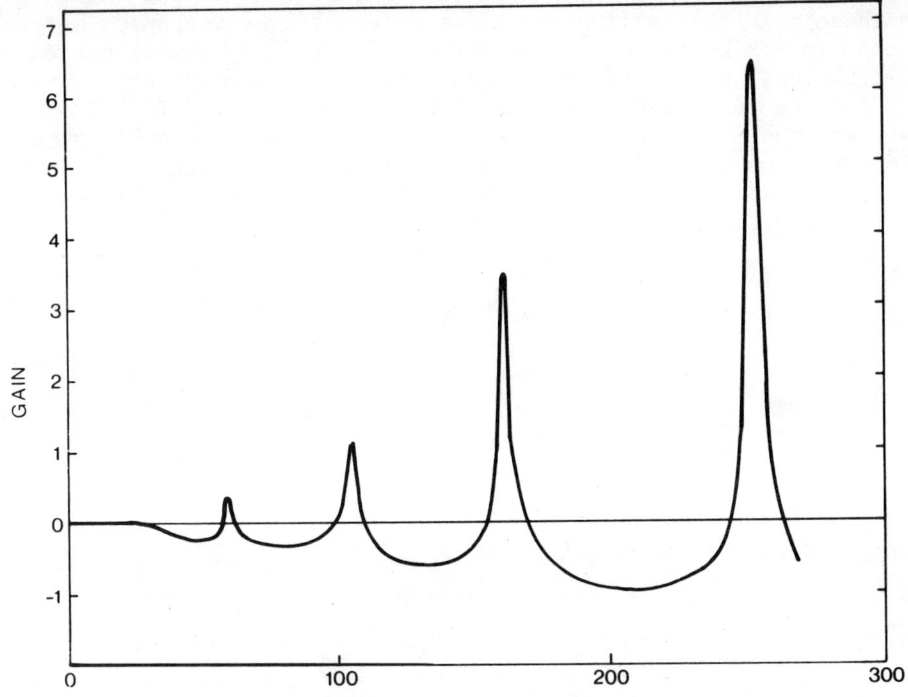

Fig. 3.   Two-beam differential gain - 'transphasor' action. The
          gain is the ratio of the power change in the weak signal
          beam to the induced change in the main pumping beam, showing
          clear peaks corresponding to the one-beam differential gain
          of Fig. 1. Sample and laser beams as in Fig. 1; laser
          beams, from separate lasers, focused coincidentally onto
          the sample face.

refractive index is $\sim 10^{-2}$esu at 5K and as high as 1 esu at 77K
at frequencies near the bandgap. Previously familiar values for
$\chi^{(3)}$ in semiconductor materials away from the bandgap are in the
range 10^{-8} to 10^{-11} esu. Resonance near the bandgap can increase
these values to about 10^{-6} esu at best but it is seen that a further
four or five orders of magnitude are required to explain the present
experimental results. From Fig. 1 we can plot the variation of n_2
with intensity (or otherwise expressed, there exists a $\chi^{(5)}$ as well
as a $\chi^{(3)}$). This plot, Fig. 4, indicates that the refractive nonlinearity saturates. Saturation of the absorption occurs in the
band-tail region[4]; we have shown that most of this effect disappears
before the onset of the nonlinear refraction responsible for the
device operation[6]. Photoconductivity measurements[8] indicate that
mobile carriers can be created by laser irradiation in the same
region. It is therefore reasonable to associate the excitation of
carriers and possible saturation of transitions with the existence
of these highly sensitive nonlinear effects. The mechanism of excitation is not so far established (we refer to this below) but we can
nevertheless deduce from the *measured* absorption coefficient at the
appropriate laser power that between 1×10^{14} and 1×10^{15} free
carriers would be created optically if we assume a recombination
time of about 200 ns (a reasonable scaling from experimental numbers)
The excitation of such a carrier density might then lead to the
following effects:

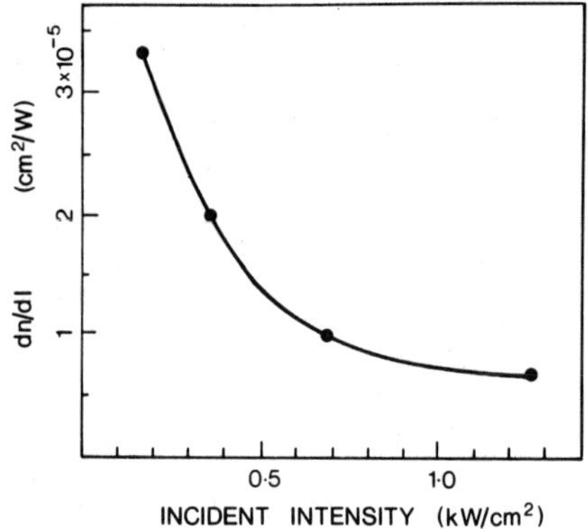

Fig. 4.   Intensity dependence of dn/dI, the nonlinear refractive
          index, derived from the separation of nonlinear inter-
          ference orders in Fig. 1.

1) Screening or saturation of exciton effects. All absorption near
    the bandgap frequency in a semiconductor is in principle modified
    by Coulomb effects. It is known that the screening of discrete
    exciton absorption is associated with the refractive nonlinearity
    in GaAs[2]. This mechanism, however, seems unlikely in InSb where
    the Bohr radius of the exciton is $\sim$ 700 Å and excitonic effects
    are totally screened at carrier densities of around $1 \times 10^{-14} \mathrm{cm}^{-3}$,
    the lowest initial value of the electron density in our material.
    In practice, exciton absorption spectra are not seen in InSb and
    both single and two photon absorption spectra near the bandgap
    energy are better fitted to experiment if excitons are neglected.
    We therefore consider significant exciton effects unlikely.

2) Nonlinearity caused by free carrier plasma creation. This
    effect has been invoked by Jain and Klein[9] to explain a value
    of $\chi^{(3)} \sim 10^{-7}$ esu in silicon. The effect is described by the
    equation

$$n_2 = - \frac{2\pi\alpha e^2 \tau}{n m_r \hbar \omega^3}$$

    (in esu) where α is the linear absorption coefficient, τ is the
    recombination time, n is the linear refractive index, m_r is the
    reduced effective mass, and $\hbar$, e and ω have their usual meanings.
    Inserting the values quoted above and noting that the absorption
    coefficient $\alpha \sim 1 \mathrm{cm}^{-1}$ this yields a value of n_2 around
    3×10^{-6} cm^2/W – 10 or 20 times too small to explain our results
    at 5K[6].

3) Excitation of carriers has been shown to saturate the absorption
    above the bandgap in InSb[10,11]. This has been attributed to a
    dynamic Burstein–Moss shift caused by filling the bottom of the
    conduction band with electrons. Equally, such an effect removes
    states contributing to dispersion and could cause nonlinear
    refraction. The published absorption work relies upon an exci-
    tation at energies *above* the bandgap and subsequent thermalisa-
    tion of the carriers. We can estimate[12] the relative values of
    this effect and the free carrier plasma

$$\frac{n_2 \ (\text{Burstein–Moss})}{n_2 \ (\text{Plasma})} = 4\left(\frac{\omega_G - \omega}{\omega_G}\right)$$

    where $\hbar\omega_G$ is the bandgap energy. In our case, however, the
    excitation energy is less than the bandgap and it is not clear
    that within the timescale of the excitation the carriers will

will distribute themselves in the same way. Thus it may not be correct to interpret this formula too literally.

4) In direct analogy with discrete two-level systems, broadened by a dephasing time T_2, it is possible to consider the band-tails of a set of two-level oscillators as capable of exciting electrons for laser photon energies less than the bandgap. This could be described as 'off-resonant pumping' and would be significant within a frequency range of $1/T_2$ of the bandgap. The suggestion of Javan and Kelley[13] that saturated absorption leads to nonlinear refraction, in principle by causality, applied to nonlinear processes can then give a contribution to nonlinear refraction. For the case of a semiconductor there is already evidence that a two-level model can reasonably explain saturable absorption in p-type Ge[14]. The Bloch states which describe interband transitions constrain the absorption to k-conserving vertical transitions so that a set of two-level oscillators is a quite reasonable model. At least it gives us a complete *a priori* model to compare with experiments. An expression for n_2 can then be obtained by summing the dispersive contributions using a density matrix formulation put forward by Miller, Smith and Wherrett[12]. The expression for n_2 is as follows:

$$n_2 = \frac{-2\pi}{15} \frac{\hbar}{n^2 c} \left| \frac{eP}{\hbar^2 \omega} \right|^4 \left(\frac{2m_r}{\hbar} \right)^{3/2} \frac{T_1}{T_2} (\omega_G - \omega)^{-3/2} .$$

Here all the parameters e, $\hbar$, ω, c take on their usual meanings, with n as the linear refractive index, m_r the reduced effective mass P and the momentum matrix element. $\hbar\omega_G$ is the bandgap energy. The parameters in this theory which require careful interpretation are the energy relaxation time T_1 and the phase relaxation time T_2, assumed constant throughout the bands for simplicity in this expression. We can interpret T_2 as being the time between collisions of excited electrons or holes giving randomisation of the quantum-mechanical phase of the wavefunctions; we therefore expect T_2 to be related to the speeds of intraband scattering mechanisms which can be on a picosecond timescale (the momentum relaxation time obtained from mobility data is typically picoseconds or less in cooled InSb). In atomic systems T_1 is normally the time taken for the excitation to relax to the ground state; in the semiconductor we might then interpret T_1 as the band-to-band recombination time. While this recombination time must set an upper limit on T_1, it is not the only process by which the excitation of a particular two-level system can be effectively relaxed. After excitation, the electron (hole) can be scattered to another state *inside* the conduction (valence) band by the fast intraband processes; the excitation of this state is then effectively relaxed because we now find this state in the condition it was initially- namely, with the electron (i.e. no hole) in the valence band and no electron in the corresponding conduction band state. Intraband energy relaxation has been measured at about $\sim$ 100 ps[15] in InSb albeit in a magnetic field. Because of the difficulty of interpreting T_1 and T_2, their ratio must remain a fitting parameter in our theory. However, we would expect to see a response of the nonlinearity on both a picosecond and a nanosecond timescale. The comparison between n_2 resonance measurements and theory with T_1/T_2 = 10 and 100 is made in reference 12 for a temperature of 5K. Such values for the ratio T_1/T_2 are not inconsistent with the result of Gornik et al.[15] Recent measurements at 77K of $n_2 \simeq 3 \times 10^{-4}$ cm^2/W at 1830 cm^{-1} suggest that using this model T_1/T_2 is as high as 10,000 at this temperature. In this case, the absorption edge is substantially more broadened, implying a shorter T_2 than the 5K case.

All the processes discussed above are in principle capable of giving large contributions to $\chi^{(3)}$. We have undertaken some experiments which give some indications of the relative importance.

Following the knowledge that the nonlinear refraction induced by one beam can be transferred to another[3] we may attempt to observe induced defocusing using two laser beams at various frequencies. If the origin of the nonlinearity was predominantly carrier plasma one would then expect the nonlinearity to increase according to $1/\omega^3$. Thus pumping at 1886 cm^{-1} near the bandgap should therefore give a large effect at say 1730 cm^{-1}, the longest probe wavelength. As shown in Fig. 5, probe laser beams at different frequencies are indeed broadened by the 1886 cm^{-1} pump but the effect falls off towards longer wavelengths in direct contradiction to this proposition. This experiment therefore favours 3 and 4.

In our original observations of nonlinear optical effects in InSb[5] we saw self-induced beam broadening at both 5K and 77K. Thus we predicted that optical bistability would be observable at 77K as well as the original temperature regime at 5K. As shown in Fig. 6 we observe both a large nonlinearity at 77K ($\chi^{(3)} \sim 10^{-1}$ esu) and very clear optical bistability. This latter was observed in a crystal 130 µm thick coated to given a reflectivity of 0.7 with layers of Ge and ZnS. This bistability occurred in first order and required an onset power of only 7.9 mW. If the mechanism were excitonic we would expect a gross reduction in the strength of the nonlinearity on increasing the temperature from 5K to 77K and a disappearance of the effect with impure samples. In fact we see a *stronger* nonlinearity at 77K, and the bistability results shown here for 77K were taken with inexpensive, impure polycrystalline InSb of the type normally used for monochromator order-blocking filters. This therefore argues strongly against excitonic effects (mechanism 1).

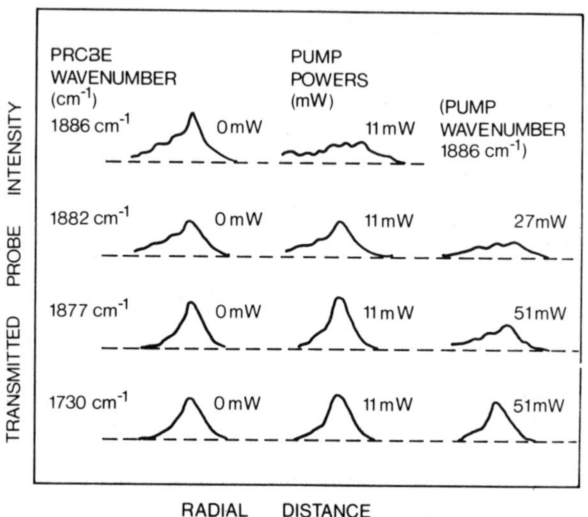

Fig. 5.   Transferred beam broadening in InSb. Two laser beams (spot size ∿ 150 μm) are coincident on the sample (5 mm thick, at 5K), a pump and a probe. At 11 mW the pump far field profile is significantly broadened by self-defocusing; this broadening is observed transferred into the weak probe beam at the same wavelength. As the probe beam wavelength is moved from the pump wavelength progressively more pump power is required to induce beam broadening in the probe beam. At wavelengths much longer than the pump wavelength (e.g. 1730 cm⁻¹) no induced effect is observed with available beam powers.

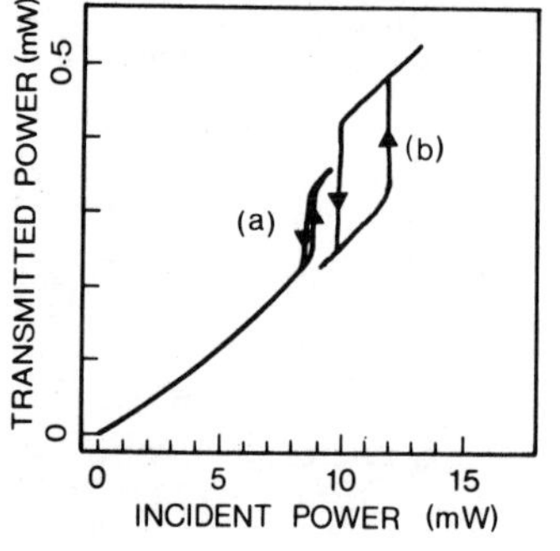

Fig. 6   Optical bistability in InSb at 77K. Crystal thickness 130 μm, surface reflectivity ∿70%, laser wavenumber 1827 cm⁻¹, spot size ∿150 μm. The onset of bistability, (a), is seen at 7.9 mW; clear, broad bistable regions are seen (e.g. (b)) as the cavity tuning is altered in ·the correct sense.

It should be possible to distinguish between the relative strengths of the refractive effects of direct saturation (i.e. population of states by direct optical excitation, mechanism (4)) and indirect saturation (i.e. population of states by scattering-in of excited carriers from other states, mechanism (3)) by making time-resolved measurements of nonlinear refraction. The total effect should be some combination of the two since one cannot exist without the other (indirect saturation can only result if there is some direct optical excitation, and intraband de-excitation of direct optical excitation must result in indirect population of other band states.).

We have made preliminary measurements of time-resolved bistable switching in InSb using the observed bistability at 77K. The laser beam intensity was modulated by $\sim$ 50% using an electro-optic modulator at a frequency of $\sim$ 1 KHz; bistable switching then occurred on a much faster timescale. The observations, currently limited by detector response times, show switching occurs in less than a microsecond and therefore conclusively eliminate any possibility of thermal origins for the process.

V.   CONCLUSIONS

The small holding powers, switching energies, size, two-dimensional character and potentiality for fast operation give considerable promise for the application of narrow gap semiconductors to optically bistable devices and their various derivatives. The new bandgap-resonant mechanism discovered in this work seems likely to be applicable to a variety of materials, possibly at higher temperatures and shorter wavelengths.

The authors acknowledge the experimental assistance of A. M. Johnston.

REFERENCES

1. D. A. B. Miller, S.D. Smith and A.M. Johnston, Appl. Phys. Lett. 35, 658 (1979).
2. H. M. Gibbs, S. L. McCall, T. N. C. Venkatesan, A. C. Gossard, A. Passner and W. Wiegmann, Appl. Phys. Lett. 35, 451 (1979).
3. D. A. B. Miller and S. D. Smith, Opt. Commun. 31, 101 (1979).
4. D. A. B. Miller and S. D. Smith, Appl. Opt. 17, 3804 (1978).
5. D. A. B. Miller, M. H. Mozolowski, A. Miller and S. D. Smith, Opt. Commun. 27, 133 (1978).
6. D. Weaire, B. S. Wherrett, D. A. B. Miller and S. D. Smith, Opt. Lett. 4, 331 (1979).
7. D. A. B. Miller, R. G. Harrison, A. M. Johnston, C. T. Seaton and S. D. Smith, Opt. Commun. 32, 478 (1980).

8.   D. G. Seiler and L. K. Hanes, Opt. Commun. _28_, 326 (1979).
9.   R. K. Jain and M. B. Klein, Appl. Phys. Lett. _35_, 454 (1979).
10. P. Lavallard, R. Bichard and C. Benoit à la Guillaume, Phys. Rev. B_16_, 2804 (1977).
11. A. V. Nurmikko, Opt. Commun. _16_, 365 (1976).
12. D. A. B. Miller, S. D. Smith and B. S. Wherrett, Optics Commun. in press.
13. A. Javan and P. L. Kelley, IEEE J. Quantum Electron. _QE-2_, 470 (1966).
14. F. Keilman, IEEE J. Quantum Electron. _QE-12_, 592 (1976).
15. E. Gornik, T. Y. Chang, T. J. Bridges, V. T. Nguyen and J. D. McGee, Phys. Rev. Lett. _40_, 1151 (1978).

OPTICAL BISTABILITY EFFECTS IN A DYE RING LASER*

L. Mandel, Rajarshi Roy, and Surendra Singh

Department of Physics and Astronomy
The University of Rochester
Rochester, New York 14627

Abstract: The theory of a homogeneously broadened two-mode ring laser is discussed. Because the mode coupling constant can be twice as great as under conditions of inhomogeneous broadening, new effects like quasi-bistability appear, and the radiation field undergoes a phase transition near threshold that is of the first order. This is reflected in a double peaked probability distribution of each mode intensity, and in alternate switching of the excitation between modes. A first passage time calculation shows that the dwell times should increase rapidly with excitation. These predictions are tested by photoelectric counting and other measurements on a dye ring laser. Although the observed photoelectric counting distributions closely resemble the theoretical ones, measurements of the dwell times reveal significant discrepancies between theory and experiment, that are believed to be due to backscattering. The adequacy of a theory based on two-level atom models in accounting for the behavior of a dye ring laser is briefly discussed.

I. INTRODUCTION

It is well known from many experiments with ring lasers, that when the laser gain medium is homogeneously broadened and the laser is operated at a single frequency, one of the two travelling wave modes tends to suppress the other one[1]. Although competition between modes is predicted[2-10] and observable[11,12] in all lasers to some extent, the effects are particularly pronounced in homogeneously broadened ring lasers,[4,6] and in Zeeman ring lasers,[5,9] in

---

*This work was supported by the National Science Foundation and by the Air Force Office of Scientific Research.

which the normalized mode coupling constant can exceed unity. However, because of the quantum fluctuations, the state in which one mode is excited and the excitation of the other mode is suppressed is not usually stable, and from time to time a fluctuation will cause the system to switch over spontaneously into the opposite configuration. Such a ring laser is therefore an example of a quasi-bistable optical system. Moreover, unlike more conventional lasers, it exhibits statistical features that are more closely associated with a first order – rather than with a second order – phase transition. Indeed, the effective Hamiltonian or potential describing the behavior of the optical field bears a striking similarity to Hamiltonians that are often used to model first order phase transitions in other systems[13,14]. We have recently treated the statistical features of the light from such a ring laser theoretically[8], and have observed some of them experimentally[15].

II. EQUATIONS OF MOTION

The simplest starting point for any theoretical treatment of the ring laser is the pair of coupled third order Lamb equations of motion[5,16] for the complex amplitudes $E_1(t)$ and $E_2(t)$ of the two counterpropagating travelling wave modes. We write them in the dimensionless form

$$\frac{dE_1}{dt} = (a_1 - |E_1|^2 - \xi|E_2|^2)E_1 + q_1(t) \tag{1}$$

$$\frac{dE_2}{dt} = (a_2 - |E_2|^2 - \xi|E_1|^2)E_2 + q_2(t) \ . \tag{2}$$

The deterministic equations, which are derivable from a semiclassical treatment of the interaction between two-level atoms and a classical field, have been supplemented by the introduction of randomly fluctuating Langevin terms $q_1(t)$ and $q_2(t)$, that represent the spontaneous emission fluctuations. Because the atomic lifetimes are typically short compared with the times in which the laser field changes, we may take $q_1(t)$ and $q_2(t)$ to be effectively δ-correlated Gaussian noise sources, with

$$\langle q_i^*(t)q_j(t')\rangle = 2\delta_{ij}\delta(t-t') \ , \qquad i,j = 1,2 \ .$$

a_1 and a_2 are dimensionless pump parameters characterizing the excitations of the two laser modes; they are positive above and negative below threshold. The third order equations do not of course apply very far above threshold, but they are usually excellent approximations up to pump parameter values of at least 50. ξ is the dimensionless mode coupling constant, which is constrained to lie between 0 and 1 for a gaseous gain medium which is inhomogeneously broadened, but can become greater than unity and as large as 2 in a homogeneously broadened laser[2-7].

Before considering the solutions of Eqs. (1) and (2), it is instructive to examine the consequences of the equations if the motion were completely deterministic. In the absence of the Langevin terms, the equations have the steady state solutions

$$(a_1 - I_1 - \xi I_2)I_1 = 0$$
$$(a_2 - I_2 - \xi I_1)I_2 = 0 , \tag{3}$$

in which $I_1 \equiv |E_1|^2$ and $I_2 \equiv |E_2|^2$ are the light intensities. The solutions for I_1, I_2 therefore have three distinct branches:

(a) $I_1 = 0$, $I_2 = a_2 = a - \dfrac{1}{2} \Delta a$

(b) $I_1 = a_1 = a + \dfrac{1}{2} \Delta a$ $I_2 = 0$ (4)

(c) $I_1 = \dfrac{a}{\xi+1} - \dfrac{\frac{1}{2}\Delta a}{\xi-1}$, $I_2 = \dfrac{a}{\xi+1} + \dfrac{\frac{1}{2}\Delta a}{\xi-1}$,

provided $- \dfrac{2a}{\xi+1} < \dfrac{\Delta a}{\xi-1} < \dfrac{2a}{\xi+1}$,

which are illustrated graphically in Fig. 1 for several different coupling constants ξ. We have used the notation

$$a \equiv \dfrac{1}{2} (a_1 + a_2)$$
$$\Delta a \equiv a_1 - a_2 . \tag{5}$$

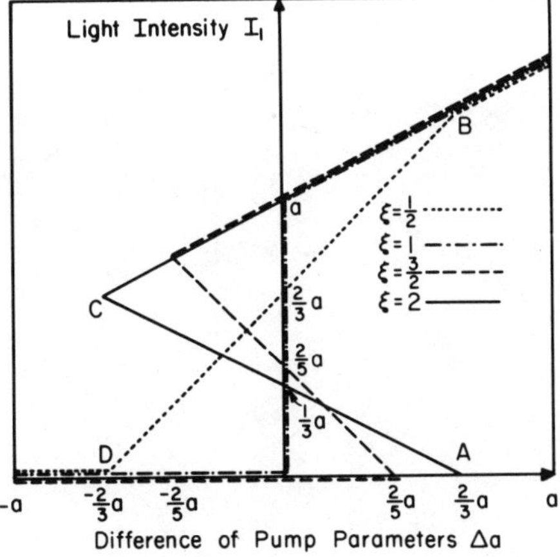

Fig. 1.   Illustrating the solution with hysteresis implied by the deterministic equations.

It will be seen that I_1 is always zero for sufficiently negative Δa and it is equal to a_1 for sufficiently positive Δa, and these solutions are connected by the third branch. However, the connecting branch has positive slope when $\xi < 1$ and negative slope when $\xi > 1$. As a branch with negative slope represents an unstable solution, it is to be expected that, as Δa increases from zero, the representative working point of the system will jump from point A to point B when $\xi = 2$. On the other hand, if Δa is decreased subsequently, the working point follows the line from B to C and then jumps to point D, after which it follows the axis. In other words, like most bistable systems the ring laser is expected to exhibit hysteresis under these conditions.

Because of the fluctuations, these simplistic solutions do not hold, and the complex field amplitudes E_1, E_2 of the two modes are describable only in statistical terms through a joint probability density $p(E_1, E_2, t)$. It is well known that, corresponding to the Langevin Eqs. (1) and (2) for E_1, E_2 there exists a Fokker-Planck equation for $p(E_1, E_2, t)$ of the form

$$\frac{\partial p}{\partial t} = - \sum_{i=1}^{4} \frac{\partial}{\partial x_i} (A_i p) + \frac{1}{2} \sum_{i=1}^{4} \sum_{j=1}^{4} \frac{\partial^2}{\partial x_i \partial x_j} (D_{ij} p) , \tag{6}$$

where we have written

$$\left. \begin{aligned} E_1 &\equiv x_1 + ix_2 \\ E_2 &\equiv x_3 + ix_4 , \end{aligned} \right\} \tag{7}$$

with drift vectors

$$\left. \begin{aligned} A_1 &= [a_1 - (x_1^2 + x_2^2) - \xi(x_3^2 + x_4^2)]x_1 \\ A_2 &= [a_1 - (x_1^2 + x_2^2) - \xi(x_3^2 + x_4^2)]x_2 \\ A_3 &= [a_2 - (x_3^2 + x_4^2) - \xi(x_1^2 + x_2^2)]x_3 \\ A_4 &= [a_2 - (x_3^2 + x_4^2) - \xi(x_1^2 + x_2^2)]x_4 , \end{aligned} \right\} \tag{8}$$

and diffusion tensor

$$D_{ij} = 2\delta_{ij} , \qquad\qquad i,j = 1,2,3,4 . \tag{9}$$

The general time dependent solution of Eq. (6) has so far only been obtained for the case $\xi = 1$.[7]

III. STEADY STATE SOLUTION

When the drift vectors obey the so-called potential condition[17]

$$\frac{\partial A_i}{\partial x_j} = \frac{\partial A_j}{\partial x_i} \quad , \tag{10}$$

the steady state solution of Eq. (6) can be written down at once, and it takes the form[2,7]

$$P_s(\vec{x}) = \text{const.} \times \exp\left[\frac{1}{2} a_1 (x_1^2 + x_2^2) + \frac{1}{2} a_2 (x_3^2 + x_4^2) - \frac{1}{4} (x_1^2 + x_2^2)^2 \right.$$

$$\left. - \frac{1}{4}(x_3^2 + x_4^2)^2 - \frac{1}{2} \xi (x_1^2 + x_2^2)(x_3^2 + x_4^2) \right] . \tag{11}$$

Alternatively, we may express the same solution in terms of the joint probability density $P_2(I_1, I_2)$ for the two mode intensities I_1, I_2 in the form

$$P_2(I_1, I_2) = Q^{-1} \exp\left[\frac{1}{2} a_1 I_1 + \frac{1}{2} a_2 I_2 - \frac{1}{4} I_1^2 - \frac{1}{4} I_2^2 - \frac{1}{2} \xi I_1 I_2\right], \tag{12}$$

where Q is a normalizing constant. This has the character of a truncated bivariate Gaussian distribution in the two variables I_1, I_2 whenever $\xi < 1$. However, when $\xi > 1$ the coefficients in the exponent do not form a positive definite matrix, and the probability density is no longer Gaussian.

From Eq. (12) we may evaluate all the moments of I_1, I_2. General expressions for the first two moments have already been given by Grossmann and Richter[2] and by M-Tehrani and Mandel[7]. Because of their length and complexities we shall not reproduce them here. A particularly interesting consequence of Eq. (12) arises when the coupling constant $\xi > 1$ and $a_1 = a_2 = a$, as for a homogeneously broadened symmetric ring laser. We then find[8] for large pump parameter a,

$$\langle I_1 \rangle = \langle I_2 \rangle = \frac{1}{2} a - \frac{1}{a} + 0\left(\frac{1}{a}\right)^3 \tag{13}$$

$$\frac{\langle I_1 I_2 \rangle}{\langle I_1 \rangle \langle I_2 \rangle} = \frac{8}{a^2 (\xi - 1)} + 0\left(\frac{1}{a}\right)^4 , \tag{14}$$

so that the relative intensity correlation tends to zero well above threshold when $\xi > 1$. Figure 2 shows the behavior of the normalized cross-correlation $\langle I_1 I_2 \rangle / \langle I_1 \rangle \langle I_2 \rangle$ as a function of a for several different values of the coupling constant ξ. It will be seen that the correlation tends to zero with increasing excitation of the laser when ξ exceeds unity. As the light intensities are necessarily non-negative and have a non-vanishing mean, this carries the implication that, although the intensities of the two-modes are equal on the average, a high intensity of one mode is always associated with a low intensity of the other one. In other words, the principal excitation switches alternately between modes.

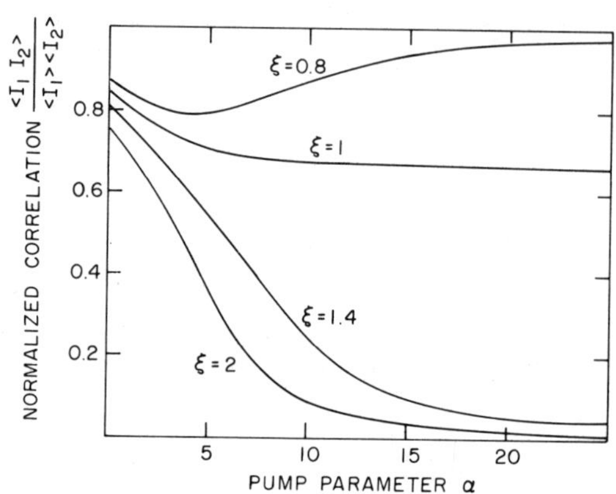

Fig. 2.　Variation of the normalized intensity cross-correlation with pump parameter a ($\Delta a = 0$), for several different coupling constants ξ.

By integrating over one of the variables I_2, in Eq. (12) we readily arrive at the probability density $P(I_1)$ of the other mode intensity I_1,[8]

$$P(I_1) = \sqrt{\pi}\, Q^{-1} \exp \tfrac{1}{4}\left(a - \tfrac{1}{2}\,\Delta a\right)^2 \exp[-V(I_1, a, \Delta a, \xi)] \tag{15a}$$

in which the 'potential' V is given by

$$V(I, a, \Delta a, \xi) = -\tfrac{1}{4}\,(\xi^2 - 1)I^2 + [\tfrac{1}{2}\,a(\xi - 1) - \tfrac{1}{4}\,\Delta a(\xi + 1)]I$$

$$- \log[1 - \mathrm{erf}(\tfrac{1}{2}\,\xi I - \tfrac{1}{2}\,a + \tfrac{1}{4}\,\Delta a)]. \tag{15b}$$

The form of this potential V as a function of I_1 is illustrated in Figs. 3 and 4 for several different pump parameters a_1 and coupling constants ξ. The difference Δa between the two pump parameters, corresponding to a small asymmetry in the reflection and diffraction losses of the two ring laser modes, is assumed to remain constant at $\Delta a = 1$.

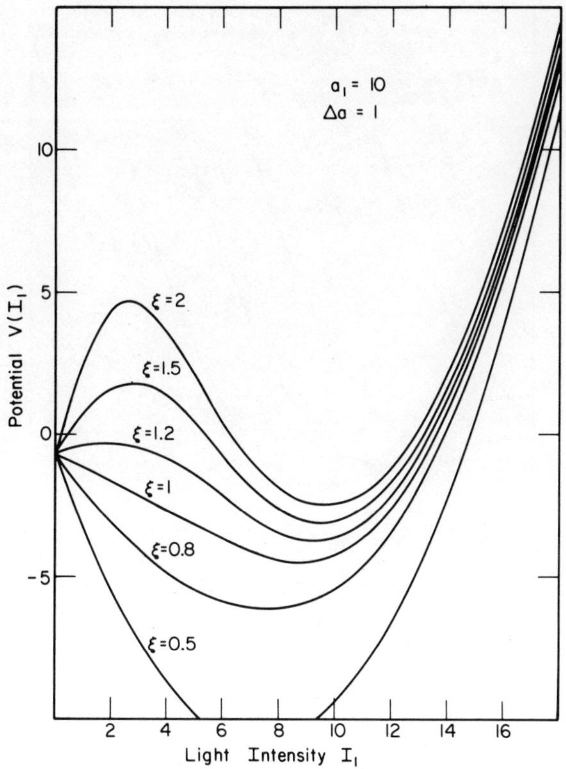

Fig. 3.   Variation of the potential $V(I_1)$ for several different
coupling constants ξ, with $a_1=10$, $\Delta a=1$.

A particularly interesting feature of the solution is the appearance of a maximum in addition to the minimum in the potential
above threshold, when the coupling constant ξ exceeds unity. It implies that the probability density $P(I)$ has two peaks; one always
occurs at $I=0$ and one at the position $I=I_B$ of the potential minimum.
This suggests that the ring laser may have two quasi-stable states
above threshold when ξ exceeds unity. In the absence of fluctuations
the states $I=0$ or $I=I_B$ would of course persist indefinitely, but, because of fluctuations, neither state is truly stable. Sooner or
later a sufficiently large fluctuation causes the system to switch
from one state to the other, at an average rate that is determined
by the coupling constant ξ and the pump parameters.

It is not too difficult to find at least approximate expressions
for the positions of the two stationary points of the potential when
$\xi > 1$. By differentiation of Eq. (15b) we find that the stationary
points are given by the intersections of the two curves

$$y = [(\xi^2-1)I - a(\xi-1) + \tfrac{1}{2}\Delta a(\xi+1)][1-\text{erf}\ \tfrac{1}{2}(\xi I - a + \tfrac{1}{2}\Delta a)] \qquad (16)$$

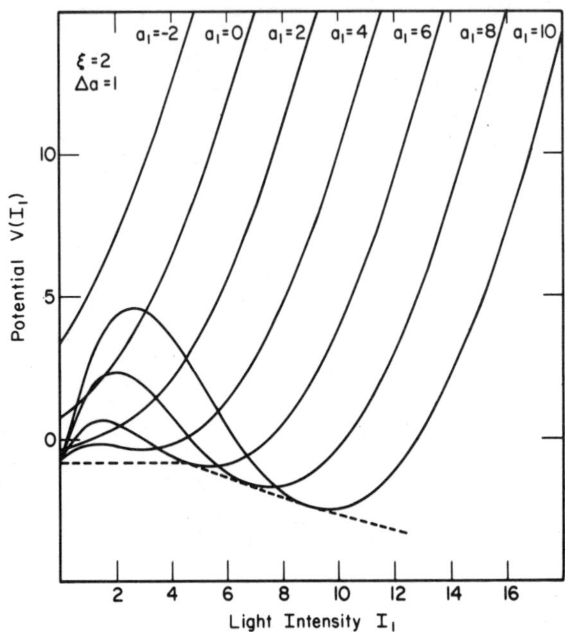

Fig. 4.   Variation of the potential $V(I_1)$ for several different
pump parameters a_1, with $\Delta a=1$, $\xi=2$.

$$y = \frac{2\xi}{\sqrt{\pi}} \exp - \frac{1}{4}(\xi I - a + \frac{1}{2} \Delta a)^2 \ . \tag{17}$$

Above threshold, when $a > 0$, and for I not too large, Eq. (16) approximately describes a straight line, but this line is effectively
cut off by the error function factor once I appreciably exceeds
$(a - \frac{1}{2} \Delta a)/\xi$. The straight line intercepts the Gaussian function
given by Eq. (17), which is peaked at $I = (a - \frac{1}{2} \Delta a)/\xi$, in the left
tail close to the I-axis. It follows from Eq. (16) that the maximum
value of the potential occurs when $I = I_A$ such that

$$(\xi^2-1)I_A - a(\xi-1) + \frac{1}{2} \Delta a(\xi+1) \approx 0$$

or

$$I_A \approx \frac{a}{\xi+1} - \frac{\frac{1}{2} \Delta a}{\xi-1} \ . \tag{18}$$

The position of the potential minimum I_B is given by the second intersection of the same two functions, and it occurs in the right
tail of the Gaussian. When I is relatively large, we may use the
asymptotic form of the error function[18] (provided $\xi I - a + \frac{1}{2} \Delta a > 4$) in
Eq. (16), and we then find that I_B is given by

$$[(\xi^2-1)I_B - a(\xi-1) + \frac{1}{2}\Delta a(\xi+1)]\ \frac{e^{-\frac{1}{4}(\xi I_B - a + \frac{1}{2}\Delta a)^2}}{\sqrt{\pi}\ \frac{1}{2}(\xi I_B - a + \frac{1}{2}\Delta a)}$$

$$\approx \frac{2\xi}{\sqrt{\pi}}\ e^{-\frac{1}{4}(\xi I_B - a + \frac{1}{2}\Delta a)^2}$$

or

$$(\xi^2-1)I_B - a(\xi-1) + \frac{1}{2}\Delta a(\xi+1) \approx \xi(\xi I_B - a + \frac{1}{2}\Delta a)$$

or

$$I_B \approx a + \frac{1}{2}\Delta a = a_1. \tag{19}$$

When $a \gg 1$ the two peaks in the probability distribution $P(I)$ at $I=0$ and at $I=I_B \approx a + \frac{1}{2}\Delta a$ are well resolved, and in between at $I=I_A$ the probability density $P(I_A)$ is close to zero. It is then convenient to define two integral probabilities

$$\left.\begin{array}{l} P_{low} \equiv \displaystyle\int_0^{I_A} P(I)dI \\[20pt] P_{high} \equiv \displaystyle\int_{I_A}^{\infty} P(I)dI \end{array}\right\} \tag{20}$$

satisfying the equality $P_{low} + P_{high} = 1$, such that P_{low} represents the probability for the light intensity to be low or near zero, while P_{high} represents the corresponding probability for it to be high or near a_1. By using the asymptotic forms of the error function, we readily find from Eqs. (15), (18) and (19),

$$P_{high} \approx \frac{4\sqrt{\pi}}{Q}\ \frac{e^{\frac{1}{4}(a-\frac{1}{2}\Delta a)^2}\ e^{\frac{1}{2}a\Delta a}}{(\xi-1)a + \frac{1}{2}(\xi+1)\Delta a}$$

$$P_{low} \approx \frac{4\sqrt{\pi}}{Q}\ \frac{e^{\frac{1}{4}(a-\frac{1}{2}\Delta a)^2}}{(\xi-1)a - \frac{1}{2}(\xi+1)\Delta a}\ ,$$

or

$$P_{high} \approx \frac{1}{1 + e^{-\frac{1}{2}a\Delta a}\ \dfrac{(\xi-1)a + \frac{1}{2}(\xi+1)\Delta a}{(\xi-1)a - \frac{1}{2}(\xi+1)\Delta a}} \tag{21}$$

$$P_{low} \approx \cfrac{1}{1 + e^{\frac{1}{2} a\Delta a} \cfrac{(\xi-1)a - \frac{1}{2}(\xi+1)\Delta a}{(\xi-1)a + \frac{1}{2}(\xi+1)\Delta a}} \tag{22}$$

It will be seen that $P_{high}(\Delta a) = P_{low}(-\Delta a)$, as was to be expected from the symmetry of the problem, and that $P_{high} = \frac{1}{2} = P_{low}$ when $\Delta a=0$. The probability P_{high} is plotted in Fig. 5 as a function of Δa for various values of the pump parameter a, with the coupling constant $\xi=2$. The curves demonstrate the effect of fluctuations on the output of the ring laser, and can be contrasted with Fig. 1, in which such fluctuations were assumed to be absent. Evidently the probability P_{high} shows no sign of hysteresis in the steady state. When Δa is large and positive it is highly probable that the light intensity I is high, and when Δa is large and negative it is highly probable that it is near zero, and vice versa for the other mode.

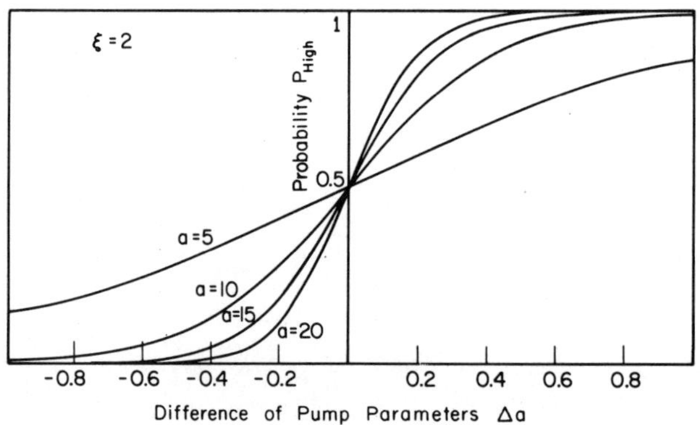

Fig. 5.   Variation of the probability P_{high} with Δa, for several different pump parameters a, and with $\xi = 2$.

IV.   ANALOGY WITH THERMODYNAMIC PHASE TRANSITIONS

The appearance of both a maximum and a minimum in the potential V(I) above threshold when $\xi > 1$ is strongly suggestive of a first order phase transition in the language of thermodynamics. So long as $\xi < 1$, the potential has only a single minimum, which lies to the left of the vertical axis below threshold, and to the right of the axis above threshold. The most probable value $I_m(a)$ of the light intensity, which is of course associated with the lowest possible value of the potential, therefore varies in a continuous manner from zero below threshold, to non-zero values above threshold. However, when ξ is greater than 1, the most probable light intensity

may vary discontinuously with pump parameter. This is illustrated
by the broken curve in Fig. 4, which gives the locus of the most
probable intensity $I_m(a)$ for different pump parameters a_1 when $\xi=2$
and $\Delta a=1$. So long as $a_1 < 4$ the most probable light intensity is
zero, and when $a_1 > 6$, the most probable value increases continuously
with a_1. But somewhere between $a_1 = 4$ and $a_1 = 6$ there is a discontinuity, as the most probable intensity jumps from a zero to a non-
zero value. The same effect is illustrated more clearly in Fig. 6
for four different coupling constants ξ, and for a range of pump
parameters, with $\Delta a = 1$. Once ξ sufficiently exceeds unity by an
amount that depends on Δa, the most probable value of the light intensity $I_m(a)$ exhibits a jump discontinuity as a function of the
pump parameter a.

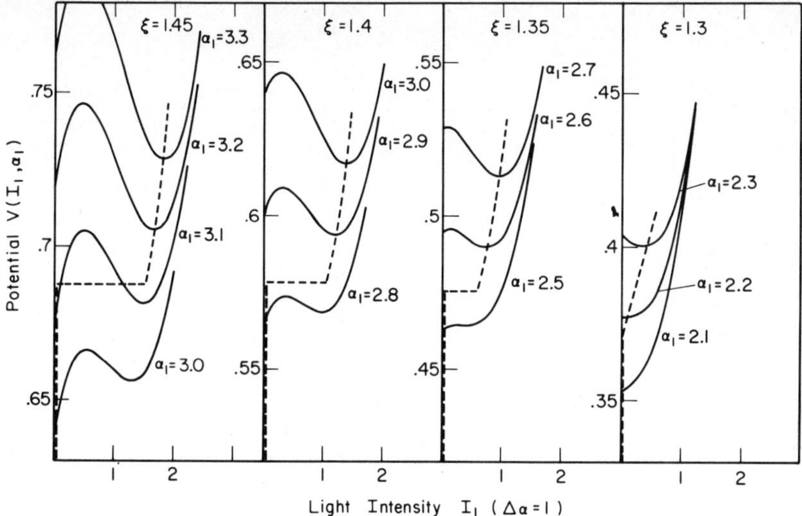

Fig. 6.   The broken lines show the locus of the most probable light
          intensity with pump parameter a, for several different
          coupling constants ξ. Note the increasing discontinuity
          as ξ increases.

     If we regard the laser field as a thermodynamic system that
undergoes a phase transition as the pump parameter is varied, we
may look on the most probable value of the field – or of the light
intensity I – as an order parameter for the phase transition[14],[19].
The order parameter is zero in the thermal or incoherent phase of
the field below threshold, and non-zero in the coherent phase above
threshold. The pump parameter plays a role that is somewhat analogous to the temperature of a thermodynamic system. Moreover, if we
substitute the most probable value of the light intensity $I_m(a)$ for
I in the potential $V(I,a)$, we obtain a quantity, the most probable
potential

$$V_m(a) \equiv V(I_m, a) \quad , \tag{23}$$

that is to some extent analogous to the free energy of a system in thermodynamic equilibrium. Figure 7 illustrates how this most probable potential $V_m(a_1)$ varies with pump parameter a_1 for several different values of the coupling constant ξ - or how the free energy varies with temperature in the analogous thermodynamic situation.

We observe that a fundamental change occurs in the behavior of $V_m(a)$ when ξ exceeds unity. Although $V_m(a)$ is always a continuous function of a, its derivative $\partial V_m(a)/\partial a$ exhibits a discontinuity when ξ lies sufficiently above 1. As the negative rate of change of the free energy with temperature is the thermodynamic entropy of an equilibrium system, we see that our system behaves as if the entropy has a discontinuity at the phase transition when ξ exceeds 1.

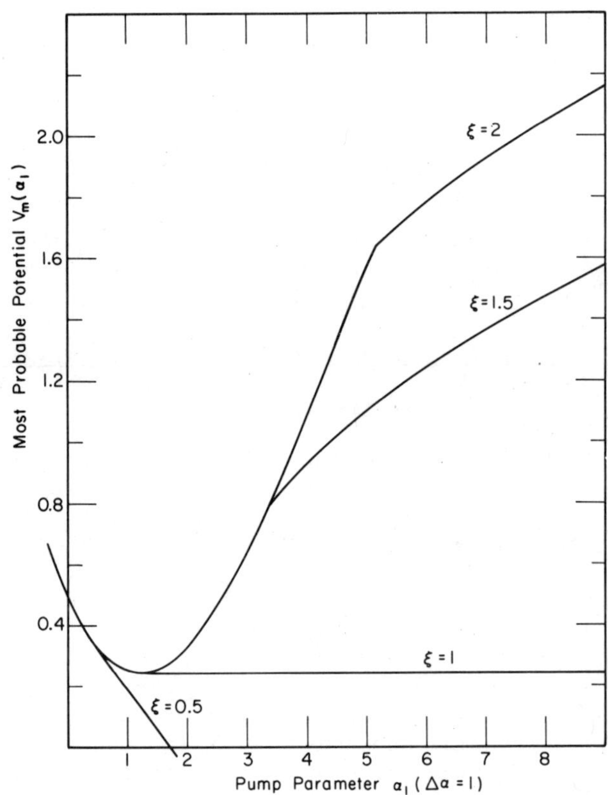

Fig. 7.   The most probable potential as a function of pump parameter a_1 with $\Delta a = 1$. Note the discontinuities of the derivative when $\xi > 1$.

In other words, the system exhibits a discontinuous – or first order – phase transition under these conditions. The second derivative $\partial^2 V_m(a)/\partial a^2$, which is analogous to the specific heat when multiplied by a, diverges at the phase transition. On the other hand, when $\xi \lesssim 1$, as in a more conventional laser, the first derivative $\partial V_m(a)/\partial a$ is continuous, and the phase transition is of the second order.

V. THE DWELL TIMES UNDER QUASI-BISTABILITY

We have already seen that when the coupling constant ξ exceeds unity, and especially when $\xi = 2$, the probability density $P(I)$ can exhibit two peaks, one at $I = 0$ and one in the neighborhood of $I = a + \frac{1}{2} \Delta a$, with a minimum in between at $I = I_A \approx a/(\xi+1) - \frac{1}{2} \Delta a/(\xi-1)$, as given by Eq. (18). These two peaks correspond to two states of the laser mode which are quasi-stable, in the sense that the mode amplitude remains in the neighborhood of one peak until a sufficiently large fluctuation causes it to switch to the other one. It is to be expected that the more highly peaked the potential is at the maximum I_A, the longer will be the time needed to tunnel through the potential maximum and to switch states. From Eqs. (15b) and (18) we readily find that when $\Delta a = 0$,

$$V(I_A) \rightarrow \frac{1}{4} a^2 \left(\frac{\xi-1}{\xi+1} \right) - \log 2 \tag{24}$$

for sufficiently large a, and this increases rapidly with increasing a. It follows that the dwell times tend to longer and longer values as the laser becomes more and more highly excited, and the system becomes bistable in the limit $a \rightarrow \infty$. However, this limit, strictly speaking, lies outside the domain of the third order laser theory.

It is sometimes possible to obtain estimates of the dwell times[8] by making use of the first passage time formalism[20] for motion in one dimension. We imagine that the domain of all values of the light intensity I for one mode is divided into two regions by the potential maximum at $I = I_A$, and we adopt the terminology of calling the mode "off" when $I < I_A$, and "on" when $I > I_A$. We then denote the average first passage time for a transition from the state $I=0$ to the "on" state by T_{off}, and the average first passage time from the state $I = I_B$ to the "off" state by T_{on}. The quantities T_{off} and T_{on} are convenient measures of the average time that the mode remains in the "off" and the "on" positions, respectively. In general the average first passage time $T(I)$ to cross the boundary at I_A when starting from state I is given by the expression[20]

$$T(I) = -2 \int_o^I dI' \; \frac{1}{P(I')D(I')} \int_o^{I'} dI''P(I'') + c_1 \int_o^I dI' \; \frac{1}{P(I')D(I')} + c_2 , \tag{25}$$

where c_1, c_2 are constants to be determined by the boundary conditions, and $D(I)$ is the diffusion constant of the Fokker-Planck equation for I, which has been shown to be $8I$ in the present problem[8]. If the initial state has $I \lesssim I_A$, then $T(I)$ should be zero when $I = I_A$, and this gives the relation

$$0 = -2 \int_0^{I_A} dI' \, \frac{1}{P(I')D(I')} \int_0^{I'} dI'' P(I'') + c_1 \int_0^{I_A} dI' \, \frac{1}{P(I')D(I')} + c_2 \ . \tag{26}$$

In addition, $T(I)$ should have its greatest value when $I = 0$, so that for all other values $I \lesssim I_A$ we have $T(I) \lesssim T(0)$. This leads to the condition

$$\int_0^{I} dI' \, \frac{1}{P(I')D(I')} \left[2 \int_0^{I'} dI'' P(I'') - c_1 \right] \geq 0 \ ,$$

which can only be satisfied for all $0 \leq I \leq I_A$ when $c_1 = 0$. Equation (26) then yields the value of c_2 immediately, and from Eq. (25) we obtain

$$T_{off} = T(0) = 2 \int_0^{I_A} dI' \, \frac{1}{P(I')D(I')} \int_0^{I'} dI'' P(I'') \ . \tag{27}$$

With the help of Eqs. (15), (18) and the value $D(I) = 8I$, we finally arrive at the equation

$$T_{off} = \int_0^{I_A} dI' \int_0^{I'} dI'' \, \frac{\exp[\tfrac{1}{4}(\xi^2 - 1)(I''^2 - I'^2 - 2I_A I'' + 2I_A I')]}{4I'}$$

$$\times \frac{1 - \mathrm{erf} \ \tfrac{1}{2}(\xi I'' - a + \tfrac{1}{2} \Delta a)}{1 - \mathrm{erf} \ \tfrac{1}{2}(\xi I' - a + \tfrac{1}{2} \Delta a)} \ . \tag{28}$$

In a similar manner we may write an expression for T_{on}.[8] It is to be expected from the symmetry of the problem that T_{off} and T_{on} will be approximately equal when $\Delta a = 0$. Some values of T_{off} given by Eq. (28) as a function of the pump parameter a, for different coupling constants ξ, and with $\Delta a = 0$, are illustrated in Fig. 8. When ξ is appreciably greater than 1, the dwell times rise very rapidly with the excitation of the laser, and the tendency towards real bistability is apparent. However, it must be remembered that we have treated the first passage problem as one-dimensional, whereas the light intensities of both modes are actually fluctuating in a correlated manner.

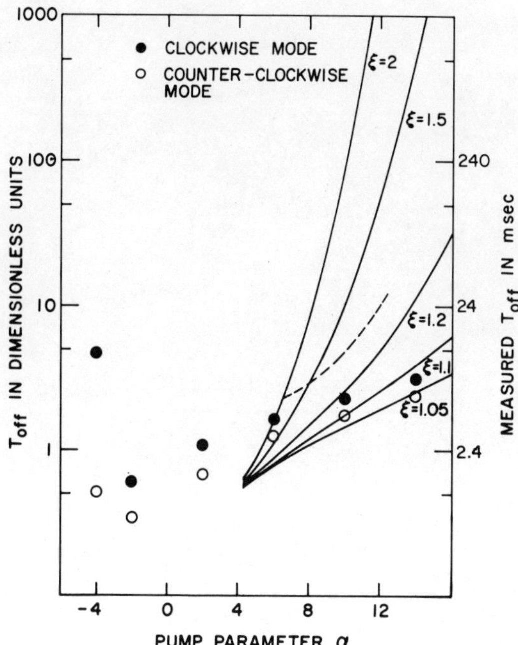

Fig. 8.   The variation of the dwell time T_{off} with pump parameter
a, for $\Delta a = 0$ and various values of ξ. The full curves are
theoretical and the experimental values found for each of
the two modes are shown superimposed. The broken curve
is obtained by averaging the values for $\xi = 2$ over a range
of pump parameters with standard deviation 4.

It is not too difficult to obtain an approximate analytic expression for T_{off} given by Eq. (28) when a is large and Δa is small.
Provided $I_A < (a - \frac{1}{2}\Delta a)/\xi$, which holds true for large a and small Δa,
the arguments of the error functions in the integrand are negative
throughout the range of integration, and both the $1 - erf \frac{1}{2}( )$ factors can be approximated by 2. The I''-integrand then has the form
$\exp[\frac{1}{4}(\xi^2-1)(I''^2-2I_A I'')]$, and for large a this is well approximated
by $\exp[-\frac{1}{2}(\xi^2-1)I_A I'']$ over the range of integration. The I''-inte-
gration is then easily performed, and the answer is independent of
I' for large I'. The numerator of the I'-integrand is a Gaussian
form in I', which is peaked at $I' = I_A$ and allows us to replace the
I' in the denominator by the peak value I_A. We then obtain approximately

$$T_{off} \approx \frac{e^{\frac{1}{4}(\xi^2-1)I_A^2}}{2(\xi^2-1)I_A^2} \int_0^{I_A} dI' \, e^{-\frac{1}{4}(\xi^2-1)(I'-I_A)^2} \left[1 - e^{-\frac{1}{2}(\xi^2-1)I_A I'}\right]$$

$$\approx \frac{\sqrt{\pi}}{2} \frac{e^{\frac{1}{4}(\xi^2-1)I_A^2}}{(\xi^2-1)^{3/2} I_A^2} \cdot \tag{29a}$$

The approximation is expected to improve as a increases. In the special case $\Delta a = 0$ this can be written

$$T_{off} \approx \frac{\sqrt{\pi}}{2} \left(\frac{\xi+1}{\xi-1}\right) \frac{e^{\frac{1}{4}a^2(\xi-1)/(\xi+1)}}{(\xi^2-1)^{\frac{1}{2}} a^2} \cdot \tag{29b}$$

VI. MEASUREMENT OF THE PHOTOELECTRIC COUNTING STATISTICS

In order to test the double peaked form of the probability density $P(I)$ given by Eqs. (15), we have measured the photoelectric counting statistics when the light beams derived from each of the two modes of a dye ring laser in turn fall on a photomultiplier tube[15]. For this purpose a dye ring laser was constructed, in which the active medium is a .01% solution (2×10^{-4} molar) of rhodamine 6G in methanol and water, that is made to flow continuously through a cell with quartz windows. As is well known, the spectrum of dye molecules is homogeneously broadened, as required. The dye is optically pumped by the light of an argon ion laser. Three etalons inserted in one arm of the ring ensure single-frequency operation in two travelling wave modes. It is found that the asymmetry of the ring laser, as measured by Δa, is strongly influenced by small angular adjustments of the etalons, and we usually operate the laser with $\Delta a \ll 1$. The geometry of the ring laser is illustrated in Fig. 11 below, but the detection system is different from that shown in the figure. The two strongly attenuated light beams emerging from the output mirror of the ring laser cavity are allowed, in turn, to strike a counting phototube, whose photoelectric pulses, after amplification and pulse shaping, are then counted by a scaler for short time intervals T of order 1 μsec. The number n registered in any one counting interval T is stored in a computer, and after many counting cycles, the number of times that the value n is encountered becomes a measure of the probability p(n) of registering n counts from the incident light beam in time T. For short intervals T, the probability p(n) is related to the probability density $P(I)$ of the incident laser beam by[21]

$$p(n) = \frac{1}{n!} \int_0^\infty (\alpha I T)^n e^{-\alpha I T} \, P(I) dI \quad . \tag{30}$$

Here α is the quantum efficiency of the detector, and the light intensity I is generally taken to be expressed in units of incident
photons per second, rather than in the dimensionless units we have
been using up to now. However, from Eq. (30) we easily find

$$\langle n \rangle = \alpha \langle I \rangle T \, , \tag{31}$$

and this allows us to continue to describe I in the dimensionless
units, provided the constant α is adjusted so as to make $\langle n \rangle$ derived from Eqs. (15) and (31) agree with the measured mean $\langle n \rangle$.

When $\langle n \rangle$ is large, the shape of the probability density $P(I)$
is generally mirrored by p(n). However, in practice it is undesir-
able to make the counting rates much higher than about 10^7/sec, because of the limitations of the counting electronics. Even at this
counting rate $\langle n \rangle$ is only of order 10, which makes p(n) a somewhat
distorted form of $P(I)$. At high counting rates the deadtime of the
counting electronics becomes a significant factor, and deadtime corrections to the measured values of p(n) have to be made by the procedure described in Ref. 22. On the other hand, the corrections for
background counts are found to be negligible.

Figure 9 gives the results of measurements of the probabilities
p(n) for each of the two ring laser modes, with the theoretical
values derived from Eqs. (15) and (30) superimposed as broken curves.
We estimate the pump parameters to be about 12, with the difference
$\Delta a \approx .135$. The results reveal the expected double peaked feature
of p(n), and they are in general qualitative agreement with the
theory, although disagreeing in detail. The agreement can be improved if it is assumed that small optical pumping fluctuations are
present, that result in effective r.m.s. fluctuations of the pump
parameter a of about 2.75. We can then generate a new probability
density from $P(I,a)$ by convolving $P(I,a)$ with a Gaussian spread
function in a with mean $\langle a \rangle = 12$ and standard deviation 2.75. When
this 'corrected' form of $P(I)$ is used in Eq. (30) to yield a corrected form of p(n), we obtain the full curves shown in Fig. 9. The
agreement between theory and experiment is now significantly improved, although discrepancies remain at $n = 0$ and $n = 1$, where the
convolution procedure is probably inappropriate. However, there
is good evidence for the existence of two peaks in each probability
distribution, and for the predicted two quasi-stable states of the
laser.

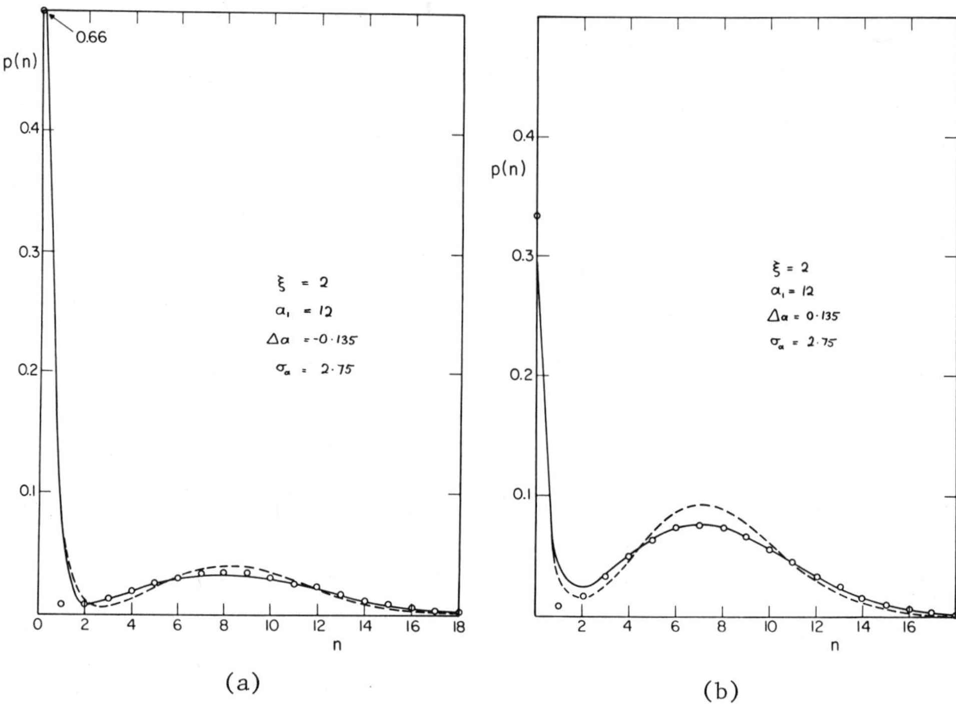

(a)   (b)

Fig. 9.   The measured photoelectric counting probabilities p(n)
           (a) for the counterclockwise mode; (b) for the clockwise
           mode. The theoretical values are shown as continuous
           curves, for clarity. The broken curve is obtained by
           putting a_1 = 12, Δa = .135, ξ = 2, α = .63, and the full
           curve when the convolution procedure is used.

VII.   MEASUREMENT OF THE DWELL TIMES T_{off}, T_{on}

The random switching of the ring laser intensity between the
two counterpropagating travelling wave modes can be demonstrated
very easily, if the outputs of two photodetectors illuminated by
light beams derived from the two modes are displayed on a double-
beam oscilloscope. Figure 10 is a photograph of an oscilloscope
display obtained in this way. The alternate switching of the excitation between the two ring laser modes is clearly visible, and
it is evident that one mode always turns off when the other one
turns on.

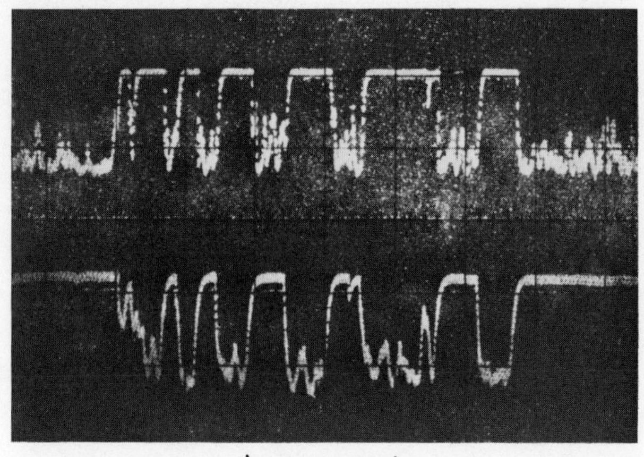

Fig. 10.   An oscilloscope photograph showing anticorrelated on-off
           switching of the two ring laser modes, (top) clockwise
           wave, (bottom) counter-clockwise wave.

In order to measure the dwell times or switching times T_{on} and
T_{off} for each of the two laser modes we make use of the apparatus
shown in Fig. 11. The photon counting photomultiplier tube is now
replaced by a much lower gain detector, that does not resolve the
individual photoelectric pulses, but generates a signal that is
nearly proportional to the incident light intensity I. After amplification, this signal is sent to a limiter that delivers rectangular
pulses of constant amplitude A_o, whose duration is equal to the
"on-time" of the laser mode. The pulses are counted by a scaler
for a few minutes, and the measured average counting rate R allows
the sum $T_{on} + T_{off}$ to be determined from the relation

$$T_{on} + T_{off} = 1/R \ . \tag{32}$$

At the same time the voltage reading A of the integrating meter
shown in Fig. 11 is related to the ratio T_{off}/T_{on} by

$$\frac{A}{A_o} = \frac{T_{on}}{T_{on} + T_{off}} \ . \tag{33}$$

From Eqs. (32) and (33) we immediately have

$$T_{on} = \frac{A}{A_o} \frac{1}{R} \quad ; \quad T_{off} = \frac{A_o - A}{A_o} \frac{1}{R} \ , \tag{34}$$

which allows the two dwell times to be determined from measurements
of A/A_o and R for various working points of the laser.

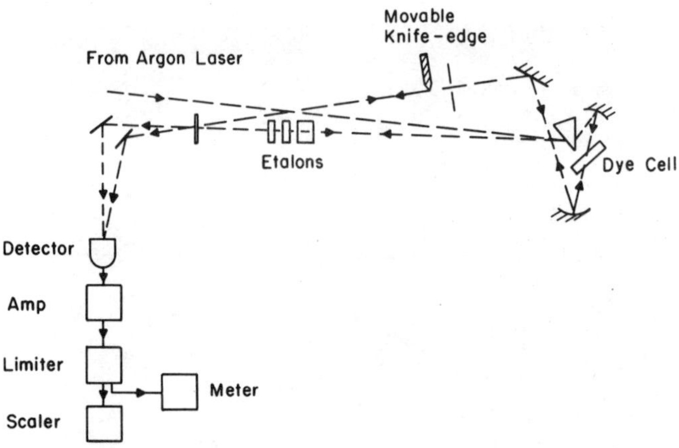

Fig. 11.   Outline of the dye ring laser and of the apparatus for
          measuring the dwell times.

In practice we vary the pump parameter a by making small displacements of a knife edge which is inserted in one arm of the ring
laser, as shown in Fig. 11. The knife-edge acts as a controlled
loss mechanism[12], and for sufficiently small displacements x, a is
a linear function of x. The laser is operated close to symmetry,
with $\Delta a \ll 1$. In order to determine a from x, we first identify the
laser threshold (a = 0) by the fact that the relative intensity
fluctuations are of order unity near threshold[7], and then use the
property that $\langle I(a)\rangle/\langle I(0)\rangle \sim a$ for large a. T_{off} and T_{on} are
then measured for different working points of the laser, as de-
scribed. As the ring laser was operated close to symmetry, the
values of T_{off} and T_{on} were found to be similar, except at the lowest intensities.

The measured values of T_{off} for the two laser modes are shown
superimposed on the dimensionless theoretical curves in Fig. 8,
with the time scale adjusted somewhat arbitrarily so that one experimental point at a = 6 falls on the ξ = 2 curve. Although T_{off}
is found to increase with pump parameter a for sufficiently large
values of a as expected, the observed variation is very much smaller
than predicted by Eq. (28) by some orders of magnitude. The experiment and the theory disagree even when the coupling constant ξ is
assumed to be only a little above unity, whereas the evidence from
the photoelectric counting experiments described above strongly
suggests that the value of ξ is close to 2. The observed minimum

of T_{off} just below the laser threshold can be understood in general terms by reference to Fig. 3. Once the maximum in the potential $V(I)$ disappears near threshold, it becomes increasingly likely that the intensity will remain near zero as a is decreased, and this is reflected in an increase of T_{off} at very low intensities, but not of T_{on}.

VIII. DISCUSSION

Although the motion of the dye molecules through the region of optical pumping may have the effect of lowering the coupling constant ξ from its ideal theoretical value $\xi = 2$,[10] the photoelectric counting measurements strongly indicate that ξ is close to 2 in our ring laser. If ξ were near unity rather than near 2, the 'on' and 'off' states of each mode would not be nearly as well defined, nor as strongly anti-correlated, as they are observed to be (see Fig. 10), and there would be strong disagreement between the theoretical and the measured probabilities p(n). It seems more plausible to accept the value $\xi = 2$, and to look for other sources of the discrepancy between the calculations and the measurements of T_{off}. Although it might appear that a one-dimensional treatment of the first passage problem is perhaps an oversimplification, there are indications that T_{off} given by Eq. (29) is at least approximately of the correct form. For large pump parameters a the mode switching rate is governed by the smallest non-vanishing eigenvalue of the Fokker-Planck operator. We have recently succeeded in making estimates of this lowest eigenvalue by considering a simplified form of the potential, and have encountered expressions that are rather similar to the reciprocal of T_{off} given by Eq. (29b).

The values of T_{off} are however extremely sensitive to small perturbations in the noise level. Any source of additional intensity fluctuations has the effect of shortening the dwell times T_{off}, and especially the longer times. The broken curve in Fig. 8 illustrates the effect of averaging the theoretical rates $1/T_{off}(a)$ for $\xi = 2$ over a Gaussian distribution in the pump parameter a with standard deviation 4, and then taking the reciprocal of the resulting average. It is clear from this rather crude model of pump parameter fluctuations that the influence of additional fluctuations can be large. Effects such as backscattering, or the presence of another very weakly excited laser mode, for example, can significantly shorten T_{off}. There is actually indirect evidence that backscattering of light from one mode into the other may be responsible for some of the discrepancies between theory and experiment. We have recently observed switching of the excitation between two longitudinal modes of a conventional dye laser, which should be governed by the same equations of motion as the ring laser. The dwell times in this case are found to increase dramatically with pump parameter, very nearly as predicted by Eqs. (29), so that the laser really exhibits bistability when it is highly excited. Apart from the fact that the two longitudinal modes of a conventional laser are less easy to separate and to investigate experimentally, the main difference between such a laser and the ring laser is that back-scattering does not play a significant role in the former, whereas it can in the latter. This suggests that backscattering could be the principal source of the discrepancy regarding T_{off}. We have recently made measurements of the moments of I_1 and I_2[23] in a ring laser and have also encountered some departures from conclusions based on Eq. (15).

Finally, we should mention a possible, rather different source of inadequacy of the ring laser theory given above. It was pointed out by Schaefer and Willis[24] and others[25] that, because of the existence of triplet states in the spectrum of the dye molecule, which compete to some extent for cavity photons against the singlet laser levels, the dye molecules used in the laser may not be adequately represented by two-level quantum systems. In that case the laser theory becomes much more complicated, and the probability distribution $P(I)$ can be double peaked even for a single-mode dye laser[24,25]. Some indirect evidence for this prediction has already been reported[26]. In addition, it has recently been pointed out that a saturable absorber in a ring laser has the effect of suppressing switching between modes[27]. As each dye molecule carries its own resonant absorber, similar suppression of the switching may also occur in a dye ring laser, and has indeed been observed under some circumstances. Our results therefore raise a number of interesting questions that await further experimental and theoretical investigation.

REFERENCES

1.   W. W. Rigrod and T. J. Bridges, IEEE J. Quant. Electron. QE-1, 298 (1965); H. W. Schröder, L. Stein, D. Fröhlich, B. Fugger and H. Welling, Appl. Phys. 14, 377 (1977); D. Kühlke, S. Schröter and W. Dietel, Sov. J. Quant. Electron. 9, 642 (1979).

2.   J. A. White, Phys. Rev. 137, A1651 (1965); F. Aronowitz, Phys. Rev. 139, A635 (1965); S. G. Zeiger and E. E. Fradkin, Opt. Spektrosk. 21, 386 (1966) [Sov. Phys. Opt. Spectros. 21, 217 (1966)]; V. S. Smirnov and B. L. Zhelnov, Zh. Eksp. Teor. Fiz. 57, 2043 (1969)[Sov. Phys. JETP 30, 1108 (1970)]; S. Grossmann and P. H. Richter, Zeits. f. Phys. 249, 43 (1971); F. Aronowitz, Appl. Opt. 11, 405 (1972); P. H. Richter and S. Grossmann, Zeits f. Phys. 255, 59 (1972); R. Graham and W. A. Smith, Opt. Comm. 7, 289 (1973); F. T. Arecchi and A. M. Ricca, Phys. Rev. A15, 308 (1977);

       D. Kühlke and W. Dietel, Opt. & Quant. Electron. $\underline{9}$, 305
       (1977); Surendra Singh, Phys. Rev. A (to be published,1980).
 3.   H. Haken, Z. Phys. $\underline{219}$, 246 (1969), and Rev. Mod. Phys. $\underline{47}$, 67
       (1975). See also H. Haken, Laser Theory from Handbuch der
       Physik, Vol. 25/2c (Springer, Berlin, 1970).
 4.   L. Menegozzi and W. E. Lamb, Jr., Phys. Rev. A$\underline{8}$, 2103 (1973).
 5.   M. Sargent, III, M. O. Scully and W. E. Lamb, Jr., Laser
       Physics (Addison-Wesley, Reading, MA., 1974) Ch. 11.
 6.   J. B. Hambenne and M. Sargent, III, IEEE J. Quant. Electron.
       QE-11, 90 (1975) and Phys. Rev. A$\underline{13}$, 797 (1976).
 7.   M. M-Tehrani and L. Mandel, Opt. Comm. $\underline{16}$, 16 (1976) and Phys.
       Rev. A$\underline{17}$, 677 (1978); F. T. Hioe, Surendra Singh and L.
       Mandel, Phys. Rev. A$\underline{19}$, 2036 (1979).
 8.   Surendra Singh and L. Mandel, Phys. Rev. A$\underline{20}$, 2459 (1979).
 9.   M. Sargent, IIII, W. E. Lamb, Jr., and R. L. Fork, Phys. Rev.
       $\underline{164}$, 436 (1967); Surendra Singh, Opt. Comm. $\underline{32}$, 339 (1980).
10.   D. Kühlke and R. Horak, Opt. and Quant. Electron. $\underline{11}$, 485 (1979).
11.   A. W. Smith and J. A. Armstrong, Phys. Lett. $\underline{16}$, 39 (1965);
       Yu. I. Zaitsev, Zh. Eksp. Teor. Fiz. $\underline{50}$, 525 (1966)[Sov.
       Phys. JETP $\underline{23}$, 349 (1966)]; T. J. Hutchings, J. Winocur,
       R. H. Durrett, E. D. Jacobs and W. L. Zingery, Phys. Rev.
       $\underline{152}$, 467 (1966); V. E. Privalov and S. A. Fridrikhov. Usp.
       Fiz. Nauk $\underline{97}$, 377 (1969)[Sov. Phys. Usp. $\underline{12}$, 153 (1969)];
       F. Aronowitz, Appl. Opt. $\underline{11}$, 405 (1972).
12.   M. M-Tehrani and L. Mandel, Opt. Lett. $\underline{1}$, 196 (1977); and
       Phys. Rev. A$\underline{17}$, 694 (1978).
13.   I. F. Lyuksyutov, Phys. Lett. $\underline{56A}$, 135 (1976); I. F. Ginzburg
       and S. L. Panfil, Phys. Lett. $\underline{72A}$, 5 (1979); A. Z.
       Patashinski and V. I. Pokrovskii, Fluctuation Theory of
       Phase Transitions (Pergamon Press, Oxford, 1979) Ch. 8.
14.   H. Haken, Synergetics (Springer, Berlin, 1978).
15.   Rajarshi Roy and L. Mandel, Opt. Comm. $\underline{34}$, 133 (1980).
16.   W. E. Lamb, Jr., Phys. Rev. $\underline{134}$, A1429 (1964).
17.   R. L. Stratonovich, Topics in the Theory of Random Noise,
       Vol. I (Gordon and Breach, New York., 1963) Ch. 4.
18.   See for example M. Abramowitz and I. A. Stegun, Handbook of
       Mathematical Functions (Dover, New York, 1965) Ch. 7.
19.   V. DiGiorgio and M. O. Scully, Phys. Rev. A$\underline{2}$, 1170 (1970);
       R. Graham and H. Haken, Zeits. f. Phys. $\underline{237}$, 31 (1970).
20.   G. H. Weiss, in Stochastic Processes in Chemical Physics,
       ed. I. Oppenheim, K. E. Schuler and G. H. Weiss (MIT,
       Cambridge, MA., 1977) p. 361.
21.   L. Mandel, Proc. Phys. Soc. (London) $\underline{74}$, 233 (1959) and in
       Progress in Optics, Vol. 2, ed. E. Wolf (North-Holland,
       Amsterdam, 1963)p. 181.
22.   L. Mandel, J. Opt. Soc. Am. $\underline{70}$, 873 (1980).
23.   R. Roy, S. Singh, J. Gordon and L. Mandel, Paper, 11'th In-
       ternational Quantum Electronics Conference 1980, to be
       published.

24. R. B. Schaefer and C. R. Willis, Phys. Rev. A13, 1874 (1976);
     C. R. Willis, in Coherence and Quantum Optics IV, eds. L.
     Mandel and E. Wolf (Plenum, New York, 1978) p. 63.
25. A. Baczinski, A. Kossakowski and T. Marzcalek, Zeits. f.
     Physik B23, 205 (1976); S. T. Dembinski and A. Kossakowski,
     Zeits. f. Physik B24, 141 (1976).
26. J. A. Abate, H. J. Kimble and L. Mandel, Phys. Rev. A14, 788
     (1976).
27. A. V. Dotsenko and E. G. Lariontsev. Sov. J. Quant. Electron.
     9, 576 (1979).

MICROWAVE DISPERSIVE BISTABILITY IN A CONFOCAL FABRY-PEROT MICROWAVE CAVITY

E. Arimondo and A. Gozzini

Istituto di Fisica dell'Università di Pisa, Italy

L. Lovitch and E. Pistelli

Istituto di Fisica dell'Università di Ferrara, Italy

Abstract: The behaviour of a Fabry-Perot transmission cavity containing a two-level absorber is derived as a function of the incident power and of the resonantor and absorber parameters by taking into account the spatially averaged absorption coefficient of the gas. The state equation so obtained exhibits a bistable behaviour. The transmittance of a microwave cavity filled with ammonia gas and tuned at the frequency of the (3,3) inversion line of ammonia is investigated experimentally as a function of the frequency of the incident em wave and gas pressure. Good quantitative agreement is found between the theoretical description and the experimental results.

I.   INTRODUCTION

The behaviour of a Fabry-Perot (FP) optical resonator filled with an absorbing (or emitting) medium, has been studied by Kastler several years ago.[1] His analysis does not consider the case of a non-linear medium. With the advent of powerful laser sources many non-linear effects have been observed in the optical range. We may cite, in particular, the effects of optical bistability caused by interaction between absorbing atoms in the interior of a resonator and the resonator itself. These effects are revealed by sudden changes of the transmitted power as a function of the incident one.

In a bistable system the output power is a two-valued function and bistable switching is realized through the continuous variation of a parameter. The bistability is a consequence of the non-linear response of one element governing the operation of the system. For instance in an FP optical cavity filled with sodium vapour, the transmitted power is governed by the non-linear absorption and dispersion of the vapour, and bistability is realized on sweeping the incident intensity, as demonstrated in an experiment by Gibbs et al[2] using a laser source. In subsequent experiments bistable operation has been observed in FP cavities containing different atomic vapours[3], Kerr liquids[4] and semiconductors[5]. Bistable operation has also been achieved by using electronic feedback in different optical and solid-state devices[6]. Very recently the bistability of an FP cavity filled with sodium vapour has been reinvestigated and the influence of the Gaussian beam distribution of the laser field has been analyzed[7]. In this paper the results of an experiment on a confocal FP microwave cavity filled with ammonia gas and irradiated with microwave radiation in resonance with an inversion line of ammonia are presented. The power transmitted by the cavity is monitored while the frequency of the incident microwave radiation is swept through the absorber resonance. The bistability occurs because of the high reflectance of the cavity mirror, i.e., the high quality factor of the cavity, and the large absorption coefficient of ammonia at the microwave inversion transition. Experimental evidence is given of the dependence of the bistability on the incident microwave power and on the pressure of the absorbing gas.

The experiment reported in this paper satisfies the conditions imposed by Bonifacio et al[8] in the treatment of absorptive and dispersive optical bistability insofar as the total absorption over the cavity length is small compared to unity and the reflectance of the cavity mirrors is high. Thus the state equation presented in Refs. 8-10 describes the main qualitative features of the experiment. However, a quantitative agreement between the theoretical and experimental results is achieved only when one relaxes the approximation assumed in the previous treatment of a plane wave electric field inside the cavity.

In Sec. II the transmittance of a cavity filled by an absorbing gas is easily derived by analyzing the response function of the resonator. In this analysis we use a spatially averaged absorption coefficient. In Sec. III the domain of bistability on sweeping the intensity or the frequency of the incident em field is analyzed and evaluated numerically under the conditions of the present experiment. In Sec. IV the experimental apparatus is described, in Sec. V the experimental and theoretical results are compared and in Sec. VI we present the conclusions.

II. TRANSMITTANCE OF A RESONATOR FILLED WITH
AN ABSORBING MEDIUM

Let us consider a high quality factor transmission cavity filled with a molecular absorber and fed by a microwave power P_i of angular frequency ω and wavelength λ. The absorber is described as a two-level system with power absorption coefficient $\alpha(\omega)$ and refractive index $n(\omega)$ given by

$$\alpha(\omega) = \alpha_o (1 + \Delta^2 + S)^{-1} \tag{1}$$

$$n(\omega) = 1 - \frac{\alpha(\omega)\lambda\Delta}{4\pi} \tag{2}$$

where

$$\Delta = \frac{\omega - \omega_a}{\Gamma}$$

and

$$S = \frac{\langle\mu^2\rangle E^2}{3\hbar^2} \frac{\tau}{\Gamma} \quad .$$

Here α_o is the intensity of the absorber resonance, i.e., the absorption coefficient at the resonance frequency ω_a and for vanishing saturation parameter S, Γ the linewidth, i.e., the transverse relaxation rate, τ the thermal (longitudinal) relaxation time, $\langle\mu\rangle$ the dipole matrix element of the transition, E the intensity of the em field at the frequency ω.

The transmission cavity is a quadrupole whose behaviour is described by its resonance frequency ω_R and the internal and external quality factors. The elements of the scattering S matrix describing the amplitude and phase relationship between the input and output waves are given by (see, for example, Ref. 11)

$$S_{11} = \frac{2}{Q_1 A(\omega)} - 1$$

$$S_{22} = \frac{2}{Q_2 A(\omega)} - 1$$

$$S_{12} = S_{21} = \frac{2}{\sqrt{Q_1 Q_2}\, A(\omega)}$$

where

$$A(\omega) = \frac{1}{Q_j} + \frac{1}{Q_a} + \frac{1}{Q_1} + \frac{1}{Q_2} + i\left(\frac{\omega}{\omega_R} - \frac{\omega_R}{\omega}\right) = \frac{1}{Q_L} + i\left(\frac{\omega}{\omega_R} - \frac{\omega_R}{\omega}\right) \quad .$$

In the latter expression

$$Q_J = \frac{\omega W}{P_J} \,, \quad Q_a = \frac{\omega W}{P_a} \,, \quad Q_1 = \frac{\omega W}{P_1} \,, \quad Q_2 = \frac{\omega W}{P_2} \,,$$

are the quality factors of the cavity walls, absorber, and input and output couplings, respectively, where W is the energy stored inside the resonator, and P_J, P_a, P_1 and P_2 are the Joule losses, the absorber losses and the power radiated from the input and output couplings, respectively, and

$$Q_L = \left(\frac{1}{Q_J} + \frac{1}{Q_a} + \frac{1}{Q_1} + \frac{1}{Q_2} \right)^{-1} \tag{3}$$

is the effective (loaded) quality factor of the resonator.

For high Q_L values and weak input and output couplings Q_1, $Q_2 \gg Q_L$, the transmitted and reflected power are given by:

$$T(\omega) = \frac{P_T}{P_i} = S_{12} S_{12}^* \approx \frac{4 Q_L^2}{Q_1 Q_2} \frac{1}{1 + 4 Q_L^2 \left(\frac{\omega - \omega_R}{\omega_R} \right)^2} \tag{4}$$

$$R(\omega) = \frac{P_R}{P_i} = S_{11} S_{11}^* \approx 1 - \frac{4 Q_L}{Q_1} \frac{1}{1 + 4 Q_L^2 \left(\frac{\omega - \omega_R}{\omega_R} \right)^2} = 1 - \frac{Q_2}{Q_L} T(\omega). \tag{5}$$

In order to calculate the transmission $T(\omega)$ of the filled resonator we must express Q_L and ω_R in terms of the absorber and resonator parameters, which in turn depend on the incident power P_i.

Let $Q_0 = \left(\frac{1}{Q_J} + \frac{1}{Q_1} + \frac{1}{Q_2} \right)^{-1}$ be the loaded Q of the unfilled resonator.   Then

$$Q_L = Q_0 \frac{1}{1 + Q_0/Q_a} \,, \tag{6}$$

where Q_a is given by

$$\frac{1}{Q_a} = \frac{P_a}{\omega W} = \frac{c}{\omega} \frac{\int \alpha \, E^2 \, dv}{\int E^2 \, dv} = \frac{\langle \alpha \rangle \lambda}{2\pi} \,. \tag{7}$$

In (7) the integral extends over the interior resonator volume and

$$\langle \alpha \rangle = \frac{\int \alpha(E^2) \, E^2 \, dv}{\int E^2 \, dv} \tag{8}$$

is an averaged absorption coefficient; we see from (7) that $\langle\alpha\rangle\lambda$ represents the fractional power losses inside the resonator due to the absorber.

In the filled resonator the cavity wavelength of the em wave is $1/n$ times that in the empty resonator, so that the resonance frequency ω_R is related to that of the unfilled one ω_o by

$$\frac{\omega - \omega_R}{\omega_R} = \frac{\langle n\rangle \, \omega - \omega_o}{\omega_o} \tag{9}$$

where $\langle n\rangle$ is the averaged refractive index

$$\langle n\rangle = \frac{\int n(E^2) \, E^2 \, dv}{\int E^2 \, dv} \, .$$

Making use of Eq. (2) which is valid at each point inside the cavity, we get

$$\langle n\rangle = 1 - \frac{\lambda\Delta}{4\pi} \, \langle\alpha\rangle \quad , \tag{10}$$

the averaged absorption coefficient $\langle\alpha\rangle$ being given by Eq. (8).

At low levels of the incident power P_i the saturation parameter S is negligible with respect to unity, $\langle\alpha\rangle = \alpha$, $\langle n\rangle = n$ and $T(\omega)$ does not depend on P_i, so that from Eqs. (2), (4), (6), (7) and (9), we have

$$T(\omega) = T_o \left[\left(1 + \frac{\alpha\lambda Q_o}{2\pi} \right)^2 + 4Q_o^2 \left(\frac{\omega - \omega_o}{\omega_o} - \frac{\alpha\lambda\omega}{4\pi\omega_o} \frac{\omega - \omega_o}{\Gamma} \right)^2 \right]^{-1} , \tag{11}$$

where $T_o = 4Q_o^2/Q_1 Q_2$ is the transmission at resonance of the empty resonator. However, for higher values of the incident power, α and n depend on P_i and vary from point to point in the interior of the resonator. The transmission $T(\omega)$ is then a function of the incident power and, depending on the absorber and resonator parameters, bistability may occur, as results from the analysis presented in Sec. III.

To calculate the expressions (8) and (10) in the general situation let us consider the electric field inside the microwave cavity. This can be written as

$$E^2 = E_o^2 \, f(x, \, y, \, z) \, ,$$

where E_o is the maximum amplitude of the electric field, and the relative field distribution $f(x,y,z)$ depends on the resonator's shape and on the resonance mode.

It has been proven[12] that in a laser medium with a non-uniform gain distribution, the field distribution $f(x,y,z)$ does not change appreciably at low gain. We assume that this result may be extended to a microwave cavity containing a non-linear absorber. In that case $\langle\alpha\rangle$ and $\langle n\rangle$ can be calculated since the field distribution in the interior of the empty resonator is known.

For a spherical mirror FP resonator the field distribution corresponding to excitation in the fundamental TEM_{oop} mode, is given by[13,14]

$$E^2(r,z) = \frac{E_o^2}{1+\xi^2} \exp\left(-\frac{A^2}{1+\xi^2}\, r^2\right) \sin^2\phi(r,z) \tag{12}$$

where:

$r = \sqrt{x^2 + y^2}$ is the distance from the interferometer axis,

$\xi = 2z/b$ is the normalized longitudinal coordinate, i.e., the distance from the interferometer center, measured along its axis in units of half of the mirror spacing b,

$A^2 = \dfrac{4\pi}{b\lambda}$

and

$$\phi(r,z) = \frac{\pi b}{\lambda}(1+\xi) + \frac{A^2 r^2}{2}\,\frac{\xi}{1+\xi^2} + tg^{-1}\left(\frac{1-\xi}{1+\xi}\right) - \frac{\pi}{2}$$

is the phase angle of the standing waves. The resonance condition for this mode with a spacing b close to the confocal spacing, such as in the present experiment, is

$$\frac{4b}{\lambda} = 2p + 1 \quad . \tag{13}$$

When the longitudinal number $p \gg 1$ one can neglect the z dependence of the cavity wavelength and write:

$$E^2(r,z) = \frac{E_o^2}{1+\xi^2} \exp\left(-\frac{A^2}{1+\xi^2}\, r^2\right) \sin\frac{p\pi\xi}{2} \quad . \tag{14}$$

This field configuration, apart from the z-dependence appearing in the amplitude and in the exponential factor, is what one has in the interior of a plane mirror FP optical resonator, excited by a laser beam.

From Eqs. (1), (7), (10) and (14) it follows that

$$\langle\alpha\rangle = \alpha_o\, \psi(\Delta, S_o) \tag{15}$$

$$<n> = 1 - \frac{\alpha_o \lambda \Delta}{4\pi} \, \psi(\Delta, S_o) \tag{16}$$

where

$$\psi(\Delta, S_o) = 2 \int_o^1 du \int_o^\infty dt \; \frac{e^{-t} \sin^2 \frac{1}{2} p\pi u}{1 + \Delta^2 + \dfrac{S_o}{1+u^2} \, e^{-t} \sin^2 \frac{1}{2} p\pi u} \; . \tag{17}$$

A good approximation to this expression is obtained by substituting
for $1/(1+u^2)$, in the denominator of the integrand, its average value
over the integration range, namely $\pi/4$. The integration may then
be performed analytically giving*

$$\psi(\Delta, S_o) = \frac{16}{\pi S_o} \log \frac{1}{2} \left(1 + \sqrt{1 + \frac{1}{4} \pi B}\right) . \tag{18}$$

In the latter expression we have written

$$S_o = \frac{<\mu^2> E_o^2}{3\hbar^2} \frac{\tau}{\Gamma} \; , \tag{19}$$

$$B = \frac{S_o}{1 + \Delta^2} \; . \tag{20}$$

Equation (16) was derived by evaluating the mean value of the refractive index throughout the whole cavity, i.e., over the whole
wavefront of the electric field. In an alternative approach we
could impose the resonance condition along the interferometer axis
that p nodes are present for the electric field amplitude in the
TEM_{oop} mode both for the empty cavity and for the cavity filled
with a saturable absorber. This resonance condition must be valid
whatever is the spatial distribution of the electric field over the
cavity volume, and may be seen to correspond to limiting the volume
integration in the expression for $<n>$ to an infinitesimal cylinder
about the interferometer axis. This alternative procedure yields

---

*It readily follows from (17) that $\psi(\Delta, S_o)$ must satisfy the in-
equality

$$\frac{4}{p S_o} \sum_{n=0}^{p-1} \left(1 + \frac{n^2}{p^2}\right) \log \frac{1}{2}\left(1 + \sqrt{1 + \frac{B}{1+n^2/p^2}}\right) < \psi(\Delta, S_o)$$

$$< \frac{4}{p S_o} \sum_{n=1}^{p} \left(1 + \frac{n^2}{p^2}\right) \log \frac{1}{2}\left(1 + \sqrt{1 + \frac{B}{1+n^2/p^2}}\right)$$

thereby confirming the validity of the approximation (18).

exactly the same expression (16) given for $\langle n \rangle$. From Eqs. (4), (6)-(10), (15) and (16) the transmission $T(\omega)$ can now be expressed as a function of E_o^2, the maximum value of the electric field in the interior of the cavity, or rather as a function of S_o:

$$T(\omega) = T_o \left\{ [1 + 2C\psi(\Delta,S_o)]^2 + [\theta - 2C\Delta \frac{\omega}{\omega_o} \psi(\Delta,S_o)]^2 \right\}^{-1} , \qquad (21)$$

where we have put

$$C = \frac{\alpha_o \lambda Q_o}{4\pi} , \qquad (22)$$

$$\theta = \frac{2Q_o}{\omega_o} (\omega - \omega_o) = \frac{\omega - \omega_o}{\gamma} ; \qquad (23)$$

θ represents the detuning from the cavity frequency in units of γ, the resonance width of the empty resonator.

Let us now express E_o^2 as a function of the incident power. To this end consider the steady state: the overall energy losses $(-dW/dt)$ must be equal to the power entering the interior of the resonantor, so that from Eq. (5)

$$- \frac{dW}{dt} = P_i - P_R = P_i \frac{Q_2}{Q_L} T(\omega) .$$

However the resonator Q-value is defined to be

$$Q_L = \frac{\omega W}{(-dW/dt)} = \frac{\omega}{(-dW/dt)} \int \frac{E^2}{4\pi} dv = \frac{\omega Q_L}{Q_2 P_i T(\omega)} \frac{E_o^2 b^2 \lambda}{32\pi} , \qquad (24)$$

where Eq. (14) has been used for the TEM_{oop} electric field. Thus

$$E_o^2 = \frac{16Q_2}{b^2 c} P_i T(\omega) . \qquad (25)$$

Introducing the expressions

$$X = P_T = P_i T(\omega) \qquad (26)$$

$$Y = P_i T_o \qquad (27)$$

$$a = \frac{\langle \mu^2 \rangle}{3\hbar^2} \frac{\tau}{\Gamma} \frac{16Q_2}{b^2 c} , \qquad (28)$$

we now have, from Eqs. (19), (25), (26) and (28),

$$S_o = aX , \qquad (29)$$

and from Eqs. (21) and (27)

$$Y = X\{[1 + 2C\psi(\Delta,X/a)]^2 + [\theta - 2C\Delta\psi(\Delta,X/a)]^2\} \quad , \tag{30}$$

which is the state equation expressing the transmitted power as a
function of the incident power and of the resonator and absorber
parameters, when one takes $\omega/\omega_o \approx 1$[15].

Using expression (18), if one expands the inverse of $\psi(\Delta,S_o)$
in a power series in $S_o = aX$ then the relation (30) is replaced to
first order by the following expression

$$Y = X\left\{\left(1 + \frac{2C}{1 + \Delta^2 + \frac{3\pi}{32}aX}\right)^2 + \left(\theta - \frac{2C\Delta}{1 + \Delta^2 + \frac{3\pi}{32}aX}\right)^2\right\} . \tag{31}$$

Equation (31) is very similar to that obtained by Bonifacio et al[8]
in the analysis of the optical bistability in a ring cavity filled
with a non-linear absorber. In fact, in the case of a homogeneously
broadened atomic system they obtain, using our notation, the relation

$$Y = X\left\{\left(1 + \frac{2C}{1 + \Delta^2 + aX}\right)^2 + \left(\theta - \frac{2C\Delta}{1 + \Delta^2 + aX}\right)^2\right\} , \tag{32}$$

which differs from Eq. (31) only in the factors $3\pi/32$ appearing
there in the denominators and may clearly be attributed to the different spatial distributions of the electric field inside the two
different cavities considered. Whereas, in principle, Eq. (31) can
be solved analytically as a cubic equation in X, the relation (30)
must be solved numerically.

Apart from a scaling of S_o by the factor $\pi/4$ in Eq. (18), Eq.
(30) is just the state equation connecting the incident and transmitted powers obtained by Ballagh et al[7] in their recent analysis
of optical bistability in an FP cavity, in which they impose a self-
consistency equation on the TEM_{oop} electric field propagating inside an FP resonator filled with a non-linear absorptive medium.

In order to exhibit the effect of the spatial distribution of
the electric field inside the resonator, we have plotted in Fig. 1
the normalized absorption coefficients α/α_o using Eq. (1) (Fig. 1a),
$\langle\alpha\rangle/\alpha_o$ from Eq. (15) (Fig. 1b) and the linear saturation parameter
approximation [as in Eq. (31)] to $\langle\alpha\rangle/\alpha_o$ (Fig. 1c) as a function of
the detuning parameter Δ. In this figure we have taken the saturation parameter to be $S_o = 10$, which in the field of magnetic resonance is situated in the high saturation region. These curves show
that $\langle\alpha\rangle$ has both a higher value and a stronger Δ dependence than
those given by Eqs. (1) and (31). This different behaviour occurs

because, when making an average over the cavity volume, the absorption coefficient $\langle\alpha\rangle$ includes both contributions from points at the center of the interferometer, where the saturation parameter is large, and contributions from points at large distances from the z axis, where the electric field is weak and the saturation is small. Summing up all these contributions we find as a result that $\langle\alpha\rangle$ is larger than α and has a strong dependence on the detuning parameter Δ.

As we shall indicate in the following section, this different behaviour of the absorption on the detuning parameter is important, because it plays a decisive role in the occurrence of bistability.

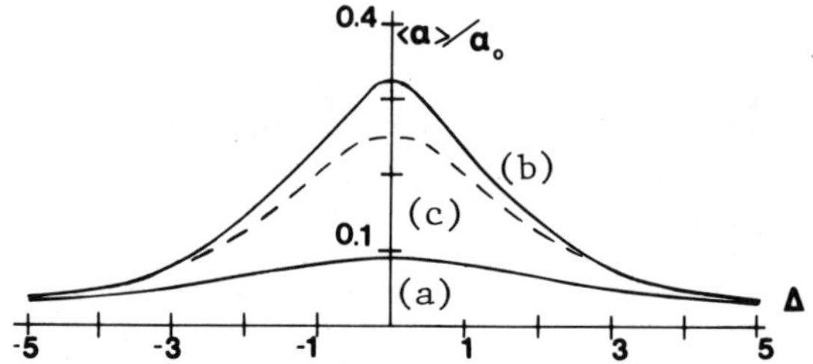

Fig. 1   The relative absorption coefficient $\langle\alpha\rangle/\alpha_o$ versus the frequency detuning Δ for saturation parameter $S_o = 10$. Curve a) is obtained from Eq. (1) and corresponds to the result for a uniform field distribution, curbe b) is obtained from Eq. (15) and curve c) is obtained from the linearized approximation in S_o to Eq. (15).

III.   BISTABILITY

The bistability of the filled resonator may be understood by studying Eqs. (30) and (31) which describe the transmission of the cavity. A different bistable behaviour results according to whether one does or does not include the spatial distribution of the electric field inside the cavity. If one takes a uniform electric field distribution inside a ring cavity then bistability occurs only if, in our notation, the following condition of Bonifacio and Lugiato[8]

$$2C = \frac{\alpha_o \lambda Q_o}{2\pi} > 8 \tag{33}$$

is valid. If, instead, a Gaussian radial dependence is assumed for the electric field in an FP cavity, on the basis of Eq. (30), Ballagh et al[7] have derived the more restricted condition for bistability:

$$2C > 20.08 \quad . \tag{34}$$

At centimeter wavelengths the latter condition requires strong absorption lines and very high Q values.

In our experiment as absorber we have used ammonia gas working at the frequency ν_a = 23.879 GHz, corresponding to λ = 1.257 cm, where the strongest line (J=K=3) of the inversion spectrum occurs and whose absorption intensity in the homogeneous broadening regime is α_o = 7.9 $\times$ 10^{-4} cm^{-1}. The collisional and Doppler line widths of ammonia are the same at a gas pressure of about 1 mTorr, while at low pressures α_o is reduced by the factor $(1 + 1.5/p^2)^{-\frac{1}{2}}$, where p is measured in mTorr. In order to obtain the highest possible Q value we used the interferometer under conditions of very weak coupling. Since the Q value of our resonator is Q_o=5 $\times$ 10^5 (see Sec. IV), the maximum value of 2C is 80 for the case of the homogeneously broadened (3,3) line of ammonia. The parameter C can be decreased by varying the pressure of the gas.

Corresponding to the (3,3) line and to our cavity, the parameters appearing in Eqs. (19), (22) and (28) have the following values:

$$b = 73 \ \underline{cm}$$

$$\tau = \Gamma_c^{-1} = 5.48 \times 10^{-6}/p \ \underline{s}$$

$$\Gamma_{Dopp} = 2.26 \times 10^5 \ \underline{s}^{-1}$$

$$\Gamma = (\Gamma_c^2 + \Gamma_{Dopp}^2)^{\frac{1}{2}} = 1.82 \times 10^5 \ (p^2 + 1.5)^{\frac{1}{2}} \ \underline{s}^{-1}$$

$$\langle\mu\rangle = 1.27 \times 10^{-18} \ \underline{cgs\ units}$$

$$\alpha_o = 7.0 \times 10^{-4} \ (1 + 1.5/p^2)^{-\frac{1}{2}} \ \underline{cm}^{-1}$$

where p is expressed in mTorr. The bistability parameters appear in Eqs. (30) and (31) with p in mTorr, $\nu-\nu_a$, $\nu-\nu_o$ in kHz, and $P_i = \tilde{Y}$, $P_T = \tilde{X}$ in mW are as follows:

$$2C = 80 \ (1 + 1.5/p^2)^{-\frac{1}{2}}$$

$$\theta = 4.2 \times 10^{-2} \ (\nu-\nu_o)$$

$$\Delta = 3.4 \times 10^{-2} \ (\nu-\nu_a) \ (p^2 + 1.5)^{-\frac{1}{2}}$$

$$S_o = aX = 1.4 \times 10^6 \ p^{-1} \ (p^2 + 1.5)^{-\frac{1}{2}} \ \tilde{X}$$

and Eq. (30) becomes

$$\tilde{Y} = 10^4 \tilde{X}\{[1+2C(p)\psi(\Delta,p,\tilde{X})]^2 + [\theta - 2C(p)\Delta \ \psi(\Delta,p,\tilde{X})]^2\} \qquad (35)$$

We may observe from the above parameter values that, with the pressure in the mTorr range and for an incident power of a few milliwatts, the saturation parameter in our experiment can reach as high a value as a few thousand. These large values, in contrast to the much lower numbers resulting in the corresponding optical experiments, arise from the microwave relaxation times which are considerably larger than the lifetime of an optical level and from the high Q value of the microwave resonator.

In order to exhibit the contribution of the spatially averaged absorption coefficient we have plotted in Figs. 2a and 2b, the transmitted power $\tilde{X}$ as a function of the incident power $\tilde{Y}$ for p=1 mTorr, i.e., $C = 25.5$, $a = 9.23 \times 10^5$, in the conditions of absorptive bistability $\Delta = \theta = 0$. The curve a) is the transmitted power given by Eq. (32) corresponding to a uniform saturation parameter S inside the resonator, while the curve b), obtained from Eq. (30), includes the spatial dependence of the absorption coefficient. In the case of a non-uniform electric field distribution the bistability occurs at larger values of the incident power. The dashed line represents the transmission of the empty cavity and at high $\tilde{Y}$ values both curves a) and b) approach the empty cavity transmission.

In Fig. 2c the transmitted power $\tilde{X}$ has been plotted as a function of the incident power $\tilde{Y}$ for $\Delta = 1.5$ and $\theta = 0.29$. On increasing Δ and θ the bistability region shifts to smaller values of the incident power $\tilde{Y}$.

In the experiment reported here, the bistability is studied taking constant vlaues for the parameters C and $\tilde{Y}$, while Δ and θ are varied by a continuous sweeping of the angular frequency of

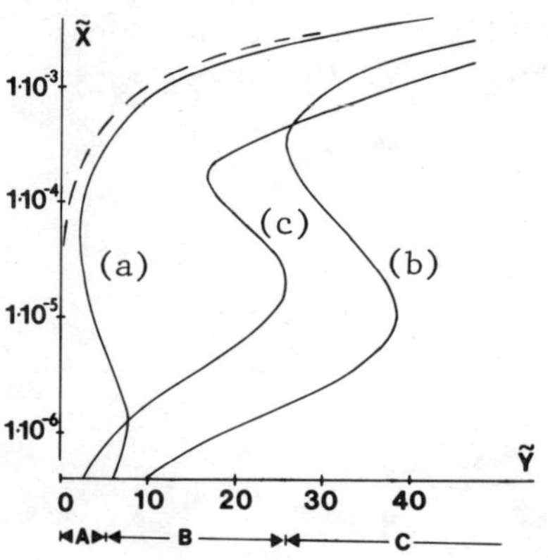

Fig. 2 The transmitted power $\tilde{X}$ on a logarithmic scale versus the incident power $\tilde{Y}$, both measured in mW, for p=1 mTorr. Curve a) describes the absorptive bistability of Eq. (32), curve b) is obtained from Eq. (30). At large incident powers $\tilde{Y}$, both curves a) and b) approach the dashed line representing the transmission of the cavity when empty. Curve c) shows the transmission in the presence of detuning, i.e., $\Delta = 1.5$ corresponding to $\nu - \nu_a = 69$ kHz.

the incident radiation. The simplest case is when the cavity and
molecular frequencies are equal, i.e., $\omega_o = \omega_a$ so that we may write

$$\theta = \frac{2Q_o \Gamma}{\omega_o} \Delta = \frac{\Gamma}{\gamma} \Delta = \rho \Delta . \tag{36}$$

Bistability curves $\tilde{X}(\Delta)$, with fixed $\tilde{Y}$, C and ρ (dispersive bistability), are shown in Fig. 3. Depending upon the incident power
$\tilde{Y}$, the transmittance has a different dependence on the detuning
parameter Δ. At low incident power, in the interval A of Fig. 2,
the transmittance is a single-valued function of Δ and no bistability occurs on sweeping the detuning Δ (Fig. 3d). Increasing the
incident power to the interval B of Fig. 2, the transmittance has
the dependence depicted in Fig. 3c with bistable behaviour. In the
latter figure the transmission signals $\tilde{X}(\Delta)$ obtained from frequency
modulating the incident power around $\Delta = 0$ are shown, and the experimental behaviour of the system through the turning points Δ_1 and
Δ_2, the zeros of $d\tilde{X}/d\Delta$, will depend on whether one is increasing
or decreasing the value of Δ as the time increases. A similar behaviour was pointed out by Carmichael and Hermann[10].

The bistable behaviour occurs because for a given value of C
the interval of $\tilde{Y}$ giving bistability is maximum for $\Delta = \theta = 0$, while
on shifting the values of Δ and θ away from zero the interval of
bistability is smaller and is centered around lower values of $\tilde{Y}$.
For an incident power $\tilde{Y}$ in the interval B of Fig. 2, the transmitted
power $\tilde{X}$ is a single-valued function of $\tilde{Y}$ when $\Delta = 0$, while between
Δ_1 and Δ_2 of Fig. 3c it becomes a three-valued function of $\tilde{Y}$. On
increasing the value of Δ from zero the value of $\tilde{X}$ varies continuously along the lower curve of Fig. 3c until the turning value Δ_2
is reached, when $\tilde{X}$ jumps to the upper branch of the curve. On the
other hand, reducing Δ from large values, the upper curve is traced
until the turning point Δ_1 is reached, when $\tilde{X}$ jumps down to the
lower branch of the curve. We have not performed a detailed numerical analysis of the onset of bistability, but it appears that condition (34) is also required when observing the bistable behaviour
as a function of Δ and θ. For higher values of $\tilde{Y}$ (region C of Fig.
2), the dependence of the transmittance on the parameter Δ is typically represented in Figs. 3a and 3b. The transmittance is a multi-
valued function at small values of Δ, but the lower branch is isolated from the upper one, so that bistable behaviour and hysteresis
are not observed on sweeping the detuning parameter Δ.

The turning points Δ_1 and Δ_2 in the bistable behaviour of Fig.
3x can be evaluated from Eq. (30) and their dependence on the incident power $\tilde{Y}$ is shown in Fig. 4 for the same parameters as used
in Fig. 3.

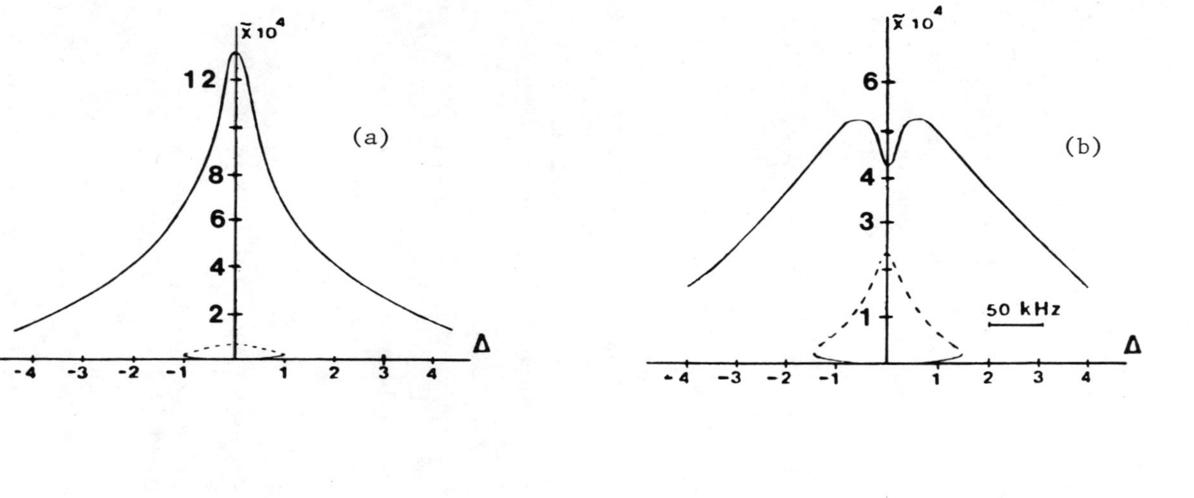
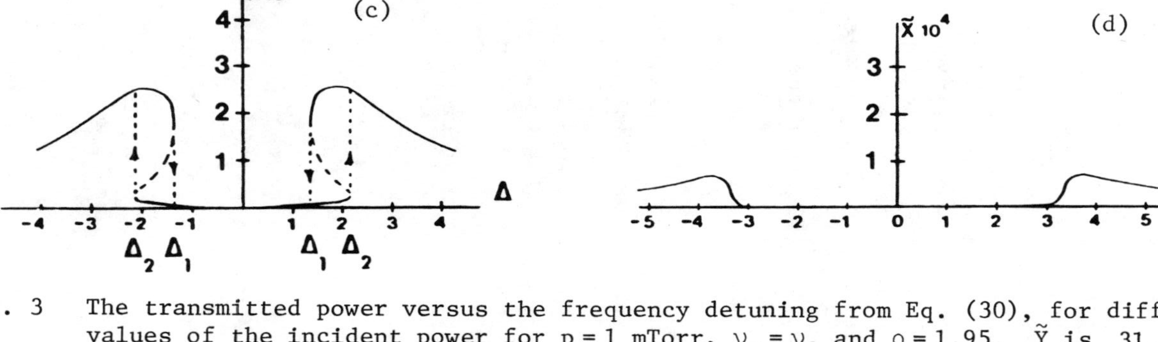

Fig. 3   The transmitted power versus the frequency detuning from Eq. (30), for different values of the incident power for $p = 1$ mTorr, $\nu_o = \nu_a$ and $\rho = 1.95$. $\tilde{Y}$ is 31.75 mW in curve a), 26 mW in curve b), 17.8 mW in curve c), 5.7 mW in curve d).

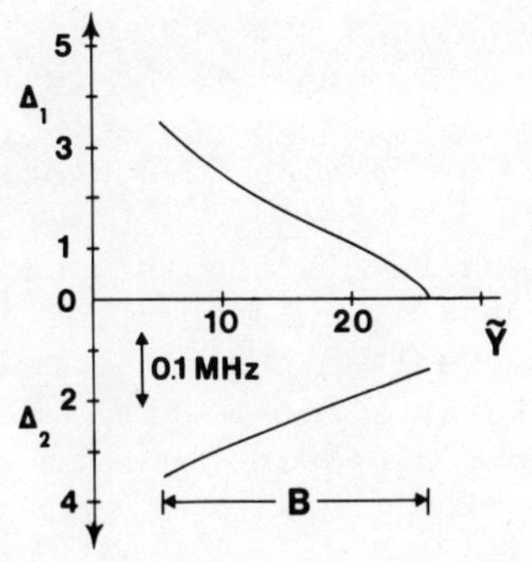

Fig. 4   The bistability turning points Δ_1 and Δ_2 versus the incident power $\tilde{Y}$ with the same parameters as in Fig. 3. Dispersive bistability is observed for those values of the incident power $\tilde{Y}$ which are inside the interval B of Fig. 2.

IV.   EXPERIMENTAL APPARATUS

The experimental apparatus is sketched in Fig. 5. The FP resonator is fed by a VA98 reflex klystron via arm 1 of a 3 port circulator. The power reflected in arm 3 is utilized to minitor the incident power, while the transmitted power, amplified by a tunnel diode X band amplifier, is detected by a 1N26 diode, and the transmission signal is observed on the screen of an oscilloscope.

The FP interferometer has already been described in Ref. (14). It is composed of two spherical mirrors mounted on a guiding system of four steel bars, at a spacing b = 72.8 cm close to the confocal spacing (73 cm). At this spacing the system resonates in the fundamental $TEM_{0,0,116}$ mode at the frequency of the (3,3) ammonia inversion line. The coupling is realized by two small holes at the center of the mirrors, which face the interiors of the input and output waveguides.

A motor driven screw allows the spacing of the mirrors to vary periodically, at a low rate, sweeping the resonator frequency ω_R around the absorber frequency ω_a over a range of several MHz.

Fig. 5   Block diagram of the microwave spectrometer: K – klystron, U – uniline, Av – variable attentuator, C – circulator with 1,2, and 3 ports, A – attenuator, X – diode, Am – amplifier, S – scope.

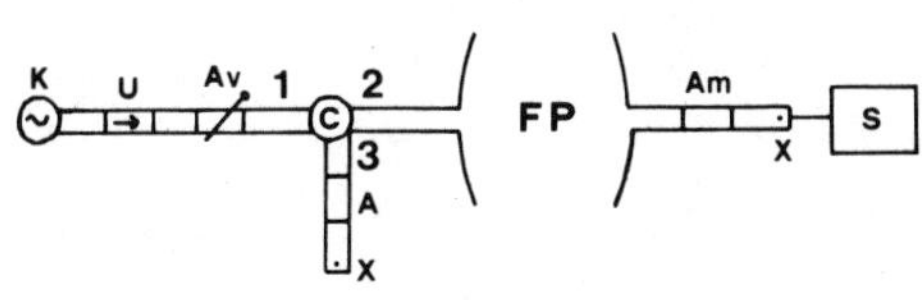

The whole system is enclosed in a stainless steel cylindrical container that can be evacuated to a pressure of $p < 10^{-6}$ Torr. The ammonia pressure is measured by means of a McLeod gauge. Owing to the large value of the Fresnel number N = 2.5 the diffraction losses

are negligible. Since the input and output couplings are very weak,
the loaded Q of the unfilled interferometer measured at $\lambda = 1.25$ cm
is $Q_o = 5 \times 10^5$, which is close to the theoretical value $Q_o = 6.4 \times 10^5$.
The measured transmission at resonance is $T_o = 10^{-4}$ giving $Q_1 = Q_2 = 10^8$.

In order to measure the incident power P_i a 1N26 diode is used
to detect the power reflected in the arm 3 of the circulator after
a 30 db attenuation. Owing to the weak input coupling the reflected
power is, in practice, equal to the incident power. For power levels
in the microwatt range the voltage diode output is proportional to
the incident power.

The absorption by the ammonia molecules contained in the FP
cavity can be observed as a function of the frequency by monitoring
the transmitted power while the interferometer spacing is slowly
varied by means of the motor driven screw, and while the klystron
is frequency modulated with a modulation depth larger than the frequency interval to be investigated. At each period of the 50 Hz
modulation frequency the transmission signal at the resonator frequency is observed and the envelope of the transmission signals
represents $T(\omega_R)$ as a function of $\omega_R - \omega_a$, the detuning between the
interferometer and absorber resonances. When the resonator fre-
quency is swept through a frequency range of several MHz around the
frequency of the (3,3) inversion line the quadrupole structure is
observed as shown in Fig. 6. The central line in this figure is
composed of the $\Delta F = 0$ lines, and the satellites are the $\Delta F = \pm 1$ components. Owing to the smaller values of the dipole matrix elements
for the $\Delta F = \pm 1$ resonances with respect to the $\Delta F = 0$ resonance, the
satellites are less saturated, and from the ratio of the intensity
of the central line to those of the satellites one obtains the
order of magnitude of the saturation parameter S_o.

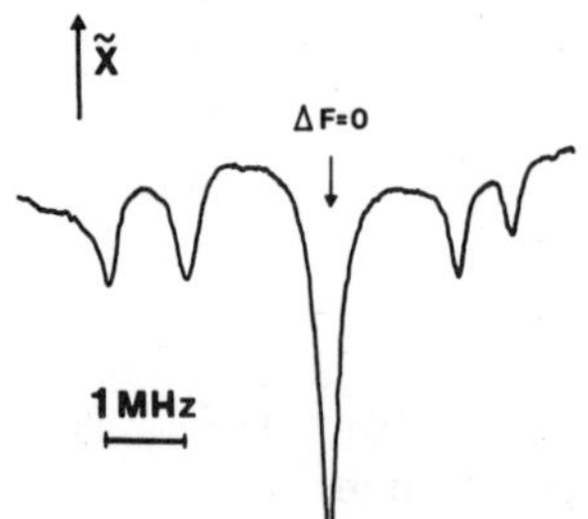

Fig. 6 The (3,3) absorption line of ammonia with the $\Delta F = 0$ central component and
$\Delta F = \pm 1$ satellites. This recording was obtained by motor scanning the cavity length
and frequency modulating the klystron, as
described in the text. Comparison of the
relative heights of the central and satel-
lite lines yields $S_o \approx 170$ for the saturation parameter.

V.   EXPERIMENTAL RESULTS

In observing the bistable behaviour of the resonator filled
with ammonia, we have taken a fixed value for the frequency detuning $\omega_o - \omega_a$ and swept the frequency of the klystron.   In this way we
have varied simultaneously the parameters Δ and θ appearing in Eqs.
(30) and (31), which describe the transmission behaviour of the
resonator.

The simplest case occurs when $\omega_o = \omega_a$, i.e., when the resonator
is tuned to the absorption frequency of ammonia, so that Δ and θ
are proportional to each other as shown in Eq. (36).   The transmittance of the interferometer, as a function of the frequency detuning $\omega - \omega_a$, is represented in Fig. 7 for different values of the incident power, and we observe that it decreases in going from Figs.
7a to 7e.   The bistability appears as a sudden change of the transmittance; this falls abruptly to an undetectable value, and rises
again after passing through the value $\omega = \omega_a$.   The behaviour reported
in Fig. 7 closely resembles the theoretical results shown in Fig. 3
for decreasing values of the incident power.

Fig. 7 Experimental recording of the
transmitted power $\tilde{X}$ as a function of
the em field frequency for p=0.5
mTorr ammonia pressure and different
values of the incident power.   The incident power $\tilde{Y}$ was estimated to be
5, 2.5, 1.5, 1 and 0.5 mW for curves
a), b), c), d), e) respectively.

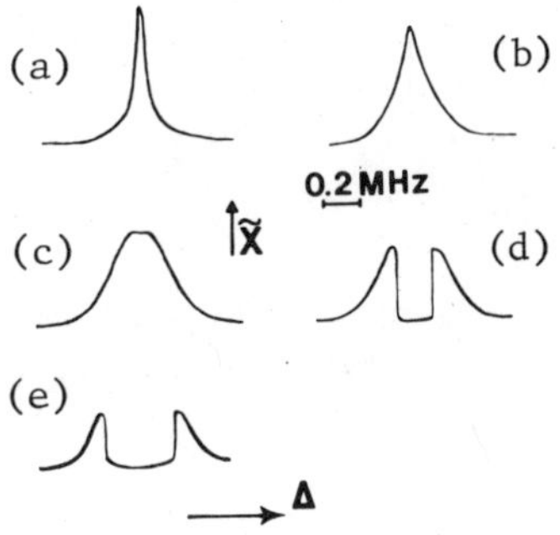

The dependence of the transmission signal on the gas pressure
is depicted in Fig. 8, where a set of curves is presented corresponding to a fixed incident power and for increasing values of the
gas pressure.   The measurement of the incident power is very approximate and our estimate gives $\tilde{Y}$ in the 20-50 mW range.   In this
set of experiments the occurrence of bistability is very critical;
in fact, on changing the pressure the parameters C, a and ρ are
simultaneously varied.   Under these conditions the pressure at
which bistability occurs cannot be measured accurately and we can
only estimate its order of magnitude.   In Fig. 9 we present a set
of theoretical curves obtained from Eq. (30) corresponding to an
incident power of $\tilde{Y} = 31.75$ mW.   The values of the pressure of these
theoretical curves are chosen to illustrate the agreement that can
be achieved with the experimental signals reported in Fig. 8.

Within the present experimental accuracy of the incident power and
ammonia pressure measurements, the agreement between the experimental and the theoretical results is certainly to be considered
good.

More accurate measurements involving both absorptive and dispersive bistability, with a careful determination of the pressure
and the incident power are in progress.

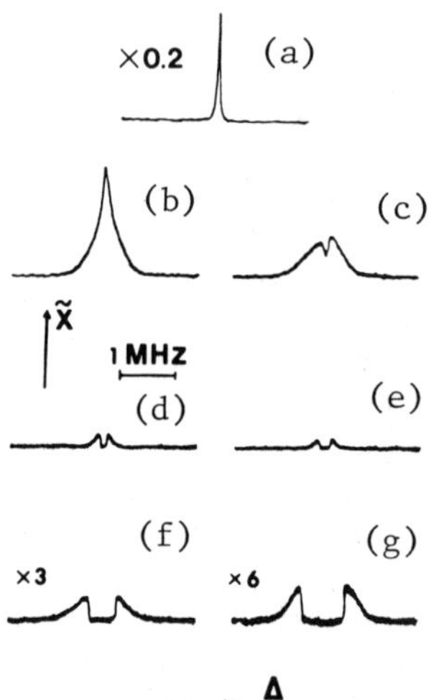

Fig. 8 Experimental curves of the
transmittance versus the frequency
at fixed incident power and different ammonia pressures. The
cavity is evacuated in curve a)
and p = 1, 1.15, 1.3, 1.4, 1.5,
1.7 mTorr for curves b), c), d),
e), f), g) respectively.

VI.   CONCLUSIONS

In the present work we have analyzed the optical bistability
in relation to an experiment in the microwave region. The starting
point of our analysis is typical of microwave studies. The behaviour of the FP cacity filled with an absorbing gas is described
from a macroscopic point of view; the FP response is expressed as
a function of the absorber and resonator parameters. Thus we are
not concerned with the details of the em field propagation. This
approach leads in a straightforward way to the state equation of
the optical bistability. An important result of our analysis is
the influence of the spatial distribution of the field on the decrease of the quality factor of the cavity and on the bistable behaviour; this influence has been neglected in most of the previous
theoretical descriptions of optical bistability. Because of the
spatial dependence of the absorber saturation the bistability occurs
for larger powers of the incident em field than those predicted by
theories which assume a uniform saturation.

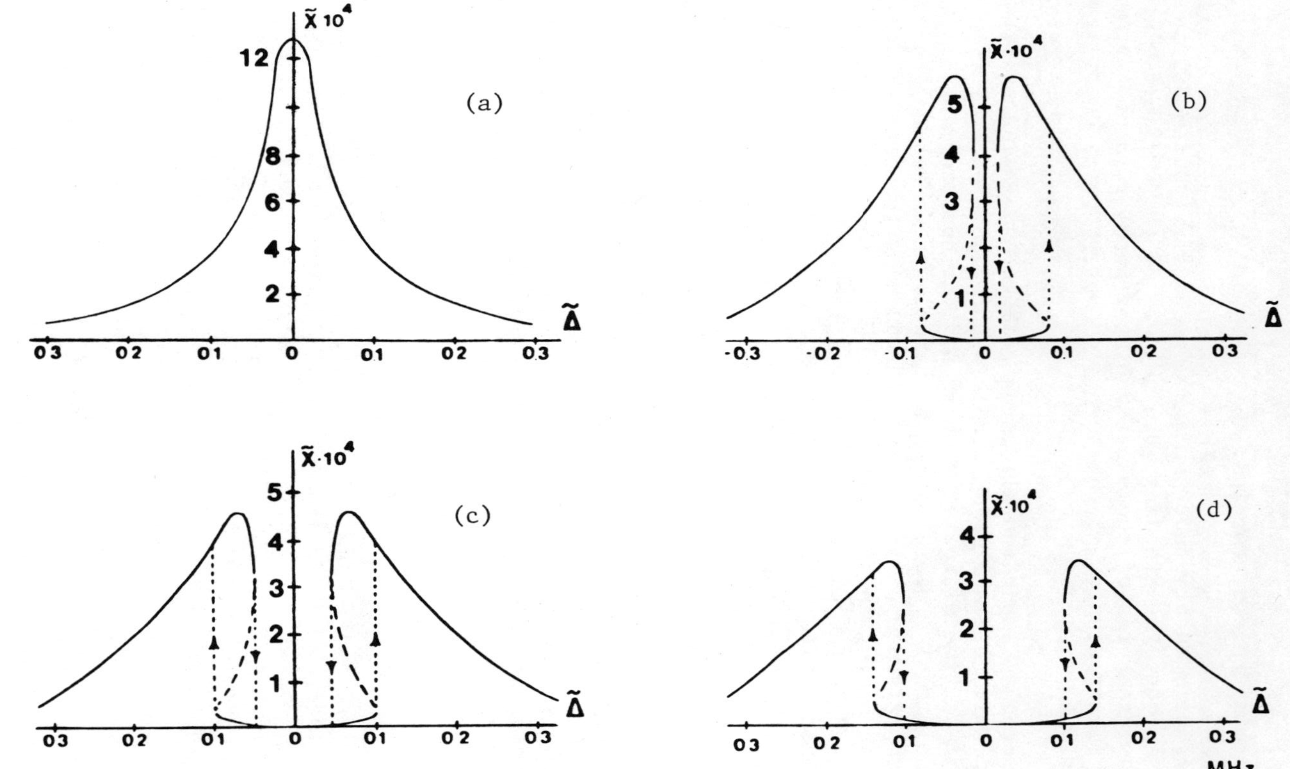

Fig. 9   Theoretical curves for the transmittance $\tilde{X}$ versus the frequency at a fixed incident power $\tilde{Y}$ = 31.75 mW and increasing values of the ammonia pressure: p = 0.8, 1,1, 1.2, 1.4 mTorr for curves a), b), c), d) respectively.

The FP response is calculated under the assumption that the spatial configuration of the em field inside the cavity is not perturbed by the absorber. At large saturation the configuration is distorted by the spatial inhomogeneity of the saturation, but we cannot determine this distortion.

We present a theory that for a homogeneously broadened line takes into account the standing wave pattern of the em field. At the very low pressures of the experiment the Doppler inhomogeneous broadening is important, but the theory cannot properly describe this condition.

The microwave apparatus allows us to make a detailed analysis of the dispersive bistability. Stable and reliable em radiation is easily obtained from microwave sources and the spectroscopic parameters of the ammonia inversion lines are well-known for making an accurate comparison with the theory. Because of the long relaxation times occurring in microwave spectroscopy the saturation parameters involved in the experiment are much larger than the corresponding values typical of the optical region.

ACKNOWLEDGMENTS

The authors wish to dedicate this work to the late Mario Gronchi who had been a student of one of them (A.G.) during his years at Pisa University and who at the initial stages of this investigation contributed to the understanding of the phenomenon with enjoyable discussions. Thanks are also due to R. Bonifacio and L. A. Lugiato for their constant interest, and to W. J. Sandle for a helpful discussion with one of them (E.A.).

REFERENCES

1.   A. Kastler, Appl. Opt. $\underline{1}$, 17 (1962).
2.   H. M. Gibbs, S. L. McCall and T. N. C. Venkatesan, Phys. Rev. Lett. $\underline{36}$, 1135 (1976); T. N. C. Venkatesan and S. L. McCall, Appl. Phys. Lett. $\underline{30}$, 282 (1977).
3.   D. Grischkowsky, J. Opt. Soc. Amer. $\underline{68}$, 641 (1978).
4.   T. Bischofberger and Y. R. Shen, Appl. Phys. Lett. $\underline{32}$, 156 (1978); Opt. Lett. $\underline{4}$, 40 (1979).
5.   H. M. Gibbs, S. L. McCall, T. N. C. Venkatesan, A. C. Gossard, A. Passner and W. Wiegman, Appl. Phys. Lett. $\underline{35}$, 451 (1979); D. A. Miller, S. D. Smith and A. Johnston, Appl. Phys. Lett. $\underline{35}$, 658 (1979).
6.   P. W. Smith and E. H. Turner, Appl. Phys. Lett. $\underline{30}$, 280 (1977); P. W. Smith, I. P. Kaminov, P. J. Maloney and L. W. Stulz, Appl. Phys. Lett. $\underline{33}$, 24 (1978) and $\underline{34}$, 62 (1979);

E. Garmire, S. D. Allen, J. Marburger and C. M. Verber, Opt. Lett. $\underline{3}$, 69 (1978); A. Feldman, Opt. Lett. $\underline{4}$, 115 (1979).

7. W. J. Sandle and A. Gallagher, Digest of Technical Papers, XI IQEC, Boston 23-26 June 1980, IEEE J. Quantum Electron. $\underline{QE-16}$, 656 (1980); R. J. Ballagh, J. Cooper, M. W. Hamilton, W. J. Sandle and D. M. Warrington, Phys. Rev. (in press); W. J. Sandle and A. Gallagher, Phys. Rev. (in press).

8. R. Bonifacio and L. A. Lugiato, Lett. Nuovo Cimento $\underline{21}$, 517 (1978); R. Bonifacio, M. Gronchi and L. A. Lugiato, Nuovo Cimento $\underline{53B}$, 311 (1979); R. Bonifacio, L. A. Lugiato and M. Gronchi, in "Laser Spectroscopy IV" Proc. Int. Conf. at Rottach-Egern, Germany, eds. H. Walther and K. W. Rothe (Springer-Verlag, Berlin, 1979) p. 426.

9. G. P. Agrawal and H. J. Carmichael, Phys. Rev. $\underline{A19}$, 2074 (1979); C. W. Bowden and C. C. Sung, Phys. Rev. $\underline{A19}$, 2392 (1979).

10. H. J. Carmichael and J. A. Hermann, Zeits Phys. (in press).

11. G. Boudouris and P. Chenevier, Circuits pour ondes guidées (Dunod, Paris, 1975).

12. H. Statz and C. L. Tang, J. Appl. Phys. $\underline{36}$, 1816 (1965); T. Li and J. G. Skinner, J. Appl. Phys. $\underline{36}$, 2595 (1965).

13. G. D. Boyd and J. P. Gordon, Bell Syst. Tech. Journ. $\underline{40}$, 489 (1961).

14. A. Battaglia, A. Gozzini and G. Boudouris, Nuovo Cimento $\underline{69B}$, 121 (1970).

15. In the now standard definitions in the literature[7-10] the state equation involves the normalized input transmitted powers, i.e., in our notation the quantities aY and aX.

BISTABILITY AND PHASE TRANSITIONS OF NUCLEAR SPIN SYSTEMS

P. Bösiger*, E. Brun, and D. Meier

Institute of Physics
University of Zurich
CH-8001 Zurich, Switzerland

Abstract: A nuclear spin system inverted by dynamic nuclear polarization shows properties similar to a laser. In addition a NMR-laser in the radio frequency region can be driven by a phase-locked external RF field. Under such conditions, bistability and hysteresis typical to a first-order phase transition can be observed. Experimental and theoretical results concerning the steady state behavior of the ruby NMR-laser and the dynamics of its phase transitions are presented. In the non-inverted spin state, a nonlinear response to an external RF field has been observed. Conditions are given for which NMR-bistability (the analog of optical bistability) becomes feasible.

I. INTRODUCTION

After the realization of the solid-state spin-flip NMR-laser [1] in the radio-frequency range (which we now call *RASER* for short) it became obvious that nuclear spins, embedded in a crystalline matrix, are extremely suitable to study the dynamic aspects of cooperative ordering processes. In particular, phase transitions far from thermal equilibrium can be demonstrated with ease, and the temporal evolution of unstable behavior may be followed with high degree of accuracy and reproducibility. We have pointed out [2,3,4] that strongly polarized spin systems may be considered as simple test systems for cooperative self-ordering which is the main theme of *Synergetics* [5].

---

*Present address: Institut für Biomedizinische Technik der
 Universität und der ETH, CH-8044 Zürich.

In this paper we review the experimental results for the ruby
raser and discuss a Bloch equation based approach to radiant nuclear
spin order and the dynamics of its instabilities. In particular, we
would like to emphasize those features of non-linear nuclear spin
response which may be directly related to the problem of optical
bistability: to the $<J^2>$-breaking situation in the small volume
limit [6,7] and the low Q-cavity [8-12]. To be specific, we want
to demonstrate the behavior of absorptive bistability [13] for positively polarized spins as well as negatively polarized ones.

Let us consider a nuclear spin system inside of a tunable rf-
coil being placed in a strong magnetic field B_O. With spins precessing about B_O, there occur reaction effects due to the coupling
between the spin system and its own radiation field. If the rf-
structure supporting the resonance field has a large Q and if the
nuclear spin polarization is large enough, then these effects have
a dramatic influence upon the response of this highly non-linear
device. This has been pointed out by Bruce et al.[14], Bloembergen
and Pound [15], and in particular by Bloom [16] for a two-state
quantum system, capable of making radiative transitions which is
driven by a high-frequency field.

For nuclear spins in solids with their broad NMR-lines and Q-
values of the order of 100, typically, the non-linear reaction
effects are difficult to observe even at lowest laboratory temperatures. Using dynamic nuclear polarization (DNP) in order to pump
the spin systems to low positive or negative spin temperatures,
enhanced nuclear magnetization may be obtained and investigations of
such effects become feasible. With negatively polarized spins we
have reached the threshold for NMR-laser action of ^{27}AL in ruby.
Similarly, Derighetti and Marxer [17] have observed NMR-laser action
of the protons in LMN and of ^{19}F in $CaF_2:Gd^{3+}$ [18]. With the same
but positively pumped samples deviations from the linear response of
driven spins have been found. However, the threshold for bistable
behavior of such a saturable NMR-absorber has not yet been reached.

II. BLOCH TYPE BEHAVIOR

We consider a situation where the rf-coil of the tank circuit
is properly tuned to a selected $\Delta m = \pm 1$ NMR-transition of frequency
ω_O of a nuclear spin system. For spins with $I > 1/2$, the states
connected by the transition form in general a fictitious spin-1/2
two-level system as has been pointed out by Abragam [19].

We assume that the corresponding cw-line is homogeneously broad
ened due to spin-spin interactions with an unique dephasing time T_2.
This leads to an exponential decay of a freely precessing transverse
magnetization M_v. We further suppose that the nuclei are not only
coupled to the resonant rf-field in the coil but also to a reservoir

of spin energy which may be pumped either to a low negative or
positive spin temperature by means of DNP. The coupling of the
spins to the reservoir is assumed to be of relaxation type. Thus,
the spins relax with a characteristic pumping time T_e towards the
longitudinal pump magnetization M_e which represents the temperature
of the reservoir. M_e may be either positive or negative. T_e is
different from what generally is called the spin-lattice relaxation
time. As we have demonstrated [2], T_e may be appreciably shorter
than T_1. This fact has a marked influence on the dynamics of nuclear
spins and is of prime importance for the understanding of their
response.

Due to the low Q of the coil, the reaction field may be adiabatically eliminated leaving only Bloch type equations for the nuclear
magnetization. The components M_v and M_z then play the role of order
parameters for the radiant transverse and the non-radiant longitudinal spin order, respectively. The detailed dynamic order parameter equations may be found in our previous publications [2,3].

III. STEADY STATES IN SPIN-1/2 SYSTEMS

In the rotating frame (u,v,z), at exact resonance $(\omega = \omega_o)$,
the steady-state values of the order parameters fulfill the conditions:

$$M_z \omega_t = M_v/T_2 \tag{1}$$

$$M_v \omega_t = (M_e - M_z)/T_e \; . \tag{2}$$

Here, the fluctuations due to the noise in the coil have been neglected. ω_t is the total rf-field (in frequency units) acting on the
spins. ω_t is determined by the reaction field ω_r of the coil and a
possible superimposed driving field ω_d. For convenience, we choose
ω_d always to point along the positive u-axis. Equations (3) and (4)
are correct only when ω_r is either in phase or 180° out of phase with
ω_d. Then, we speak of a cooperative and a competitive field configuration, respectively. Under these circumstances the total field

$$\omega_t = \omega_d + \omega_r \tag{3}$$

falls into the u-axis but may be either positive or negative. This
way we avoid a dispersive response in the u,v-plane. Hence, the
transverse spin order can be treated as a one-dimensional problem.
Finally, the adiabatic elimination of the reaction field can be
accomplished by the condition

$$\omega_r = -\frac{1}{2}\,\eta\gamma Q M_v \quad , \tag{4}$$

where η is the filling factor of the rf-coil. To discuss the non-linear response due to the expressions (1) through (4) we set

$$M_K = \frac{2}{\mu_o\gamma\eta Q T_2} \tag{5}$$

and introduce the relative pump strength by defining the pump parameter

$$p = \frac{M_e}{M_K} \quad . \tag{6}$$

In passing, we note that M_K is always positive, M_e is either positive or negative. Now, we replace the total rf-field and the driving field by X and Y, respectively:

$$X = \omega_t\,\sqrt{T_e T_2} \quad , \tag{7}$$

$$Y = \omega_d\,\sqrt{T_e T_2} \quad . \tag{8}$$

Thus, we obtain the general steady state relation:

$$Y = X + \frac{pX}{1 + X^2} \quad . \tag{9}$$

For positive pump parameters, setting p = 2C, Eq. (9) is identical with the one derived by Bonifacio and Lugiato [13] for optical bistability of a saturable absorber inside a cavity. However, (9) is more general as it gives the steady state fields for inverted systems also, and this not only for the driven spin system but also for the free-running raser.

If the spin system is driven with an external field, we have bistability for p > 8 and p < -1. In the region p > 8 we have the analogue of optical bistability. For p < -1 we speak of raser (or laser) bistability. For -1 < p < 0 we have subcritical behavior of an inverted system, where critical narrowing (or slowing down) can be observed. In contrast, for 0 < p < 8 we have enhanced saturation with a corresponding reactive line broadening. The non-linear steady-state response in accord with Eq. (9) is given for different pump parameters p by the lines in Fig. 1. Those sections of the response curves with negative slope are unstable. They give rise to first order phase transitions with the well-known hysteresis effects.

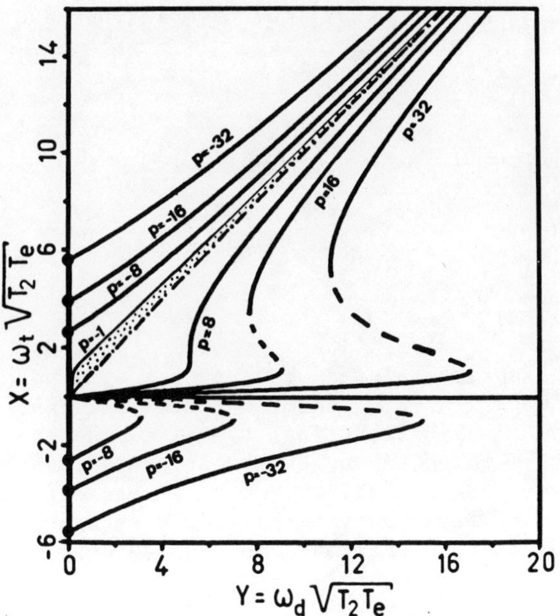

Fig. 1.   Total field versus driving field in steady-states of pumped
spin systems with pump parameter p.

To obtain the steady states of the free-running raser we set
$Y = 0$ and $X = X_r \neq 0$. From (9) we obtain

$$X_r = \sqrt{-1 - p} \ . \tag{10}$$

Hence, $p = -1$ or $M_e = -M_K$ is the threshold condition for raser activity. Since the raser output is proportional to X_r, we have the
mean field behavior $(-p - 1)^{\frac{1}{2}}$ of a Landau type second order phase
transitions with values as indicated by the dots in Fig. 1.

IV.   STEADY STATES IN RUBY

So far, we have considered the non-linear response of true
spin-1/2 particles. However, Eq. (9) holds also for fictitious spin-
1/2 two state systems when proper scaling of the physical parameters
is taken into account. This requires the computation of spin factors
which are necessary to relate the fictitious variables of (9) to the
real physical quantities. The spin factors depend on the actual
spin I and the magnetic quantum numbers of the Zeeman level partici-
pating in a radiative transition. For ^{27}Al in ruby ($Al_2O_3:Cr^{3+}$) with
$I = 5/2$, and the $(1/2, - 1/2)$-transition (central line of a

quadrupolar split NMR-spectrum) for example, we have to set

$$M_K = \frac{2}{9\mu_o \eta\gamma Q T_2}$$ (5')

$$X = 3\omega_t \sqrt{T_e T_2}$$ (7')

$$Y = 3\omega_d \sqrt{T_e T_2} \; .$$ (8')

in order to preserve Eq. (9).

With our experimental set-up we obtain spin polarizations within the limits $-2 < p_{exp} < +2$. Hence, critical line narrowing, free raser activity, raser bistability, reactive line broadening, and non-linear absorptive response may be investigated with relative ease, as one can infer from Fig. 1. However, we are off by a factor of 4 to observe bistability and hysteresis of a saturable NMR-absorber. Adding a Q-amplifier to the tank could bring us into the bistable region.

To illustrate critical narrowing in the response of a subcritically inverted spin system ($-1 < p < 0$), we have plotted in Fig. 2 the conventional 5-line NMR-absorption spectrum of ^{27}AL in ruby as a function of the inverse spin temperature (which is a direct measure of p). As we have shown [3] the linewidth or the inverse effective dephasing time $T_{2,eff}$ can be related directly to the pump parameter:

$$\frac{1}{T_{2,eff}} = \frac{1}{T_2} (1 + p) \; .$$ (11)

Since each of the 5 NMR-lines has its individual p-parameter (due to different spin factors and the non-linearity of the Boltzmann distribution function), the lines narrow differently when the common spin temperature is lowered as can be seen in Fig. 2.

To demonstrate the Landau behavior of the phase transition which leads to a spontaneous raser activity, we have plotted in Fig. 3 the steady state output voltage versus M_K while keeping M_e constant. The best fit is obtained with $\beta_{exp} = 0.56$. The discrepancy between theory ($\beta = 0.5$) and experiment is not yet fully understood. Presumably, it has its roots in the neglected inhomogeneous broadening mechanisms which contribute to the linewidth. This situation cannot be handled with simple Bloch type equations.

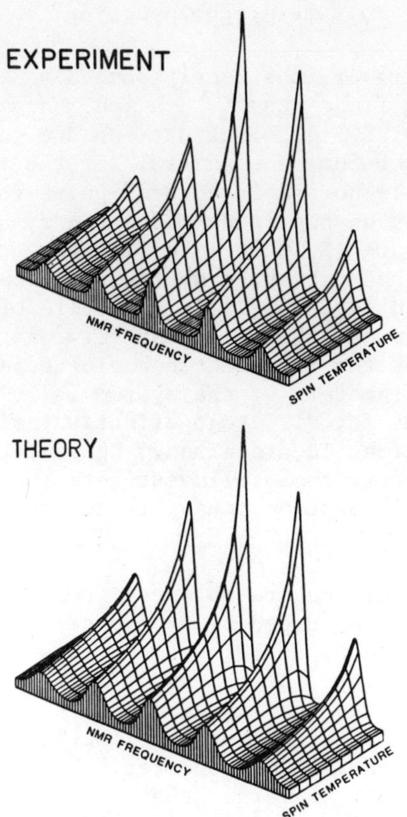

Fig. 2.   Critical narrowing (growing) of the 5 NMR-lines of ^{27}Al
          versus spin temperature for pump parameter $p \to -1$.

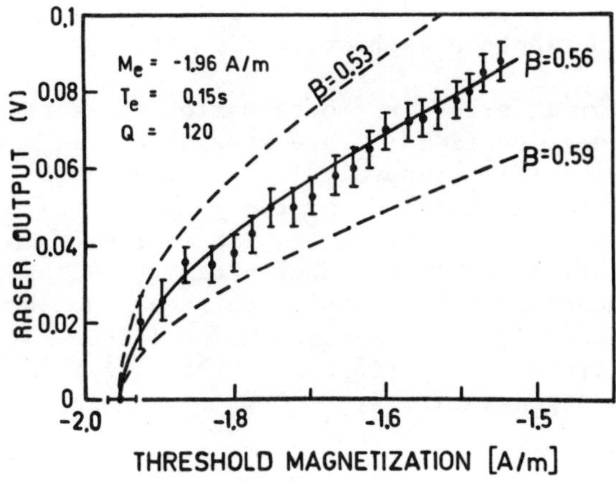

Fig. 3.   Raser output proportional to X_r for different $p < -1$.

V.   TRANSIENT BEHAVIOR

If a de-tuned raser is strongly pumped and subsequently tuned
with a Q-switch, it finds itself in a non-radiant unstable equilibrium (X = 0; Y = 0). Field fluctuations due to the noise current
in the coil are now becoming essential for the ignition. The raser
responds with a well-known delayed Sech-pulse followed after a
typical dead time by an amplitude-modulated relaxation oscillation
towards a steady state ($X_r \neq 0$). In Fig. 4 we compare the observed
experimental response with the computer solutions of the general
raser equation given in Ref. [2,3]. In spite of the simplicity of
the theoretical approach the agreement between theory and experiment
is gratifying. That the transient behavior depends strongly on
various physical parameters of the system is illustrated in Fig. 5.
Computer simulations for different effective noise strengths M_{vn} and
different pumping times T_e are given. We see that the transient
response is a sensitive tool to investigate the spin dynamics of
solids, namely the relaxation behavior and the response to external
fluctuations.

Remarkable effects can be observed when a free-running raser
(p < -1) is suddenly perturbed by a competitive field (Y antiparallel
X_r). If Y for a fixed p, is below the corresponding critical value
Y_c for raser bistability, then a damped amplitude-modulated oscillation leads from the free-running raser state (dot in Fig. 1) to a
new stable state (on the dot-crossing line in Fig. 1). This is
demonstrated experimentally and theoretically in Fig. 6a. If,
however, Y is chosen to be larger than Y_c then a growing amplitude-
modulated oscillation sets in which finally quenches the coherent
radiative state. There, the spins loose their phase memory in a
time of the order of T_2. Now, the driving field induces the
reorganization of spins such that a cooperative field configuration
(Y parallel X_r) results. A stable state can ultimately be reached.
The system thus has performed a first order phase transition. The
corresponding transients are shown in Fig. 6b.

Of particular interest is the situation where the driving field
Y > Y_c and the reaction field X_r are kept always in a competitive
field configuration (X_r antiparallel Y). This can be accomplished
by means of an electric feedback system. Again, the spin system
responds after suddenly turning on Y with a growing amplitude-
modulated oscillation until the raser activity ceases. From there
on, as a consequence of the interplay between pumping, cooperative
self-ordering, and driving field in a competitive state, the raser
emits a regular train of superradiant bursts. Figure 7 shows the
behavior of this regenerative system, both experimentally and
theoretically.

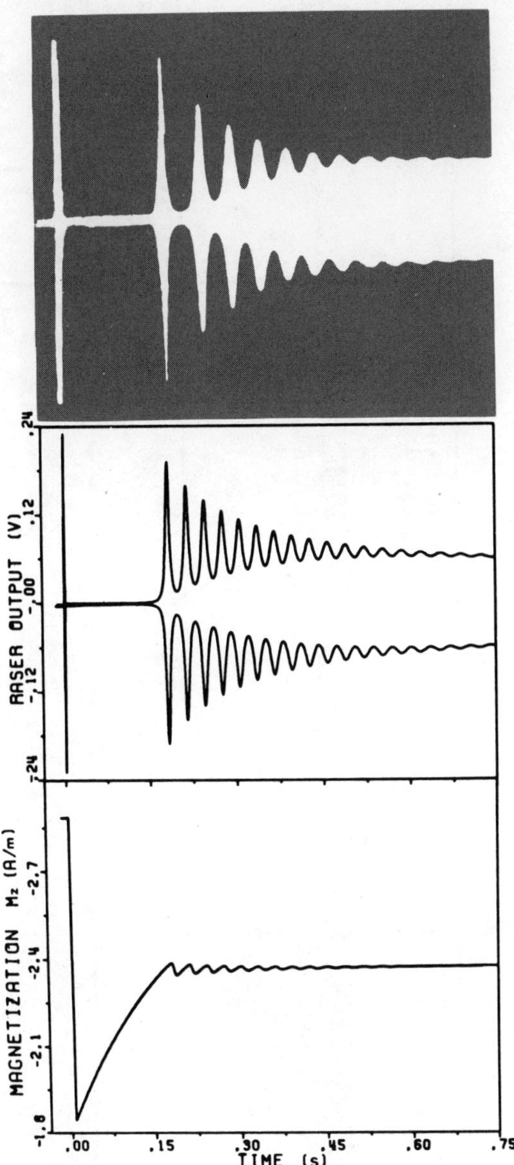

Fig. 4. Experimental and computed transients from $X = Y = 0$, $p < -1$ to $X_r \neq 0$; $Y = 0$.

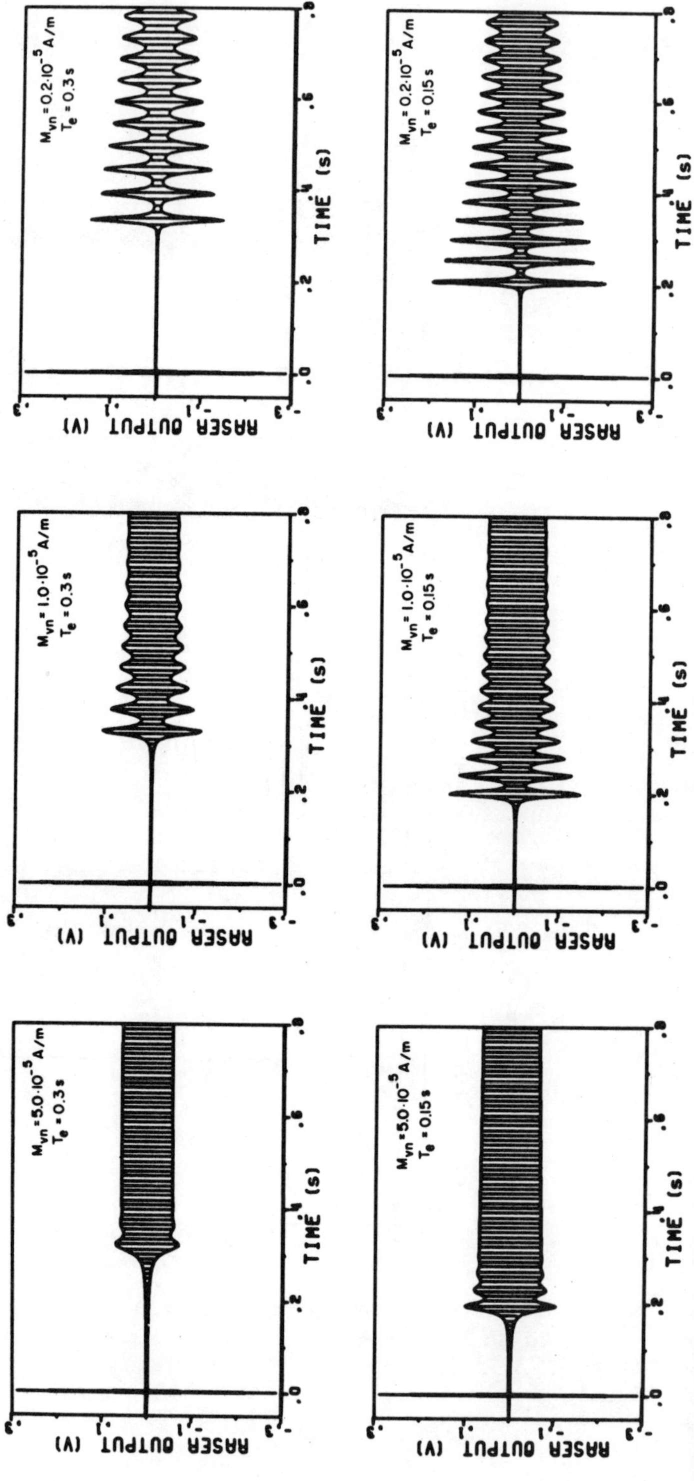

$M_{vn} = 5.0 \cdot 10^{-5}$ A/m
$T_e = 0.3$ s
RASER OUTPUT (V)
TIME (s)
$M_{vn} = 1.0 \cdot 10^{-5}$ A/m
$T_e = 0.3$ s
RASER OUTPUT (V)
TIME (s)
$M_{vn} = 0.2 \cdot 10^{-5}$ A/m
$T_e = 0.3$ s
RASER OUTPUT (V)
TIME (s)
$M_{vn} = 5.0 \cdot 10^{-5}$ A/m
$T_e = 0.15$ s
RASER OUTPUT (V)
TIME (s)
$M_{vn} = 1.0 \cdot 10^{-5}$ A/m
$T_e = 0.15$ s
RASER OUTPUT (V)
TIME (s)
$M_{vn} = 0.2 \cdot 10^{-5}$ A/m
$T_e = 0.15$ s
RASER OUTPUT (V)
TIME (s)

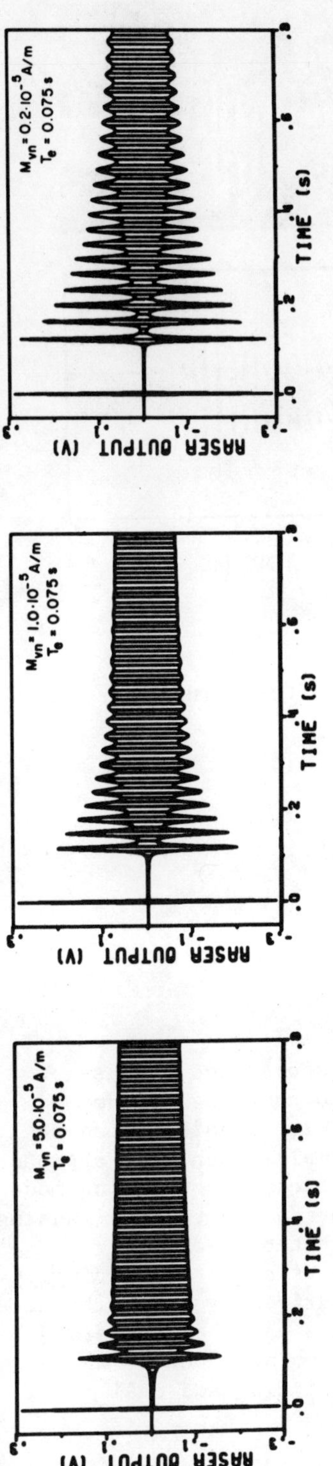

Fig. 5. Computed raser response after Q-switch procedure with different system parameters M_{vn} and T_e.

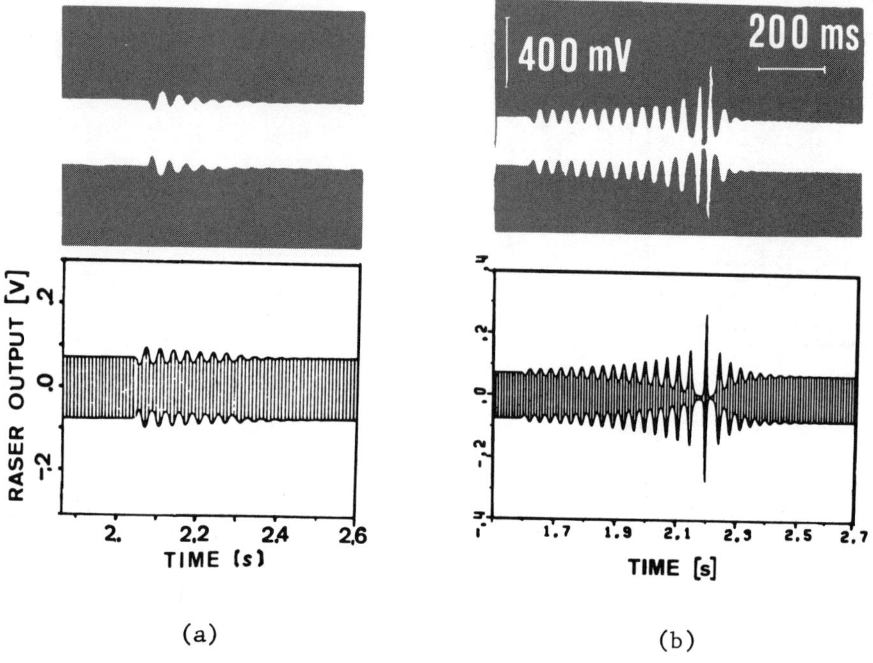

(a)   (b)

Fig. 6.   Response of the free-running raser after turning on Y for
          $Y < Y_c$ (a) and $Y > Y_c$ (b).

VI.   FINAL REMARKS

The material presented on non-linear response and on raser
bistability of pumped nuclear spins refers to the single-mode
behavior only. With strong inversion, however, transients can be
observed which cannot be explained with a simple Bloch type approach.
Multi-mode excitation leads to mode interference or to hopping modes
[2,4] as well as to single-mode pulse modulation due to spin fanning
[18]. These effects require a more sophisticated treatment.
Further, in de-tuned systems dispersion becomes essential giving
rise to complicated transient patterns. In driven systems without
phase locking of ω_d and ω_r interesting effects have been observed
which may be due to phase drifts as has been pointed out by Walls
[20]. The experimental observations are intriguing and call for a
systematic investigation of these exciting new effects.

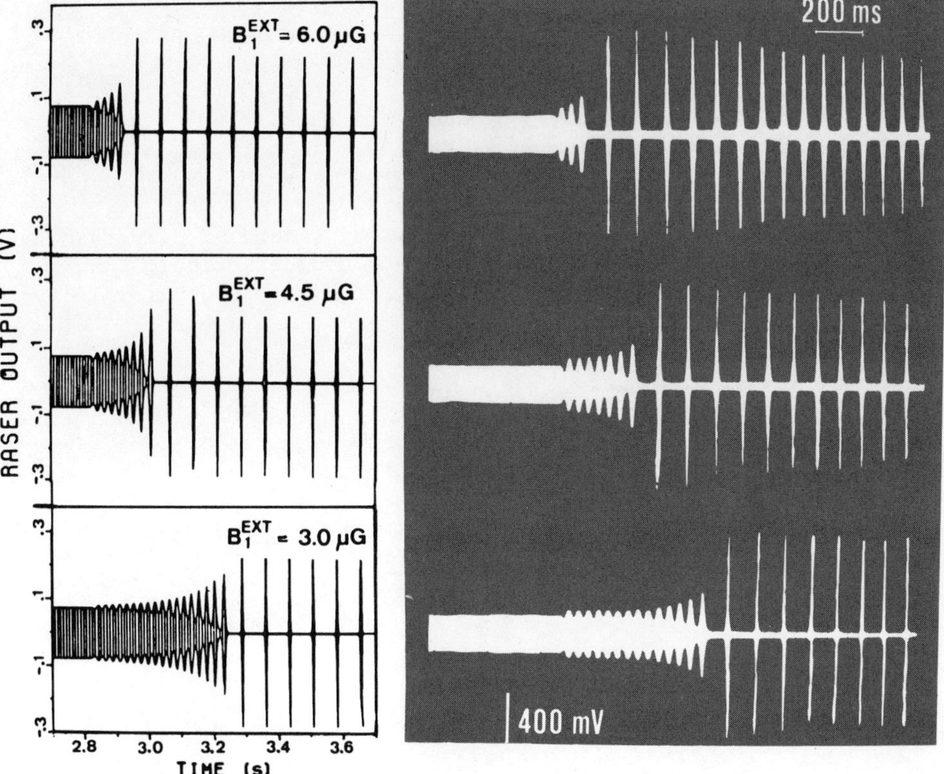

Fig. 7.   Response of the free-running raser after turning on $Y > Y_c$ keeping Y antiparallel X_r by electronic means.

REFERENCES

1. P. Bösiger, E. Brun, and D. Meier, Phys. Rev. Lett. __38__ (1977) 602.
2. P. Bösiger, E. Brun, and D. Meier, Phys. Rev. __A18__ (1978) 671.
3. P. Bösiger, E. Brun, and D. Meier, Phys. Rev. __A20__ (1979) 1073.
4. In *Dynamics of Synergetic Systems*, ed. H. Haken, Springer, 1980.
5. H. Haken, *Synergetics*, Berlin, Springer, 1977.
6. H. J. Carmichael and D. F. Walls, J. Phys. __10B__ (1977) L685.
7. S. S. Hassan and D. F. Walls, J. Phys. __11A__ (1978) L87.
8. G. S. Agarwal, L. M. Narducci, R. Gilmore, and D. H. Feng, Phys. Rev. __A18__ (1978) 620.
9. L. M. Narducci, R. Gilmore, D. H. Feng and G. S. Agarwal, Optics Lett. __2__ (1978) 88.
10. R. Bonifacio, M. Gronchi, and L. A. Lugiato, Phys. Rev. __A18__ (1978) 1129.

11. R. Bonifacio, M. Gronchi, and L. A. Lugiato, Phys. Rev. A18
     (1978) 2266.
12. L. A. Lugiato, Nuovo Cim. 50B (1979) 89.
13. R. Bonifacio and L. A. Lugiato, Opt. Comm. 19 (1976) 172.
14. C. R. Bruce, R. E. Norberg, and G. E. Pake, Phys. Rev. 104
     (1956) 419.
15. N. Bloembergen and R. V. Pound, Phys. Rev. 95 (1954) 8.
16. S. Bloom, J. Appl. Phys. 28 (1957) 800.
17. B. Derighetti and H. Marxer, Helv. Phys. Acta 52, (1979) 374.
18. B. Derighetti and H. Marxer, private communication.
19. A. Abragam, *The Principles of Nuclear Magnetism*, Oxford,
     Clarendon Press, 1961, p. 36.
20. D. Walls, private communication.

TRANSIENT PHENOMENA IN BISTABLE OPTICAL DEVICES

J. A. Goldstone, P.-T. Ho, and E. Garmire

Center for Laser Studies
University of Southern California
University Park
Los Angeles, California 90007

Abstract: We demonstrate both experimentally and theoretically
a new mode of transient switching using rapid onset light signals in
a bistable optical device with two comparable time constants. Four
distinct transmission modes for input pulse trains have been observed,
one of which produces a subharmonic of the input pulse train by
alternate overshoot switching.

I.   INTRODUCTION

There has been considerable recent interest in bistable optical
devices (BOD's), nonlinear optical devices whose output may be multi-
valued with respect to input. This phenomenon is typically obtained
with a nonlinear Fabry-Perot[1] or with a hybrid electrical/optical
device.[2,3] The steady state characteristics of these devices are now
fairly well understood and recent emphasis has been directed toward
the transient response of these devices.[4,5] In this paper we turn
our attention to BOD's which have two comparable time constants. We
will show that it is possible for a BOD to switch at incident light
levels considerably smaller than the critical incident signals in
steady state operation. This effect occurs for rapid onset input
signals when the device has two time constants of roughly equal value.
The discovery of overshoot switching will have a significant impact
on BOD's whose feedback time constants approach their internal time
constants. Both hybrid BOD's and nonlinear Fabry-Perots may show
overshoot switching. In this paper we emphasize the hybrid BOD[3],
since this device is most easily amenable to experimental verification. We will show that the theory and experiment are in excellent
numerical agreement. Similar analyses of other hybrid configurations

and of the nonlinear Fabry-Perot lead to similar results.[6]

When a train of pulses is incident on a BOD, there is a regime
of operation in which the device switches alternately yielding an
output which is a subharmonic of the input pulse train. We demonstrate this effect both experimentally and theoretically in hybrid
devices. In addition, we predict its existence in the all-optical
BOD's such as the nonlinear Fabry-Perot.

<h2 style="text-align:center">II.   OVERSHOOT SWITCHING</h2>

The hybrid BOD used for our theoretical analysis and experimental studies is the same as that used previously.[5] It consists of
a LiNbO$_3$ polarization modulator driven by the output of a photoconductor which monitors the transmission of the BOD. The time
constants in the device come from the rise-time of the photoconductor
and from a variable capacitance placed across the biasing resistor.
Figure 1 shows the steady-state transfer curve of the experimental
BOD used to demonstrate overshoot switching. The critical incident
light level required for switching to the high transmission state
under steady-state operation is indicated by I_c. I_s represents the
experimentally measured critical incident signal for switching with
a rapid onset input. The fact that this incident intensity is considerably below the steady-state value is the result of overshoot
switching.

It can be shown that the dynamic response of the hybrid BOD
with two time constants may in general be represented by an equation
for the voltage across the modulator,[5]

$$m \frac{d^2 v}{dt^2} = T(v)\ I_i\ -\ v\ -\ \alpha\ \frac{dv}{dt} \tag{1}$$

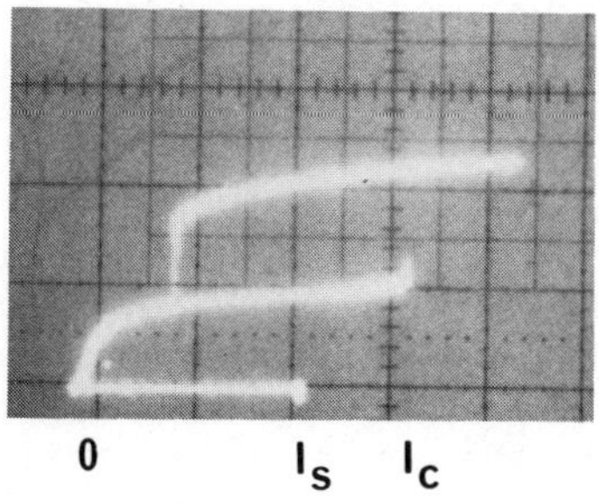

Fig. 1.   Voltage across the modulator of the BOD (vertical) vs.
incident light intensity (horizontal). I_c is the steady-
state switching intensity, I_s is the minimum overshoot
switching intensity.

and an equation for the transmitted light intensity,

$$I_t = T(v)\, I_i \tag{2}$$

where $m = \tau_1 \tau_2$, $\alpha = \tau_1 + \tau_2$; I_i and I_t are the incident and transmitted light intensities, respectively, measured in voltage units
and equal to the photoconductor voltages produced by the incident
and transmitted intensities; v is the instantaneous voltage across
the modulator; τ_1 and τ_2 are the two device time constants; and $T(v)$
is the transmission of the modulator as a function of the applied
voltage.

For step function inputs, $I_i = I_o$ for $t \geq t_o$, Eq. (1) may be
written as

$$m\,\frac{d^2v}{dt^2} = -\,\frac{\partial U(v)}{\partial v} - \alpha\,\frac{dv}{dt} \tag{3}$$

where

$$U(v) = -\int^v \{T(v')\, I_o - v'\}\, dv' \tag{4}$$

represents a conservative potential. In this form, Eq. (3) represents
the motion of a particle of "mass" m and "position" v moving in a one
dimensional potential $U(v)$ with an additional friction term $-\,\alpha dv/dt$.

Figure 2 shows $U(v)$ for various input intensities and for the
transmission function $T(v) = \frac{1}{2}[1-0.7 \cos\{\pi(v + 6.8)/30\}]$, which
closely matches our experimental modulator. Points of stable and
unstable equilibrium can be seen immediately in the potential diagram.
Steady-state operation occurs at stable equilibrium points, and the
existence of more than one stable point represents the possibility
of multistable operation. The final steady-state output depends on
the initial output, input step intensity, BOD characteristics, and
on the ratio of "mass" to "friction coefficient", $m/\alpha = \tau_1\tau_2/(\tau_1 + \tau_2)$.

The time constants τ_1 and τ_2 appear symmetrically in Eq. (1).
We define $\tau_1 < \tau_2$. As $\tau_1/\tau_2 \to 1$, m/α increases and the system becomes
underdamped. Thus, voltage overshoots occur. Critical slowing down,
demonstrated previously,[5] occurs when the voltage overshoots a point
of stable equilibrium and approaches the adjacent point of unstable
equilibrium with zero velocity. Overshoot switching occurs when the
voltage reaches an unstable equilibrium point with nonzero velocity
and continues to the next stable point. The analog in the nonlinear
Fabry-Perot is the overshoot of the nonlinear index.

Theoretical analysis shows that hybrid overshoot switching for
rapid onset signals can reduce the input intensity necessary to reach
the high transmission state by nearly 50% in some cases. Figure 1
shows that for our experimental setup switching occurred for a rapid
onset signal at an input intensity 32% smaller than that required
for steady-state operation. The difference between steady-state
critical input intensity and that obtained by overshoot switching
plotted as a function of τ_1/τ_2, is shown in Fig. 3. The theoretical
curve was obtained from solutions of Eq. (3). The experimental points
were obtained by varying an adjustable capacitance C_2 across the bias
resistor, R_2, with BOD time constants τ_1 and τ_2 determined by comparing Eq. (1) and the hybrid circuit equation. The photoconductor
and feedback circuits had resistances and capacitances R_1C_1 ($\sim$5 msec)
and R_2C_2 respectively. For this hybrid BOD, $\tau_1\tau_2 = R_1C_1R_2C_2$ and
$\tau_1 + \tau_2 = R_1C_1 + R_2C_2 + R_2C_1$. The experimental values fall on the
theoretical curve within experimental error (20%), with no adjustable or normalizing parameters. It was not possible in this experiment to obtain larger values of τ_1/τ_2 because a smaller R_2 did not
provide sufficient bias for switching. Nonetheless, the experiment
provides quantitative confirmation of the theory.

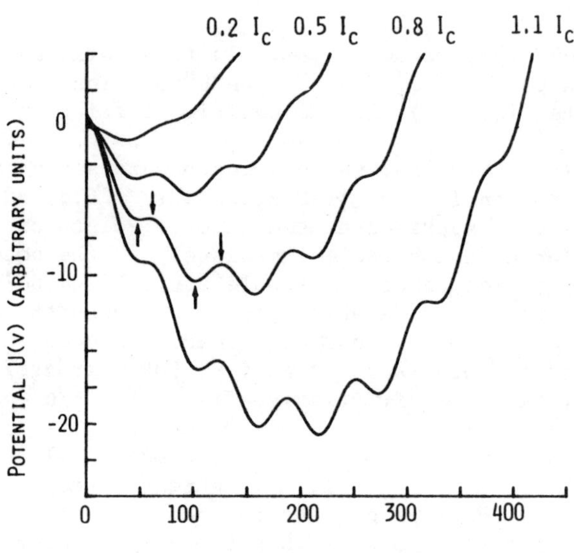

Fig. 2. Effective potential wells used to describe the time dependence of the voltage across the modulator for various input
intensities. Upward arrows indicate stable equilibrium
points for steady-state intensities. Downward arrows indicate unstable equilibrium points.

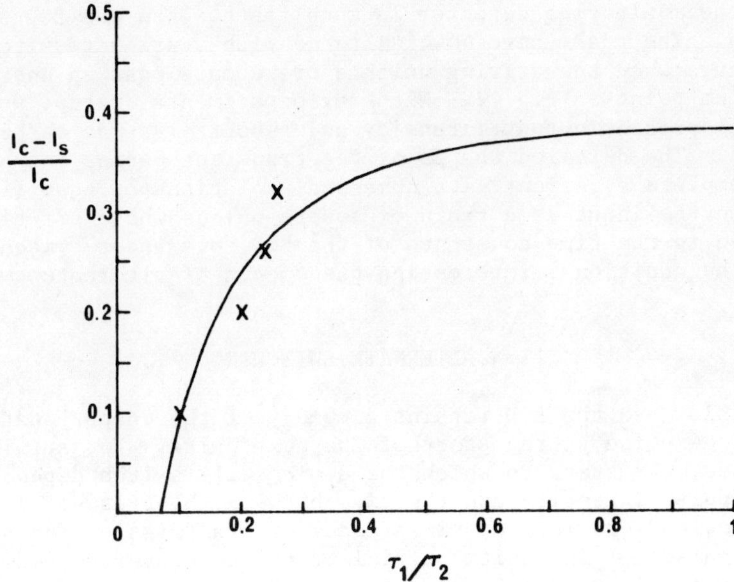

Fig. 3. Overshoot switching as a function of the ratio of the system
         time constants. The vertical axis represents the fractional
         difference between steady-state and overshoot switching
         intensities.

The final output intensity of the BOD depends not only on τ_1/τ_2
and $T(v)$, but also on the rise-time of the input signal. If the rise
time is slow enough, the device will stop at the first stable equilibrium point. As the input rise-time decreases, the device may overshoot switch to the second stable voltage point. For a given ratio
τ_1/τ_2 there is a critical input rise-time below which the device will
overshoot switch to the second (high transmission) output and above
which the device will not. This critical rise-time can be determined
from numerical integrations of Eqs. (1) and (2), which apply for time
dependent inputs.

The rise-time effect was observed by chopping the incident beam
with a rotating blade. The transit time of the blade edge across the
incident beam cross-section determined the effective rise-time of the
incident signal. Overshoot switching to the high transmission mode
was observed at high chopping rates. As the motor speed was
decreased, the onset time of the incident beam was effectively
increased and the device transmitted in the low mode for the same
input intensity. These results are shown in Fig. 4, where the initial
peaks in the output light intensity are due to tracing out peaks of
the transmission curve, as indicated by Eq. (2). These peaks are commonly seen in all BOD's,[2,4] and are observed only when the response

time of the monitoring detector is much faster than the BOD time
constants. The peaks have nothing to do with overshoot switching,
which occurs when the driving voltage overshoots past an unstable
equilibrium point in Fig. 2. With our experimental setup, voltage
overshoots produce output intensity undershoots, as may be observed
in Fig. 4. The detailed shapes of the transient output intensities
are in complete agreement with numerical solutions of Eqs. (1) and
(2). When the input is a train of square pulses whose off-time is
comparable to the time constants of the BOD, overshoot switching may
lead to the additional interesting phenomenon of alternate switching.

III. ALTERNATE SWITCHING

In this case the BOD retains a memory of the output state of
the previous pulse at the start of the next pulse. For sufficiently
long pulses, the state to which the device will switch depends on
both the pulse intensity and the time between the pulses. This
dependence leads to four distinct modes of transmission for a train
of input pulses of intensity sufficient to cause overshoot switching
but less than the critical value for steady-state switching.

(a)   (b)

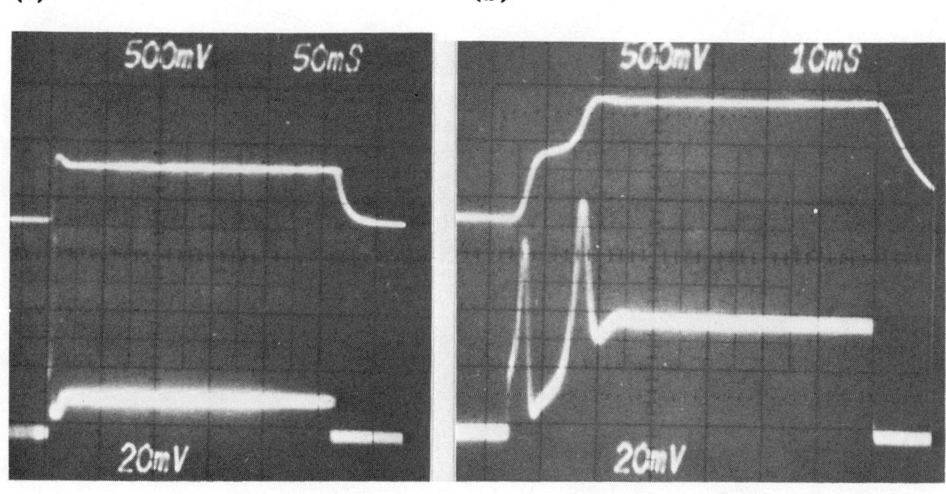

Fig. 4.   Transient response of BOD to a square pulse of incident
          light. Upper traces, voltage across the modulator; lower
          traces, transmitted light. a) Horizontal scale, 50 ms/div;
          input rise time $\sim$1.4 ms. b) Horizontal scale, 10 ms/div;
          input rise time $\sim$0.3 ms. Except for the chopper rate, all
          experimental conditions were identical in both cases. Note
          that the BOD did not switch in a), but did switch in b).

We have demonstrated these new effects with the same experimental hybrid BOD. The input light beam was chopped by a motor-driven rotating blade, producing a train of square pulses whose input intensity was chosen to be below the steady-state level required to switch the BOD but above that required for overshoot switching. The speed of the motor was varied, holding all other parameters constant. Four distinct transmission modes corresponding to four ranges of chop frequencies were observed, as shown in Fig. 5. For the slowest chop rates the device transmitted in the high mode for all input pulses (Fig. 5a). As the chop frequency increased, the device alternated between high and low transmission modes for successive input pulses (Fig. 5b). This alternating behavior occurs even though the incident intensity goes to zero between the pulses, and the alternation produces a subharmonic of the incident pulse train. Increasing the chop rate beyond this range produced low mode transmission for all input pulses (Fig. 5c). The fastest chop rates produced high transmission of all input pulses (Fig. 5d).

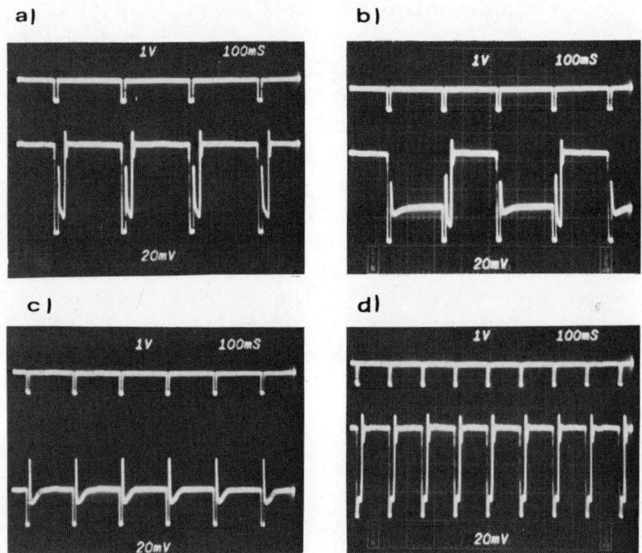

Fig. 5. Four transmission modes of a BOD occurring from overshoot switching in response to an input of square pulse train. Upper traces: input light intensity. Lower traces: transmitted light intensity. Horizontal scale = 100 ms/div. (a) All input pulses transmitted in high mode. Input pulse train off-time τ_{off} ~16 ms. (b) Alternate input pulses transmitted in high mode. τ_{off} ~12.5 ms. (c) Input light pulse transmitted in low mode τ_{off} ~10 ms. (d) Input pulses transmitted in high mode. τ_{off} ~7.0 ms. Except for the input pulse chop rate, experimental conditions were identical in all 4 cases.

It can be seen in Fig. 5 that the BOD light output displays rapid variations at the beginning of each pulse, before settling into its equilibrium value. These transients result from the modulator tracing out its transmission curve as the voltage rapidly increases, and are not related to overshoot switching.

The four transmission modes may be understood within the theory of overshoot switching using the model of a particle moving with friction in a potential well.

In addition to Eqs. (2-4), we include the fact that when the input is shut off, $I_o = 0$, Eq. (1) reduces to that of a linear over-damped oscillator,

$$m \frac{d^2 v}{dt^2} + \alpha \frac{dv}{dt} + v = 0 \quad . \tag{5}$$

Figure 6 shows the potential well $U(v)$, for the experimental BOD, which applies during a square pulse of intensity I_o, chosen to be in the regime of overshoot switching. Characteristic voltages which are used to describe the behavior of the device are v_B and v_D, the low and high steady-state transmission voltages respectively, v_C, the point of unstable equilibrium where overshoot switching and critical slowing down occur, and v_A, the maximum starting voltage for overshoot switching to the high transmission state. For square pulses Eqs. (2-4) apply during each pulse. Between pulses the voltage across the modulator decays with time according to Eq. (5) (Fig. 7). The off-time between square pulses is the crucial factor in determining the transmission response of the device. Four regions of off-time which correspond to the four transmission modes shown in Fig. 5 can be identified.

For the longest off-times (slowest chop rates), the pulses act nearly independently, and each successive pulse causes overshoot switching into the high transmission state. The condition that each pulse cause overshoot switching is that the off-time be sufficient for v_D to decay to below v_A (T_3 in Fig. 7). This transmission mode in which all pulses switch into the high state is shown in Fig. 5(a).

If the off-time lies between T_2 and T_3 (Fig. 7), the device over shoot switches to v_D on the first pulse, but during the off-time the voltage decays to some value such that $v_C > v > v_A$ at the start of the second pulse. Since $v > v_A$, the device cannot overshoot switch and settles into the low transmission state v_B during that pulse. During the next off-time, the voltage decays from v_B to a value below v_A (Fig. 7) thereby making it possible for the device to overshoot switch to the high state on the third input pulse. This alternation of high and low transmission continues ad infinitum, and a second subharmonic of the incident pulse train is produced (Fig. 5b).

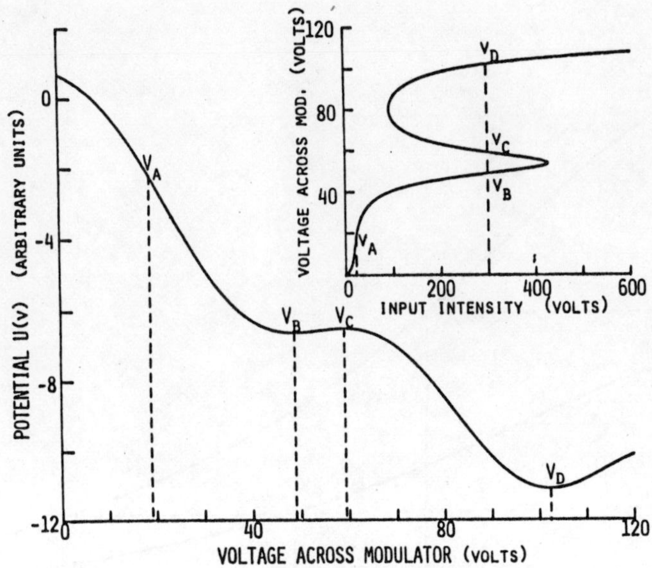

Fig. 6.   Conservative potential well for the experimental BOD; v_B and
          v_D are the low and high steady state voltages, v_A the maximum
          starting voltage for overshoot switching, v_C the unstable
          equilibrium voltage. Inset shows comparison with steady-
          state transfer curve.

Consider the case when the time between pulses is further decreased. As before, the first input pulse causes overshoot switching
to the high mode, but in this case the voltage decays during the off
time so that at the start of the second pulse, $v_C > v > v_A$, and the
second pulse is transmitted in the low state. Referring to Fig. 7,
for $T_1 < T < T_2$, the voltage decay from the low state voltage v_B does
not drop below v_A before the start of the next pulse, so overshoot
switching cannot take place and all succeeding pulses are also transmitted in the low state (Fig. 5c).

The fourth distinct region in off-times occurs for $T < T_1$: This
region is unique, since it has two possible modes of operation depending on the history of the device. If $T_1 < T < T_2$, and the train of
pulses is being transmitted in the low mode, increasing the chop rate
so that $T < T_1$ will not allow the voltage to decay below v_A, for
example by interrupting the beam for a time longer than T_3, then the
first pulse causes overshoot switching as usual into the high state.
Since the time between pulses is now insufficient for the voltage to
decay below v_C, all pulses are transmitted in the high state. (Fig.
5d).

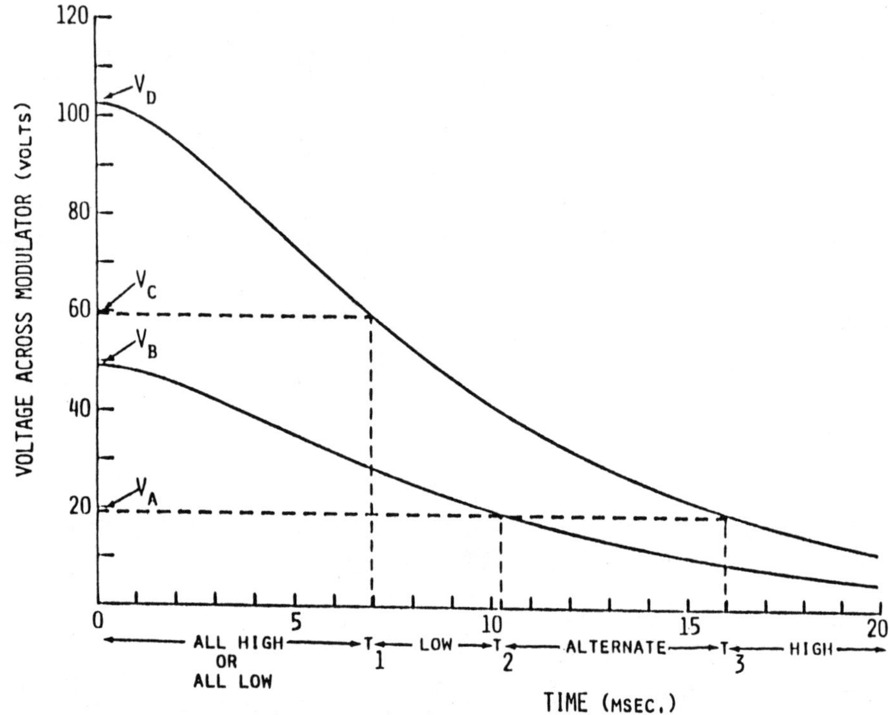

Fig. 7.   Zero input time decays which define the four transmission
mode regions. Times T_1, T_2, and T_3 correspond to the critical times between transmission mode regions.

Comparison of the theoretical and experimental ranges of off-
times which produce the four modes of transmission can be made directly, since the theoretical predictions are based on a knowledge of
R_1C_1 and R_2C_2, easily measureable electrical quantities. The experimental ranges of off-times for the four modes of transmission were
within 15% of the theoretical predictions, easily within experimental
uncertainties.

Although, the theoretical discussion and experimental results on
alternate switching and subharmonic pulse generation presented in this
paper refer to the hybrid BOD, we predict the same behavior for the
all-optical BOD's such as the nonlinear Fabry-Perot. We have performed numerical calculations on a theoretical model of the nonlinear
Fabry-Perot and observed the same overshoot and alternate switching.
This will be the subject of a forth-coming paper.

IV.   CONCLUSION

In conclusion, we have shown that when a BOD has two comparable
time constants it is possible to cause switching with input pulses
whose intensity level is up to 50% less than that required for switching in the steady-state. In addition, since the BOD retains a memory
of its output state for a time comparable to the time constants, it
is possible to obtain four distinct modes of transmission for rapid
onset input pulse trains depending on the amount of off-time between
pulses. The most dramatic of these is an alternating of the transmission between the high and low states, with the output going to
zero between each of these states. This alternate switching can be
used to generate subharmonics. The four distinct transmission modes
for input pulse trains discussed above open up new possibilities in
optical signal processing and for logic operations utilizing both
the hybric BOD and the nonlinear Fabry-Perot.

This work was supported by AFOSR Grant #78-3686.

REFERENCES

1.   H. M. Gibbs, S. L. McCall, and T. N. C. Venkatesan, Phys. Rev.
     Lett. 36, 1135 (1976); T. N. C. Venkatesan and S. L. McCall,
     Appl. Phys. Lett. 30, 282 (1977); H. M. Gibbs, S. L. McCall,
     T. N. C. Venkatesan, A. C. Gossard, A. Passner, and W.
     Weigmann, Appl. Phys. Lett. 35, 658 (1979).
2.   P. W. Smith, I. P. Kaminow, P. J. Maloney, and L. W. Stutz, Appl.
     Phys. Lett. 34, 24 (1979); P. S. Cross, R. V. Schmidt, R. L.
     Thornton, and P. W. Smith, IEEE J. Quant. Electr. QE-14, 577
     (1978).
3.   E. Garmire, J. H. Marburger, and S. D. Allen, Appl. Phys. Lett.
     32, 320 (1978); E. Garmire, S. D. Allen, J. H. Marburger, and
     C. M. Verber, Opt. Lett. 3, 69 (1978).
4.   T. Bischofberger and Y. R. Shen, Phys. Rev. A 19, 1169 (1979).
5.   E. Garmire, J. H. Marburger, S. D. Allen and H. G. Winful, Appl.
     Phys. Lett. 34, 374 (1979).
6.   F. A. Hopf, P. Meystre, P. D. Drummond, and D. F. Walls, Opt.
     Commun. 31, 101 (1979); J. A. Goldstone, and E. Garmire,
     unpublished.

ANALYTIC DESCRIPTION OF MULTIPHOTON OPTICAL BISTABILITY IN A

RING CAVITY

J. A. Hermann

Optics Section Department of Physics
Imperial College, London SW7 2AZ, U.K.

B. V. Thompson

Mathematics Department, University of Manchester
Institute of Science and Technology, P. O. Box 88
Manchester M60, 1QD, U.K.

Abstract: A theory of two-photon optical bistability with
two distinct fields is described, within the context of a ring
cavity geometry. A number of specific physical features are considered in the analysis, in particular the incorporation of:
(a) Phase and Stark effects; (b) Spatial effects; (c) Cavity and
Atomic detunings; (d) Ionization from the upper excited level.
Calculations are performed in the slowly-varying envelope approximation. Some of the results are extended to describe multiphoton
interactions more generally.

I.   INTRODUCTION

The relatively recently-investigated optical phenomena of
resonant (or near-resonant) multiphoton processes, and also of
optical bistability (OB) with associated hysteresis, have provided
fruitful areas for the development and elaboration of theoretical
models. Particularly notable and instructive in the multiphoton
category are the works of Takatsuji[1], Brewer and Hahn[2], Grischkowsky
et al.[3], and of Elgin, New and Orkney[4], which adopt and explore
the density-matrix formalism; Wilson-Gordon and Friedmann[5] have
alternatively used a projection-operator formalism, while Narducci
et al.[6] utilize the probability-amplitude approach. The original
theoretical papers of Bonifacio and Lugiato[7,8] on absorptive OB

were soon followed by a flood of papers incorporating atomic detunings (dispersive effect)[9-13], as well as spatial and propagation
effects[17-29], the inclusion of higher harmonic components of the
polarization where standing waves were present[20-22,28,29] cavity
detuning[13-16,21-29], inhomogeneous (Doppler) dephasing[12,28,29], and
ionization losses[30].

Almost all investigations of OB to date have been concerned
with two-level single field ("one-photon") processes. For various
reasons to be elaborated, a marriage between the above two phenomena
might prove very useful. Arecchi and Politi[31], motivated by the
possibility of examining Doppler-free two-photon OB, have investigated the degenerate case (one distinct field only) for a ring
cavity. A schematic diagram of such a geometry is shown in Fig. 1.
Very recently, Agrawal and Flytzanis[32] have studied the resonant
absorption which may be induced by two distinct and appropriately
tuned beams of radiation in a Fabry-Perot interferometer. We will
develop and extend this early work in various directions, based
upon analogies to be found with the single-photon theory. One
valuable advantage of using two distinct fields to resonantly
couple the end states via an intermediate level (or levels), is
the potential for controlling switching intensities of one of the
fields by varying the other, as we will show.

The theory of two-photon optical bistability is to be developed
for an absorptive medium in which dispersive effects, cavity detunings, and spatial effects are taken into account. Some mean-
field results will be generalized to the case where an arbitrary
number of photons are involved in the net resonance.

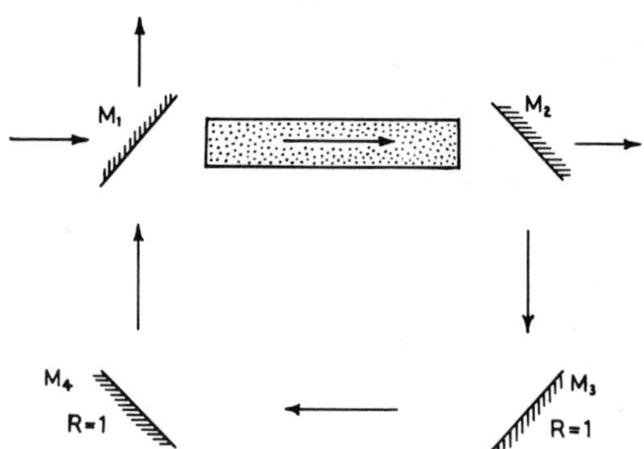

Fig. 1.   The Ring-Cavity Arrangement.

II.   THE EQUATIONS OF MOTION FOR TWO-PHOTON ABSORPTION

We consider firstly the equations of motion for a resonant absorption process in which two end states $|a\rangle$ and $|b\rangle$ of like parity are effectively coupled by two-photon processes involving a set of intermediate states of opposite parity to $|a\rangle$, $|b\rangle$. The approach of Narducci et al.[6] may be employed, using two distinct field amplitudes. All terms which do not oscillate at the frequencies of the two input fields are discarded when calculating the polarization density envelope, so that the type of nonlinearity we consider is limited.

The complex field equations are:

$$\frac{\partial E_1}{\partial z} + \frac{\eta_1}{c}\frac{\partial E_1}{\partial t} = -\alpha\{P^*E_2^* - iE_1(\hat{S}_1 D + \hat{S}_1' N) + \gamma_1 E_1(D-N)\} \qquad (1a)$$

$$\frac{\partial E_2}{\partial z} + \frac{\eta_2}{c}\frac{\partial E_2}{\partial t} = -\alpha\{P^*E_1^* - iE_2(\hat{S}_2 D + \hat{S}_2' N) + \gamma_2 E_2(D-N)\} \qquad (1b)$$

while the matter equations can be expressed as:

$$\frac{\partial P}{\partial t} = \lambda E_1^* E_2^* D + iP\Omega - \Gamma_\perp P \qquad (1c)$$

$$\frac{\partial D}{\partial t} = -\tfrac{1}{2}\lambda(E_1 E_2 P + E_1^* E_2^* P^*) - \Gamma_\parallel (D-N) \qquad (1d)$$

$$\frac{\partial N}{\partial t} = \tfrac{1}{2}[\gamma_1|E_1|^2 + \gamma_2|E_2|^2](D-N) \qquad (1e)$$

where z,t are the laboratory-frame space and time coordinates; $\omega_1^{\frac{1}{2}}E_1, \omega_2^{\frac{1}{2}}E_2$ are envelopes of the complex electric fields, ω_1 and ω_2 being their respective frequencies; P is the complex polarization envelope; N is essentially the number density of active atoms (in the absence of atomic relaxation we find that $D^2 + |P|^2 - N^2 = 0$, even when N is not constant). The incoherent loss parameters are defined as

$$\Gamma_\perp = \gamma_\perp + \tfrac{1}{2}[\gamma_1|E_1|^2 + \gamma_2|E_2|^2]; \quad \Gamma_\parallel = \gamma_\parallel + \tfrac{1}{2}[\gamma_1|E_1|^2 + \gamma_2|E_2|^2] \qquad (2)$$

where $\gamma_\perp, \gamma_\parallel$ are the transverse and longitudinal relaxation rates, while $\gamma_{1,2}$ are ionization rates. The other quantities are

$$\alpha = (2\varepsilon_o c)^{-1} \cdot \lambda \qquad (3)$$

where ε_o is the permittivity of free space (MKS units are used throughout) and c is the velocity of light; also $\lambda = (\omega_1\omega_2)^{\frac{1}{2}} K_{ab}$ where

$$K_{ab} = \hbar^{-1} \sum_j \mu_{aj} \mu_{jb} [(\omega_{ja} - \omega_1)^{-1} + (\omega_{ja} - \omega_2)^{-1}] \qquad (4)$$

is the compound matrix element for the near-resonant two-photon transitions[1-6], while the atomic detuning and Stark shifts are represented by the parameter

$$\Omega = \Delta + (2\hbar)^{-1} \lambda [(\omega_1/\omega_2)^{\frac{1}{2}} S_1 |E_1|^2 + (\omega_2/\omega_1)^{\frac{1}{2}} S_2 |E_2|^2] \; ; \qquad (5)$$

$$S_k = \frac{1}{2} K_{ab}^{-1}[K_{aa}(\omega_k) - K_{bb}(\omega_k)], \quad S_k' = \frac{1}{2} K_{ab}^{-1}[K_{aa}(\omega_k) + K_{bb}(\omega_k)],$$
$$k = 1,2; \qquad (6a)$$

with $\hat{S}_k = (\omega_k/\omega_\ell)^{\frac{1}{2}} - S_k$, $\hat{S}_k' = (\omega_k/\omega_\ell)^{\frac{1}{2}} - S_k'$, $\ell(\neq k) = 1,2;$ $\qquad (6b)$

$$K_{aa}(\omega_k) = 2\hbar^{-1} \sum_j |\mu_{aj}|^2 \omega_{ja} / (\omega_{ja}^2 - \omega_k^2) \; , \qquad (7a)$$

$$K_{bb}(\omega_k) = 2\hbar^{-1} \sum_j |\mu_{bj}|^2 \omega_{jb} / (\omega_{jb}^2 - \omega_k^2) \qquad (7b)$$

being the real parts of the linear susceptibilities for the a and b levels with frequencies ω_k. The η_i are linear refractive indices.

The equations (1a)-(1e) are a natural extension of those given by Takatsuji[1], Brewer and Hahn[2], Elgin, New, and Orkney[4], as well as by Narducci et al.[6] (apart from a sign change in the field equations in the latter case). To establish the connection, let $\omega_1^{\frac{1}{2}} e_1$, $\omega_2^{\frac{1}{2}} e_2$ be the real field amplitudes, while r_1, r_2, r_3 are the Bloch-vector components. Then we set

$$E_1 = e_1 \exp(-i\phi_1), \quad E_2 = e_2 \exp(-i\phi_2),$$

$$P = (r_2 - ir_1) \exp i(\phi_1 + \phi_2), \quad D = r_3 \; .$$

The basis for including the ionization loss terms has been discussed by Eberly et al.[33], Elgin[34] and Elgin et al.[35] Both fields are associated with the loss processes in this case. We have consistently assumed, in both the field and matter equations, that $\hbar\gamma_k$ is the imaginary part of the susceptibility $\chi_b(\omega_k) = K_{bb}(\omega_k) - i\hbar\gamma_k$.

The bistability effects in ionization loss from a one-photon resonant process have also been treated by Armstrong[30], however the loss term was omitted here from the field equation. For the one-photon system with ionization, a rate equation for N can be obtained by making[33] the adiabatic approximation $\dot{P} = \dot{D} = 0$; it is reasonable to assume that a similar approximation is valid also for the two-photon process described here. Equations (1c) and (1d) at steady state become

$$D = N(1 + \Omega^2/\Gamma_\perp^2)\{1 + \Omega^2/\Gamma_\perp^2 + \lambda^2|E_1|^2|E_2|^2/\Gamma_\parallel\Gamma_\perp\}^{-1} , \tag{8a}$$

$$P = E_1^* E_2^* D/(\Gamma_\perp - i\Omega) . \tag{8b}$$

For convenience, we set $x_1 = \lambda^{\frac{1}{2}}(\Gamma_\parallel\Gamma_\perp)^{-\frac{1}{4}}E_1$ and $x_2 = \lambda^{\frac{1}{2}}(\Gamma_\parallel\Gamma_\perp)^{-\frac{1}{4}}E_2$. Substituting (8a), (8b) into (1a), (1b) and putting $\dot{E}_1 \simeq 0 \simeq \dot{E}_2$, we find that the quasi steady-state field equations reduce to the following:

$$\frac{dx_k}{dz} \simeq$$

$$-\alpha N x_k \left\{ \frac{(\Gamma_\parallel/\Gamma_\perp)^{\frac{1}{2}}(1-i\Omega/\Gamma_\perp)|x_\ell|^2 - \hat{S}_k (1+\Omega^2/\Gamma_\perp^2) - \gamma_k|x_1|^2|x_2|^2}{1 + \Omega^2/\Gamma_\perp^2 + |x_1|^2|x_2|^2} \right.$$

$$\left. - i\hat{S}_k' \right\} \quad ; \quad k,\ell = 1,2 \ (\ell \neq k) . \tag{9}$$

It should be noted that the dynamical variables, including the fields, are still to be regarded as slowly-varying in time, and hence Eqs. (9) are only crudely valid unless γ_k is extremely small; in the limit of $\gamma_k \to 0$ we find that N becomes constant as it must and that Eqs. (9) become exact total-differential equations. It is convenient to assume as we have done, that stimulated radiative recombination is not an important effect. Although this may not always be a reasonable assumption, it is clear from the preceding that a non-trivial modification of the bistability equation relating the incident field to the transmitted field is an unavoidable consequence of retaining terms proportional to γ_k in the field equations. Other sources of field loss could be incorporated in the model, i.e. further damping terms could be introduced into the field equations. Where the losses are linear the effect upon the transmission curve is to re-scale the incident field and some of the fixed parameters, so that the functional relationship between incident and transmitted fields remains unchanged. It is not our intention to pursue problems of ionization and other losses in a detailed manner however, and the present treatment of ionization should be regarded as only a qualitative guide to a small effect.

III. THE EQUATIONS OF STATE IN MEAN-FIELD APPROXIMATION

In passing from the state of affairs described by Eqs. (9) to the so-called mean-field limit (MFL), we impose by analogy with the one-photon resonant counterpart the limiting processes

$$g' = (\Gamma_\parallel / \Gamma_\perp)^{\frac{1}{2}} g = \alpha N (\Gamma_\parallel / \Gamma_\perp)^{\frac{1}{2}} c^{-1} \to 0 ; \qquad T \to 0;$$

$$C_1 = 2g'L/T \to \text{positive constant} \tag{10}$$

where T is the mirror transmission coefficient $(0 < T < 1)$, assumed the same for both of the semi-transmitting mirrors at each end of the active region; L is the effective length of this region. When $\gamma_k = 0$, C_1 is the cooperativity parameter defined previously in reference 31. It is also necessary to impose the following conditions for the transmitted fields x_{1T}, x_{2T} and incident fields x_{1I}, x_{2I}:

$$|x_{kT}| \to 0 ; \quad |x_{kI}| \to 0$$

$$|x_{kT}|/T^{\frac{1}{2}} \to \text{constant}; \quad |x_{kI}|/T^{\frac{1}{2}} \to \text{constant} \quad (k=1,2) . \tag{11}$$

The mean field approximation has been analyzed for one-photon OB as a limiting case of a more general situation, where the fields are spatially dependent, by various authors[8,18-28]. It has been rigorously shown[21] that low resonant absorption, an essential requirement for the mean-field limit, is one of three possible physical limits which lead to linearized solutions for the fields. This conclusion applies to both ring and Fabry-Perot geometries. It is plausible to infer that, when gL and T are sufficiently small, the fields in the analogous two-photon resonant system are again essentially linear. The proof can be established in principle by methods similar to those utilized in reference 21, but is not given here. It should also be added that for one-photon OB the low-absorption limit has been shown[8] to be compatible with taking a spatial average of each envelope $F(z)$, so that $\bar{F} = L^{-1} \int_0^L F(z) dz$ and $\overline{F_1 F_2} = L^{-2} \int_0^L F_1(z) dz \int_0^L F_2(z') dz'$. With this assumption then, the fields x_1 and x_2 are found to satisfy

$$x_k(o) - x_k(L) \simeq \bar{x}_k \left\{ \frac{\bar{C}_1 |\bar{x}_\ell|^2 (1 - \bar{\beta}_k |\bar{x}_k|^2)}{1 + \bar{\delta}^2 + |\bar{x}_1|^2 |\bar{x}_2|^2} - i \left[\frac{\bar{C}_1 (\bar{\delta} |\bar{x}_\ell|^2 + (\bar{\Gamma}_\perp / \bar{\Gamma}_\parallel)^{\frac{1}{2}} \hat{S}_k (1 + \bar{\delta}^2))}{1 + \bar{\delta}^2 + |\bar{x}_1|^2 |\bar{x}_2|^2} + \alpha N \hat{S}_k' L c^{-1} \right] \right\} \tag{12}$$

where $\delta = \Omega / \Gamma_\perp$ and $\beta_k = (\Gamma_\perp / \Gamma_\parallel)^{\frac{1}{2}} \gamma_k$ $(k, \ell = 1, 2)$. Consistent with linearity, we replace fields $\bar{x}_k$ on the right of Eq. (12) with $x_k(L)$.

Boundary conditions are now required, in order to construct a set of coupled equations of state. These are analogous to those given in reference 21 (providing that the ionization processes are ignored), thus we have:

$$x_k(L) = x_{kT}/T^{\frac{1}{2}} \; ; \tag{13a}$$

$$T^{\frac{1}{2}}x_{kI}e^{i\theta_{kT}} = x_k(0) - (1-T)x_k(L)e^{2i\theta_{kR}}e^{iK_k(L+L')} \tag{13b}$$

where θ_{kT} and θ_{kR} are the phase changes accompanying transmission
and reflection, L' is the optical path length *outside* the active
region. The initial cavity detunings are $\mathcal{E} = 2\pi - [K_k(L+L')+2\theta_{kR}]_{\mathrm{mod}\ 2\pi}$,
and to complete the requirements for the mean-field approximation
we assume

$$\xi_k \to 0; \quad \phi_k = \frac{(1-T)}{T}\,\xi_k \to \mathrm{constant} \; . \tag{14}$$

Writing $Y_k = |x_{kI}|/T^{\frac{1}{2}}$ and $X_k = |x_{kT}|/T^{\frac{1}{2}}$, and employing (10),(12),(13)
and (14), we eventually obtain the coupled set of double-absorption
equations

$$Y_1^2 = X_1^2 \left\{ \left[1 + \frac{C_1\,X_2^2}{1+\delta^2+X_1^2X_2^2} \right]^2 \right.$$
$$\left. + \left[\frac{C_1(\delta X_2^2 + (\Gamma_\perp/\Gamma_\parallel)^{\frac{1}{2}}\,\hat{S}_1(1+\delta^2))}{1+\delta^2+X_1^2X_2^2} - \Phi_1 \right]^2 \right\} \; ; \tag{15a}$$

$$Y_2^2 = X_2^2 \left\{ \left[1 + \frac{C_1\,X_1^2}{1+\delta^2+X_1^2X_2^2} \right]^2 \right.$$
$$\left. + \left[\frac{C_1(\delta X_1^2 + (\Gamma_\perp/\Gamma_\parallel)^{\frac{1}{2}}\,\hat{S}_2(1+\delta^2))}{1+\delta^2+X_1^2X_2^2} - \Phi_2 \right]^2 \right\} \tag{15b}$$

with $\Phi_k = \phi_k - \alpha N L c^{-1}\,\hat{S}_k'$.   These equations are similar in many
respects to the corresponding equations for the Fabry-Perot system
given by Agrawal and Flytzanis[32]; note however, that they have
omitted terms containing S_1', S_2'.

   Although a general analysis of the functional relationships
contained in Eqs. (15a) and (15b) is difficult, a considerable simplification occurs when one field, say X_2, is sufficiently large
for the approximation $Y_2 \sim X_2(1 + \Phi_2^2)^{\frac{1}{2}}$ to hold. We then find
that Eq. (15a) becomes

$$Y_1'^2 = X_1'^2 \left\{ \left[1 + \frac{C_1'}{1+\delta^2+X_1'^2} \right]^2 + \left[\frac{C_1'\delta'}{1+\delta^2+X_1'^2} - \Phi_1 \right]^2 \right\} \tag{16}$$

where $C_1' = C_1 Y_2^2 (1 + \Phi_2^2)^{-\frac{1}{2}}$, $Y_1' = Y_1 Y_2 (1 + \Phi_2^2)^{-\frac{1}{2}}$, $X_1' = X_1 Y_2 (1 + \Phi_2^2)^{-\frac{1}{2}}$ and
$\delta' = \delta + (\Gamma_\perp / \Gamma_\parallel)^{\frac{1}{2}} \hat{S}_1 (1 + \delta^2)(1 + \Phi_2^2)^{\frac{1}{2}} Y_2^{-2}$. This is now in a form
similar to the equation of state for one-photon OB in a ring cavity.
In order to obtain a tractable analysis, we assume that (a) the
Stark terms may be neglected, at least to a first-order approximation; (b) the incident field Y_2 is large and held constant, so that
X_2 is also approximately constant. With these restrictions, it is
possible to establish a condition for the turning points in the
Y_1' (X_1') curve, i.e. from $dY_1'/dX_1' = 0$ we find

$$U[U^2(1 + \delta^2)(1 + \Phi_1^2)(1 + \Phi_2^2) + C_1 Y_2^2(1 - \delta'\Phi_1)] =$$
$$= C_1^2 Y_2^4[(1 + \delta'^2)/(1 + \delta^2)(1 + \Phi_2^2)](U-2) \qquad (17)$$

with $U = 1 + X_1^2 Y_2^2/[(1 + \delta^2)(1 + \Phi_2)^2]$. This may be interpreted as a
quadratic equation in the C_1 such that for fixed δ, Φ_1 and Φ_2 the
solutions for a chosen $U=U_S$ produce bistability curves for the turning points versus C_1. It may also be regarded as a quadratic equation in Φ_1, for fixed C_1, Φ_2 and δ, so that solutions for U_S versus
Φ_1 are obtainable. Note that a further simplification occurs when
$\delta=0$, $\Phi_1=0$, $\Phi_2=0$: Eq. (17) then factorizes to

$$(U^2 - C_1 Y_2^2 U + 2C_1 Y_2^2)(U + C_1 Y_2^2) = 0 \qquad (18)$$

which implies that $C_1 > 8/Y_2^2$ is a necessary condition for bistability.
This shows that switching values for field Y_1 can be adjusted over
a wide range of values by an appropriate choice of Y_2, and demonstrates a definite advantage over the two-level, single field counterpart. Note however that the constraints leading to Eq. (18) are
not always valid, since it is not permissible generally to treat
Stark effects as negligibly small. However, if $\gamma_\perp >> \Delta >> |\Omega - \Delta|$ and
$\Phi_1 \simeq 0 \simeq \Phi_2$, then it is readily shown that the bistability condition
becomes $C_1 > 8(1 + \delta^2)/Y_2^2$, which suggests that other degrees of freedom could be exploited in adjusting the switching intensities. In
Fig. 2 some restricted cases of Eqs. (15a,b) are shown: in particular, we show the transmission curves for X_1 and X_2 as functions
of Y_1, when Y_2 is fixed and of the same order of magnitude as the
parameter C_1. One important effect arising by increasing Y_2 is
the appearance of a dramatic increase in the switching intensity at
the lower branch. Note also that $X_2 \lesssim Y_2$; thus the two fields tend
to switch in opposing directions.

In the Raman case, the mean-field bistability equations for
$Y_k(X_1, X_2)$ are the same as in Eqs. (15a,b) except that the sign of
C_1 in (15b) is reversed. The behavior of X_1 and X_2 with respect to
Y_2 is shown in Fig. 3, where different constant values of Y_2 are
used. It will be observed that both fields switch in the same

direction, and that bistable behavior is still in evidence when Y_2 is very small ($Y_2 \ll 1$).

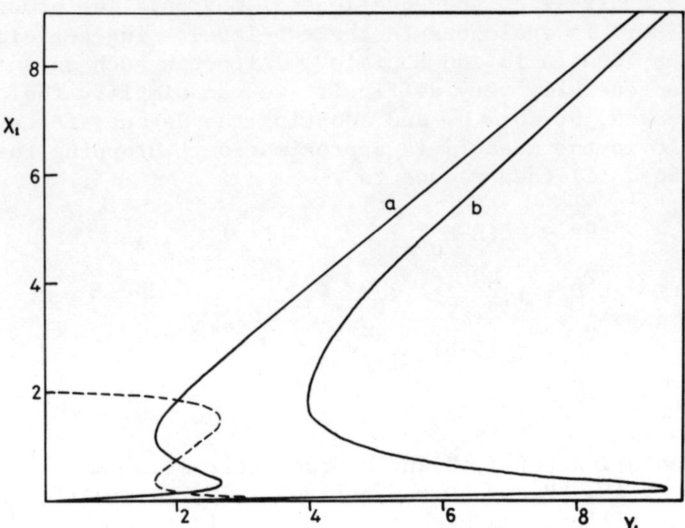

Fig. 2. Double absorption using two different fields in the MFL. Transmitted field X_1 as a function of incident field Y_1 when Y_2 is given the constant values (a) $Y_2=2$; (b) $Y_2=5$. The broken line depicts X_2 as a function of Y_1 when $Y_2=2$. The cooperativity parameter is $C_1=4$.

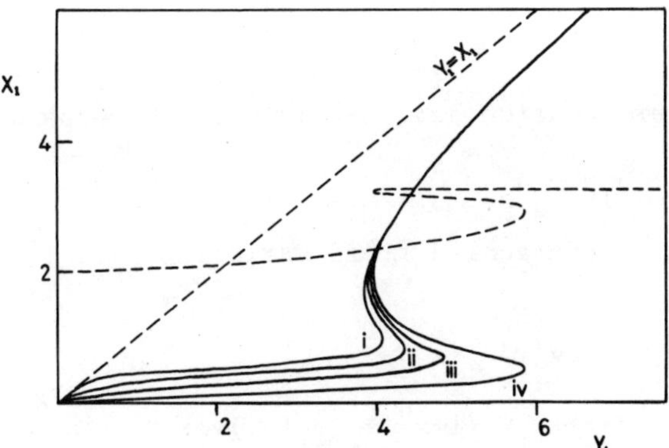

Fig. 3. Two-photon Raman process for $C_1=4$. Transmitted field X_1 versus Y_1 when Y_2 is fixed at (i) 0.1; (ii) 0.5; (iii) 1.0; (iv) 2.0. The broken line shows X_2 versus Y_1 when $Y_2= 2$.

IV. LINEAR STABILITY ANALYSIS

It has been anticipated that regions of the curves $Y_k(X_1, X_2)$ with negative slope will be unstable, and the formal proof requires a stability analysis of the equations of motion. The procedure to be adopted here is analogous to the two-level, single-field calculation given by Bonifacio and Lugiato[8]. Although such an analysis of Eqs. (1) is generally very difficult, we can simplify the problem by putting $\Delta = 0$, $S_k = S_k' = 0$, and adopting the degenerate case of a single field in the mean-field approximation. Dropping the 1,2 suffices, Eqs. (1) then reduce to

$$dE/dt = -\alpha cEP - \kappa c(E - E_o) \; ; \tag{19a}$$

$$dP/dt = \lambda E^2 D - \gamma_\perp P \; ; \tag{19b}$$

$$dD/dt = -\lambda E^2 P - \gamma_\parallel (D - N) \; . \tag{19c}$$

where $\kappa^{-1} = L/c$.

The deviations of E, P and D from their stationary values will be denoted by δE, δP and δD respectively. We retain linear terms in these quantities, and seek nontrivial solutions of the consequent set of differential equations of the form

$$(\delta E, \delta P, \delta D) = e^{\mu t} \cdot (\delta E_{(o)}, \delta P_{(o)}, \delta D_{(o)}). \tag{20}$$

These solutions occur if μ is an eigenvalue of

$$M = \begin{bmatrix} -\alpha cP_{(o)} - \kappa c & -\alpha cE_{(o)} & 0 \\ 2\lambda E_{(o)} D_{(o)} & -\gamma_\perp & \lambda E^2_{(o)} \\ -2\lambda E_{(o)} P_{(o)} & -\lambda E^2_{(o)} & -\gamma_\parallel \end{bmatrix} . \tag{21}$$

The cubic secular equation det.$(M - \mu I) = 0$ can be expanded into

$$\mu^2 + a_2 \mu^2 + a_1 \mu + a_o = 0 \tag{22}$$

where, in terms of the scaled field variables $y = |E_o| \{\lambda(\gamma_\parallel \gamma_\perp)^{-\frac{1}{2}}\}^{\frac{1}{2}}$ and $x = |E| \{\lambda(\gamma_\parallel \gamma_\perp)^{-\frac{1}{2}}\}^{\frac{1}{2}}$,

$$a_2 = \gamma_\perp + \gamma_\parallel + \kappa y/x \tag{23a}$$

$$a_1 = 2\kappa\gamma_\perp (y/x - 1) + \gamma_\perp \gamma_\parallel (1 + x^4) + \kappa(\gamma_\perp + \gamma_\parallel) y/x \tag{23b}$$

$$a_o = \kappa\gamma_\perp\gamma_\parallel \{(3 - x^4)y/x - 2(1 - x^4)\} = \kappa\gamma_\perp\gamma_\parallel (1 + x^4) dy/dx . \tag{23c}$$

The stability condition according to the Routh–Hurwitz theorem is $a_o > 0$, thus from Eq. (23c) we have the condition $dy/dx > 0$.

The same argument as has been used in Ref. 8 also establishes the existence of a critical slowing down at the transition points (where $dy/dx = 0$); for a "good" cavity in which $\kappa \ll \gamma_\perp, \gamma_\parallel$ we find the damping constant to possess the general form $\bar{\mu} = -\kappa\{1 + C(3-x^4)x^2/(1+x^4)^2\}$ where $C = (\gamma_\parallel/\gamma_\perp)^{\frac{1}{2}} \alpha N\kappa^{-1}$.

V. SPATIAL EFFECTS

Let us again consider Eq. (9), and ignore ionization loss, so that in the steady-state a pair of exact total differential equations forms:

$$\frac{dx_k}{dz} = -2gx_k\left\{\frac{|x_\ell|^2(1-i\delta) - i\hat{S}_k(1 + \delta^2)\,(\Gamma_\perp/\Gamma_\parallel)^{\frac{1}{2}}}{1 + \delta^2 + |x_1|^2|x_2|^2}\right.$$

$$\left. - i\hat{S}'_k(\gamma_\perp/\gamma_\parallel)^{\frac{1}{2}}\right\}, \quad k,\ell = 1,2 \ (\ell \neq k) . \tag{24}$$

It is convenient to separate out the real and imaginary parts at this stage, with $x_k = |x_k|e^{-i\phi_k}$. Defining an effective phase $\bar{\phi}_k$ as $\bar{\phi}_k = \phi_k + 2g\,\hat{S}'_k(\gamma_\perp/\gamma_\parallel)^{\frac{1}{2}}z$ we have

$$\frac{d|x_1|}{dz} = -2g\,\frac{|x_1||x_2|^2}{1+\delta^2 + |x_1|^2|x_2|^2} \tag{25a}$$

$$\frac{d|x_2|}{dz} = -2g\,\frac{|x_2||x_1|^2}{1+\delta^2 + |x_1|^2|x_2|^2} \tag{25b}$$

$$\frac{d\bar{\phi}_1}{dz} = -2g\,\frac{|x_2|^2\delta + S_1(1 + \delta^2)(\gamma_\perp/\gamma_\parallel)^{\frac{1}{2}}}{1+\delta^2 + |x_1|^2|x_2|^2} \tag{25c}$$

$$\frac{d\bar{\phi}_2}{dz} = -2g\,\frac{|x_1|^2\delta + \hat{S}_2(1 + \delta^2)(\gamma_\perp/\gamma_\parallel)^{\frac{1}{2}}}{1+\gamma^2 + |x_1|^2|x_2|^2} . \tag{25d}$$

Equations (25a) and (25b) imply a conservation law, namely

$$|x_1(z)|^2 - |x_2(z)|^2 = |x_1(o)|^2 - |x_2(o)|^2 . \tag{26}$$

Since the fields have been previously scaled (i.e. divided by the square roots of their respective frequencies) we recognize Eq.(26)

as an expression of the Manley–Rowe formula for conservation of photon number. Note that transient, simulated Raman scattering described by Elgin and O'Hare[36], Carman et al.[37] and Raymer et al.[38] is also a two-photon process, with the same field equations as for the resonant absorption process except that the spatial derivatives in Eq. (24) have opposing signs, and that the corresponding frequency factors also possess opposing signs. As a consequence, the Manley–Rowe relation appropriate to the Raman case becomes

$$\left|x_1(z)\right|^2 + \left|x_2(z)\right|^2 = \left|x_1(o)\right|^2 + \left|x_2(o)\right|^2 . \tag{27}$$

The conservation relations (26,27) can be utilized for the purpose of integrating (25a-d) exactly. For (25a) and (25b) we obtain

$$
\begin{aligned}
-C_1 T = -2gL = &\frac{1}{2}\left[\left|x_1(L)\right|^2 - \left|x_1(o)\right|^2\right] \\
&\pm\left[\left|x_1(o)\right|^2 \mp \left|x_2(o)\right|^2\right] \ln\left|\frac{x_2(L)x_1(o)}{x_2(o)x_1(L)}\right| \\
&+ \int_{z=o}^{z=L} \frac{\delta^2(\beta)d\beta}{\beta(\beta-\beta_o)}
\end{aligned}
\tag{28}
$$

with $\beta = \left|x_i(z)\right|^2$ and $\beta_o = \left|x_1(o)\right|^2 \mp \left|x_2(o)\right|^2$, where i=1 for double absorption (upper sign) and i=2 for the Raman case (lower sign). Because of the logarithmic term in Eq. (28), bistability curves for arbitrary values of T require greater computational effort than for the mean-field case. These numerical results will be presented elsewhere. The last term in (28) may be evaluated without difficulty if required. Equations (26,27) are also employed when integrating (25c) and (25d), thus with $d\bar{\phi}_k/dz = (d\bar{\phi}_k/d\left|x_k\right|)(d\left|x_k\right|/dz)$ and using (25), (26) and (27) we have

$$\bar{\phi}_k(L) - \bar{\phi}_k(o) = \frac{1}{2}\int_{z=o}^{z=L} d\beta'\left\{\frac{\delta(\beta')}{\beta'} + \frac{\varepsilon_k \hat{S}_k(1+\delta^2(\beta'))}{\beta'(\beta'-\beta_o)}\right\} \tag{29}$$

with $\beta' = \left|x_(z)\right|^2, \varepsilon_k = (\Gamma_\perp/\Gamma_\parallel)^{\frac{1}{2}}$ (k=1,2).

VI.   EQUATIONS OF STATE FOR ARBITRARY TRANSMISSIVITY

Passing beyond the mean-field approximation, we can show that a set of equations of state may be constructed using the boundary conditions (13a), (13b) together with the integrated solutions (28) and (29). The definitions of X_k, Y_k and ϕ_k used in the mean-field theory apply in these circumstances too, with the result

$$Y_k^2 = T^{-2}\left\{|x_k(o)|^2 + R^2 X_k^2 - 2R|x_k(o)|X_k \cos[\bar{\xi}_k + \bar{\phi}_k(L) - \bar{\phi}_k(o)]\right\} \tag{30}$$

in which $\bar{\xi}_k = \xi_k - 2gL\hat{S}'_k \lambda^{-1}(\gamma_\perp/\gamma_\parallel)^{\frac{1}{2}}$ is a new effective detuning parameter, and $R = 1-T$ is the reflectivity. The procedure outlined here, and the formal structure of the equations of state, are very similar to their single-photon counterparts derived very recently by several authors[21,23,26]. The additional feature of Stark shifts, which of course is not present in the one-photon model, complicates the analysis considerably.

When the stark terms and detunings may be considered negligible, it will be found that the phases $\bar{\phi}_k(z)$ may be consistently set equal to zero, so that all of the dynamical variables become real quantities. Equation (30) then becomes

$$x_k(o) = T Y_k + R X_k . \tag{31}$$

The conservation identities (26,27) evaluated at Z=L may now be rewritten using Eq. (31) as

$$X_1^2 \mp X_2^2 = (TY_1 + RX_1)^2 \mp (TY_2 + RX_2)^2 . \tag{32}$$

For given T and input fields (Y_1, Y_2) the transmitted field amplitudes X_1, X_2 lie on a hyperbola for the double-absorption resonance and on an ellipse for the Raman resonance. This equation is to be used in conjunction with Eq. (28) in the form

$$-C_1 T = \frac{1}{2}\left\{X_1^2 - (TX_1 + RY_1)^2\right\} \pm \left\{(TY_1 + RX_1)^2 \mp\right.$$

$$\left. (TY_2 + RX_2)^2\right\}^{-1} \ln\left|\frac{X_2(TY_1 + RX_1)}{X_1(TY_2 + RX_2)}\right| \tag{33}$$

Rearranging Eq. (32), we obtain

$$(Y_2 - X_2)[Y_2 + X_2(1+R)/T] = \pm(Y_1 - X_1)[Y_1 + X_1(1+R)/T] . \tag{34}$$

Assuming that $Y_1 \geqslant X_1$, we have immediately that $Y_2 \geqslant X_2$ for double-absorptive resonance, while $Y_2 \leqslant X_2$ for Raman resonance. This implies that switching processes in each case are fundamentally different: in double absorption the fields switch in opposing directions but in the Raman case the fields switch in the same direction (see Figs. 2 and 3).

VII.   DEGENERATE TWO-PHOTON ABSORPTION

The degenerate case of doubly-absorptive OB will occur when

a single field, with an energy which is approximately half the separation of the end states, can support the two-photon effect in the bistable regime. Assuming this behavior, we drop all 1,2 indices from Eqs. (25) which are respectively integrated to yield

$$-2C_1 T = -4gL = |x(L)|^2 - |x(o)|^2 - |x(L)|^{-2} + |x(o)|^{-2} +$$

$$+ \int_{z=o}^{z=L} \frac{\delta^2(\beta)}{\beta^2}\, d\beta \tag{35}$$

$$\bar{\phi}(L) - \bar{\phi}(o) = \frac{1}{2} \int_{z=o}^{z=L} d\beta \left\{ \frac{\delta(\beta)}{\beta} + \frac{\varepsilon S(1 + \delta^2(\beta))}{\beta^2} \right\} \tag{36}$$

To demonstrate that a closed expression for $Y(X)$ may appear, we drastically simplify the calculation by setting δ equal to a constant. Equation (35) now rearranges to

$$|x(o)|^2 = X^2 \eta(X, \delta) \tag{37}$$

with

$$2\eta(X,\delta) = 1 + \frac{2C_1 T}{X^2} - \frac{1+\delta^2}{X^4} + \left\{ \left(1 + \frac{2C_1 T}{X^2} - \frac{1+\delta^2}{X^4}\right)^2 + \frac{4(1+\delta^2)}{X^4} \right\}^{\frac{1}{2}} . \tag{38}$$

The function η is plotted against X, in Fig. 4, for various values of δ. Note that $\eta \geqslant 1$ and that $\eta \to 1$ whenever (a) $T \to 0$; (b) $X \to 0$; (c) $X \to \infty$. The equation of state (30) becomes

$$Y = XT^{-1} \{\eta + R^2 - 2R\eta^{\frac{1}{2}} \cos[\bar{\xi} + \bar{\phi}(L) - \bar{\phi}(o)]\}^{\frac{1}{2}} \tag{39}$$

with

$$\bar{\phi}(L) - \bar{\phi}(o) = -\frac{1}{2} \delta \ln \eta - \frac{1}{2} \varepsilon S(1 + \delta^2)(1 - \eta^{-1})/X^2 . \tag{40}$$

The fact that η approaches unity as T becomes smaller is consistent with the requirement that $\phi(L) - \phi(o)$ should vanish in the mean-field limit. No logarithmic terms appear in either the spatial dependence of the internal field or in the equation of state; in this respect the degenerate problem presents an analytic advantage by comparison with either the previous two-fields problem or the single-field two-level problem. Note that the parameter K_{ab} in the degenerate case has been shown[1] to differ from the corresponding quantity obtained in the two-fields case by a factor of ½; this similarly affects the respective definitions of C_1. The MFL of Eq. (39) can be derived by evaluating the series expansion of Eq. (38) for small T as far as the T^2 term, substituting into (39), and cancelling powers of T throughout. The result is

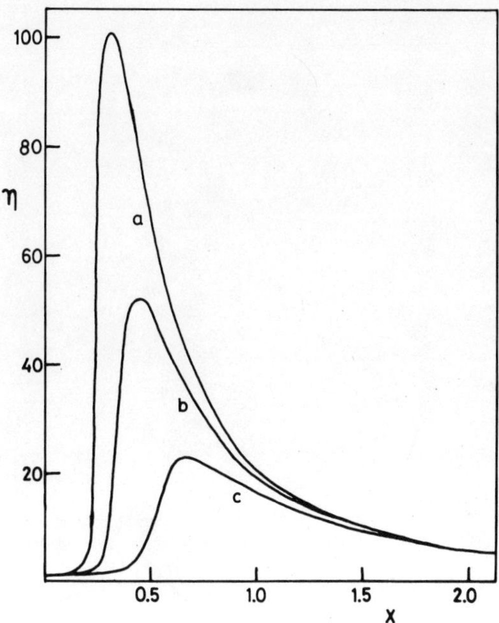

Fig. 4.   Graphs of $\eta(X,\delta)$ when T=0.25, C_1=40, and (a) $\delta = 0$; (b) $\delta = 1$; (c) $\delta = 2$.

$$Y = X\left\{\left[1 + \frac{C_1 X^2}{1 + \delta^2 + X^4}\right]^2 + \left[\frac{-C_1(X^2\delta + \varepsilon S(1 + \delta^2))}{1 + \delta^2 + X^4} - \Phi\right]^2\right\}^{\frac{1}{2}} \tag{41}$$

where $\varepsilon = (\Gamma_\perp/\Gamma_\parallel)^{\frac{1}{2}}$ and $\Phi = T^{-1}\xi + C_1\varepsilon\hat{S}'$. Transmission curves have been plotted from Eq. (41) in Fig. 5; they are evidently very sensitive to the parameter Φ, but are closely similar when the cavity is perfectly tuned. The inset shows that $\bar{\phi}(L)-\bar{\phi}(o)$, as a function of Y, also exhibits bistability (see Ref. 21).   Spatial effects have been taken into account in Fig. 6; Eq. (39) has been plotted for different values of T, with $\delta = \xi = 0$.

At no stage (as shown in Figs. 5 and 6) do the transmission curves approach the Y-axis asymptotically; this is in contradistinction to the behavior of the one-photon OB counterpart[17-21]. The effect of including both dispersive components and spatial effects is exhibited in Fig. 7. In particular, bistability would seem to disappear at quite low values of T when some representative values are given to C_1, δ, Φ, S', S and ε, suggesting that it is very likely to be necessary to employ a high-Q ring cavity if bistability is to be observed with this system. The full transmissivity situation (T=1) is simply described by the equation

$$Y = X \, \eta^{\frac{1}{2}}, \tag{42}$$

which is plotted in Fig. 5 with $\delta = 0$.

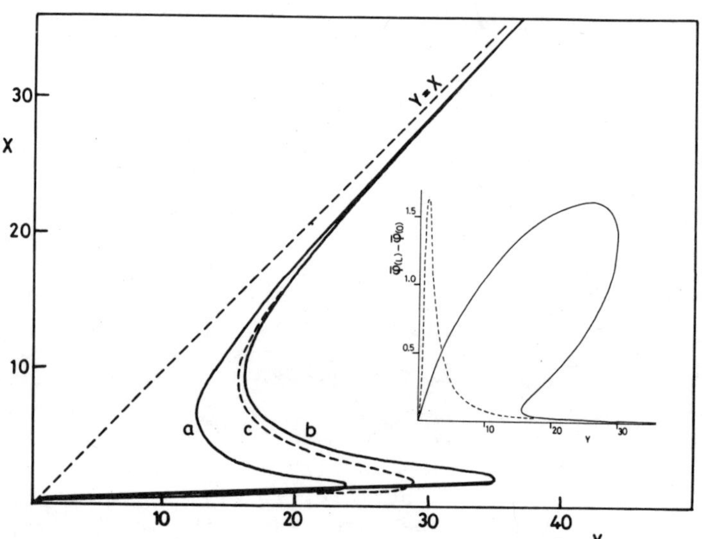

Fig. 5. Transmission curves for degenerate two-photon absorption
         in MFL with $C_1 = 40$, $S' = 0.5$, $S = 0$, $\varepsilon = 0.01$ and
         (a) $\delta = 0$, $\Phi = 0$; (b) $\delta = 2$, $\Phi = 0$. Curve (c) depicts
         the $T = 0.1$ transmission curve for $\delta = 2$, $\Phi = 0$.
         INSET: Phase changes in the degenerate case. Plots of
         $\bar{\phi}(L) - \bar{\phi}(o)$ against Y (full line) and X (broken line)
         for fixed $\delta(= 2)$ with $T = 0.1$. It is assumed that
         $(S/S') \ll 1$.

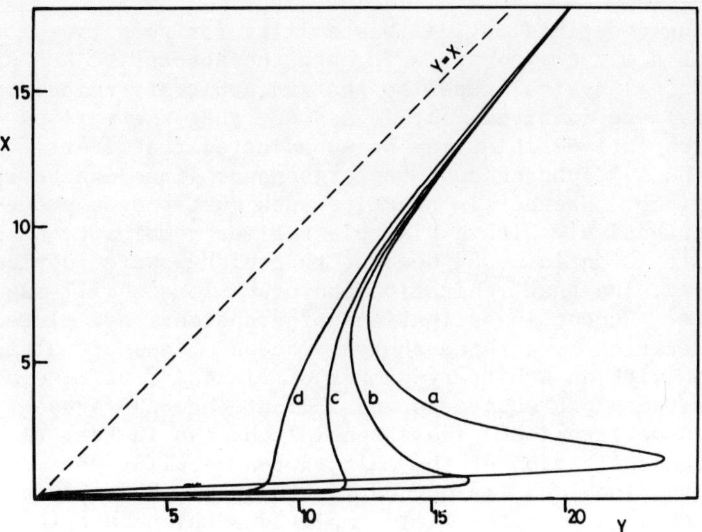

Fig. 6.   Graphs of X against Y for degenerate two-photon absorption
          when $C_1 = 40$, $\delta = 0$, $\xi = 0$, $S' = 0.5$, $S = 0$ and
          (a) $T \simeq 0$; (b) $T = 0.25$; (c) $T = 0.5$ (d) $T = 1$.

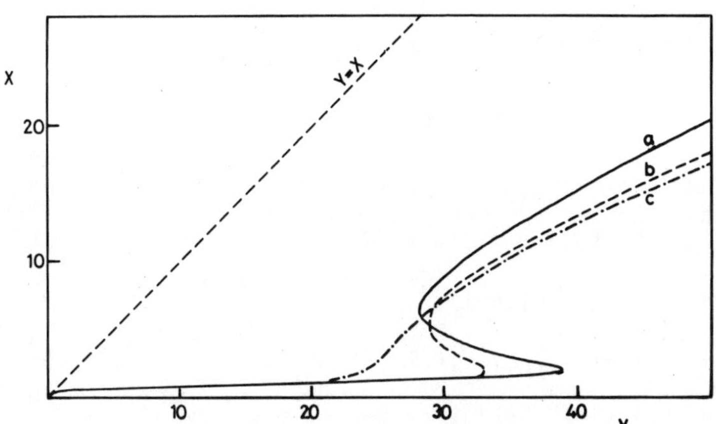

Fig. 7.   Degenerate two-photon absorption with dispersive components.
          The parameters are: $C_1 = 40$, $\delta = 2$, $\Phi = -2$, $S' = 0.5$,
          $S = 0$, $\varepsilon = 0.01$ when (a) $T \simeq 0$, (b) $T = 0.1$; (c) $T = 0.25$.

VIII. SOME GENERALIZATIONS

Having described optical bistability for some two-photon processes, it now seems possible in both the absorptive and Raman cases to extend the physical model so that an arbitrary number of photons (or fields) are involved. It is assumed that the various detunings for the respective fields may be so adjusted that overlap effects between the multiphoton and other resonance lines can be regarded as negligible. Whether in practice such multiphoton processes can operate effectively within bistable regimes remains to be demonstrated; it is anticipated however that high-powered devices will be required, and that ionization and other losses will assume greater importance. Recent investigations of transient, stimulated hyper-Raman scattering as a three-photon process[39], and of multiphoton absorption with an arbitrary number of fields[40], utilize a vector model for the Bloch equations and indicate how the present description could be extended. The essential changes in Eqs. (1) should be: (a) Generalization of the multiphoton coupling parameter λ; (b) In both Bloch and field equations λ should be associated with the product of all of the field amplitudes which contribute to the multiphoton process.

By way of demonstration that bistable regimes exist *in principle*, we consider degenerate n-photon absorption (n=1,2,3,...) in the mean-field approximation. The equations of motion are essentially

$$\dot{E} + \kappa(E - E_o) = - E^{n-1}P \; ; \tag{43}$$

$$\dot{P} + \gamma_\perp P = \lambda E^n D \; ; \tag{44}$$

$$\dot{D} + \gamma_\parallel (D-N) = -\lambda E^n P \; . \tag{45}$$

All detunings, Stark terms, and ionization losses are ignored in these equations. The linear stability analysis proceeds in the manner indicated in Section IV, and produces a characteristic equation of the form (22), with however

$$a_2 = \gamma_\perp + \gamma_\parallel + \kappa[(n-1)y/x-(n-2)] \tag{46}$$

$$a_1 = n\kappa\gamma_\perp (y/x-1) + \gamma_\perp\gamma_\parallel (1+x^{2n}) + \kappa(\gamma_\perp+\gamma_\parallel)[(n-1)y/x-(n-2)] \tag{47}$$

$$a_0 = \kappa\gamma_\perp\gamma_\parallel (1+x^{2n})dy/dx = \kappa\gamma_\perp\gamma_\parallel[(2n-1-x^{2n})y/x-2(n-1-x^{2n})] \tag{48}$$

where

$$x = [\lambda(\gamma_\perp\gamma_\parallel)^{-\frac{1}{2}}]^{1/n}E, \quad y = [\lambda(\gamma_\perp\gamma_\parallel)^{-\frac{1}{2}}]^{1/n}E_o \; . \tag{49}$$

This result is a generalization of the calculation of Bonifacio and Lugiato[7].

Again we observe the features that $dy/dx > 0$ for stability, and that critical slowing down occurs where $dy/dx = 0$.

Two specific examples of multiphoton resonances which might exhibit bistable behavior are three-photon absorption with three independent fields, and hyper-Raman resonance with two independent fields. For the former case the (grossly simplified) equations of motion are

$$\dot{E}_k + \kappa(E_k - E_k^0) = - E_\ell E_m P \qquad (k \neq \ell \neq m, \text{ permutation over } 1,2,3); \tag{50}$$

$$\dot{P} + \gamma_\perp P = \lambda E_1 E_2 E_3 D; \quad \dot{D} + \gamma_\parallel (D-N) = -\lambda E_1 E_2 E_3 P .$$

It has again been assumed that each field is predominantly associated with one particular (component) transition. In an obvious scaling, the mean-field equations of state may be cast into the form

$$Y_k = X_k \left\{ 1 + \frac{C_1 X_\ell^2 X_m^2}{1 + X_1^2 X_2^2 X_3^2} \right\} \tag{51}$$

An approximate solution of this set is effected when two of the input fields, say Y_2 and Y_3, are held constant; for sufficiently large fields $X_2 \simeq Y_2$ and $X_3 \simeq Y_3$, and so we can put $\bar{C}_1 = C_1 Y_2^2 Y_3^2$, $X_1' = X_1 Y_2 Y_3$, $Y_1' = Y_1 Y_2 Y_3$. Equation (51) becomes equivalent to the single-photon equation, and the constraint on C_1 for bistable behavior is $C_1 > 8/(Y_2^2 Y_3^2)$. We see that the third field has conferred an additional advantage over two-field OB, in that an additional degree of freedom is available for the purpose of controlling or adjusting the switching intensities.

The hyper-Raman resonance is described by the equations
$$\dot{E}_1 + \kappa(E_1 - E_1^0) = - \alpha E_1 E_2 P; \quad \dot{E}_2 + \kappa(E_2 - E_2^0) = \tfrac{1}{2} \alpha E_1^2 P; \quad \dot{P} + \gamma_\perp P = E_1^2 E_2 D;$$
$$\dot{D} + \gamma_\parallel (D-N) = - \lambda E_1 E_2 P .$$
A rigorous derivation is provided elsewhere[39]. This time the appropriately scaled mean-field equations of state become

$$Y_1 = X_1 \left\{ 1 + \frac{C_1 X_1^2 X_2^2}{1 + X_1^4 X_2^2} \right\} , \tag{52a}$$

$$Y_2 = X_2 \left\{ 1 - \frac{\tfrac{1}{2} C_1 X_1^4}{1 + X_1^4 X_2^2} \right\} . \tag{52b}$$

In common with the ordinary Raman case we find $Y_1 \gtrless X_1$, $Y_2 \lessgtr X_2$, thus the fields switch in opposing directions. The assumption $X_2 \simeq Y_2$ (large, constant) reduces Eq. (52a) to the degenerate double-absorption form.

The mechanisms of the above forms of absorptive bistability depend (as in two-level, single-field absorption) upon the possibility of saturating the particular atomic resonance, although this is not a necessary requirement for OB. A scheme involving a Raman-type resonance has been proposed as a mechanism for a novel OB device by Walls and Zoller[41]. Their predicted effect depends critically upon detunings; it cannot be obtained from our Raman-case formalism as we have effectively assumed a vector model, whereas they consider density matrix elements coupling states $|1\rangle$ and $|2\rangle$ to the intermediate state $|3\rangle$, and ρ_{33} explicitly.

IX. SUMMARY AND CONCLUSIONS

A theory of optical bistability for the ring-cavity geometry has been developed employing multilevel atoms in which two particular eigenstates are effectively coupled by two-photon processes involving the other eigenstates as intermediates. The Bloch-vector model used by Takatsuji, and Brewer and Hahn is adopted, it being assumed that the optical electrons spend a negligible time in the intermediate states. Generally, two distinct fields of frequencies ω_1 and ω_2 can be tuned so that either $\omega_1 + \omega_2$ or $\omega_1 - \omega_2$ is close to a resonance of the medium. It has been demonstrated that conditions for bistability to occur may be obtained and modified. Dispersive and spatial effects are taken into account, and detailed calculations involving these features have been performed for a degenerate doubly-absorptive process. The two-photon Stark terms enter into the analysis in an unavoidable manner, and present an additional complication; it is anticipated that future work will lead to a better understanding of their effects.

Generalizations to processes involving more than two photons have been discussed. Doppler broadening has been ignored since in the absorptive cases it can be substantially reduced by using counterpropagating beams, while in the Raman case there is a natural cancelling effect between processes of emission and absorption. It is premature at this stage to discuss possible device applications; a more detailed analysis of transients is required, and experimental work is also desirable. However, as a consequence of the conservation relations regulating multi-field processes, we find regimes in which novel switching effects occur, involving substantial changes in the different interacting fields.

REFERENCES

1. M. Takatsuji, Phys. Rev. $\underline{A4}$, 808 (1971); Phys. Rev. $\underline{A11}$, 619 (1975).
2. R. G. Brewer and E. L. Hahn, Phys. Rev. $\underline{A11}$, 1641 (1975).

3. D. Grischkowsky, M. M. Loy and P. F. Liao, Phys. Rev. $\underline{A12}$, 2514 (1975).
4. J. N. Elgin, G. H. New and K. E. Orkney, Optics Comm. $\underline{18}$, 250 (1976).
5. H. Friedmann and A. D. Wilson-Gordon, Optics Comm. $\underline{24}$, 5 (1978).
6. L. M. Narducci, W. W. Eidson, P. Furcinitti and D. C. Eteson, Phys. Rev. $\underline{A16}$, 1665 (1977); L. E. Estes, L. M. Narducci and B. Shammas, Lett. Nuovo Cim. $\underline{19}$, 775 (1971).
7. R. Bonifacio and L. A. Lugiato, Optics Comm. $\underline{19}$, 172 (1976).
8. R. Bonifacio and L. A. Lugiato, Phys. Rev. $\underline{A18}$, 1129 (1978).
9. R. Bonifacio and L. A. Lugiato, Phys. Rev. Lett. $\underline{40}$, 1023 (1978).
10. R. Bonifacio, M. Gronchi, and L. A. Lugiato, Phys. Rev. $\underline{A18}$, 2266 (1978).
11. L. A. Lugiato, Nuovo Cimento $\underline{B50}$, 89 (1979).
12. R. Bonifacio, and L. A. Lugiato, Lett. Nuovo Cimento $\underline{21}$, 505, 517 (1978).
13. R. Bonifacio, M. Gronchi and L. A. Lugiato, Nuovo Cimento (to appear).
14. S. S. Hassan, P. D. Drummond and D. F. Walls, Opt. Comm. $\underline{27}$, 480 (1978).
15. F. S. Felber and J. H. Marburger, Appl. Phys. Lett. $\underline{28}$, 732 (1976).
16. J. H. Marburger and F. S. Felber, Phys. Rev. $\underline{A17}$, 335 (1978).
17. E. Abraham, R. K. Bullough and S. S. Hassan, Optics Comm. $\underline{29}$, 109 (1979).
18. P. Meystre, Optics Comm. $\underline{26}$, 277 (1978).
19. G. P. Agrawal and H. J. Carmichael, Phys. Rev. $\underline{A9}$, 2074 (1979).
20. J. A. Hermann, Optica Acta $\underline{27}$, 159 (1980).
21. H. J. Carmichael and J. A. Hermann, Zeitschrift für Physik $\underline{B38}$, 365 (1980).
22. S. L. McCall and H. M. Gibbs, Optics Comm. $\underline{33}$, 335 (1980).
23. R. Bonifacio, M. Gronchi and L. A. Lugiato, Theory of Optical Bistability, in "Laser Spectroscopy IV," Proc. Fourth Conf. on Laser Spectroscopy, 1979; Eds. Walther and Rothe, Springer-Verlag 1979.
24. M. Gronchi and L. A. Lugiato, preprint (to appear in Optics Letters).
25. R. Roy and M. S. Zubairy, Optics Comm. $\underline{32}$, 163 (1980).
26. R. Roy and M. S. Zubairy, Phys. Rev. $\underline{A21}$, 274 (1980).
27. E. Abraham, S. S. Hassan and R. K. Bullough, preprint (to be published).
28. G. P. Agrawal and H. J. Carmichael, Optica Acta (to appear); H. J. Carmichael, Optica Acta (to appear).
29. H. J. Carmichael and G. P. Agrawal, preprint (to be published).
30. L. Armstrong, J. Phys. B: Atom. Molec. Phys. $\underline{12}$, L719 (1979).
31. F. T. Arecchi and A. Politi, Lett. al Nuovo Cimento, $\underline{23}$, 65 (1978).
32. G. P. Agrawal and C. Flytzanis, Phys. Rev. Lett. $\underline{44}$, 1058 (1980).

33. J. H. Eberly and S. V. O'Neil, Phys. Rev. $\underline{A19}$, 1161 (1979); J. L. F. de Meijere and J. H. Eberly, Phys. Rev. $\underline{A17}$, 1416 (1978).

34. J. N. Elgin, J. Phys. B: Atom. Molec. Phys., $\underline{12}$, L261 (1979).

35. J. N. Elgin, G. H. C. New, and T. B. O'Hare, (to be published).

36. J. N. Elgin, and T. B. O'Hare, J. Phys. B: Atom. Molec. Phys., $\underline{12}$, 159 (1979).

37. R. L. Carman, F. Shimizu, C. S. Wang and N. Bloembergen, Phys. Rev. $\underline{A2}$, 60 (1970).

38. M. G. Raymer, J. Mostowski and J. L. Carlsten, Phys. Rev. $\underline{A19}$, 2304 (1979).

39. J. A. Hermann and B. V. Thompson (to be published).

40. J. A. Hermann and B. V. Thompson (to be published).

41. D. F. Walls and P. Zoller, preprint (to be published).

ACTIVE TWO-BEAM OPTICAL BISTABILITY

G. P. Agrawal
Quantel, 17 avenue de l'Atlantique
Z.I. 91400 Orsay, France

C. Flytzanis
Laboratoire d'Optique Quantique*, Ecole Polytechnique
91128 Palaiseau, France

Abstract: The principle of bistable operation with two independent input optical beams is presented. The nonlinear medium inside a Fabry-Perot resonator is in two-photon resonance with the beams. The steady-state transmission characteristics of the bistable device are derived in the absorptive and the dispersive regimes. One has an active control over the switching intensities through the second beam.

I. INTRODUCTION

Optical bistability by which a light transmitting system is made to operate in only two stable states, namely "opaque" and "transparent", and exhibit a hysteresis has now been proposed and observed for a number of schemes in transmission[1-8] or reflection[9-10]. Some of these schemes show features that can be practically exploited and may give birth to useful optical bistable devices.

The central idea of the one beam optical bistability is the self-action of an optical beam on its own optical path

$$\delta = \frac{2\pi \ell \sqrt{\varepsilon}}{\lambda} \tag{1}$$

inside a limited region filled with a nonlinear medium and under appropriate boundary conditions (for instance a cavity); ℓ is the longitudinal dimension of the region and ε is the beam-intensity

---

*Laboratoire propre du Centre National de la Recherche Scientifique.

dependent complex dielectric constant of the nonlinear medium where
the self-action takes place, λ is the wavelength of the light. Since
in general

$$\varepsilon = \varepsilon' + i\varepsilon'', \tag{2}$$

one may distinguish two regimes of optical bistability, the dispersive (a Kerr type nonlinearity) and the absorptive (saturated absorption type nonlinearity).

However, the very same nonlinearity that underlies the self-
action of the beam on its own optical path can be used to concur-
rently modify this optical path with a second beam of different
wavelength and this is the central idea[8] of the active or two-beam
bistability we shall deal with in the following. We anticipate that
this scheme has features and applications drastically different from
the ones associated with the passive one-beam bistability.

After a short disgression on the two-photon resonantly enhanced
Kerr effect we proceed to set up the physical basis of the active
optical bistability using an equivalent two-photon Hamiltonian to
describe the evolution of the material in the presence of the two
optical beams. The resulting equations for the density matrix
together with the propagation equation for the electromagnetic
fields inside the cavity completely describe the dynamics of the
active bistability. The characteristics of the bistable operation
are then derived for the stationary regime. A short discussion of
the stability conditions follows and we conclude with some concrete
proposals for active bistable devices. In the course of this presentation we also discuss the possibility to study the optical bistability through a generalized phase-conjugation method.

II.   TWO-PHOTON RESONANTLY ENHANCED KERR EFFECT

The possibility of constructing a practical optical bistable
device consisting of a multiple pass interferometer (Fabry-Perot
cavity) containing a nonlinear dispersive material and operating
over a large frequency range very much relies on the access to a
material with a large optical Kerr coefficient; further the material
must be transparent over the desired frequency range.

Here we wish to point out that these two requirements can be
met in some materials by operating with frequencies ω well below
their absorption threshold but with 2ω near a two-photon transition.
Indeed the physical origin of the dispersive optical bistability is
the third order optical polarisation

$$P^{(3)}(\omega) = \chi^{(3)}(\omega,-\omega,\omega)\, |E(\omega)|^2\, E(\omega), \tag{3}$$

where $E(\omega)$ is the electric field inside the cavity and $\chi^{(3)}$ is the
third order susceptibility; it is related to the optical Kerr coefficient n_2 at the frequency ω defined by

$$n(\omega) = n_o(\omega) + n_2(\omega)I, \tag{4}$$

where n_o is linear refraction index and I the beam intensity $I = cE^2/2\pi$, through the relation

$$n_2(\omega) = \frac{8\pi^2}{n_o c} \chi^{(3)}(\omega,-\omega,\omega). \tag{5}$$

The expression for $\chi^{(3)}(\omega,-\omega,\omega)$ for ω well below the onset of electronic transitions but 2ω near a two photon transition frequency
Ω_{eg} between the ground state g and an excited state e is[11]

$$\chi^{(3)}(\omega,-\omega,\omega) \simeq \frac{N}{\hbar(\Omega_{eg}-2\omega-i\Delta\omega)} \left| \sum_i \left(\frac{p_{gi}\,p_{ie}}{\hbar\Omega_{ig}}\right)\right|^2 , \tag{6}$$

where N is the number density of the polarisable units, p is the
dipole moment operator and $\Delta\omega$ the linewidth of the transition Ω_{eg}.
Under certain conditions, generally met in most materials, this
can also be written

$$\chi^{(3)} \simeq \frac{\Omega_{eg}}{\Omega_{eg}-2\omega-i\Delta\omega} \chi_o^{(3)} , \tag{7}$$

where $\chi_o^{(3)}$ is the zero frequency limit of the third order susceptibility which is related to the conventional Kerr coefficient $n_2(o)$
through $n_2(0) = 8\pi^2\chi_o^{(3)}/n_o c$ or

$$n_2(\omega) = \frac{\Omega_{eg}}{\Omega_{eg}-2\omega-i\Delta\omega} n_2(o) . \tag{8}$$

Thus the Kerr coefficient at frequency ω is resonantly enhanced
although ω is far below the onset of optical transitions and thus
the absorption losses can be kept quite low.

Near or on the two photon resonance the absorption loss competes
with the two-photon enhancement of the dispersion of n_2 and one
must proceed beyond the two-photon saturation regime to achieve
bistable behavior.

Clearly the intensity I occurring in (4) does not need to be
that of the beam at frequency ω and the two-photon resonant enhancement of $n_2(\omega)$ can also occur with a second beam present of different

frequency ω' such that $\omega'+\omega\simeq\Omega_{eg}$. This introduces three more degrees of freedom, the intensity I', the frequency ω' and the polarization direction of the second beam with respect to the first. The two beams will mutually interact with each other and the one will affect the bistable behavior of the other. This is the essence of the active optical bistability[8] which will be presented below.

It is appropriate to stress here that the two frequencies can be tuned over the whole transparency range of the nonlinear material and yet as long as their sum or difference (but not both) is near a two-photon or Raman active resonance of the medium one can exploit the enhancement of the nonlinearity to achieve bistable operation. Otherwise stated, one can be in the dispersive bistable regime with respect to each beam and exploit the advantages of resonant enhancement of the nonlinearity.

III.   TWO BEAMS INSIDE A CAVITY

The Fabry-Perot resonator we consider is axial to the z direction and is bounded at $z = 0$ and $z = d$ by plane parallel mirrors of the intensity reflection coefficient R. The resonator is filled with an isotropic nonlinear medium. Two light beams of the electric field strengths E_1^{in} and E_2^{in} and of frequencies ω_1 and ω_2 respectively, copropagating along the z-axis are incident at the plane $z=0$. The frequencies are such that either $\omega_1+\omega_2$ or $|\omega_1-\omega_2|$ can be near or coincide with the atomic transition frequency Ω_{eg} between the ground state $|g>$ and an excited state $|e>$, while ω_1 and ω_2 are kept away from any resonances of the medium. In the plane wave approximation the transverse variation of the electric field is ignored. Using Maxwell's equations the total electric field $\varepsilon(z,t)$ at any point z inside the cavity satisfies the wave equation

$$\frac{\partial^2 \vec{\varepsilon}}{\partial z^2} - \frac{1}{c^2}\frac{\partial^2 \vec{\varepsilon}}{\partial t^2} = \frac{4\pi}{c^2}\frac{\partial^2 \vec{P}}{\partial t^2}, \tag{9}$$

where

$$\vec{\varepsilon}(z,t) = \hat{x}\ \mathrm{Re}\left[E_1\ e^{-i\omega_1 t} + E_2\ e^{-i\omega_2 t}\right], \tag{10}$$

$$\vec{P}(z,t) = \hat{x}\ \mathrm{Re}\left[\tilde{\chi}_1\ E_1\ e^{-i\omega_1 t} + \tilde{\chi}_2 E_2\ e^{-i\omega_2 t}\right] \tag{11}$$

and $\hat{x}$ is a unit vector along the direction of polarization of two beams. Here,

$$\tilde{\chi}_j \equiv \tilde{\chi}(\omega_j) = \chi_j^L + \chi_j^{NL} \tag{12}$$

is the effective susceptibility at the frequency ω_j. Equation (9) is to be solved subject to the boundary conditions appropriate for the Fabry-Perot cavity. In what follows we make the mean-field approximation to simplify the propagation problem. Its use can be justified for a high Q cavity and when $\chi_j^{NL} \ll 1$. Under these conditions the forward and backward waves associated with each beam have equal amplitudes and z-dependence of each beam can be taken to be of the form

$$E_j(z,t) = 2 A_j(t) \sin (k_j z), \tag{13}$$

where $k_j = \sqrt{\varepsilon_j}\, \omega_j/c$ and $\varepsilon_j = 1 + 4\pi \chi_j^L$ is the linear dielectric constant. Further $A_j(t)$ is a slowly varying function of time. Using Eqs. (9)-(13) we obtain

$$\frac{dA_j}{dt} = 2\pi i\, \omega_j\, \chi_j^{NL}\, A_j. \tag{14}$$

This can be interpreted as a rate equation for the field amplitude A_j. To account for the boundary conditions we introduce gain and loss terms. We then obtain

$$\frac{dA_j}{dt} = 2\pi i\, \omega_j\, \chi_j^{NL}\, A_j + \frac{c}{2d\sqrt{\varepsilon_j}}\left[\sqrt{1-R}\, E_j^{in} - (\frac{1-R}{R} + i\, \theta_j)\, A_j\right], \tag{15}$$

where $2d\sqrt{\varepsilon_j}/c$ is the roundtrip time, $\sqrt{1-R}\, E_j^{in}$ is the gain term because of the incident field, and θ_j is the phase shift arising due to initial cavity detuning. In the steady state Eqs. (13) and (15) combine to give the cavity field associated with the j-th beam and we have

$$E_j(z) = \frac{2RE_j^{in}}{\sqrt{1-R}}\left[1 - \frac{iR}{1-R}(4\pi\, k_j d\, \chi_j^{NL} - \theta_j)\right]^{-1} \sin (k_j z). \tag{16}$$

IV. MATTER DYNAMICS AND INDUCED POLARIZATION

The matter-radiation interaction is governed through the time evolution of the density operator $\rho(t)$ which satisfies the equation of motion

$$i\hbar \frac{\partial \rho}{\partial t} = [H_o + H',\rho] \ , \tag{17}$$

where H_o is the free Hamiltonian and $H' = - \vec{p}\cdot\vec{E}$ in the dipole approximation. Here $\vec{p}$ is the dipole moment operator. The induced polarization is then given by

$$\vec{P}(t) = N \ \text{Tr} \ [\vec{p} \ \rho(t)], \tag{18}$$

where N is the atomic density. In Eq. (18) it is convenient to evaluate the trace in a transformed space[12-15]. First a unitary transformation U is made to convert the dynamics of the multilevel system to that of an effective two-level system[12-14]. The diagonal elements of the effective Hamiltonian in the transformed frame contain the intensity dependent optical Stark shift while the off-diagonal elements contain the terms oscillating at frequencies $2\omega_1$, $2\omega_2$, $\omega_1+\omega_2$ and $\omega_1-\omega_2$. Here we shall explicitly consider the case $\omega_1+\omega_2 \simeq \Omega_{eg}$; the Raman case $|\omega_1-\omega_2| \simeq \Omega_{eg}$ can be treated along the same lines replacing ω_2 by $-\omega_2$. Under these conditions only terms oscillating at $\omega_1+\omega_2$ are retained in the off-diagonal elements of the density operator. For the degenerate case $\omega_1=\omega_2=\omega$, the terms oscillating at 2ω should be retained. A second transformation R is then made to a "doubly rotating" frame. Details can be found in Ref. 14. The transformed density matrix is written in the form

$$\rho''(t) = R \ U \ \rho(t) \ U^{-1} \ R^{-1} = \frac{1}{2}\begin{pmatrix} 1+r_3 & r_1-ir_2 \\ r_1+ir_2 & 1-r_3 \end{pmatrix} \ , \tag{19}$$

where the components r_i satisfy the equations

$$\dot{r}_1 = - \gamma_3 r_2 - r_1/T_2$$

$$\dot{r}_2 = - \gamma_1 r_3 + \gamma_3 r_1 - r_2/T_2 \tag{20}$$

$$\dot{r}_3 = \gamma_1 r_2 - (r_3 + \frac{1}{2})/T_1$$

and we have introduced the phenomenological relaxation times T_1 and T_2 in analogy with the Bloch vector model for the case of one-photon resonance. Their introduction allows the system to achieve a non-equilibrium steady state. The other parameters in Eqs. (20) are:

$$\gamma_1 = - \kappa |E_1 \ E_2| \tag{21}$$

$$\gamma_3 = (\Omega_{eg}-\omega_1-\omega_2) + (\Delta E_e - \Delta E_g)/\hbar \tag{22}$$

$$\kappa = \left| \sum_n \frac{P_{en} P_{ng}}{2\hbar^2} \left(\frac{1}{\Omega_{ng}-\omega_1} + \frac{1}{\Omega_{ng}-\omega_2} \right) \right| \qquad (23)$$

$$\Delta E_i = \frac{1}{4} [\alpha_i(\omega_1)|E_1|^2 + \alpha_i(\omega_2)|E_2|^2], \qquad (24)$$

$$\alpha_i(\omega) = \frac{2}{\hbar} \sum_n \frac{|P_{ni}|^2 \Omega_{ni}}{(\Omega_{ni}^2-\omega^2)} , \quad i = g,e \quad . \qquad (25)$$

Here κ is the two-photon gyroelectric ratio, ΔE_i is the optical
Stark shift for level i=g and e; and $\alpha_i(\omega)$ is the corresponding
linear electronic polarizability. The sum in Eqs. (23) and (25) is
over all the intermediate states and $p_{ij} = \langle i|p|j\rangle$ is the matrix
element of the x-component of the dipole moment operator.

Using Eqs. (19) and (20) the induced polarization $\vec{P}(t)$ is found
to contain terms oscillating at ω_1, ω_2, $2\omega_1+\omega_2$ and $2\omega_2+\omega_1$. For the
analysis of the two-beam optical bistability, only the first two
terms are relevent and we obtain

$$P(t) = \frac{N}{2} \, \mathrm{Re} \left\{ \sum_{j=1}^{2} [\alpha_g(\omega_j)(1-r_3) + \alpha_e(\omega_j)(1+r_3)] \, E_j \, e^{-i\omega_j t} \right\}$$

$$+ N\hbar\kappa \, \mathrm{Re}\left[(r_1-ir_2) \left(\left|\frac{E_2}{E_1}\right| E_1 e^{-i\omega_1 t} + \left|\frac{E_1}{E_2}\right| E_2 e^{-i\omega_2 t} \right) \right] . \qquad (26)$$

The first term in Eq. (26) contributes mainly to χ^L. Its contribution to χ^{NL} will be neglected. On comparing Eq. (11) and Eq. (26)
we obtain the nonlinear susceptibility

$$\chi^{NL}_j = N\hbar\kappa(r_1-ir_2)|E_{3-j}/E_j|, \quad j=1,2. \qquad (27)$$

Equations (15) and (20) together with (27) constitute a set of seven
coupled equations and completely describe the dynamics of the system.
Their solution is desired if one is interested in the switching
time of the bistable device. However, these equations are highly
nonlinear and a numerical approach appears to be necessary to obtain
transient solutions. In what follows we consider the steady-state
solutions.

V.   STEADY STATE BISTABLE OPERATION

By setting $\dot{r}_j = 0$ in Eqs. (20) we obtain

$$\vec{r} = \frac{1}{2}\left(\frac{T_2}{T_1}\right)^{1/2} \frac{1}{1+\Delta^2+I_1I_2}\left[\Delta\sqrt{I_1I_2}, \; -\sqrt{I_1I_2}, \; -\left(\frac{T_1}{T_2}\right)^{1/2}(1+\Delta^2)\right], \quad (28)$$

where we have introduced the intensity dependent detuning parameter

$$\Delta = \gamma_3 T_2 = \Omega + \frac{1}{2}(\delta_1 I_1 + \delta_2 I_2), \tag{29}$$

and,

$$\Omega = (\Omega_{eg}-\omega_1-\omega_2)T_2 \tag{30}$$

is the detuning of the two laser frequencies from the two-photon
resonance and

$$\delta_j = (T_2/T_1)^{1/2}[\alpha_g(\omega_j) - \alpha_e(\omega_j)]/(2\hbar\kappa) \tag{31}$$

governs the intensity dependent optical Stark shift.   Further
$I_j = |E_j|^2/I_s$, where $I_s = (\kappa^2 T_1 T_2)^{-1/2}$ is the normalized beam
intensity.

We now substitute Eq. (28) in Eq. (27) and obtain the following
expression for the nonlinear susceptibility:

$$\chi_j^{NL} = \chi_o \frac{(\Delta+i)I_{3-j}}{1+\Delta^2+I_1I_2}, \quad \chi_o = \frac{1}{2}N\hbar\kappa(T_2/T_1)^{1/2}. \tag{32}$$

We remark that Δ is intensity dependent and is given by Eq. (29).
Equation (32) clearly exhibits coupling of two beams arising due
to their interaction with the nonlinear medium. It is to be noted
that $I_j = |E_j|^2$ is a sinusoidal function of z as indicated in
Eq. (13). In the mean field approximation adopted here we average
out the rapid oscillation of χ_j^{NL} arising from the standing wave
nature of the cavity fields. We note that such an averaging is not
required for the ring cavity case. Assuming that the two beams are
statistically independent we take

$$\langle\sin^2\theta_1\sin^2\theta_2\rangle = \langle\sin^2\theta_1\rangle\langle\sin^2\theta_2\rangle,$$

$$\langle\sin^2\theta_1\rangle = \frac{1}{2}, \quad \langle\sin^4\theta_1\rangle = \frac{3}{8}, \tag{33}$$

where angular brackets denote average over θ. The transmitted field $E_j^{Trans} = (1-R)^{1/2} A_j$ where A_j is the steady state solution of Eq. (15). Using Eqs. (15), (32) and (33) we obtain the following set of two coupled state equations relating the transmitted field amplitudes of those of the incident ones:

$$Y_1 = X_1 [(1 + 2C_1 X_2^2/D)^2 + (\phi_1 - 2C_1 X_2^2 N_2/D)^2]^{1/2}$$

$$\qquad\qquad\qquad\qquad\qquad\qquad\qquad\qquad\qquad\qquad\qquad\qquad (34)$$

$$Y_2 = X_2 [(1 + 2C_2 X_1^2/D)^2 + (\phi_2 - 2C_2 X_1^2 N_1/D)^2]^{1/2},$$

where

$$N_j(X_1, X_2) = \Omega + \frac{3}{2} \delta_j X_j^2 + \frac{1}{2} \delta_{3-j} X_{3-j}^2 \qquad\qquad (35)$$

represents the detuning and the Stark-shift effects in the dispersive contribution and

$$D(X_1, X_2) = 1 + \Omega^2 + \Omega(\delta_1 X_1^2 + \delta_2 X_2^2) + \frac{3}{8} (\delta_1^2 X_1^2 + \delta_2^2 X_2^2) +$$

$$(1 + \frac{1}{2} \delta_1 \delta_2) X_1^2 X_2^2 \qquad\qquad (36)$$

reflects the dependence of the saturation effects on Ω and δ. Further we introduced the dimensionless field amplitudes

$$Y_j = \left(\frac{2}{1-R}\right)^{1/2} \frac{|E_j^{in}|}{E_s} \; , \quad X_j = \left(\frac{2}{1-R}\right)^{1/2} \frac{|E_j^{trans}|}{E_s} , \qquad (37)$$

where $E_s = (\kappa^2 T_1 T_2)^{-1/4}$ is the two-photon saturation amplitude. The cooperativity parameters C_j and the cavity detuning parameters ϕ_j are defined as follows:

$$C_j = 2\pi\chi_o k_j Rd/(1-R), \quad \phi_j = \theta_j R/(1-R) \; . \qquad (38)$$

A general treatment of active two beam optical bistability requires the numerical solution of the set (34). For given values of the parameters ϕ_j, δ_j, C_j, Ω and Y_2 one can obtain X_1 and X_2 as a function of Y_1. The essential features of the bistable operation, however, can be seen in the special case $\phi_j = 0$, $\delta_j = 0$ and $\Omega = 0$. It corresponds to the case of the absorptive bistability in a medium on exact two-photon resonance in a tuned cavity and where the linear polarizability $\alpha_g(\omega_j) \simeq \alpha_e(\omega_j)$. For simplicity we further choose $C_1 = C_2 = C$. The set (34) now reduces to

$$Y_j = X_j [1 + 2C \, X^2_{3-j} / (1 + X^2_1 X^2_2)], \; j = 1,2. \tag{39}$$

On eliminating X_2 we obtain a ninth degree polynomial in X_1 where coefficients depend on C, Y_1 and Y_2. For a given value of C and Y_2 we obtain the real positive roots of this polynomial as a function of Y_1.

In Fig. 1 we have plotted X_1 and X_2 as a function of Y_1 for $C = 2$ and $Y_2 = 10$. Similar to the case of one beam bistability we obtain three branches of the steady state solutions. The middle branch in Fig. 1 is unstable and the device will switch "on" and "off" at the switching amplitudes $Y_1{}^{max}$ and $Y_1{}^{min}$ respectively. The essential difference with the one beam optical bistability is that we now have external control over the switching amplitudes through the second beam. To obtain an estimate of the switching amplitudes we note from Fig. 1 that for a fixed Y_2, X_2 does not change appreciably with Y_1 and to a good degree of approximation we may replace X_2 by Y_2 in Eq. (39). We then obtain a single state equation

$$Y_1 = X_1 [1 + 2C \, Y^2_2 / (1 + X^2_1 Y^2_2)] \quad , \tag{40}$$

which can be analyzed in a straightforward manner similar to the one-beam bistability case. One finds that for the observation of two-beam bistablility $C > 4/Y^2_2$, and the switching amplitudes, assuming $Y_2 > 1$, are given by

$$Y_1{}^{min} \simeq 2\sqrt{2C} , \qquad Y_1{}^{max} \simeq CY_2 . \tag{41}$$

Thus by proper choice of C and Y_2, two-beam bistability can be actively controlled.

We remark that in the general case described by Eqs. (34), when a complete exploration of the parameters C_j, ϕ_j, δ_j, Ω and Y_2 is undertaken, a rich variety of effects takes place. For instance, under certain conditions as much as seven steady-state branches are found to occur. To look for the stable part of the transmission curve it is necessary to obtain the stability conditions, and this will be done in the next section. One more remark is in order. In the Raman case when $|\omega_1 - \omega_2|$ is in two-photon resonance, one should replace ω_2 by $-\omega_2$ as noted earlier. This implies that C_2 is negative in Eqs. (34). If an analysis similar to Fig. 1 is carried out, we find that both beams switch on in the direction of increasing intensities. This is expected, as the situation corresponds to the case of stimulated Raman scattering and the Stokes field $X_2 \gtrsim Y_2$ experiences a net gain in the nonlinear medium.

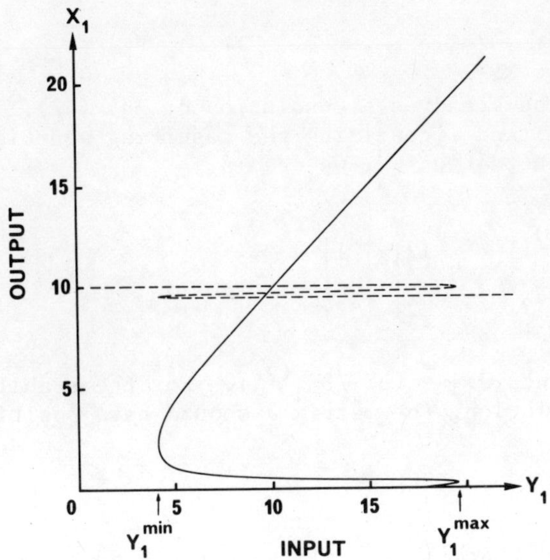

Fig. 1. Variation of the transmitted amplitudes X_1 (full line)
and X_2 (dashed line) of the two beams with the incident
amplitude Y_1. The parameters are $C = 2$ and $Y_2 = 10$. The
bistable device turns on and off at $Y_1{}^{max}$ and $Y_1{}^{min}$, respectively, which are functions of the second beam amplitude Y_2.

VI.   STABILITY CONDITIONS

A detailed study of the linear stability analysis for the case
of two-beam bistable operation is given in Ref. 10. Here we content ourselves in reporting the results for the case when the cavity
decay rate is much shorter than the atomic relaxation rates $T_1{}^{-1}$
and $T_2{}^{-1}$. In this case the atomic system follows the optical
fields adiabatically and one may use steady-state solutions, given
by Eq. (28), for the atomic variables. For simplicity, we further
consider the case of absorptive bistability and set $\Omega = 0$, $\phi_j = 0$
and $\delta_j = 0$. This makes the field amplitudes in Eq. (15) real.
Noting $E_j{}^{trans} = (1-R)^{1/2}A_j$ and using Eqs. (15), (32)-(38), we
obtain

$$\dot{X}_j = Y_j - X_j[1 + 2CX_{3-j}^2/(1+X_1^2X_2^2)] \qquad (42)$$

where $j = 1, 2$ and a dot on X_j represents derivative with respect
to $\tau = ct(1-R)^{1/2}/(2d)$. For the linear stability analysis we write

$$X_j = X_j^{(o)} + \chi_j \tag{43}$$

where $X^{(o)}$ is the steady-state solution of Eq. (42). On substituting Eq. (43) in (42) and linearizing the resulting equations with respect to the small fluctuation χ_j, we obtain

$$\begin{pmatrix} \chi_1 \\ \chi_2 \end{pmatrix} = - \begin{pmatrix} J_{11} & J_{12} \\ J_{21} & J_{22} \end{pmatrix} \begin{pmatrix} \chi_1 \\ \chi_2 \end{pmatrix} \tag{44}$$

where the elements $J_{ij} = (\partial Y_j / \partial X_j^{(o)})$. For the stability of the steady-state solution, the matrix J should have positive eigenvalues for which

$$\text{Tr } J > 0, \qquad \text{Det } J > 0. \tag{45}$$

It clearly shows that the stability criteria are more involved for the two-beam case. In the special case corresponding to Fig. 1, when $X_2 \simeq Y_2 \gg 1$, Eq. (45) is equivalent to $J_{11} > 0$. This implies that the negative slope regions of Fig. 1 are unstable as one should expect. Similar to the one-beam case critical slowing down[3] will occur when one of the eigenvalues of J is zero.

VII. DEGENERATE CASE: ONE BEAM TWO-PHOTON BISTABILITY

We briefly consider the degenerate case $\omega_1 = \omega_2 = \omega$. The details are similar to Sec. IV. In the off-diagonal elements of the effective Hamiltonian, one should now keep the terms oscillating at 2ω. Equations (20) still hold and the analysis of Secs. IV and V can be used by simply dropping the beam index j. We remark that κ given by Eq. (23) should be multiplied by $1/2$ now. Using Eqs. (29) and (32), the nonlinear susceptibility is given by

$$\chi^{NL}(\omega) = \chi_o \left[\frac{(\Omega+i)\, I + \delta I^2}{1+\Omega^2+2\Omega\delta I+(1+\delta^2) I^2} \right], \tag{46}$$

where χ_o, Ω and δ are obtained by putting $\omega_1 = \omega_2 = \omega$ in their previous expressions. After averaging out the rapid oscillations of I we finally obtain a single state equation

$$Y = X\left\{ (1 + 2CX^2/D)^2 + [\phi - 2C(\Omega X^2 + \tfrac{3}{2}\,\delta X^4)/D]^2 \right\}^{1/2} \tag{47}$$

where

$$D = 1 + \Omega^2 + 2\Omega\delta X^2 + \frac{3}{2}(1 + \delta^2)X^4. \tag{48}$$

An equation similar to (47) was also obtained by Arecchi and Politi[7] for the ring-cavity case.

In Fig. 2 we have plotted the transmission characteristics of a two-photon single-beam bistable device for several values of the parameters Ω, ϕ and δ with C = 20. The curves (a), (b) and (c) correspond to the case of absorptive bistability ($\Omega = \phi = 0$) and show the effect of the Stark-shift parameter δ; the curves for δ and $-\delta$ are identical. As is evident, the bistability effect diminishes with increasing $|\delta|$ and disappears for $|\delta| = 0.2$. In curves (d) and (e), $\Omega = \phi = 1$ and dispersive effects are included. We further note that the sign of δ matters. In particular for negative δ the switching amplitudes are lower.

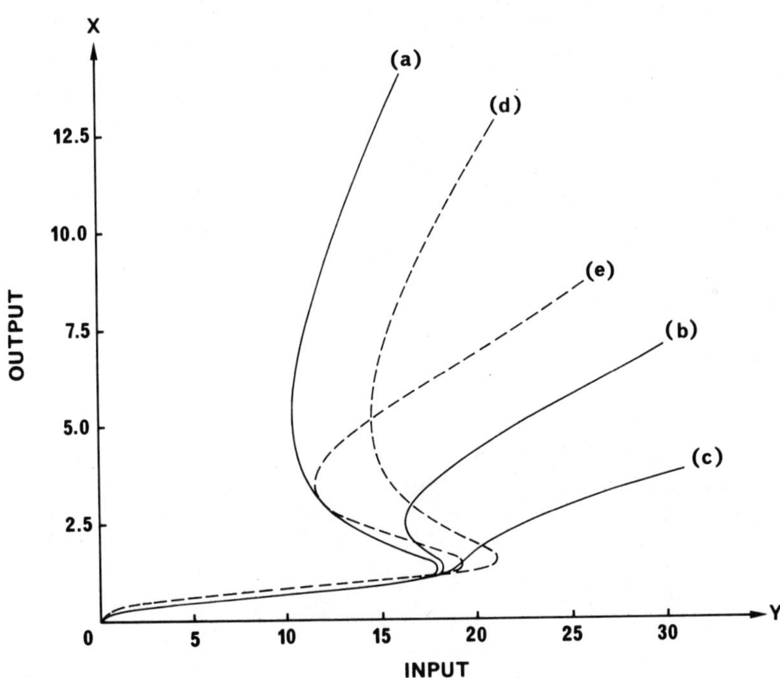

Fig. 2. Degenerate case when a single beam is in two-photon resonance. Variation of the transmitted amplitude X with the incident amplitude Y is shown for various values of Ω, ϕ and δ with C=20. For curves (a), (b) and (c), $\Omega=\phi=0$ and $\delta=0$, 0.1 and 0.2 respectively. For curves (d) and (e), $\Omega=\phi=1$, and $\delta=0.05$ and $\delta=-0.05$ respectively.

VIII. DISCUSSION AND CONCLUSIONS

A very convenient material to study two-beam two-photon optical bistability is InSb. The band gap of this semiconductor is at approximation $5\mu m$ and using two mode-frequencies of a CO_2 laser (each nearly at $10\mu m$), one can easily satisfy the resonance condition and observe bistable behaviour in thin samples. The nonlinear susceptibility of the material is indeed large, $\chi^{(3)}I$ easily reaching ~ 1 e.s.u. with a few KW/cm^2 intensities. Incidently, an indication that the use of InSb is feasible is given by the hugh non-linearities observed in one-beam bistability experiments by Miller et al.[16,17] In their experiment the transmission of a CO laser beam ($\lambda \sim 5\mu m$) through a InSb crystals (with polished plane parallel faces to form a Fabry-Perot cavity) is studied. The large non-inearity found here to account for the experimental results can well be due to a simultaneous one-photon (interband) and two-photon (intraband) resonant enhancement. We mention that the use of two CO laser beams in Ref. 17 does not correspond to the case of active two-beam optical bistability considered here. Their two beams are in one-photon resonance while use of two CO_2 laser beams, as we are proposing here, will correspond to two-photon resonance.

Quite remarkable is also the possibility to achieve bistable behaviour with two beams such that their frequency difference $\omega_1-\omega_2$ is near a Raman active vibrational mode. As mentioned earlier, this case corresponds to the stimulated Raman scattering taking place inside a cavity. The Stokes field at ω_2 amplifies as it propagates. Optical bistability may be observed in transmission of the pump field or the Stokes field while the other is kept fixed in intensity at the input.

Mutual coupling of two beams shown by Eq. (34) manifests through a variety of rich behaviour. In this article we considered the specific situation when the intensity of one beam was kept fixed at the input. In general one may vary the input intensities of both beams in some predetermined manner and study the transmission characteristics. In particular, the transmitted intensity of one beam may change by a large amount with a small change in the intensity of the other beam. Another generalization of the present analysis is to treat the case of optical double resonance. This corresponds to a three level atomic system which is simultaneously in one and two-photon resonances. The steady-state solution for the density matrix was obtained by Brewer and Hahn[13]. The case of double resonances provides extra control as there are now two detuning parameters.

In the present treatment two beams are incident on one of the mirrors of the Fabry-Perot cavity. A possible variation is to inject the second beam through a sidewall of the cavity. The beam incident along the axis of the cavity (pump beam) will exhibit usual one-beam bistability features in transmission when the input intensity is varied. Two counter propagating pump beams and the second beam will generate a phase conjugated beam inside the non-linear medium through the four wave mixing process[18-19]. However, as the input pump intensity is varied, the phase-conjugated signal will also display bistable behaviour.

REFERENCES

1. J. Seidel, U.S. Patent N° 3610731; A. Szoke, V. Daneu, J. Goldhar, and N. A. Kurnit, Appl. Phys. Lett. 15, 376 (1969); J. W. Austin and L. G. DeShazer, J. Opt. Soc. Am. 61, 650 (1971); E. Spiller, ibid. 61, 699 (1971); J. Appl. Phys. 43, 1673 (1972); S. L. McCall, Phys. Rev. A9, 1515 (1974).

2. H. M. Gibbs, S. L. McCall, and T. N. C. Venkatesan, Phys. Rev. Lett. 36, 1135 (1976). T. N. C. Venkatesan and S. L. McCall, Appl. Phys. Lett. 30, 282 (1977); H. M. Gibbs, S. L. McCall, A. C. Gossard, A. Passner, W. Wiegmann, and T. N. C. Venkatesan in Laser Spectroscopy IV, H. Walther and K. W. Rothe, Eds. (Springer-Verlag, Berlin, 1979).

3. R. Bonifacio and L. A. Lugiato, Opt. Commu. 19, 172 (1976); Phys. Rev. A18, 1129 (1978); Lett. Nuovo Cimento 21, 517 (1978).

4. F. S. Felber and J. H. Marburger, Appl. Phys. Lett. 28, 732 (1976); J. H. Marburger and F. S. Felber, Phys. Rev. A17, 335 (1978).

5. P. W. Smith and E. H. Turner, Appl. Phys. Lett. 30, 280 (1977); P. W. Smith, J. P. Kaminov, P. J. Maloney, and L. W. Stulz, ibid. 33, 24 (1978) and 34, 62 (1979).

6. G. P. Agrawal and H. J. Carmichael, Phys. Rev. A19, 2074 (1979); G. P. Agrawal and H. J. Carmichael, Optica Acta (in press); H. J. Carmichael and G. P. Agrawal, Opt. Commun. (to be published).

7. F. T. Arecchi and A. Politi, Lett. Nuovo Cimento 23, 65 (1978); M. Devaud, G. Grynberg, C. Flytzanis and B. Cagnac, J. de Physique (to be published).

8. G. P. Agrawal and C. Flytzanis, Phys. Rev. Lett. 44, 1058 (1980); G. P. Agrawal, A. De Martino, I. Abram, and C. Flytzanis, Phys. Rev. A (to be published).

9. A. E. Kaplan, JETP. Lett. 24, 114 (1976); Sov. Phys. JETP. 45, 896 (1977).

10. P. W. Smith, J. P. Hermann, W. J. Tomlinson, and P. J. Maloney, Appl. Phys. Lett. 35, 846 (1979).

11. C. Flytzanis in Quantum Electronics Treatise, Vol. 1, Part A, A. Rabin and C. L. Tang, eds. (Academic Press, New York, 1975).

12. M. Takatsuji, Phys. Rev. A4, 808 (1971).

13. R. G. Brewer and E. L. Hahn, Phys. Rev. A11, 1641 (1973).

14. D. Grischkowsky, M. M. T. Loy, and P. F. Liao, Phys. Rev. A12, 2514 (1975).

15. L. M. Narducci, W. W. Eidson, P. Furcinitti, and D. C. Eteson,
     Phys. Rev. A16, 1665 (1977).
16. D. A. B. Miller, M. H. Mozolowski, A. Miller, and S. D. Smith,
     Opt. Commun. 27, 133 (1978).
17. D. A. B. Miller and S. D. Smith, Opt. Commun. 31, 101 (1979).
18. R. W. Hellwarth, J. Opt. Soc. Am. 67, 1-3 (1977).
19. For a review, see the article by J. AuYeung and A. Yariv in
     Laser Spectroscopy, H. Walther and K. W. Rothe, eds.
     (Springer-Verlag, Berlin, 1979).

ABSORPTIVE AND DISPERSIVE BISTABILITY FOR A DOPPLER-BROADENED

MEDIUM IN A FABRY-PEROT: STEADY-STATE DESCRIPTION*

H. J. Carmichael

Center for Studies in Statistical Mechanics
University of Texas at Austin
Austin, Texas 78712

and

G. P. Agrawal

Quantel, 17 av. de l'Atlantique
Z. I. 91400 Orsay
France

Abstract: We present a steady-state theory for absorptive and
dispersive bistability using a Doppler-broadened two-level medium
in a Fabry-Perot.   Details in the saturating Doppler line reflect
the mode structure of the standing-wave cavity.   However the qualitative effects of Doppler-broadening on transmission characteristics
are understood in general terms which apply also to non-Doppler
inhomogeneous broadening in a Fabry-Perot and inhomogeneous broadening in a ring cavity.   The approximate treatment of Doppler-broadening using a truncated Bloch hierarchy is compared with our theory
which treats the Bloch hierarchy to all orders.

I.   INTRODUCTION

The first proposals for optical bistability exploited the absorptive nonlinearity in a saturable medium[1-5].   Optical bistability is
now recognised as a more general phenomenon which may be engineered
around a variety of optical nonlinearities.   Following the experimental realization of dispersive bistability by Gibbs et al[6] this

*Work supported in part by a grant from the Robert A. Welch founda-
 tion.

diversification has proceeded apace. With the potential for device
applications as a primary stimulus numerous schemes have now been
proposed[7-16]. This paper is concerned with both absorptive and
dispersive bistability in a saturable two-level medium. More particularly, we consider a Doppler-broadened medium set within a Fabry-
Perot.

McCall[5] gave the first detailed analysis of absorptive bistability using a homogeneously broadened two-level medium in a Fabry-Perot
In conjunction with the experiments of Gibbs et al[6] this description
was subsequently extended to include dispersive effects. Much
theoretical work has been conducted since using this saturable two-
level model. In its simplest formulation[17-21] a cubic state equation
characterises bistable steady states. Bonifacio and Lugiato[17]
initiated interest in this simplified approach with their discussion
of the fluorescent spectrum for absorptive bistability. It has now
been employed extensively for theoretical studies of quantum-statistical[22-26], stochastic[27-30], and transient[31-34] features in optical
bistability. Cubic state equations arise with the spatial averaging
of cavity fields, and at appropriate levels of approximation,
essentially equivalent[35] models may be built around either a ring
cavity[8,19,21] or a Fabry-Perot[17,20]. However, more correctly, the
standing-wave mode structure in a Fabry-Perot distinguishes these
geometries. If a uniform assumption of weak absorption and high
cavity Q (mean-field approximation, MFA) is applied to both, only
the ring-cavity model yields a cubic state equation[36-42]. For a
dispersive bistability well below saturation (Kerr medium limit) the
cubic form is of course regained in a Fabry-Perot after expansion of
the nonlinear susceptibility to first order.

The mode structure in a Fabry-Perot also carries direct consequences for the treatment of Doppler-broadening; the subject of
special interest in this paper. In a ring-cavity model Doppler and
non-Doppler inhomogeneous broadening are included in a single formulation. The nonlinear susceptibility is now defined by integrating
contributions over a distributed atomic detuning. Both gaussian[19]
and lorentzian[18,43] distributions have been discussed in the recent
literature. Non-Doppler inhomogeneous broadening in a Fabry-Perot
follows a similar prescription[38]. However a Doppler-broadened mediur
meets with added complexity due to atomic motion through the standing-wave field. Each atom sees frequency shifts associated with
both forward and backward waves. In the linear regime this merely
gives rise to two tuned populations contributing forward and backward polarizations respectively. However, nonlinearities couple the
counter-propagating fields. At this level features are introduced
which have no counterpart in a ring cavity. Treatments have been
given by Stenholm and Lamb[44] and Feldman and Feld[45] in theories of
the high intensity gas laser. Lax[46] also discusses this problem in
the context of laser theory. His analysis includes nonlinearities

up to first order in the field intensities and will be adequate here
only for a purely dispersive bistability achieved well below saturation. The consequences of Doppler-broadening in the presence of
counter-propagating waves are also met, and exploited, in saturation
spectroscopy[47].

     In order to achieve a general formulation for optical bistabil-
ity it is necessary to extend the theories worked out for the
laser[44-46] to allow unequal forward and backward wave amplitudes.
We present our approach to such a description for the medium in the
following section. Although developed independently, our formalism
is equivalent to that previously reported by Shirley[48]. We proceed
then to a steady-state formulation for absorptive and dispersive
bistability in a Fabry-Perot. Most generally, all spatial effects
(other than transverse effects) are included. However, specific
results and discussion will be presented for the mean-field limit
(MFL)[36,42] where required numerical procedures are more economical.

     Treatments for Doppler-broadening and standing waves in a Fabry-
Perot are intimately connected. In both, special features enter
through the coupling of forward and backward waves in the nonlinear
susceptibility. An approximate procedure has been used by a number
of authors to include standing waves in an homogeneously broadened
medium[17,49-52]. In this approach the Bloch hierarchy defining the
polarization and inversion is truncated[37] and a cubic state equation
resembling that for a ring cavity is recovered. While at present it
does not seem possible to remove this approximation in studies of
transient effects[31,50], it need not be made in a steady-state
theory[37]. The steady-state polarization calculated from a truncated
Bloch hierarchy departs from the correct value through the saturation
region and is in error by 33% in a well saturated medium. Differences in bistable transmission characteristics and related steady-
state features have been found to be generally less than ~20%. This
same approximation may be employed in the treatment of Doppler-
broadening[53], and is here closely related to the rate equation
approach in laser theory[44,45]. Again, for the steady-state truncation of the Bloch hierarchy is not necessary. The detailed picture
for the saturating Doppler line is quite different in this approximate scheme. In a final section we discuss treatments for a Doppler-
broadened medium using both truncated and full Bloch hierarchies.
We look at differences in the calculated polarization for various
values of Doppler linewidth and relate these to the differences
which follow for transmission characteristics. Errors of a similar
order to those for an homogeneously broadened medium are found.

II.   COUNTER-PROPAGATING WAVES IN A TWO-LEVEL DOPPLER-BROADENED MEDIUM

In a recent publication[54] we outlined a treatment for Doppler broadening in the presence of counter-propagating waves. We will briefly review this formalism in the first part of this section. Further details in the physical background and mathematical development may be sought in Refs. 44-46 and 48, each of which bears close relationship to the presentation here. Shirley's treatment[48] corresponds most closely to our own, his being the only one of these theories to include unequal forward and backward wave amplitudes.

We consider a system of N two-level atoms distributed uniformly throughout a volume V and having a Maxwell distribution of velocities. For running waves in the z-direction, only the z-component of velocity is important. We denote this by v. Then if n(v) gives the density of atoms in both space and velocity we write

$$n(v) = \frac{N}{V} \frac{1}{\sqrt{\pi}u} e^{-\frac{v^2}{u^2}} \tag{2.1}$$

where u is the most probable speed for an atom ($u/\sqrt{\pi}$ is the average atomic speed). Each atom has a resonant frequency ω_a and a dipole moment with modulus μ.

A classical field of frequency ω_0 is assumed to propagate in this medium as

$$E(z,t) = \mathcal{E}(z,t)e^{-i\omega_0 t} + c.c., \tag{2.2}$$

$$\mathcal{E}(z,t) = \mathcal{E}_f(z,t)e^{ik_0 z} + \mathcal{E}_b(z,t)e^{-ik_0 z}, \tag{2.3}$$

where $\mathcal{E}_f(z,t)$ and $\mathcal{E}_b(z,t)$ are slowly varying amplitudes polarized in the direction of the atomic dipoles and $k_0 = \omega_0/c$. Optical Bloch equations now describe internal atomic dynamics within the rest frame of each atom. First consider only those atoms with a z-component of velocity in the interval v to v + dv. We may view the response of this subpopulation either within a frame moving with a velocity v[44], where standing waves are encountered as an excitation of periodic intensity, or in the laboratory frame[45,48]. The latter choice facilitates a more direct presentation and we define a macroscopic polarization p(v,z,t) and inversion d(v,z,t) which then satisfy the equations[45,48,54]

$$(\frac{\partial}{\partial t} + v\frac{\partial}{\partial z})p = -2i\frac{\mu}{\hbar}\,Ed - \gamma_\perp\,(1 + i\Delta)p, \qquad\qquad (2.4)$$

$$(\frac{\partial}{\partial t} + v\frac{\partial}{\partial z})d = i\frac{\mu}{\hbar}\,(Ep^* - E^*p) - \gamma_\parallel\,(d + \frac{n(v)}{2}). \qquad (2.5)$$

Here $\gamma_\perp$ and $\gamma_\parallel$ are transverse and longitudinal atomic relaxation rates and $\Delta = (\omega_a - \omega_0)/\gamma_\perp$. If τ_{coll} is the time between velocity changing collisions we have assumed $\gamma_\parallel^{-1} \ll \tau_{coll}$ so that velocity subpopulations evolve to independent steady states. We denote the net polarisation and inversion by $P(z,t)$ and $D(z,t)$ respectively. With

$$P(z,t) = \mu P(z,t)e^{-i\omega_0 t} + c.c. \qquad\qquad (2.6)$$

we have

$$P(z,t) = \int_{-\infty}^{\infty} dv\; p(v,z,t), \qquad\qquad (2.7)$$

$$D(z,t) = \int_{-\infty}^{\infty} dv\; d(v,z,t)\;. \qquad\qquad (2.8)$$

Our concern in the following section is with the steady state for optical bistability. The field $E(z,t)$ is then defined by time-independent amplitudes $E_f(z)$ and $E_b(z)$. Setting time derivatives in Eqs. (2.4) and (2.5) to zero, their steady-state solution may be written in terms of a continued fraction. We introduce Fourier expansions for $p(v,z) \equiv p(v,z,\infty)$ and $d(v,z) \equiv d(v,z,\infty)$ with

$$p(v,z) = \sum_{n\ \mathrm{odd}} p_n(v,z)e^{ink_0 z}, \qquad\qquad (2.9)$$

$$d(v,z) = \sum_{n\ \mathrm{even}} d_n(v,z)e^{ink_0 z}\;. \qquad\qquad (2.10)$$

Here the summation covers both positive and negative n, and since d must be real, $d_n = d^*_{-n}$. Then, if $E_s = (\hbar/2\mu)(\gamma_\parallel\,\gamma_\perp)^{\frac{1}{2}}$ is the saturation amplitude and

$$X_f(z) = E_f(z)/E_s, \quad X_b(z) = E_b(z)/E_s\;, \qquad (2.11)$$

we find[54]

$$p_n = -i \left(\frac{\gamma_\parallel}{\gamma_\perp}\right)^{\frac{1}{2}} L_n \left(X_f d_{n-1} + X_b d_{n+1}\right) , \quad n \text{ odd} , \tag{2.12}$$

$$\frac{d_{n+2}}{d_n} = R_n , \quad \text{even } n \geq 0 , \tag{2.13}$$

$$R_n = -\left(X_f^* X_b \frac{a_{n+1}}{A_n}\right)^{-1} \left(1 + X_f X_b^* \frac{a_{n-1}}{A_n} \frac{1}{R_{n-2}}\right) , \quad \text{even } n \geq 2 , \tag{2.14}$$

with

$$d_0 = -\frac{n(\nu)}{2} \frac{1}{1+S} , \tag{2.15}$$

$$R_0 = -X_f X_b^* F , \tag{2.16}$$

where

$$S(|X_f|^2, |X_b|^2, \Delta, \nu) = |X_f|^2 |L_1|^2 + |X_b|^2 |L_{-1}|^2 - |X_f|^2 |X_b|^2$$

$$\times \text{Re}[F(|X_f|^2, |X_b|^2, \Delta, \nu)(L_1 + L_{-1}^*)] , \tag{2.17}$$

$$F(|X_f|^2, |X_b|^2, \Delta, \nu) = \frac{a_1}{A_2}$$

$$\frac{1}{1 - |X_f|^2 |X_b|^2 \dfrac{a_3^2}{A_2 A_4} \dfrac{1}{1 - |X_f|^2 |X_b|^2 \dfrac{a_5^2}{A_4 A_6} \dfrac{1}{1 - \cdots}}} , \tag{2.18}$$

and

$$A_n = 1 + in \frac{\gamma_\perp}{\gamma_\parallel} \nu + \tfrac{1}{2} |X_f|^2 (L_{n+1} + L_{-n+1}^*)$$

$$+ \tfrac{1}{2} |X_b|^2 (L_{n-1} + L_{-n-1}^*) , \quad n \text{ even} , \tag{2.19}$$

$$a_n = \tfrac{1}{2}(L_n + L_{-n}^*) , \quad n \text{ odd} , \tag{2.20}$$

$$L_n = [1 + i(\Delta + n\nu)]^{-1} \, , \quad n \text{ odd} \, . \tag{2.21}$$

In Eqs. (2.17) – (2.19) and (2.21) we have introduced the scaled velocity

$$\nu = \frac{k_0 v}{\gamma_\perp} \, . \tag{2.22}$$

The steady-state medium response is fully determined by Eqs. (2.12) – (2.21) once the continued fraction F is evaluated. Generally we must accomplish this task numerically. However, for two special cases a compact analytical form may be given. We will simply note these results here without giving the details of their proof. First note that as a generalization of Eq. (2.18) all of the $R_n = d_{n+2}/d_n$ may be written as continued fractions:

$$R_n = -X_f X_b^* \frac{a_{n+1}}{A_{n+2}}$$

$$\cfrac{1}{1 - |X_f|^2 |X_b|^2 \dfrac{a_{n+3}^2}{A_{n+2} A_{n+4}} \cfrac{1}{1 - |X_f|^2 |X_b|^2 \dfrac{a_{n+5}^2}{A_{n+4} A_{n+6}} \cfrac{1}{1 - \cdots}}} \, ,$$

$$\text{even } n \geq 0 \, . \tag{2.23}$$

Then for $\nu = 0$ we employ the continued fraction representation for $(1 + x)^{\frac{1}{2}}$ [54,55] to find

$$R_n = e^{i(\phi_f - \phi_b)} \frac{1 + \Delta^2 + |X_f|^2 + |X_b|^2}{2 |X_f| |X_b|} \left\{ 1 - \left[1 - \frac{4 |X_f|^2 |X_b|^2}{(1 + \Delta^2 + |X_f|^2 + |X_b|^2)^2} \right]^{\frac{1}{2}} \right\}$$

$$\text{even } n > 0 \, . \tag{2.24}$$

We have written $X_{f,b} = |X_{f,b}| \exp(i\phi_{f,b})$. This expression reproduces results obtained by more direct procedures for a homogeneously broadened medium[5,37,54]. The second special case takes $\Delta = 0$, $\gamma_\parallel = \gamma_\perp$ and $|X_f| = |X_b| = X/\sqrt{2}$, where we introduce a scaling by $\sqrt{2}$ in anticipation of our definition for X in the following section. After some algebraic manipulation a standard continued fraction representation for the ratio of two Bessels functions[56] leads to the result

$$R_n = e^{i(\phi_f - \phi_b)} \frac{J_{n+2-i\frac{1}{\nu}}\left(\frac{\sqrt{2}X}{\nu}\right)}{J_{n-i\frac{1}{\nu}}\left(\frac{\sqrt{2}X}{\nu}\right)}, \quad \text{even } n > 0 . \qquad (2.25)$$

In the context of laser theory this solution has been derived in different approaches by Stenholm and Lamb[44], and Feldman and Feld[45]. Its use in optical bistability will be restricted to the absorptive case ($\Delta = 0$) and the MFL ($|X_f| = |X_b| = X/\sqrt{2}$).

In Eqs. (2.17) - (2.19) and (2.21) velocity dependence enters the solutions for $p(v,z)$ and $d(v,z)$ through the variable ν. This scaled velocity is then the natural choice for the variable of integration in Eqs. (2.7) and (2.8). We therefore introduce a density $n(\nu)$ with $n(\nu)d\nu = n(v)dv$. Explicitly we write

$$n(\nu) = \frac{N}{V}\frac{1}{\sqrt{\pi}\sigma} e^{-\frac{\nu^2}{\sigma^2}} \qquad (2.26)$$

where $\sigma = k_0 u/\gamma_\perp$ gives the ratio of inhomogeneous to homogeneous line-width. The corresponding polarization $p(\nu,z)$ and inversion $d(\nu,z)$ are also distinguished in script notation.

We may resolve $n(\nu)$ into separate components $n_1(\nu,z)$ and $n_2(\nu,z)$ corresponding to the occupation of lower and upper atomic states respectively; $n = n_1 + n_2$ and $(n_1 - n_2)/2 = -d$. In particular consider these densities averaged over a wavelength. We have $n = \bar{n}_1 + \bar{n}_2$, $(\bar{n}_1 - \bar{n}_2)/2 = -d_0$ (for a discussion of the spatial structure associated with higher harmonics in Eq. (2.10) see Refs. 44 and 45) and from Eq. (2.15) we may write

$$\bar{n}_{1,2}(\nu,z) = \frac{n(\nu)}{2}\left[1 \pm \frac{1}{1 + S(|X_f|^2, |X_b|^2, \Delta, \nu)}\right]; \qquad (2.27)$$

$\bar{n}_1$ and $\bar{n}_2$ retain any z-dependence entering through X_f and X_b. In Fig. 1 we plot numerical results for $\bar{n}_1(\nu)$ with $|X_f| = |X_b| = X/\sqrt{2}$. This serves to illustrate general features in the saturation of the Doppler line. For relatively weak fields symmetric hole burning (where $|X_f|$ and $|X_b|$ are unequal $\bar{n}_1(\nu)$ is no longer symmetric) at $\nu = -\Delta$ and $\nu = +\Delta$ corresponds to resonant interaction with forward and backward waves respectively. At higher intensities these holes are deepened and power broadened. An overlap eventually occurs where the atomic population at line center is saturated by both forward and backward waves. The characteristic oscillatory structure

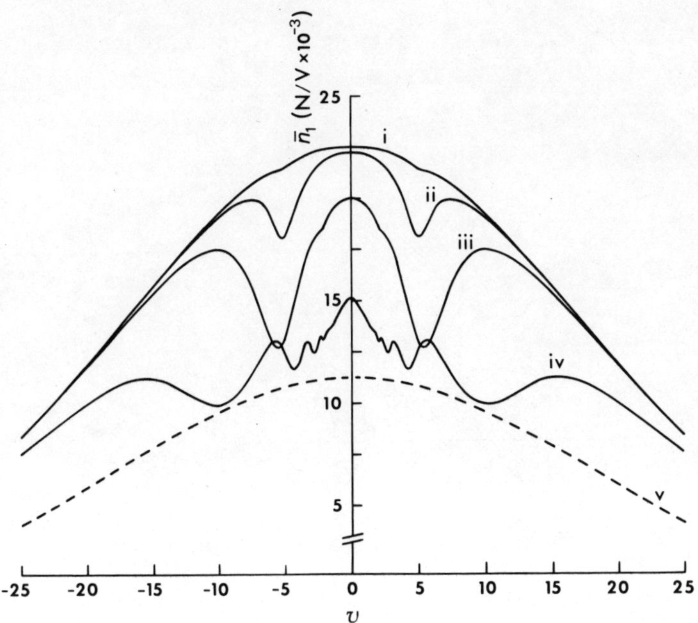

Fig. 1.   Saturation of the Doppler line by forward and backward waves
of equal amplitude. $\bar{n}_1(\nu)$ is plotted for $\Delta = 5$, $\sigma = 25$,
$\gamma_\perp/\gamma_\parallel = 0.5$, and (i) $X = 0.2$, (ii) $X = 1$, (iii) $X = 3$,
(iv) $X = 10$, and (v) full saturation ($X \rightarrow \infty$).

which emerges is both a standing-wave and nonlinear effect[44,45]. It
arises in the modulated intensity experienced by atoms moving through
the standing-wave field.

We will let $P_n(z)$ and $D_n(z)$ denote the integrated Fourier coefficients which enter expansions of the net polarization and inversion
via Eqs. (2.7) - (2.10). As has been the case in earlier discussions
of optical bistability we assume that only $P_{\pm 1}(z)$ couple to significant cavity fields, and write $P_{\pm 1}(z) \equiv P_{f,b}(z)$. This is consistent
with the form we have taken for $E(z,t)$. From Eqs. (2.17) - (2.21)
it is useful to note that

$$F(|X_f|^2, |X_b|^2, \Delta, \nu) = F(|X_b|^2, |X_f|^2, \Delta, -\nu)^* ,$$

$$\tag{2.28}$$

$$S(|X_f|^2, |X_b|^2, \Delta, \nu) = S(|X_b|^2, |X_f|^2, \Delta, -\nu) .$$

Then Eq. (2.12) gives

$$p_{f,b}(\nu,z) = p_{\pm 1}(\nu,z) = \frac{n(\nu)}{2}\left(\frac{\gamma_{\parallel}}{\gamma_{\perp}}\right)^{\frac{1}{2}} X_{f,b}$$

$$\times\ a(|X_{f,b}|^2,\ |X_{b,f}|^2,\ \Delta,\ \pm\nu),\qquad\qquad (2.29)$$

with

$$a(|X_f|^2,\ |X_b|^2,\ \Delta,\ \nu) = \frac{i + (\Delta + \nu)}{1 + (\Delta + \nu)^2}$$

$$\frac{1 - |X_b|^2\ F(|X_f|^2,\ |X_b|^2,\ \Delta,\ \nu)}{1 + S(|X_f|^2,\ |X_b|^2,\ \Delta,\ \nu)}\ .\qquad\qquad (2.30)$$

For the net forward and backward polarization we may write

$$P_{f,b}(z) = P_{\pm 1}(z) = \varepsilon_0\left(\frac{\alpha c}{\mu\omega_0}\right) E_{f,b}(z)$$

$$\times\ A(|X_{f,b}|^2,\ |X_{b,f}|^2,\ \Delta,\ \sigma)\ ,\qquad\qquad (2.31)$$

where

$$A(|X_f|^2,\ |X_b|^2,\ \Delta,\ \sigma) = \frac{1}{\sqrt{\pi}\sigma}\int_{-\infty}^{\infty} d\nu\ e^{-\frac{\nu^2}{\sigma^2}}$$

$$a(|X_f|^2,\ |X_b|^2,\ \Delta,\ \nu)\qquad\qquad (2.32)$$

and $\alpha = N\mu^2\omega_0/\varepsilon_0\hbar V\gamma_{\perp}c$ (mks units) is the resonant absorption coefficient for an homogeneously broadened medium of density N/V. In
addition to the symmetry of Eq. (2.28), complex conjugation for F
and S (S is of course real) corresponds to the transformation $\Delta \rightarrow -\Delta$
$\nu \rightarrow -\nu$. We then show

$$A(|X_f|^2,\ |X_b|^2,\ \Delta,\ \sigma) = -A(|X_f|^2,\ |X_b|^2,\ -\Delta,\ \sigma)^*\ .\qquad (2.33)$$

For an homogeneously broadened medium

$$A(|X_f|^2, |X_b|^2, \Delta, 0) = a(|X_f|^2, |X_b|^2, \Delta, 0) , \qquad (2.34)$$

and using Eqs. (2.16), (2.17) and (2.24) we find agreement with previous results[39].

In Eq. (2.31) $(\alpha c/\omega_0) A(|X_f|^2, |X_b|^2, \Delta, \sigma)$ and $(\alpha c/\omega_0) A(|X_b|^2, |X_f|^2, \Delta, \sigma)$ are nonlinear susceptibilities governing the steady-state response to forward and backward waves respectively. They are resolved into contributions from separate velocity subpopulations in $(\alpha c/\omega_0) a(|X_{f,b}|^2, |X_{b,f}|^2, \Delta, \pm v)$. Figures 2 and 3 illustrate the saturation of $a(|X_f|^2, |X_b|^2, \Delta, v)$ for $|X_f| = |X_b| = X/\sqrt{2}$. Corresponding curves for the backward wave are simply reflected about $v = 0$. For weak intensities $(|X_{f,b}|^2 \ll 1)$ we may write

$$a(|X_f|^2, |X_b|^2, \Delta, v) \approx a(0,0,\Delta,v) = \frac{i + (\Delta + v)}{1 + (\Delta + v)^2} ; \qquad (2.35)$$

$A(0,0,\Delta,\sigma)$ is given by the plasma dispersion function $Z(\xi)$[54]:

$$A(0,0,\Delta,\sigma) = \frac{1}{\sigma} Z(\frac{i-\Delta}{\sigma}) . \qquad (2.36)$$

In Figs. 2 and 3 $a(0,0,\Delta,v)$ is shown inset together with the early stages of saturation. Note that saturation initially proceeds more slowly for $\Delta \neq 0$ (Fig. 3) where only the forward wave is tuned at $v = -\Delta$. Detuning of the backward wave is greater for $v < -\Delta$ than for $v > -\Delta$ and this gives an asymmetry to the curves of Fig. 3. At higher intensities we again observe the structure which we met in Fig. 1 (the symmetric picture of Fig. 1 would correspond here to curves for $a(X^2/2, X^2/2, \Delta, v) + a(X^2/2, X^2/2, \Delta, -v)$ which add contributions from forward and backward waves).

III.   STEADY-STATE OPTICAL BISTABILITY
IN A FABRY-PEROT

We consider a Fabry-Perot interferometer filled with a Doppler-broadened medium. The cavity axis is aligned in the z-direction, where, in the plane-wave and slowly-varying-amplitude approximations, cavity fields are specified by Eqs. (2.2) and (2.3) (field contributions which vary as $\exp(\pm nik_0 z)$ $n = 3,5...$ are neglected). Maxwells equations for forward and backward wave amplitudes read

$$(\frac{\partial}{\partial t} \pm c \frac{\partial}{\partial z}) E_{f,b} = i \frac{\mu\omega_0}{2\varepsilon_0} P_{\pm 1}. \qquad (3.1)$$

                                    H. J. CARMICHAEL AND G. P. AGRAWAL

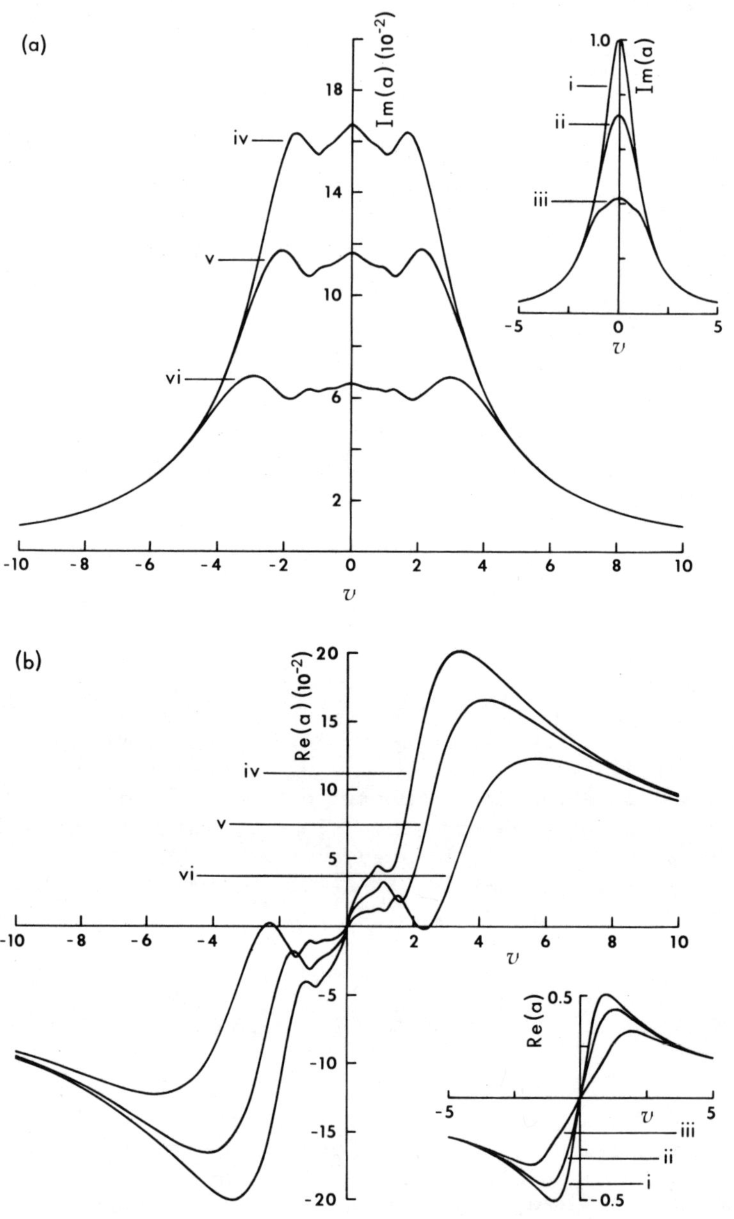

Fig. 2. Saturation of $a(X^2/2, X^2/2, \Delta, \nu)$ for a tuned medium. (a) $\mathrm{Im}(a)$ and (b) $\mathrm{Re}(a)$ for $\Delta = 0$, $\gamma_\perp/\gamma_\parallel = 0.5$, and (i) $X = 0$ (Eq. (2.35)), (ii) $X = 0.5$, (iii) $X = 1.0$, (iv) $X = 2.0$, (v) $X = 2.5$, and (vi) $X = 3.5$.

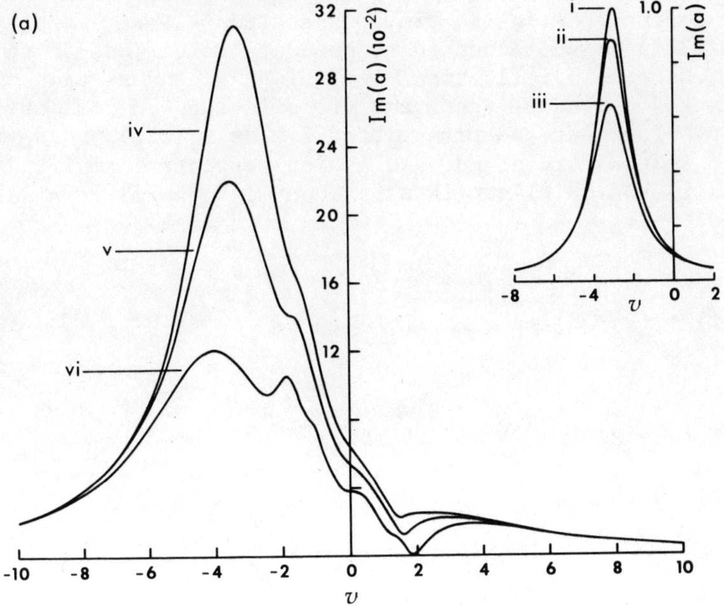

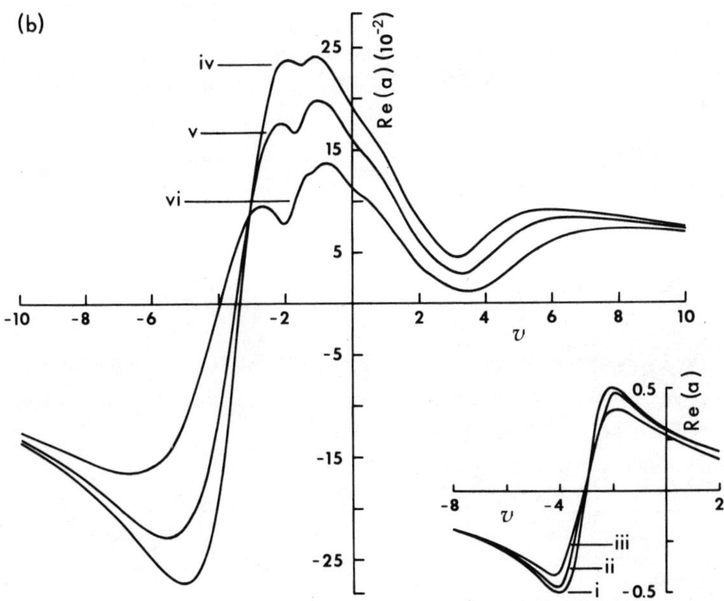

Fig. 3. Saturation of $a(X^2/2, X^2/2, \Delta, \nu)$ for a detuned medium. (a) Im(a) and (b) Re(a) for $\Delta = 3$, $\gamma_\perp/\gamma_\parallel = 0.5$, and (i) X = 0 (Eq. (2.35)), (ii) X= 0.5, (iii) X = 1.0, (iv) X = 2.0, (v) X = 2.5, and (vi) X = 3.5.

Here $P_{\pm 1}(z,t)$ are coefficients in the Fourier expansion of $P(z,t)$ and from definitions in the previous section $P_{\pm 1}(z,\infty) \equiv P_{\pm 1}(z)$. Equation (3.1) is supplimented by boundary conditions at the cavity mirrors. We take a reflection coefficient R, and at each mirror phase changes ϕ_T and ϕ_R accompany transmission and reflection respectively. Incident and transmitted fields $E_i(z,t)$ and $E_t(z,t)$ are expressed in the form of Eq. (2.2) with respective amplitudes $E_i \exp(ik_0 z)$ and $E_t(t) \exp(ik_0 z)$. Then with the mirrors set at $z = 0$ and $z = L$:

$$E_t(t) = (1 - R)^{\frac{1}{2}} e^{i\phi_T} E_f(L,t) \; , \tag{3.2}$$

$$E_b(L,t) = R^{\frac{1}{2}} e^{i\phi_R} e^{2ik_0 L} E_f(L,t) \; , \tag{3.3}$$

$$E_f(0,t) - R^{\frac{1}{2}} e^{i\phi_R} E_b(0,t) = (1 - R)^{\frac{1}{2}} e^{i\phi_T} E_i \; . \tag{3.4}$$

For the steady state $P_{+1}(z) \equiv P_f(z)$ and $P_{-1}(z) \equiv P_b(z)$ are given by Eq. (2.31). Nonlinear susceptibilities $(\alpha c/\omega_0) \, A(|X_f|^2, \, |X_b|^2, \, \Delta, \, \sigma)$ and $(\alpha c/\omega_0) \, A(|X_b|^2, \, |X_f|^2, \, \Delta, \, \sigma)$ then provide the central ingredient for a steady-state description of optical bistability. Writing $X_{f,b} = |X_{f,b}| \exp(i\phi_{f,b})$, and setting $I_{f,b} = |X_{f,b}|^2$, $\beta_{\pm} = \phi_f \pm \phi_b$, from Eq. (3.1) we find

$$\frac{dI_{f,b}}{dz} = \mp \alpha I_{f,b} \; \mathrm{Im} A(I_{f,b}, \, I_{b,f}, \, \Delta, \, \sigma) \; , \tag{3.5}$$

$$\frac{d\beta_{\pm}}{dz} = \frac{\alpha}{2} \; \mathrm{Re} \, [A(I_f, \, I_b, \, \Delta, \, \sigma) \mp A(I_b, \, I_f, \, \Delta, \, \sigma)] \; . \tag{3.6}$$

Using these equations, together with boundary conditions at $z = 0$ and $z = L$, we may calculate the input amplitude E_i corresponding to a given output amplitude E_t. We define

$$Y = \frac{\sqrt{2}}{(1-R)^{\frac{1}{2}}} \frac{|E_i|}{E_s} \; , \qquad X = \frac{\sqrt{2}}{(1-R)^{\frac{1}{2}}} \frac{|E_t|}{E_s} \; , \tag{3.7}$$

and Eqs. (3.2) – (3.4) give

$$X^2/2 = I_f(L) = R^{-1} I_b(L) \; , \tag{3.8}$$

$$Y^2/2 = (1-R)^{-2} \left\{ (I_f(0)^{\frac{1}{2}} - R^{\frac{1}{2}}I_b(0)^{\frac{1}{2}})^2 - 2R^{\frac{1}{2}}I_f(0)^{\frac{1}{2}}I_b(0)^{\frac{1}{2}} \right.$$

$$\left. x \; [\cos(\theta + \beta_-(0) - \beta_-(L)) - 1] \right\} \; , \qquad\qquad (3.9)$$

where $\theta = 2\pi - (2k_0L + 2\phi_R)_{\text{mod } 2\pi}$ is an initial cavity detuning. Further, if $E_i = |E_i| \exp(i\phi_i)$ and $E_t = |E_t| \exp(i\phi_t)$, the phase change between incident and transmitted fields is obtained with

$$\phi_t - \phi_i = 2\phi_T + \phi_f(L) - \phi_f(0) - \xi \; , \qquad\qquad (3.10)$$

where ξ gives the phase shift due to cavity detuning:

$$\tan[\xi + (\theta + \beta_-(0) - \beta_-(L))/2] = \frac{I_f(0)^{\frac{1}{2}} + R^{\frac{1}{2}}I_b(0)^{\frac{1}{2}}}{I_f(0)^{\frac{1}{2}} - R^{\frac{1}{2}}I_b(0)^{\frac{1}{2}}}$$

$$x \; \tan[(\theta + \beta_-(0) - \beta_-(L))/2] \; , \qquad\qquad (3.11)$$

and the angles $\xi + (\theta + \beta_-(0) - \beta_-(L))/2$ and $(\theta + \beta_-(0) - \beta_-(L))/2$ lie in the same quadrant. Note that $\phi_f(L) - \phi_f(0) = [(\beta_+(L) - \beta_+(0)) + (\beta_-(L) - \beta_-(0))]/2$ and only the phase differences $\beta_\pm(L) - \beta_\pm(0)$ enter Eqs. (3.9) and (3.10).

Most generally Eqs. (3.5) and (3.6) may be integrated numerically to obtain transmission characteristics for arbitrary values of the parameters αL, R, Δ, θ and σ (for $\sigma \neq 0$ evaluation of the susceptibility A also requires that the ratio $\gamma_\perp/\gamma_\parallel$ be specified). Such a program is readily accomplished, although with the continued fraction F appearing in the integrand of Eq. (2.32) the repeated evaluation of this integral can be quite time consuming. In this paper numerical results are given only for the MFL. For weak absorption ($\alpha L \ll 1$) we may integrate Eqs. (3.5) and (3.6) taking the right hand sides constant. From Eqs. (3.8) and (3.9) we find

$$Y^2 = X^2 \left\{ \left[1 + \frac{\alpha L}{2(1-R)} \; \text{Im}\left(A\left(\frac{X^2}{2}, \; R\frac{X^2}{2}, \; \Delta, \; \sigma\right) + RA\left(R\frac{X^2}{2}, \; \frac{X^2}{2}, \; \Delta, \; \sigma\right)\right) \right]^2 \right.$$

$$+ R\left[\frac{\theta}{1-R} - \frac{\alpha L}{2(1-R)} \; \text{Re}\left(A\left(\frac{X^2}{2}, \; R\frac{X^2}{2}, \; \Delta, \; \sigma\right)\right.$$

$$\left.\left. + A\left(R\frac{X^2}{2}, \; \frac{X^2}{2}, \; \Delta, \; \sigma\right)\right)\right]^2 \right\} \; , \qquad\qquad (3.12)$$

where we require $\theta \ll 1$. This equation holds for all values of R.
For the MFL:

$$\alpha L \to 0, \ (1 - R) \to 0, \ \theta \to 0, \ |E_i| \to 0, \ |E_t| \to 0;$$

$$\frac{\alpha L}{1 - R}, \ \frac{\theta}{1 - R}, \ \frac{|E_i|}{(1 - R)^{\frac{1}{2}}}, \ \frac{|E_t|}{(1 - R)^{\frac{1}{2}}} \ \text{constant},$$

the state equation then reads

$$Y^2 = X^2 \left\{ \left[1 + 2C \mathrm{Im} A \left(\frac{X^2}{2}, \frac{X^2}{2}, \Delta, \sigma \right) \right]^2 \right.$$

$$\left. + \left[\phi - 2C \mathrm{Re} A \left(\frac{X^2}{2}, \frac{X^2}{2}, \Delta, \sigma \right) \right]^2 \right\} , \tag{3.13}$$

where

$$\phi = \theta / (1 - R), \quad C = \alpha L / 2 (1 - R) \ . \tag{3.14}$$

Further, Eqs. (3.10) and (3.11) give

$$\phi_t - \phi_i = 2\phi_T + \theta/2 - \tan^{-1} \left[\phi - 2C\mathrm{Re} A \left(\frac{X^2}{2}, \frac{X^2}{2}, \Delta, \sigma \right) \right] \tag{3.15}$$

where the inverse tangent lies in the range $\pm(\pi/2)$. Using Eqs.
(2.16), (2.17) and (2.24) to evaluate $A(X^2/2, X^2/2, \Delta, 0)$ (Eq.
(2.34)), Eq. (3.13) reduces to the mean-field state equation for an
homogeneously broadened medium[39,40,42].

Absorptive and dispersive effects enter Eq. (3.13) via the terms
$2C\mathrm{Im}(A)$ and $2C\mathrm{Re}(A)$ respectively. In Eq. (2.32) $A(X^2/2, X^2/2, \Delta, \sigma)$
is evaluated by integrating $a(X^2/2, X^2/2, \Delta, \nu)$ against a gaussian
velocity distribution. Clearly, the symmetry of Fig. 2(b) then
implies that $\mathrm{Re} A(X^2/2, X^2/2, 0, \sigma) = 0$, and with $\Delta = 0$ Eq. (3.13)
describes purely absorptive bistability. More generally, for $\Delta = 0$
Eq. (2.33) requires

$$\mathrm{Re} A(I_f, I_b, 0, \sigma) = 0 \tag{3.16}$$

and Eqs. (3.5) and (3.6) also give absorptive bistability outside
the MFL. Transmission characteristics for absorptive bistability in
the MFL are plotted in Fig. 4. Here C remains constant and we have
varied the Doppler linewidth σ. Gibbs et al[6] commented on the difficulty of observing absorptive bistability in the presence of inhomogeneous broadening, and the loss of bistability for large σ, as

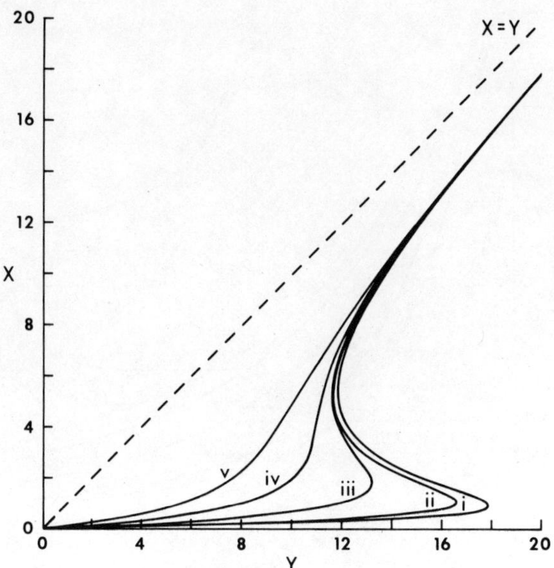

Fig. 4.   The effects of Doppler broadening on absorptive bistability
in the MFL.   Steady-state transmission characteristics for
$C = 20$, $\Delta = \phi = 0$, $\gamma_\perp/\gamma_\parallel = 0.5$, and (i) $\sigma = 0$, (ii) $\sigma = 1$,
(iii) $\sigma = 3$, (iv) $\sigma = 6$, and (v) $\sigma = 10$.

illustrated here, has been noted previously in discussions of inhomogeneous broadening in a ring cavity[18,19,40] and both Doppler[54] and
non-Doppler[39] broadening in a Fabry-Perot.

The curves in Fig. 4 do not distinguish the two separate effects
which are responsible for the reduction of absorptive bistability by
inhomogeneous broadening.   At constant C the atomic density N/V is
fixed.   However, absorption decreases with increasing σ as more of
the atomic population becomes detuned from line center.   Using Eq.
(2.36), absorption at weak fields is reduced from 2C at $\sigma = 0$ to
$(2C/\sigma)\,\mathrm{Im}\,Z(i/\sigma)$.   For $C = 20$ this effect is illustrated by the inset
to Fig. 5.   If this curve is taken as a definition of an effective C,
for $\sigma = 6$, C_{eff} is already only slightly above the critical value
$2C = 8$[2,17].

If decreased absorption is offset by an increase in atomic
density a second effect of inhomogeneous broadening may be identified.
In Fig. 5, saturation of the nonlinear susceptibility is illustrated
for various values of σ.   In each case C has been chosen to give
$2C\,\mathrm{Im}\,A(0,0,0,\sigma) = (2C/\sigma)(\mathrm{Im}\,Z(i/\sigma) = 40$.   We note a 'weakening' of the
absorptive nonlinearity with increasing σ, where the absorption due
to atoms detuned from line center saturates more slowly than that due
to the resonant atoms[6] (a power-broadening effect).   Transmission
characteristics for Doppler-broadened systems which show equal

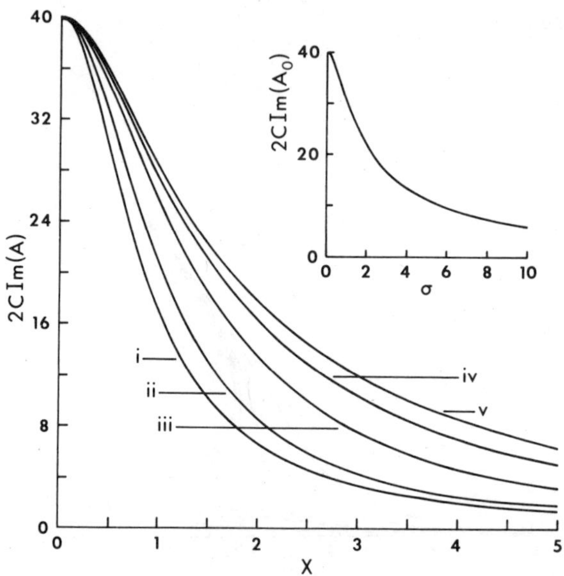

Fig. 5.   Saturation of the nonlinear susceptibility $2CImA(X^2/2, X^2/2,$
           $0,\sigma)$ for (i) $\sigma = 0$ (C = 20), (ii) $\sigma = 1$ (C = 26.39), (iii)
           $\sigma = 3$ (C = 47.53), (iv) $\sigma = 6$ (C = 80.93), and (v) $\sigma = 10$
           (C = 125.9).   The inset gives $2CIm(A_0) = 2CImA(0,0,0,\sigma)$
           $= (2C/\sigma)Im\ Z(i/\sigma)$(Eq. (2.36)) and determines the scaling
           of C in curves (i) – (v). $\gamma_\perp/\gamma_\parallel = 0.5$.

weak-field absorption are plotted in Fig. 6.   For $\sigma = 30$ bistability
is virtually eliminated.   Note that it will always be possible to
regain absorptive bistability with a further increase in C, although
higher incident intensities will then be required.

The nonlinearity which underlies dispersive bistability is
derived from the tuning of an intensity-dependent phase shift through
a cavity resonance.   The specific form of nonlinear response from
the medium itself does not have the importance it has in absorptive
bistability (phase shifts vary linearly with intensity in a Kerr
medium[7,13]).   In Eq. (3.13) the dispersive phase shift is given by
$2CReA(X^2/2, X^2/2, \Delta, \sigma)$.   Then the range of cavity detuning scanned
by $\phi - 2CReA(X^2/2, X^2/2, \Delta, \sigma)$ is the underlying determinant for
bistability.   For a saturable medium ϕ sets the detuning at high
intensities, and the weak-field phase shift $2CReA(0,0,\Delta,\sigma) =$
$(2C/\sigma)Re\ Z((i-\Delta)/\sigma)$ fixes the second limit to this range.   In Fig. 7
we plot transmission characteristics for dispersive bistability in
the MFL.   Note that this is not the purely dispersive limit[6,7,18,20];
residual absorption may be detected where curves (i) – (iv) (all tuned
through the resonance peak) remain below the line X = Y.   However, the
bistability is itself a dispersive feature since $2C/(1+\Delta^2) = 4$ is only
half the critical value for absorptive bistability.

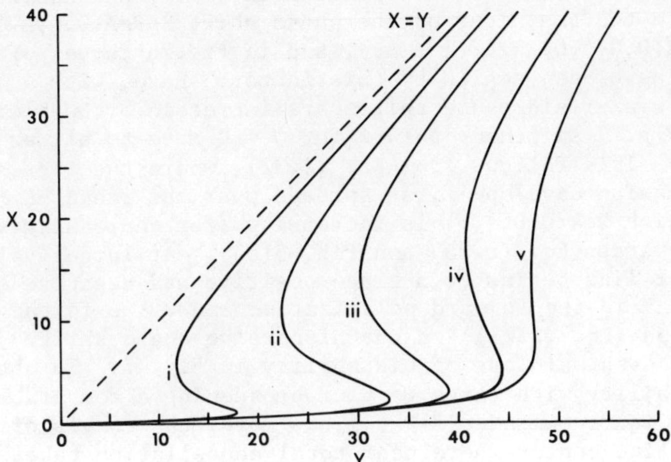

Fig. 6.   The effects of Doppler broadening on absorptive bistability
in the MFL for systems showing equal weak-field absorption.
Steady-state transmission characteristics for $\Delta = \phi = 0$,
$2CImA(0,0,0,\sigma) = 40$, $\gamma_\perp/\gamma_\parallel = 0.5$, and (i) $\sigma = 0$ $(C = 20)$,
(ii) $\sigma = 5$ $(C = 69.74)$, (iii) $\sigma = 10$ $(C = 125.9)$, (iv)
$\sigma = 20$ $(C = 238.6)$, and (v) $\sigma = 30$ $(C = 251.3)$.

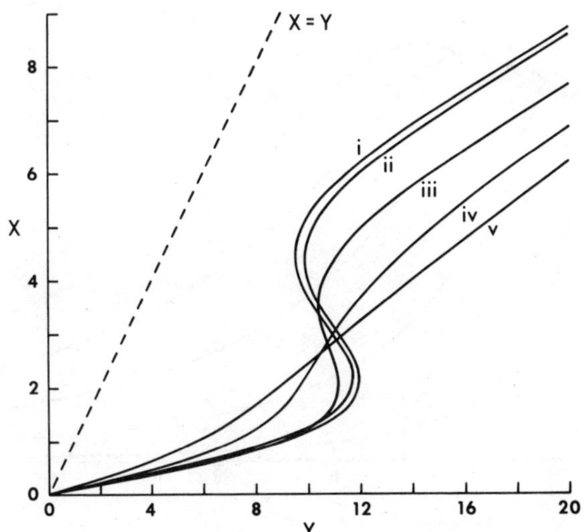

Fig. 7.   The effects of Doppler broadening on dispersive bistability
in the MFL.   Steady-state transmission characteristics for
$C = 20$, $\Delta = \phi = 3$, $\gamma_\perp/\gamma_\parallel = 0.5$, and (i) $\sigma = 0$, (ii) $\sigma = 1$,
(iii) $\sigma = 3$, (iv) $\sigma = 6$, and (v) $\sigma = 10$.

The transmission characteristics in Fig. 7 are understood in
relation to the behaviour of the phase shift $2CReA(0,0,\Delta,\sigma)$. For
$\sigma = 0$, $ReA(0,0,\Delta,0) = \Delta/(1 + \Delta^2)$, and in Fig. 8 curves of $ReA(0,0,\Delta,\sigma)$
against σ have been scaled by this factor. Here, with $\Delta > 1$ and
$\sigma \lesssim \Delta - 1$ weak-field phase shifts are increased by Doppler broadening. In Fig. 7 switching points for $\sigma = 0$ move to higher intensities
for $\sigma = 1$. This follows from Eq. (2.32), where the increasing width
of the gaussian envelope first spreads over the range $-\Delta + 1 \leq \nu \leq \Delta - 1$
through which $Rea(0,0,\Delta,\nu)$ is increasing (for decreasing ν) and
concave upwards (Eq. (2.35) and Fig. 3(b)). At larger values of σ
the Doppler line begins to average positive and negative contributions
(for $\nu \gtrsim -\Delta$) in the forward polarization and $\nu \lesssim \Delta$ in the backward
polarization (Eq. (2.29)) to the dispersive phase shift. This ex-
plains the eventual loss of bistability in Fig. 7. To obtain dispersive bistability with large Doppler broadening a comparably large
detuning Δ is recommended[19,40]. This moves the curves of Fig. 3(b)
away from line center where near total cancellation takes place. Of
course for large Δ an increase in C is required. With $\Delta \gg \sigma$ weak-field
phase shifts fall off as $1/\Delta$, is in an homogeneously broadened medium.
If $\Delta \gtrsim \sigma$ the fall off is slower, as in this region the curves of Fig.
8 have $ReA(0,0,\Delta,\sigma) > \Delta/(1+\Delta^2)$. If in the limit of large Δ we expand
$A(I_f,I_b,\Delta,\sigma)$ to first order in forward and backward intensities, a
generalization of the method used by Lax[46] for the laser gives the
integral in Eq. (2.32) in terms of the plasma dispersion function[57]
and its derivative (also see Ref. 40). We will not give details here

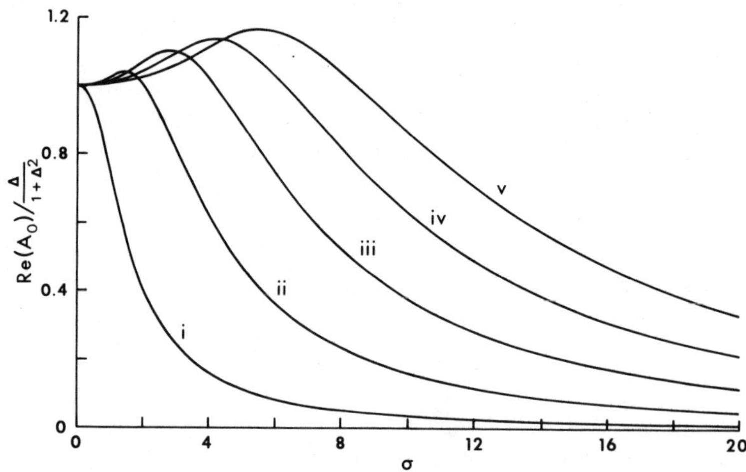

Fig. 8.   Behaviour of the weak-field phase shift $2CRe(A_0) = 2CReA(0,0$
$\Delta,\sigma) = (2C/\sigma)Re\ Z((i - \Delta)/\sigma)$ (Eq. (2.36)) as a function of
Δ and σ. Curves are normalized to the value $2C\Delta/(1 + \Delta^2)$
for homogeneous broadening: (i) $\Delta = 1$, (ii) $\Delta = 3$, (iii)
$\Delta = 5$, (iv) $\Delta = 7$, and (v) $\Delta = 9$.

Our comments in the last few paragraphs apply in qualitative respects to inhomogeneous broadening in both Fabry-Perot (Doppler and non-Doppler) and ring-cavity geometries. The detailed features illustrated by Figs. 1-3 have few consequences after they are integrated against the Doppler line. However, saturation of the dispersive phase shift does exhibit a unique property which deserves comment. This has been illustrated in Fig. 9. For homogeneous broadening $Im(A)$ (Fig. 5) and $Re(A)$ (Fig. 9) are saturated identically. If $\sigma \neq 0$, and $\sigma > \Delta$, $ReA(X^2/2, X^2/2, \Delta, \sigma)$ averages positive and negative contributions as mentioned above. In relation to Fig. 3(b) we have also pointed to the asymmetric saturation of $a(X^2/2, X^2/2, \Delta, \nu)$. These two effects may combine to change the sign of the dispersive phase shift (Fig. 9). For weak fields the positive contribution to Eq. (2.32) dominates ($\nu > -\Delta$ in Fig. 3(b)) as this lies nearer line center. However, $a(X^2/2, X^2/2, \Delta, \nu)$ saturates more quickly for $\nu >- \Delta$. It is then possible for positive and negative contributions to cancel giving $ReA(X^2/2, X^2/2, \Delta, \sigma) = 0$. We have yet to determine whether or not bistability may be obtained using this effect. It does seem however that generally absorption will supress this possibility.

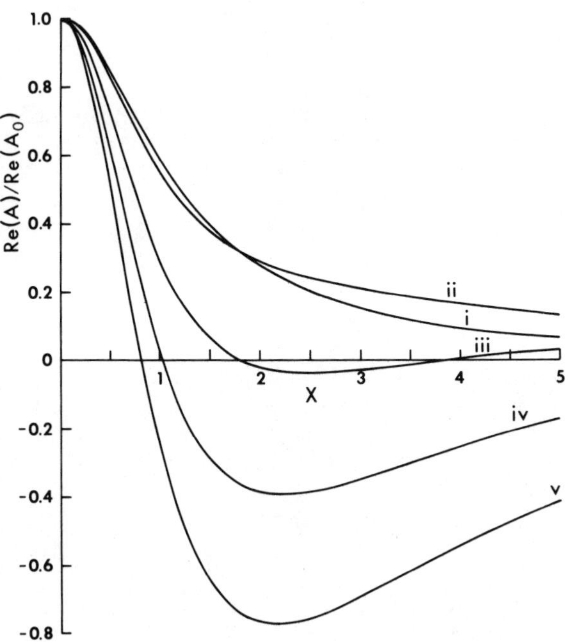

Fig. 9.   Saturation of the dispersive phase shift $2CReA(X^2/2, X^2/2, \Delta, \sigma)$ for $\Delta = 1$ and (i) $\sigma = 0$, (ii) $\sigma = 5$, (iii) $\sigma = 10$, (iv) $\sigma = 15$, and (iv) $\sigma = 20$. Curves are normalized by $2CRe(A_0) = 2CReA(0,0,\Delta,\sigma)$. $\gamma_\perp/\gamma_\parallel = 0.5$.

IV. COMPARISONS WITH A TRUNCATED BLOCH HIERARCHY

For homogeneously broadened systems a number of authors have treated standing-wave effects in an approximate scheme based on a truncated Bloch hierarchy[17,49-52]. The same approximation has been used in a recent discussion of Doppler broadening in a Fabry-Perot[53]. For transient studies[50,53] the use of a truncated system of equations seems unavoidable. However, as we have seen in the previous sections, in the steady state this approximation is not necessary, and in this context its accuracy may then be assessed. For absorptive bistability in an homogeneously broadened medium a comparison of results obtained with truncated and full Bloch hierarchies has been made in an earlier publication[37]. To conclude this paper we compare results from sections II and III with those obtained within a truncated scheme.

In the present formulation a truncation is introduced in Eqs. (2.9) and (2.10) where we write

$$p(v,z) = p_1(v,z)\, e^{ik_0 z} + p_{-1}(v,z)\, e^{-ik_0 z} , \qquad (4.1)$$

$$d(v,z) = d_0(v,z) + (d_2(v,z)\, e^{2ik_0 z} + c.c.) , \qquad (4.2)$$

and $d_2(v,z)^* = d_{-2}(v,z)$. These are substituted in Eqs. (2.4) and (2.5) and terms varying as $\exp(\pm 3ik_0 z)$ are neglected. Then steady-state solutions for $p_{\pm 1}(z)$, $d_0(z)$ and $d_2(z)$ may be taken from Eqs. (2.12) – (2.21), with two adjustments:

$$F(|X_f|^2,\ |X_b|^2,\ \Delta,\ \nu) = a_1/A_2 , \qquad (4.3)$$

$$A_2 = 1 + 2i\,\frac{\gamma_\perp}{\gamma_\parallel}\,\nu + \frac{1}{2}|X_f|^2\,L_{-1}^* + \frac{1}{2}|X_b|^2\,L_1 . \qquad (4.4)$$

The development of sections II and III may now be carried over with this new definition for F. This approximation is similar, although not identical, to the rate equation approximation[44,45]. In the latter case d_2 is also set to zero and Eq. (4.3) is replaced by F = 0.

The pressure of space will allow us only a brief discussion of the results presented in Figs. 10-13. Figure 10 illustrates differences in the saturation of $a(X^2/2,\ X^2/2,\ \Delta,\ \nu)$ arising with the use of Eq. (4.3). The oscillatory structure observed in Figs. 1-3 is not reproduced in this approximation. Beyond the loss of this detail significant quantitative deviations appear in the strongly saturated region around $\nu = 0$. While detailed oscillatory features are averaged in Eq. (2.32), these larger differences are carried through to the

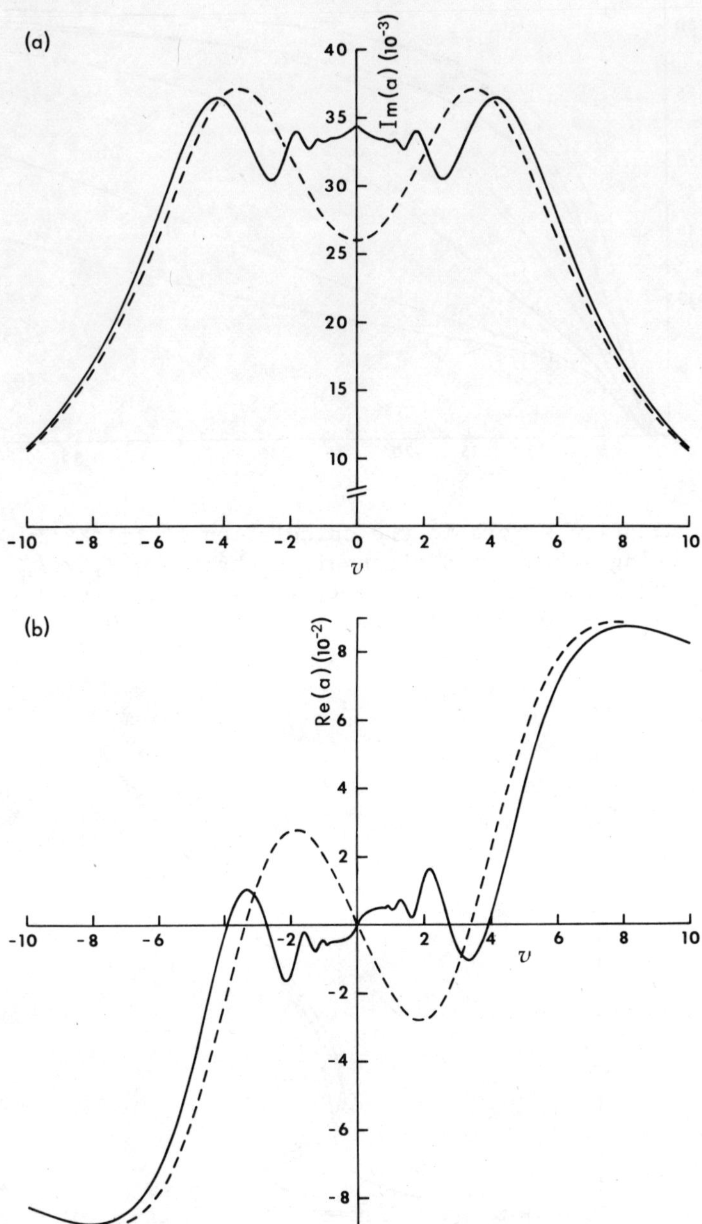

Fig. 10.   (a) Ima$(X^2/2, X^2/2, \Delta, \nu)$ and (b) Rea$(X^2/2, X^2/2, \Delta, \nu)$ calculated using full (solid curve) and truncated (dashed curve) Bloch hierarchies: $\Delta = 0$, $\gamma_\perp/\gamma_\parallel = 0.5$, and $X = 5$.

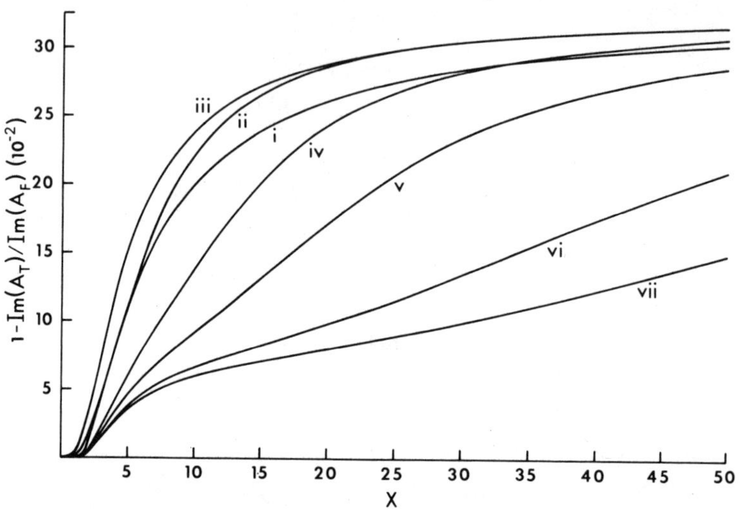

Fig. 11. Relative errors in the calculation of $\mathrm{Im}A(X^2/2,\ X^2/2,\ \Delta,\ \sigma)$
using a truncated Bloch hierarchy: $\Delta = 3$, $\gamma_\perp/\gamma_\parallel = 0.5$,
and (i) $\sigma = 0$, (ii) $\sigma = 1$, (iii) $\sigma = 3$, (iv) $\sigma = 6$,
(v) $\sigma = 10$, (vi) $\sigma = 20$, and (vii) $\sigma = 30$.

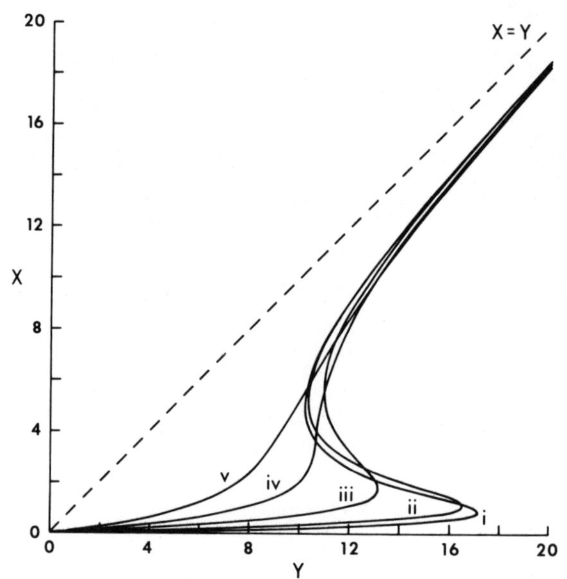

Fig. 12. Steady-state transmission characteristics in the MFL using
a truncated Bloch hierarchy: $C = 20$, $\Delta = \phi = 0$, $\gamma_\perp/\gamma_\parallel =$
0.5, and (i) $\sigma = 0$, (ii) $\sigma = 1$, (iii) $\sigma = 3$, (iv) $\sigma = 6$,
and (v) $\sigma = 10$.

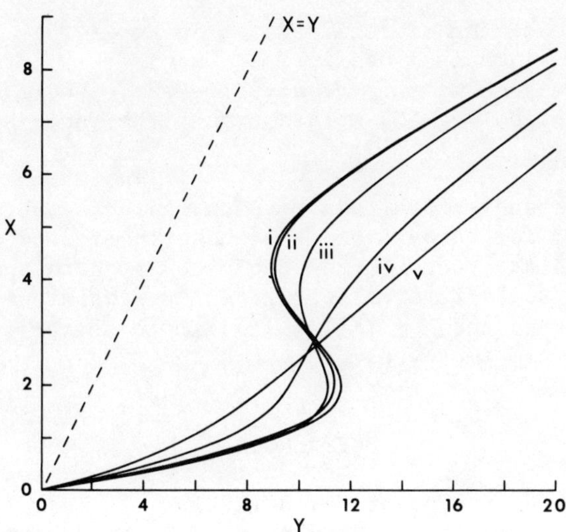

Fig. 13.   Steady-state transmission characteristics in the MFL using
           a truncated Bloch hierarchy: $C = 20$, $\Delta = \phi = 3$, $\gamma_\perp/\gamma_\parallel =$
           0.5, and (i) $\sigma = 0$, (ii) $\sigma = 1$, (iii) $\sigma = 3$, (iv) $\sigma = 6$,
           and (v) $\sigma = 10$.

susceptibility.   For $\Delta = 3$ the relative errors introduce to $\text{ImA}(X^2/2,$
$X^2/2, \Delta, \sigma)$ are shown in Fig. 11.   We use the notation A_T and A_F to
distinguish truncated and full Bloch hierarchies respectively, and
for homogeneous broadening (curve (i))

$$\frac{\text{Im}(A_T)}{\text{Im}(A_F)} = \frac{\left(1 + \Delta^2 + \frac{3}{2} X^2\right)^{-1}}{\frac{1}{X^2}\left[1 - \left(\frac{1 + \Delta^2}{1 + \Delta^2 + 2X^2}\right)^{\frac{1}{2}}\right]} . \tag{4.5}$$

Well below saturation ($X^2 \ll 1 + \Delta^2$), $\text{Im}(A_T)/\text{Im}(A_F) = 1$, and in the
limit $X \to \infty$, $\text{Im}(A_T)/\text{Im}(A_F) \to 2/3$.   With $\Delta = 3$ small values of σ introduce atoms which are closer to resonance with either the forward or
backward wave than those at line center.   For a given intensity these
atoms saturate more strongly.   This explains the curves for $\sigma = 1$
and $\sigma = 3$ in Fig. 11.   Larger values of σ introduce atoms which are
further from resonance, and then the trend is for Doppler broadening
to reduce the error in $\text{Im}(A_T)$.   At a fixed intensity the Doppler line
is now less saturated as σ is increased.

     We do not make a similar comparison for $\text{ReA}(X^2/2, X^2/2, \Delta, \sigma)$
here.   Note however, that with $\Delta \neq 0$ the relative differences for

homogeneous broadening are the same as in Eq. (4.5). For $\sigma \neq 0$ relative errors in $\mathrm{Re}(A_T)$ can in fact become very large, but this is not always a good measure of disagreement. Where $\mathrm{ReA}(X^2/2, X^2/2, \Delta, \sigma)$ changes sign (Fig. 9) we can expect the ratio $\mathrm{Re}(A_T)/\mathrm{Re}(A_F)$ to diverge.

Finally, steady-state transmission characteristics are given in Figs. 12 and 13 for comparison with those in Figs. 4 and 7. Here the differences are reduced from those observed in the susceptibility, since, as $A(X^2/2, X^2/2, \Delta, \sigma)$ saturates its contribution to Eq. (3.13) is also diminished and $X \rightarrow Y/(1 + \phi^2)$. Note that for large values of σ the agreement is improved.

REFERENCES

1. H. Seidel, U. S. Patent No. 3610731.
2. A. Szoke, V. Daneu, J. Goldhar, and N. A. Kurnit, Appl. Phys. Lett. <u>15</u>, 376 (1969).
3. J. W. Austin and L. G. DeShazer, J. Opt. Soc. Am. <u>61</u>, 650 (1971).
4. E. Spiller, J. Opt. Soc. Am. <u>61</u>, 669 (1971); J. Appl. Phys. <u>43</u>, 1673 (1972).
5. S. L. McCall, Phys. Rev. <u>A9</u>, 1515 (1974).
6. H. M. Gibbs, S. L. McCall, and T. N. C. Venkatesan, Phys. Rev. Lett. <u>36</u>, 1135 (1976); T. N. C. Venkatesan, Ph.D. thesis, City University of New York (1977) (unpublished).
7. F. S. Felber and J. H. Marburger, Appl. Phys. Lett. <u>28</u>, 732 (1976); J. H. Marburger and F. S. Felber, Phys. Rev. <u>A17</u>, 335 (1978).
8. T. N. C. Venkatesan and S. L. McCall, Appl. Phys. Lett. <u>30</u>, 282 (1977); H. M. Gibbs, T. N. C. Venkatesan, S. L. McCall, A. Passner, A. C. Gossard, and W. Weigmann, Appl. Phys. Lett. <u>34</u>, 511 (1979); <u>35</u>, 451 (1979).
9. P. W. Smith, E. H. Turner, and P. J. Maloney, IEEE J. Quantum Electron. <u>QE14</u>, 207 (1978); P. W. Smith, I. P. Kaminow, P. J. Maloney, and L. W. Stulz, Appl. Phys. Lett. <u>33</u>, 24 (1978); <u>34</u>, 62 (1979); P. W. Smith, J. P. Hermann, W. J. Tomlinson, and P. J. Maloney, Appl. Phys. Lett. <u>35</u> 846 (1979).
10. E. Garmire, J. H. Marburger, and S. D. Allen, Appl. Phys. Lett. <u>32</u>, 320 (1978); E. Garmire, S. D. Allen, J. Marburger, and C. M. Verber, Opt. Lett. <u>3</u>, 69 (1978).
11. F. T. Arecchi and A. Politi, Lett. Nuovo Cimento <u>23</u>, 65 (1978).
12. P. D. Drummond, K. J. McNeil, and D. F. Walls, Opt. Commun. <u>28</u>, 255 (1979).
13. T. Bischofberger and Y. R. Shen, Opt. Lett. <u>4</u>, 40 (1979); Phys. Rev. <u>A19</u>, 1169 (1979).
14. D. A. B. Miller and S. D. Smith, Opt. Commun. <u>31</u>, 101 (1979); D. A. B. Miller, S. D. Smith, and A. Johnston, Appl. Phys. Lett. <u>35</u>, 658 (1979).
15. A. Feldman, Opt. Lett. <u>4</u>, 115 (1979).

16. W. Sohler, Appl. Phys. Lett. $\underline{36}$, 351 (1980).

17. R. Bonifacio and L. A. Lugiato, Opt. Commun. $\underline{19}$, 172 (1976);
       Phys. Rev. $\underline{A18}$, 1129 (1978).

18. R. Bonifacio and L. A. Lugiato, Lett. Nuovo Cimento $\underline{21}$, 517
       (1978); R. Bonifacio, M. Gronchi, and L. A. Lugiato, Nuovo
       Cimento $\underline{B53}$, 311 (1979).

19. S. S. Hassan, P. D. Drummond, and D. F. Walls, Opt. Commun. $\underline{27}$,
       480 (1978).

20. G. P. Agrawal and H. J. Carmichael, Phys. Rev. $\underline{A19}$, 2074 (1979).

21. P. Schwendimann, J. Phys. $\underline{A12}$, L39 (1979).

22. H. J. Carmichael and D. F. Walls, J. Phys. $\underline{B10}$, L685 (1977);
       S. S. Hassan and D. F. Walls, J. Phys. $\underline{A11}$, L87 (1978); D. F.
       Walls, P. D. Drummond, S. S. Hassan, and H. J. Carmichael,
       Prog. Theor. Phys. Suppl. $\underline{64}$, 307 (1978); P. D. Drummond,
       D. Phil. thesis, University of Waikato (1979) (unpublished).

23. R. Bonifacio and L. A. Lugiato, Phys. Rev. Lett. $\underline{40}$, 1023 (1978);
       M. Gronchi and L. A. Lugiato, Lett. Nuovo Cimento $\underline{23}$, 593
       (1978); L. A. Lugiato, Nuovo Cimento $\underline{B50}$, 80 (1979).

24. L. M. Narducci, R. Gilmore, D. H. Feng and G. S. Agarwal, Opt.
       Lett. $\underline{2}$, 88 (1978), G. S. Agarwal, L. M. Narducci, R. Gilmore,
       and D. H. Feng, Phys. Rev. $\underline{A18}$, 620 (1978); Phys. Rev. $\underline{A21}$
       1029 (1980).

25. C. R. Willis, Opt. Commun. $\underline{23}$, 151 (1977); $\underline{26}$, 62 (1978); C. R.
       Willis and J. Day, Opt. Commun. $\underline{28}$, 137 (1979).

26. F. Casagrande and L. A. Lugiato, Nuovo Cimento (to be published).

27. A. R. Bulsara, W. C. Schieve, and R. F. Gragg, Phys. Lett. $\underline{68A}$,
       294 (1978); Phys. Rev. $\underline{A19}$, 2052 (1979).

28. A. Schenzle and H. Brand, Opt. Commun. $\underline{27}$, 485 (1978); $\underline{31}$, 401
       (1979).

29. F. T. Arecchi and A. Politi, Opt. Commun. $\underline{29}$, 361 (1979).

30. K. Kondo, M. Mabuchi, and H. Hasegawa, Opt. Commun. $\underline{32}$, 136
       (1980).

31. R. Bonifacio and P. Meystre, Opt. Commun. $\underline{27}$, 147 (1978); $\underline{29}$,
       131 (1979).

32. F. A. Hopf and P. Meystre, Opt. Commun. $\underline{29}$, 235 (1979).

33. F. A. Hopf, P. Meystre, P. D. Drummond, and D. F. Walls, Opt.
       Commun. $\underline{31}$, 245 (1979).

34. V. Benza and L. A. Lugiato, Lett. Nuovo Cimento $\underline{26}$, 405 (1979).

35. If quantitative comparisons are to be made it is important to
       keep track of the notational conventions of different authors.
       One possible source of confusion is different definitions for
       normalized incident and transmitted field amplitudes all
       designated by Y and X. For a clarification see Eqs. (54),
       (74), (83) and their context in ref. 42. Different definit-
       ions for the absorption coefficient α are also used; see the
       footnote in ref. 42.

36. R. Bonifacio and L. A. Lugiato, Lett. Nuovo Cimento 21, 505
     (1978); R. Bonifacio, L. A. Lugiato, and M. Gronchi, Theory
     of Optical Bistability, in Laser spectroscopy IV, Proceedings
     of the Fourth Conference on Laser Spectroscopy, 1979, H.
     Walther and W. K. Rothe, eds. (Springer, Berlin, 1979).
37. H. J. Carmichael, Optica Acta 27, 147 (1980).
38. J. A. Hermann, Optica Acta 27, 159 (1980).
39. G. P. Agrawal and H. J. Carmichael, Optica Acta 27, 651 (1980).
40. P. D. Drummond, D. Phil, thesis, University of Waikato (1979)
     (unpublished).
41. R. Roy and M. S. Zubairy, Phys. Rev. A21, 274 (1980).
42. H. J. Carmichael and J. A. Hermann, Z. Phys. B38, 365 (1980).
43. M. Gronchi and L. A. Lugiato, Opt. Lett. 5, 108 (1980).
44. S. Stenholm and W. E. Lamb, Jr., Phys. Rev. 181, 618 (1969);
     Phys. Rev. B1, 15 (1970).
45. B. J. Feldman and M. S. Feld, Phys. Rev. A1, 1375 (1970).
46. M. Lax, Fluctuations and Coherence Phenomena in Classical and
     Quantum Physics, in Brandeis University Summer Institute in
     Theoretical Physics 1966, Statistical Physics, Phase Transi-
     tions and Superfluidity, M. Chretien ed. (Gordon and Breach,
     New York, (1968)).
47. V. S. Letokhov and V. P. Chebotayev, Nonlinear Laser Spectroscopy
     (Springer, Berlin, 1977).
48. J. H. Shirley, Phys. Rev. A8, 347 (1973).
49. P. Meystre, Opt. Commun. 26, 277 (1978).
50. E. Abraham and R. K. Bullough, Opt. Commun. 29, 109 (1979).
51. R. Roy and M. S. Zubairy, Opt. Commun. 32, 163 (1980).
52. E. Abraham, S. S. Hassan, and R. K. Bullough, Opt. Commun. 33,
     93 (1980).
53. E. Abraham and S. S. Hassan (unpublished).
54. H. J. Carmichael and G. P. Agrawal, Opt. Commun. 34, 293 (1980).
55. A. N. Khovanskii, The Application of Continued Fractions and
     their Generalizations to Problems in Approximation Theory
     (P. Noordhoff, Groningen, 1963) p. 101.
56. Handbook of Mathematical Functions, M. Abramowitz and I. A.
     Stegun eds. (Dover, New York, 1965) p. 363.
57. B. D. Fried and S. D. Conte, The Plasma Dispersion Function,
     (Academic Press, New York, 1961).

THE ROLE OF PHASES IN THE TRANSIENT DYNAMICS OF NONLINEAR

INTERFEROMETERS*

J. D. Cresser and P. Meystre

Max Planck Gesellschaft zur Förderung der
Wissenschaften e.V.
Projektgruppe für Laserforschung

Abstract: Under appropriate conditions, the light intensity
transmitted by an interferometer filled with a non-linear medium
and irradiated by a resonant or near-resonant driving field exhibits one or many hysteresis cycles, and optical bi- or multistability.

In this paper, we analyze the response of such systems to
sudden changes in the driving field. The existence of anomalous
thresholds shows that it is not sufficient in general to consider
the standard intensity-out vs. intensity-in curves. Rather, a detailed analysis of the phase-space of the system must be carried
out. We discuss the cases of absorptive and dispersive bistability.
We show that in all cases, the phase of the driving field plays an
essential role.

This high sensitivity to phases leads in particular to the
question of the effect of phase-noise in the driving field. We
present preliminary results on this problem.

I. INTRODUCTION

Under appropriate conditions, the light intensity transmitted
by an interferometer filled with a non-linear medium and irradiated
by a resonant or near-resonant driving field exhibits one or many
hysteresis cycles, and bi- or multistability. This effect was suggested about ten years ago[1]. The first detailed theoretical analysis, due to McCall[2], its experimental evidence by Gibbs et al.[3],

---

*Research supported by the Bundesministerium fur Forschung und
Technologie and Euratom.

and the theoretical developments of Bonifacio and Lugiato[4], have led to an "explosion" of this subject, both experimentally and theoretically[5].

Over the last few years, and in collaboration with the Milano and the Hamilton groups, we have studied theoretically the transient response of bistable systems[6]. In two recent papers[7,8], we analyzed the response of a dispersive non-linear interferometer to an instantaneous change of the driving field. We found that the system switched sometimes "too soon," and sometimes in what looked like an erratic way (anomalous thresholds). These anomalies are easily understood once one realizes that the standard intensity-out vs. intensity-in curves usually considered when discussing bistability give only part of the story. To achieve a complete understanding of the transient dynamics of bistable and multistable systems, it is necessary to perform a detailed phase-space analysis.

In this paper, we generalize and extend the results of Ref. 7 and 8. We consider a Fabry-Perot interferometer filled with two-level atoms, and allow for cavity mistuning and atomic detuning with respect to the frequency of the driving laser field. In the mean-field limit, and after adiabatic elimination of the material variables, we obtain a single equation for the field, whose relation to the model of Ref. 7 is straightforward.

In Section II, we briefly summarize the main features of the model.

In Section III, we analyze the phase-space of this system. We recall how one can go about determining for which class of initial conditions one winds up on which steady-state. Phase-switching[7] is discussed in the dispersive and absorptive cases. We show that it is *not* limited to dispersive media. This is the major new result of this paper.

This result is of considerable practical relevance, since a laser operating well above threshold has an almost constant amplitude, but is not fully phase coherent. The phase of the field undergoes random variations in time in a manner which is analogous to one-dimensional Brownian motion.

The natural question which arises is whether these random phase fluctuations are sufficient to produce changes of the switching behavior of a bistable device away from those predicted on the basis of a fully coherent model for the field. In Section IV, we discuss briefly our numerical simulation of a phase diffusion model of the driving laser, and present preliminary results of the effects of such a laser on optical bistability. Finally, Section V is a summary and conclusion.

II. THE MODEL

We consider an ensemble of N two-level atoms placed inside a Fabry-Perot (or ring) cavity, and irradiated by an incident laser of frequency ω_L. We assume the medium to be homogeneously broadened, and allow for a frequency detuning $\delta_A = \omega_o - \omega_L$ between the atomic frequency ω_o and ω_L. We label by δ_F the mistuning between a cavity mode frequency ω_o and ω_L. The equations of motion of this system are known[9]. In this paper, we limit our discussion to the case where the atomic relaxation times are short compared to all other times, and adiabatically eliminate the atomic variables. In the mean field limit, the equation of motion for the transmitted field reduces then to

$$\dot{E}_T = \kappa E_I/T - \left[\kappa + i\delta_F + \frac{\alpha}{1 + \beta|E_T|^2}\right]E_T, \tag{2.1}$$

where E_I (E_T) is the incident (transmitted) field, and dot means time derivative. $\kappa = cT/L$ is the inverse cavity bandwidth, c the speed of light, L the cavity length, and T the mirror transmission.

The coefficients α and β are given by

$$\alpha \equiv \frac{\omega N\mu'^2(\gamma_\perp - i\delta_A)}{2\varepsilon_o V\hbar(\gamma_\perp^2 + \delta_A^2)} , \tag{2.2}$$

$$\beta \equiv 3\left(\frac{\mu'}{\hbar}\right)^2 \frac{\gamma_\perp}{\gamma_\parallel(\gamma_\perp^2 + \delta_A^2)T} , \tag{2.3}$$

respectively, where $1/\gamma_\parallel$ is the lifetime of the upper level and $\gamma_\perp$ the decay rate of the atomic polarization. μ' is the dipole matrix element of the atomic transition, N the number of atoms, and V the active volume.

It is useful to express Eq. (2.1) in dimensionless units. In analogy with Ref. 3, we let

$$x \equiv E_T(\mu'/\hbar)\sqrt{3\gamma_\perp/T\gamma_\parallel} / (\gamma_\perp + i\delta_A) , \tag{2.4}$$

and

$$y \equiv E_I(\mu'/\hbar)\sqrt{3\gamma_\perp/T\gamma_\parallel} / (\gamma_\perp + i\delta_A) . \tag{2.5}$$

Furthermore, we rewrite α as

$$\alpha \equiv 2\kappa(C_a - iC_d) . \tag{2.6}$$

The field equation of motion then becomes

$$\dot{x} = \kappa y - \left[\kappa + i\delta_F + \frac{2\kappa(C_a - iC_d)}{|+|x|^2}\right] x \ . \tag{2.7}$$

In Ref. 7 and 8 we used a different model, based on a non-linear susceptibility approach, to describe dispersive bistability and multistability. In the bistability limit, the field equation was

$$\dot{x} = \kappa y - (\kappa + i\Delta\omega)x - 2i\varepsilon|x|^2 x \ , \tag{2.8}$$

which can obviously be obtained from (2.7) by expanding the denominator, readjusting the constants, and setting $C_a = 0$. Although Eqs. (2.7) and (2.8) apply to different systems, their close analogy will allow us to use most of the results of Ref. 7, with minor alterations. However, Eq. (2.7) gives both absorptive and dispersive bistability.

Before proceeding with the analysis of the transient behavior of this system, we make a brief comment on its steady-state properties, in order to emphasize the fundamental difference between absorptive and dispersive bistability. This can be seen best by looking at the transmission $\tau = I_T/I_I$, where I_I (I_T) is the incident (transmitted) intensity.

In Fig. 1 we show the Fabry-Perot transmission for both the absorptive case (with and without cavity mistuning), and the dispersive case.

For absorptive systems, the transmission grows monotonically in the upper branch, while it goes through a sharp maximum in the dispersive case. The reason is that in the first case, the mechanism leading to bistability is the bleaching of the atomic medium, while in the second case, the index of refraction is changed by the presence of the field. Under these conditions, the system becomes transparent when the non-linear index compensates exactly for the cavity mistuning, which (if one considers only one cavity mode) occurs for one value of the driving field only.

III. PHASE-SPACE ANALYSIS

As discussed in Ref. 7 and 8, intensity curves such as shown in Fig. 1 are insufficient to analyze the transient dynamics of bistable systems, since they give no information on the phases of the fields. Rather, one must consider four-dimensional curves (functions of both amplitude and phase of the fields), or, equivalently, the phase-space of the output field for various values of the driving field.

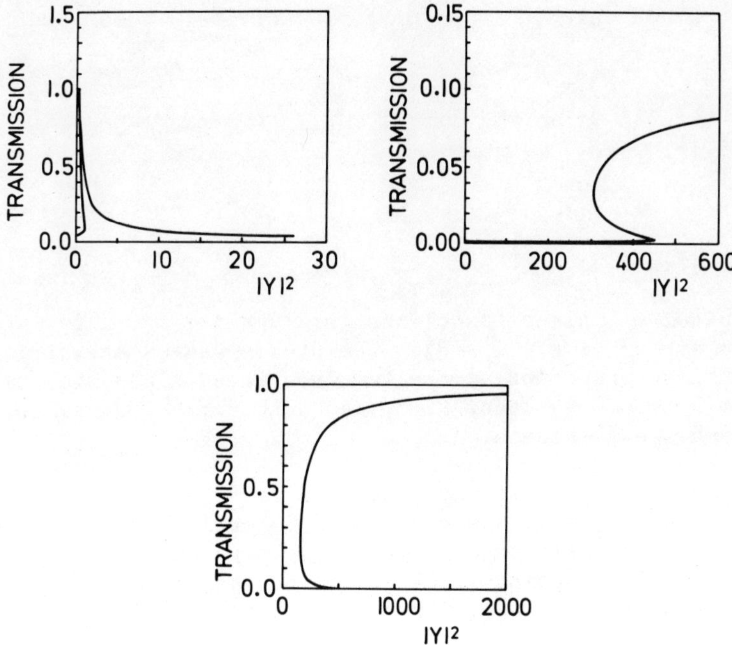

Fig. 1. Interferometer transmission as function of incident intensity (in dimensionless units) for (a) dispersive sys-
tem with $\kappa = 1$, $\delta_F = 15$, $C_a = 0$ and $C_d = -10$; (b) absorptive
system with $\kappa = 1$, $\delta_F = 3$, $C_a = 20$, and $C_d = 0$; (c) same as
(b) but $\delta_F = 0$.

When analyzing the response of the system to step-function
changes in the driving field, however, the situation is drastically
simplified, since one need only consider the phase-space corresponding to its final value. We further use the freedom of choice
of one phase and take the driving field to be real.

The problem is then to determine for which set of initial conditions one winds up on the lower or upper steady-state branch, respectively. As shown in Ref. 7, this question is easily answered
provided that the isoclines of the system are known (i.e., the
curves along which the real, imaginary part, respectively, of the
transmitted field remains constant). In Ref. 8, we presented a
general procedure to determine the isoclines, in the case where
the equation of motion of the field is of the general form

$$\dot{x} = \kappa(y + Q(|x|^2))x \ . \tag{3.1}$$

The curves along which $\mathcal{R}e(\dot{x})$, $\mathcal{I}m(\dot{x})$, respectively equals zero are given by,

$$x = \frac{y\,Q*}{|Q|^2} \pm i \left\{ \frac{|x|^2 - y^2/|Q|^2}{|Q|^2} \right\}^{\frac{1}{2}} Q* ,$$

(3.2)

and

$$\dot{x} = \pm |x| Q*/|Q| .$$

(3.3)

A typical example of isoclines is given in Fig. 2 for a purely dispersive system (i.e., $C_a = 0$). The steady-states are of course given by the intersection of the two isoclines, since one has then $\mathcal{R}e(\dot{x}) = \mathcal{I}m(\dot{x}) = 0$. We label L, M, and U the lower, unstable, and upper branch steady-states.

The general aspect of the isoclines is very similar to that obtained in Ref. 7, where we considered a non-linear susceptibility model. The major difference is that for large amplitudes of the transmitted field, the isoclines do not become horizontal ($\mathcal{R}e(\dot{x})= 0$) and vertical ($\mathcal{I}m(\dot{x}) = 0$). Rather, they become straight lines of slopes κ/δ_F and $-\delta_F/\kappa$, respectively.

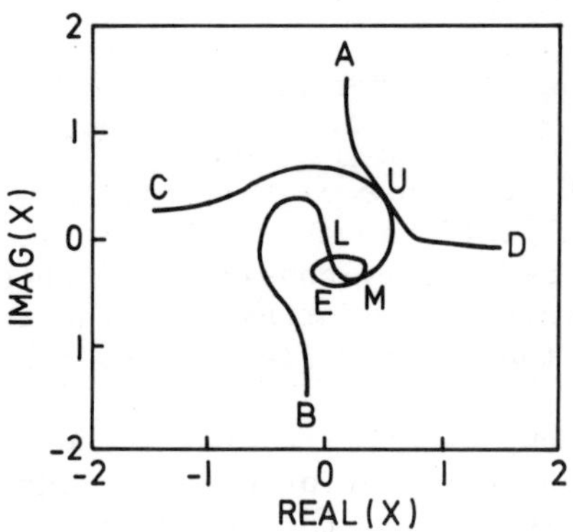

Fig. 2.   Curve AB: isocline along which $\mathcal{I}m(\dot{x}) = 0$. Curves CD and E: isocline along which $\mathcal{R}e(\dot{x}) = 0$. The points L, M and U label the lower branch, unstable, and upper branch steady-state, respectively. Isoclines obtained for the model of Eq. (2.7), for a purely dispersive system with $\kappa = 1$, $\delta_F = -15$, $C_a = 0$, and $C_d = -10$, and a real driving field $y = 0.8$.

The similarity between the two models for small values of the field is of course easy to understand, since the equations of motion (2.7) and (2.8) approach each other for $|x| < 1$. For large $|x|$, the denominator in Eq. (2.7) leads to a saturation, and is the cause of the "tilt" of the isoclines when $|x| \to \infty$. Except for these minor differences, however, the behavior of the model discussed here, in the dispersive case, is qualitatively very similar to that discussed in Ref. 7, and the reader is referred to this paper for further details.

We now turn to a discussion of the absorptive case ($C_d = 0$). In Fig. 3, we show how the isoclines vary as a function of the amplitude of the (real) driving field. Obviously, for y=0 the isocline $Re(\dot{x}) = 0$ is the same as $Im(\dot{x}) = 0$, but rotated by $\pi/2$. As the field strength is increased, it starts deforming, until an "ellipse" separates, leading to three steady-states. When y is further increased, the "ellipse" shrinks, reduces to a single point, and eventually disappears, so that one is again left with one steady-state only. It is worth noting that in this example, we have allowed for a cavity mistuning. Thus, the steady states do not have the same phase as the (real) driving field.

For absorptive bistability without mistuning, the behavior of the system is similar, except that the $Re(x)$ axis remains an axis of symmetry, and all steady-states lie on it. That is, the steady-states have the same phase as the driving field, as should be expected. This is illustrated in Fig. 4 for a value of the driving field such that the system is in the bistable region.

In order to determine for which initial conditions the system evolves towards U after a step-function change in the driving field, we proceed along the lines discussed in detail in Ref. 7 and 8. We consider initial conditions very close to the unstable steady-state M and integrate the equations of motion (2.7) backward in time. This yields two trajectories dividing the phase-space of the system into two regions. For a deterministic system, trajectories are not allowed to cross each other, except at the steady-states L, U and M. Furthermore, the system does not exhibit any limit cycles. We conclude that all initial conditions within the region containing the lower branch steady-state must evolve toward it, and all others evolve towards U (except for an ensemble of points of measure zero which wind up on the unstable state).

In Fig. 5, we show the result of such an analysis, for the cases of an absorptive system with and without cavity mistuning. We see that as in the case with mistuning, the ensemble of initial conditions leading to one or the other of the possible steady-states is rather complex. In particular, we observe the existence of anomalous thresholds[7] for $\delta_F \neq 0$.

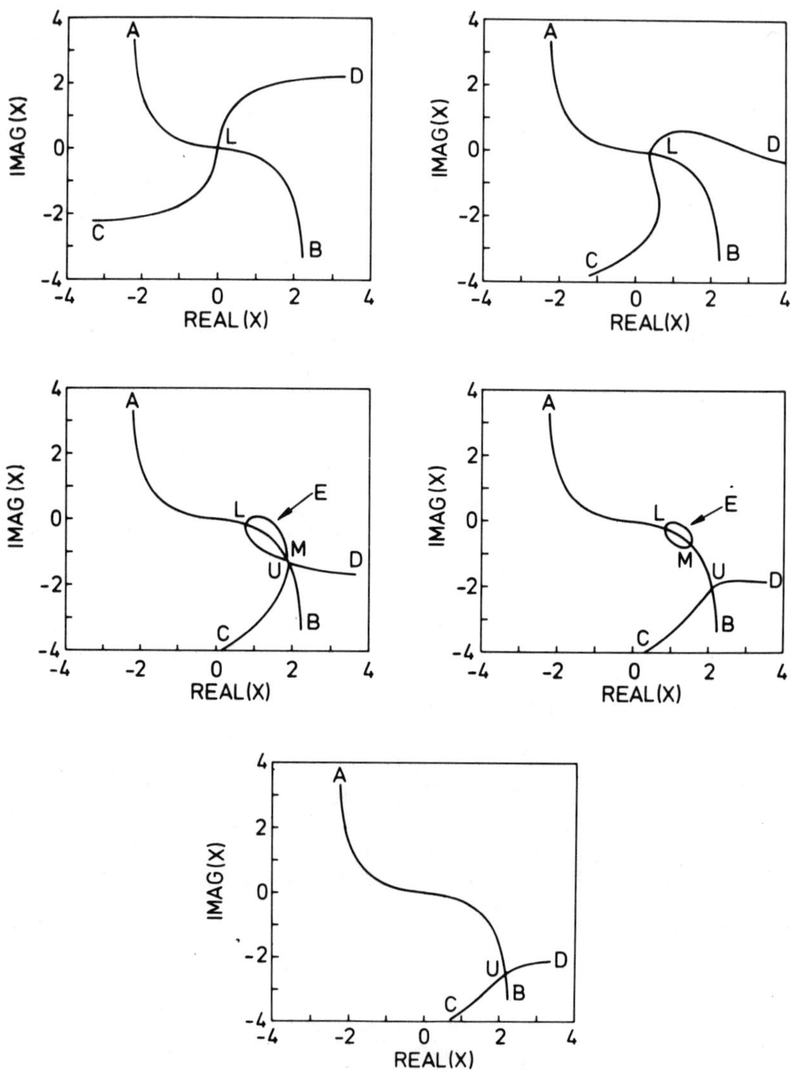

Fig. 3.   Isoclines $\mathrm{Im}(\dot{x})=0$ (curve AB) and $\mathcal{R}e(\dot{x})=0$ (curves CD and
E) for an sbsorptive system with detuning. L, M, and U
label the lower branch, unstable, and upper branch steady-
states respectively. The system parameters are $\kappa = 1$,
$\delta_F = 5$, $C_d = 0$, and $C_a = 20$. The driving field is taken to
be real and its value is (a) 0, (b) 15.0, (c) 20.0,
(d) 21.0 and (e) 22.0.

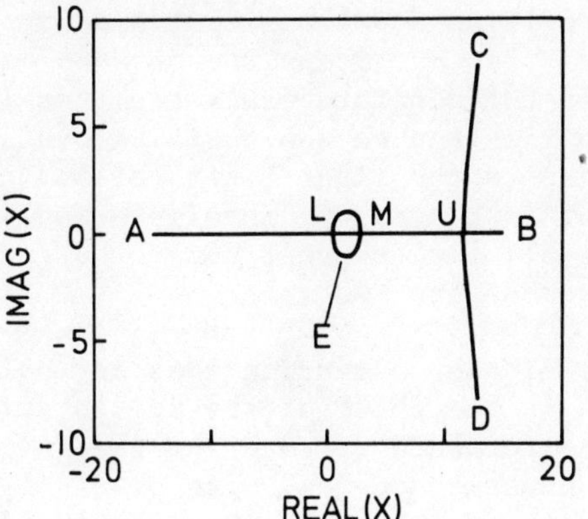

Fig. 4.   Same as Fig. 3, but for $\delta_F = 0$ (no mistuning). The value
of the driving field is $y = 15$.

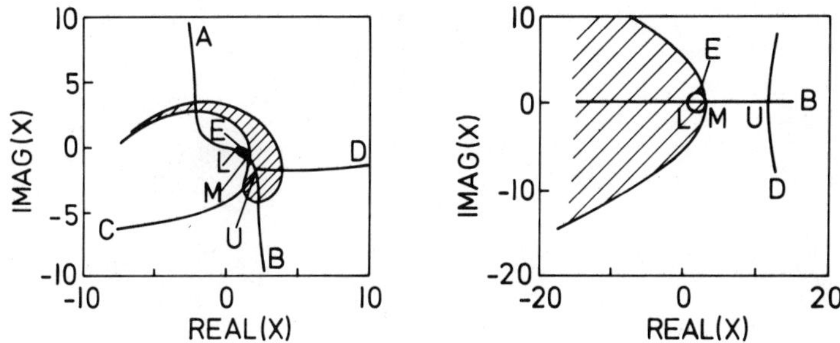

Fig. 5.   (a) Absorptive system with mistuning, same parameters as
Fig. 3d. The shaded area gives the region of phase-space
evolving towards the *upper branch* steady-state U.
(b) Absorptive system without mistuning. Same parameters
as Fig. 4. The shaded area gives the region of phase-
space evolving towards the *lower branch* steady-state L.

That is, when moving the initial condition along, say, the real
axis, one crosses regions of phase-space leading alternatively to
the lower and upper branch steady-states. Consequently, the final
state of the system cannot be predicted simply on the basis of the
standard intensity-out vs. intensity-in curve. Rather, a detailed
phase-space analysis is necessary.

However, there is no anomalous switching in the case of zero mistuning, *provided that one keeps the phase of the driving field constant*. Consider for concreteness y real, as in Fig. 5. All initial conditions lie then on the positive branch of the real axis. Clearly, if they are to the right (left) of M, they must evolve towards U (L). This explains why in simple models of absorptive bistability, where all fields are taken to be real, no anomalous transient behavior was observed[6].

However, this simple conclusion does not hold any more if one allows for changes in the phase of the driving field. It is obvious[7] that this corresponds to rotating the initial condition about the origin of the phase-space (or, more precisely, rotating the phase-space around the origin while keeping the initial point fixed). Clearly, such rotations can bring U into the region of phase-space leading to L. Thus, one can switch the system from the upper to the lower branch steady-state merely by changing the phase of the driving field (phase switching). Note however that phase up-switching (i.e., switching from L to U) is not possible in this case.

In conclusion, phase-switching is a very general feature, present in absorptive, dispersive[7], and mixed bistability, and in multistability[8]. Because lasers well above threshold exhibit significant phase fluctuations, this effect may have important practical implications. In Section IV, we give a preliminary discussion of this point.

IV.   EFFECTS OF LASER PHASE-FLUCTUATIONS

It is well known[10] that an amplitude stabilized laser well above threshold still presents significant phase fluctuations, which are the major cause of its finite linewidth. They arise as a result of the presence of various noise sources, such as spontaneous emission, thermal noise, etc. In the preceding section, we have seen that phase changes in the driving field can be sufficient to switch a bistable (or multistable) device. The natural question that arises then is to determine whether the phase-fluctuations present in any real laser are sufficient to produce changes in the switching properties of the device away from those predicted on the basis of a fully coherent model for the field.[14]

In the case of a dispersive device or of an absorptive system with cavity mistuning, the effects of laser phase fluctuations are quite difficult to analyze, since both phase up- and down-switching are possible. As a result, and for large enough fluctuations, the system hops at random between L and U. We are still in the process of studying this problem.

For absorptive systems without mistuning, however, the situation is much simpler, since only down-switching is possible. Thus, if the device jumps from U to L, it will then remain there for all times. We now restrict our discussion to this case.

We consider the phase diffusion model of the laser[10], in which the phase of the field undergoes a one-dimensional Brownian motion. For the numerical integration of Eq. (2.7) one needs to know the phase at various times t_0, t_1, t_2 ... . Rather than solving the Langevin or Fokker-Planck equation for the phase, we find it more convenient to use directly the conditional probability of the random phase $\phi(t_{m+1})$ having a certain value at some instant, given its value at an earlier instant in time:

$$P(\phi(t_{m+1}):\phi(t_m)) = \sqrt{\frac{\tau_c}{2\pi(t_{m+1}-t_m)}} \ \exp-\left\{\frac{[\phi(t_{m+1})-\phi(t_m)]^2}{2(t_{m+1}-t_m)/\tau_c}\right\}$$

$$(4.1)$$

This probability distribution is obviously a Gaussian of mean $\phi(t_m)$ and standard deviation $[2(t_{m+1}-t_m)/\tau_c]^{\frac{1}{2}}$, where τ_c is the inverse of the spectral width of the laser.

A realization of the Brownian motion of the phase is then modelled numerically in the following way: The phase at $t_0 = 0$ is chosen arbitrary.   In practice, we chose an initial, real, value of the output field, and from the steady-state form of Eq. (2.7) determine the corresponding complex input field. Its phase is then taken as the initial phase $\phi(t_0)$. By means of the probability distribution (4.1), the phase at the next instant required by the integration routine is chosen by calling a random number from a Gaussian distribution of mean $\phi(t_1)$ and standard deviation $[2(t_1-t_0)/\tau_c]^{\frac{1}{2}}$. With this new $\phi(t_1)$, we can then determine $\phi(t_2)$ from a random distribution of mean $\phi(t_1)$ and standard deviation $[2(t_2-t_1)/\tau_c]^{\frac{1}{2}}$, etc.   The successive values $\phi(t_0)$, $\phi(t_1)$, $\phi(t_2)$, then devine one realization of the Brownian motion of the field phase. In order to obtain useful statistical information on the dynamics of the bistable system, the integration has to be repeated for various choices of random numbers.

In Fig. 6, we show a typical computer run giving the dynamics of the system. Initially, the device is on the upper branch steady-state, and at time $t = 0$, we turn on the phase fluctuations of the driving field. After some time, the system switches down to the lower branch, and remains there for all ulterior times. In this example, the laser linewidth is taken to be the same as the resonator linewidth κ. We have performed a limited statistical analysis of this case, with 20 runs, and find that the average switching

time from U to L is very short, as illustrated by the histogram in
Fig. 7.

   As one would expect, the average switching time increases for
narrower laser lines. We have performed a limited series of runs
for $\Delta\omega = \kappa/10$ and $\kappa/100$. We have let the system evolve for a time
$t = 100/\kappa$, and sometimes $t = 400/\kappa$ after the turning on of the
phase fluctuations, and have not found a single run in which it
equilibrated down to the lower branch. Although our data are too
limited to draw definite conclusions, this clearly indicates that
the average switching (or equilibration) time of the system decreases dramatically as the laser linewidth is increased.

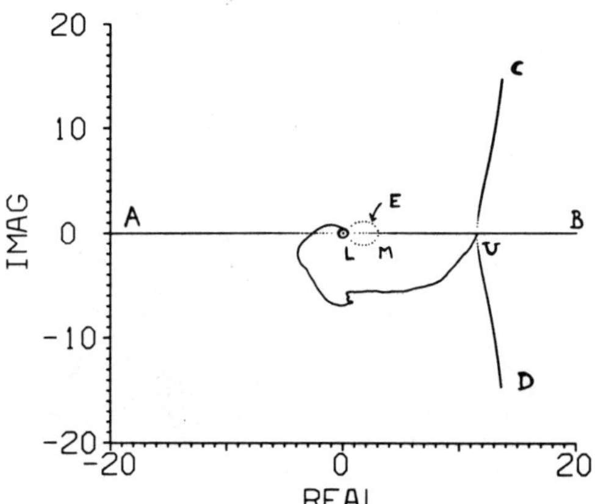

Fig. 6.   Typical phase-space trajectory of a purely absorptive
           system initially on the high transmission branch and
           driven by a field exhibiting phase fluctuations. The
           system parameters are the same as in Fig. 5b. The
           curves AB, CD, and E, and the points U, M, and L repre-
           sent the isoclines and steady-states of the system at
           $t = 0$. For later times, they rotate at random around
           the origin. The correlation time of the phase fluctua-
           tions is $\tau_c = 1$, that is, the laser bandwidth is the
           same as the resonantor bandwidth.

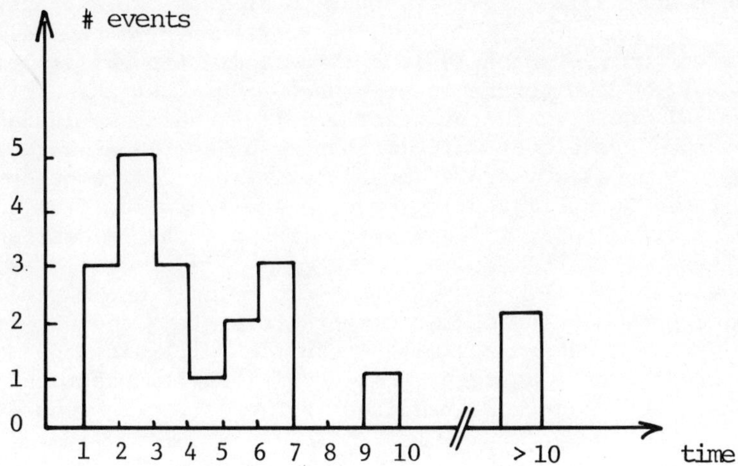

Fig. 7.   Histogram of the switching time of a purely absorptive
           system initially on the upper bistability branch and
           driven by a laser field of constant amplitude but ex-
           hibiting phase fluctuations. Parameters are the same
           as in Fig. 6. Time in units of $1/\kappa$. The histogram is
           the result of 20 different realizations of the phase
           fluctuations.

V.   CONCLUSION

From the preliminary results presented here, we can already
draw the conclusion that the phase fluctuations of the driving
field (i.e., its finite linewidth) are in general far from being
negligible, even in purely absorptive bistability. For instance,
if the laser linewidth is of the same order as the resonator
bandwidth, the effects are quite dramatic, and have serious practical implications.

This shows the need to describe the complete hysteresis cycle
of the device when phase fluctuations are present. This problem
can be tackled quite simply by describing the driving field as a
slow ramp in amplitude with random phase. What we expect is that
the switching from upper to lower bistability branch will then
occur, on the average, at input field strengths higher than that
for the perfectly coherent situation. The net effect would then
be a narrowing of the hysteresis curve.

One must realize, however, that there is no unique answer to this problem. Clearly, if one waits long enough, even small phase fluctuations will eventually lead to a finite probability of switching the system down to the lower branch. One must thus introduce some kind of "quality factor" of the device, which would have to do with the time scale over which it has to behave as a bistable system with given "steady-state" characteristics. For some applications, it may be sufficient to have an equilibration time on the order of microseconds, while minutes or hours may be desirable in other cases.

As already mentioned, for dispersive systems and absorptive systems with mistuning, a similar reduction will also appear on the lower branch: on the average, the transition will occur for field strengths lower than those required in the perfectly coherent case.

For very broad laser linewidths, we expect that bistability will disappear altogether, for all time scales of practical interest. This is in agreement with recent results of Drummond et.al., who have shown[11] that this is precisely what happens in the case of a non-linear interferometer driven by a broadband (delta-correlated) field. As pointed out by these authors, the crucial point now is that bistability is a transient effect, that one would like to exist on relatively long time scales. In order for this to occur, the switching (or equilibration) time should be slow in comparison to all times of practical interest. It is clear that the effects of quantum fluctuations are extremely small in optical bistability[12,13,15]. However, the noise in the driving field is certainly a problem which must be considered seriously.

A complete analysis of this problem is currently in progress, and will be reported elsewhere.

ACKNOWLEDGMENTS

Numerous discussions with Profs. J. H. Eberly, F. A. Hopf, W. E. Lamb, Jr., W. H. Louisell, L. A. Lugiato, M. Milani, M. Sargent III, M. O. Scully, and D. F. Walls are gratefully acknowledged. We thank Prof. H. Walther for his interest in this problem and his constant support.

REFERENCES

1.   H. Seidel, U. S. Patent No. 3610731 (1971); A. Szoke, V. Daneu and N. A. Kurnit, Appl. Phys. Letters $\underline{15}$, 376 (1969).
2.   S. L. McCall, Phys. Rev. $\underline{A9}$, 1515 (1974).
3.   H. M. Gibbs, S. L. McCall, and T. N. C. Venkatesan, Phys. Rev. Letters $\underline{36}$, 1135 (1976).

4. R. Bonifacio and L. Lugiato, Optics Commun. 19, 172 (1976); Phys. Rev. A18, 1129 (1978).

5. For recent lists of references, see in particular the review papers by H. M. Gibbs et.al., and by R. Bonifacio et.al., in "Laser Spectroscopy IV," H. Walther and K. Rothe, eds., Springer-Verlag, Berlin (1979); and H. M. Gibbs, S. L. McCall and T. N. C. Venkatesan, Optics News, Summer 1979.

6. See in particular, R. Bonifacio and P. Meystre, Optics Commun. 29, 131 (1979); F. A. Hopf and P. Meystre, Optics Commun. 29, 235 (1979), and references therein.

7. F. A. Hopf, P. Meystre, P. D. Drummond and D. F. Walls, Optics Commun. 31, 245 (1979).

8. F. A. Hopf and P. Meystre, Optics Commun. 33, 225 (1980).

9. R. Bonifacio and L. Lugiato, Lett. Nuovo Cimento 21, 517 (1978). At steady-state, Eq. (2.1) is a particular case of their Eq. (21). See also L. Lugiato, M. Milani, W. H. Louisell and P. Meystre, in preparation.

10. See, for instance, M. Sargent III, M. O. Scully and W. E. Lamb, Jr., "Laser Physics," Addison-Wesley (1974).

11. P. D. Drummond and D. F. Walls, preprint.

12. L. A. Lugiato, Nuovo Cimento 50B, 89 (1979).

13. K. Kondo, M. Mabuchi and B. Hasegawa, Optics Commun. 32, 136 (1980).

14. Note added in proof: in real lasers, frequency fluctuations due to the jitter of the resonator are likely to give the main contribution to the linewidth (S. F. Jacobs, private communication). In this context, see the analysis of frequency switching by F. A. Hopf and S. Shakir, this volume.

15. See also the Panel Discussion, this volume and in particular the comments of J. Farina.

FREQUENCY SWITCHING IN DISPERSIVE OPTICAL BISTABILITY

Frederic A. Hopf and Sami A. Shakir

Optical Sciences Center
University of Arizona
Tucson, Arizona 85721

Abstract: The theory of dispersive optical bistability is considered with special attention to the cavity dynamics when the input
frequency is changed. Changes in the conventions used in the slowly
varying amplitude and phase approximation (SVEA) are necessary. The
distinction between phase- and frequency-switching is shown. Preliminary results are discussed.

I.   INTRODUCTION

Considerable attention has been given recently to the subject
of optical bistability. The problem has proved interesting from
diverse points of view ranging from practical devices to a model
problem of nonequilibrium statistical mechanics. Past theories[1] have
been extensive, with parallel experimental efforts,[2] and while the
static behavior has been extensively studied, the time dynamics has
received less exhaustive treatment. In this paper we consider the
problem of frequency-switching of a dispersive bistable device. The
model used here is "exact" in the sense that no approximation is made
to the mode-pulling,[3,5] and hence we can deal with resonators that
are tuned far from the small-signal resonance, and we can look for
the dynamics of multi-stable (three or more stable states) configurations. This study is directly motivated by earlier studies on phase-
switching,[4,5] which showed interesting potential for switching "near-
resonance" or "third-order" devices,[4] but was otherwise disappoint-
ing in the "exact" limit.[5] Frequency switching is the logical next
step in looking for fast, low power switching techniques, since phase
switching is just a frequency switch that is accomplished "instantaneously" (This is developed in more detail later on).

Primarily, however, this paper is concerned with finding a natural theoretical approach to bistability that can be applied to frequency switching. The difficulty is that the theory in Ref. 5 simply fails to switch the device, even when it is self-evident that the device should switch, and if used without care, it even fails to predict the correct steady state behavior.

The difficulty lies directly in the effort to write a simple, comprehensible and generalizable theory of the bistable device. One uses the ring geometry illustrated in Fig. 1 which eliminates standing-wave effects and hence allows the use of propagation formalism which is the simplest version of the slowly varying amplitude and phase approximation (SVEA).[6] The problem lies in the fact that if the incident frequency changes, this SVEA neglects the change in the optical path length due to the frequency change. This change is, of course, vital to the proper description of the Fabry-Perot. The goal of this development is to generate a definition of the amplitude of the field that can be used in the SVEA in such a manner that includes the change in the optical phase in a satisfactory fashion.

To accomplish this task, we start by dealing with the linear but non-empty Fabry-Perot. This case will serve to illustrate the problem with the SVEA, and will help in setting up a procedure that enables us to discuss phase switches as a limiting case of frequency switches. Next, the nonlinearity is included and preliminary results on frequency switching are described. The final section is a conclusion.

II. THE DYNAMICS OF A LINEAR FABRY PEROT

Let us begin the discussion of the linear Fabry-Perot by writing down the small-signal SVEA and seeing what the difficulty is. We write the electric field E in terms of a slowly varying complex amplitude E as

$$E = \frac{1}{2} \left(E e^{i(kz-\omega t)} + c.c. \right) \ , \tag{1}$$

with an analagous expression for the amplitude P of the polarization P. From any standard text,[6] one gets the SVEA, written here as a function of retarded time $\mu = t - z/c$ as

$$\frac{\partial E}{\partial z} = \frac{4\pi\omega^2}{(2ik)c^2} P \ . \tag{2}$$

Here we have taken the medium to have a very low density so the linear effect of the medium can be taken as a perturbation. In the next section we write the theory in the case when only the nonlinear term is treated as a perturbation. Using $P = \chi E$, Eq. (2) is readily solved, giving

$$E(\mu,L) = E(\mu,0)e^{-i\frac{2\pi\omega}{c}\chi L} \; . \tag{3}$$

In the standard usage of the SVEA, ω is a constant even when the incident frequency changes. Hence the optical phase, which is the imaginary part of the argument of the exponential, is constant. Frequency changes appear in the argument of $E(\mu,0)$, but these do not affect the optical phase. This is fatal to describing the Fabry-Perot. What is needed is to have the time-dependent frequency appear on the RHS of Eq. (2). This is accomplished easily enough, but requires some careful consideration to how one treats the limits of frequency and phase switching.

Let us ignore, for the moment, the mirrors in Fig. 1 and consider only the propagation through the medium. Let us define the incident field amplitude at $z = 0$ as

$$E_{in} = \frac{1}{2} \left(|E_{in}| e^{i\theta_{in}(t)} + c.c. \right) \tag{4}$$

and the field amplitude and polarization inside the medium as

$$E = \frac{1}{2} \left(E e^{i\theta_{in}(\mu)} + c.c. \right) \; , \tag{5a}$$

$$P = \frac{1}{2} \left(P e^{i\theta_{in}(\mu)} + c.c. \right) \; . \tag{5b}$$

Substituting these into Maxwell's equations gives

$$\frac{\partial E}{\partial z} = -\frac{2\pi}{c} \frac{\partial \theta}{\partial \mu} P \tag{6}$$

with a solution for $P = \chi E$ given by

$$E(\mu,L) = e^{-i\frac{2\pi}{c}\frac{\partial \theta_{in}}{\partial \mu} L} E(\mu,0) \; . \tag{7}$$

This makes the optical phase dependent on the instantaneous frequency. The "slowly varying" requirements now read $(\partial E/\partial z) \ll (\partial \theta/\partial z)E$ etc.

This construction of the amplitudes in Eq. (5) has some subtle-ties when one considers the dynamics of the cavity. Since these

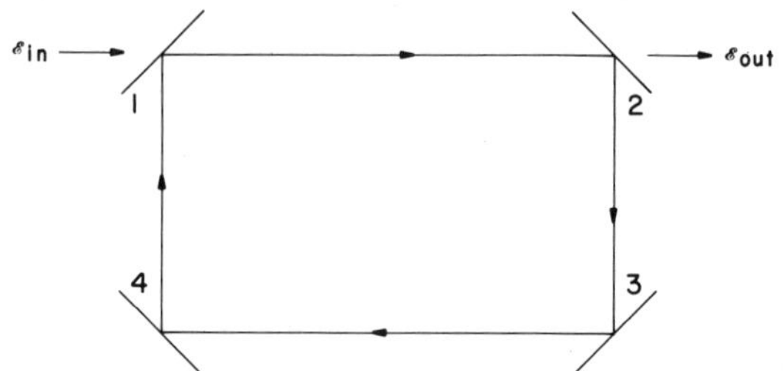

Fig. 1.   The ring cavity set up.   The nonlinear medium fills the
         entire ring.

subtleties are essentially part of the empty cavity dynamics we
discuss that case next.   The boundary conditions for the field E
defined just to the right of mirror #1 in Fig. 1 reads

$$E(t) = \sqrt{T}\, E_{in} + R\, E(t - L/c) \tag{8}$$

where T and R are the transmission and reflection coefficients.
Substituting the fields in Eq.'s (4) and (5a) gives

$$E(t) = \sqrt{T}\, |E_{in}| + R\, E(t - L/c)\, e^{i[\theta_{in}(t - \frac{L}{c}) - \theta_{in}(t)]} \,. \tag{9}$$

This "exact" equation can be used as the working equation of the
Fabry-Perot, and can be generalized to include nonlinearities.
However, in the "Good Cavity" (High finess) limit, in which the cavity
dynamics plays an interesting role, the electric field varies on a
timescale $L/c(1 - R)$ which is much longer than L/c.   Hence it is
uneconomical to use such a short step.   In this limit it is customary
to write Eq. (9) as a differential equation.   To do this we must
differentiate between a phase and frequency-switch.

Let us define $\theta_{in}(t) = \omega t + \phi(t)$ where $\phi(t)$ is the switch (i.e.
we consider only the case $|E_{in}|$ constant) and has the property that
$\dot{\phi}(t)$ is zero except on an interval $0 \le t \le t_s$.   Let us consider,
first, the "frequency switch" limit which is defined by

$$t_s \gtrsim L/c(1 - R) \quad , \tag{10}$$

$$\phi(t) - \phi(t - \frac{L}{c}) \approx \frac{L}{c} \left. \frac{\partial \phi}{\partial t'} \right|_{t' = t - \frac{L}{c}} \,. \tag{11}$$

The second approximation is really the critical one, since otherwise the amplitude varies rapidly because of the definition in Eq. 5a

Physically it is $E \exp(i\phi)$ that varies slowly on the scale L/c.
Using these plus

$$E(t) \approx E(t - \frac{L}{c}) + \frac{L}{c} \frac{\partial E}{\partial t'}\Big|_{t'=t-L/c} \tag{12}$$

and redefining the zero of time as $t - L/c \to t$ gives

$$\frac{\partial E}{\partial t} = \frac{c}{L}\left(\sqrt{T}|E_{in}| + \left[\mathrm{Re}\, e^{-i\frac{\partial\theta_{in}}{\partial t}L/c} - 1\right]E\right) \tag{13}$$

when the medium is included. The form of Eq. (7) with $\partial/\partial\mu \to \partial/\partial t$
implies that the length L is replaced by the optical path length.

In the phase-switch limit, we have $t_s \ll L/c(1 - R)$. In this
case the definition of Eq. (5a) must be abandoned if one wants to
convert Eq. (9) to differential form. The definition of Eq. (5a)
causes E to be rapidly varying in the interval $0 \leq t \leq t_s$ and Eq.
(12) cannot be applied. In this case it is best to use the standard
definitions of the SVEA used Eq. (1). Taking the phase switch to
occur instantaneously and applying the expansion in Eq. (12) to the
amplitude in Eq. (1) we have

$$\frac{\partial E}{\partial t} = \frac{c}{L}\,(\sqrt{T}\,|E_{in}|e^{i\theta_{in}(t_s)} + (R\,e^{-i\omega L/c} - 1)E) \quad . \tag{14}$$

This is the result of Ref. 5, and to include the effect of the medium,
one replaces L by the optical path length.

Thus far, we have seen that the set of Equations (5) - (7) and
(9) constitute a basis upon which one can construct a general version
of the bistable system using the SVEA. The difference equation is
quite suitable, but it is not the conventional form of the theory.
The differential form is more economical, from a numerical standpoint, in the "good" cavity limit that we treat here. To go over to
the differential equation requires two separate approaches. The
frequency switch used Eq.'s (5) - (7) and (13), and the phase switch
uses Eq.'s (1) - (3) and (14). In the adiabatic limit it is evident
from Eq. (14) that the output field amplitude is independent of the
phase of the input field. In the next section, we write the equations
and solve Eq. (13) for the dynamics of the frequency switch.

III.   FREQUENCY SWITCHING

In the purely dispersive limit, the inclusion of the nonlinearity
is straightforward, since one replaces L by the optical path $n(I)L$ in

Eq.'s (13) and (14). In the overall multiplier the factor of nL can be replaced by $n^{(1)}L$, where $n^{(1)}$ is the linear index, since it involves just a small change in the overall timescale. The full nonlinear behavior must be kept intact in the phasor. We particularize the discussion to the cubic nonlinearlity, i.e., $n \approx n^{(1)} + n^{(2)}I$, for simplicity in discussion. In the limit in which $n^{(1)} - 1$ is not small, the linear part of the polarization must be solved first. This is quite standard, and gives

$$\frac{\partial E}{\partial z} = -i \, \frac{\pi \chi^{(3)}}{n^{(1)} c} \, |E|^2 E \quad , \tag{15}$$

where the cubic polarization reads $P^{(3)} = \chi^{(3)}E^3$ and $\mu = t - n^{(1)}z/c$.

Using the nonlinear index notation, this reduces to

$$\frac{\partial E}{\partial z} = -i \, \frac{n^{(2)}I}{c} \, \frac{\partial \theta}{\partial \mu} \, E \quad , \tag{16}$$

from which one immediately finds that $I = nc|E|^2/4\pi$ is a function of μ only. Hence

$$E(\mu,L) = E(\mu,0)e^{-i \, \frac{n^{(2)}I}{c} \, \frac{\partial \theta}{\partial \mu} \, L} \quad . \tag{17}$$

Defining $\ell = (n^{(1)} + n^{(2)}I)L$ and $\ell^{(1)} = n^{(1)}L$ one finds directly that the argument leading from Eq. (9) to Eq. (13) gives

$$\frac{\partial E}{\partial t} = \frac{c}{\ell^{(1)}} \, (\sqrt{T}|E_{in}| + (R \, e^{-i \, \frac{\ell}{c} \, (\omega + \dot{\phi})} - 1)E) \quad . \tag{18}$$

At any time t, the intensity I and hence ℓ is computed using the field at that instant at steady state, where $\dot{\phi}$ and E are constant. Equation (18) leads to a cavity transmission $\tau = S_{out}/S_{in}$ as

$$\tau = \frac{T^2}{1 + R^2 - 2R \cos[\frac{\ell}{c} \, (\omega + \dot{\phi})]} \quad . \tag{19}$$

The transmission must, of course, be solved self-consistantly with $I = (\tau/T)I_{in}$. The transmission is shown as a function of $\dot{\phi}$ in Fig. 2. The role of the nonlinearity is to tilt the peak of the transmission curve by an amount

$$\Delta\omega_{NL} = -\frac{n^{(2)}_I}{n^{(1)}_T}\omega_o \; .$$

(20)

The standard procedure in the numerical calculations is to start with the system in one of the stable states of Fig. 2. For convenience we show the standard "s" curve in Fig. 3 (power out vs power in) for the operating points in Fig. 2. We then change the frequency such that $\dot{\phi} = \delta t$ for $0 \le t \le t_S/2$, $\dot{\phi} = (t_S-t)\delta$ for $t_S/2 \le t \le t_S$. We define as "fast" any case where $t_S \le L/c(1-R)$, i.e. the switch takes place over a time scale faster than or comparable to the cavity response, vs."slow" or "adiabatic" in which $t_S > L/c(1-R)$, i.e. the system adiabatically follows the frequency. We take $R = .7$ as the working example, since it is the case discussed in Ref. (5) and we label $n = 1,2,\ldots$, the stable states of the device, which are indicated by circles in Fig.'s 2 & 3. We have considered only switching operations among the states one through 4. The states $n = 3$ and $n = 4$ behave the same way, and there is no reason to suppose $n > 4$ is any different. We find that all up-switching operations (i.e. that increase n) can be performed adiabatically. All adiabatic down-switching operations return the system to $n = 1$. Furthermore, adiabatic down-switching requires extremely large frequency shifts. This is a case in which phase-switching has been shown to be effective, and is clearly simpler and more flexible.

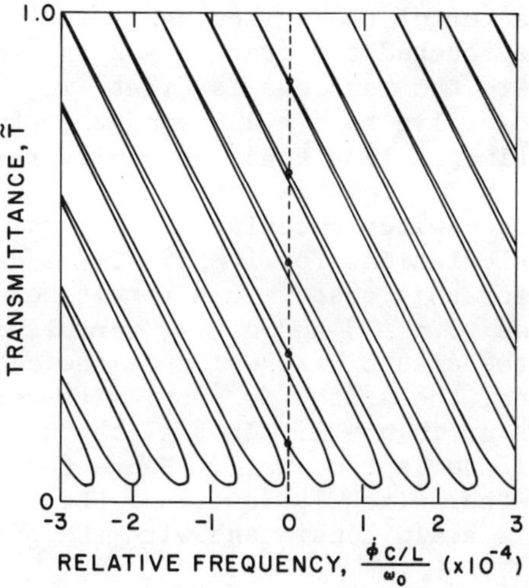

Fig. 2.   Transmittance vs. optical frequency. The operating points are indicated by circles. The zero of the frequency axis corresponds to the steady state frequency ω_0. The curve corresponds to $R = 0.7$, $T = 0.3$ and $|E_{out}| = 0.1$.

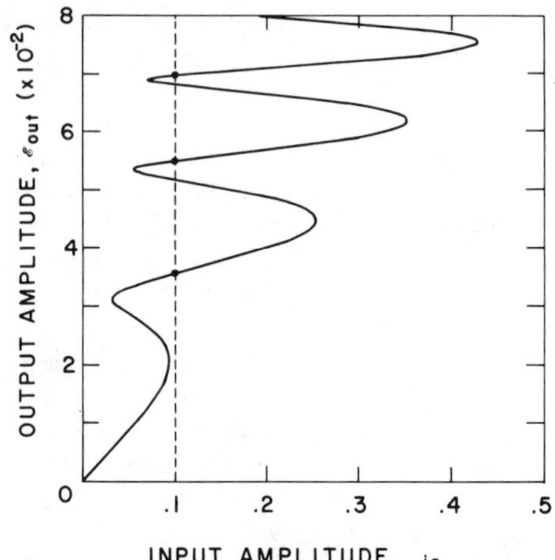

Fig. 3.   The bistability "S" curve.   The operating points corresponding to the ones in the transmittance curve are indicated by
circles.

In Fig. 4 we show an example of a "slow" up-switching $1 \to 2$.
We plot $\dot{\phi}$ and I_{out} as a function of time where the cavity response
time is 2.5 units.   In Fig. 4a, t_S is longer than this time (t_S = 10)
The adiabatic character of the switching is indicated by the fact
that the system has reached the state n = 2 at t = t_S.   In Fig. 4b,
we have t_S = 4.   Here the response is "fast" in the sense that the
system continues to evolve to n = 2 beyond the time t = t_S.   In Fig.
4c, the switching time is less than the cavity response time and the
system returns to n = 1.   This is consistant with the negative conclusions of the phase-switch calculations for this case.   In Fig. 5
we show the same calculations for $1 \to 3$.   One sees that only the
slow switch works for this case.   This result holds for all up-
switching operations to n = 3 and n = 4, for all initial states.
In Fig. 6 we show the case $3 \to 2$, which is the case alluded to
earlier.   In Fig. 6a, the switch is slow, with very large amplitude,
and the system ends up in n = 1.   In Fig. 6b the switch is again
slow, but the amplitude is too small.   Hence the system returns to
n = 3.   In Fig. 6c the switch is fast, and the system moves to n = 2
as desired.   This is again consistent with the phase-switch analysis.

IV.   CONCLUSION

In summary, we have considered the problem of frequency switching in dispersive bistability.   In order to describe frequency switching, which involves changing the frequency of the incident field, one

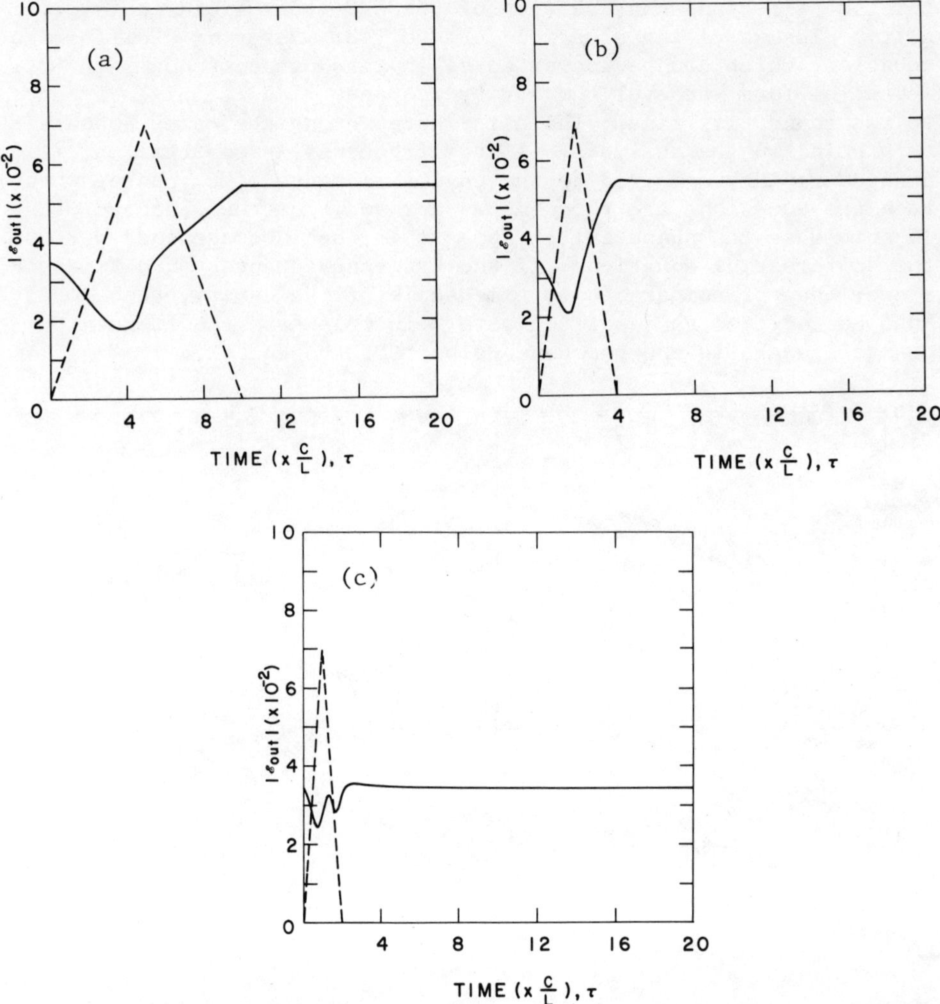

Fig. 4.   Time dynamics for up-switching.   The initial operating point
is the first point in Fig. 2.   The discontinuous curves
represent the frequency modulation.   The scale for the
frequency curve is $10^{-5}(\omega_0 L/c)$.   (a) "slow" up-switching
$1 \rightarrow 2$.   (b) "fast" up-switching $1 \rightarrow 2$.   (c) no switching.

must treat the medium equations with some care.   The SVEA, which we
use because it is the most commonly used and most generalizable
version of the theory of the interaction of light with matter,
neglects the change in optical path length that that occurs when the
frequency is changed.   This is an unacceptable approximation when
describing the dynamics of a Fabry-Perot resonator.   We describe,

here, a self-consistant version of the SVEA that avoids this difficulty. The resulting equation for the Fabry Perot is a difference
equation, which can be converted to an inhomogeneous nonlinear differential form whenever the cavity response time is large compared
to the round trip time. The differences among the cases appears at
this point in the analysis. If the frequency is constant, or if it
changes rapidly compared to the cavity response time (phase switch),
then the equations are the same as derived from the standard SVEA.
In this case the phase switch appears in the inhomogeneous part of
the differential equation. If the frequency changes on a time scale
slower than or comparable to time scale of the cavity, then the SVEA
must be modified as described here. In this case the frequency
change appears in the homogeneous part of the equation.

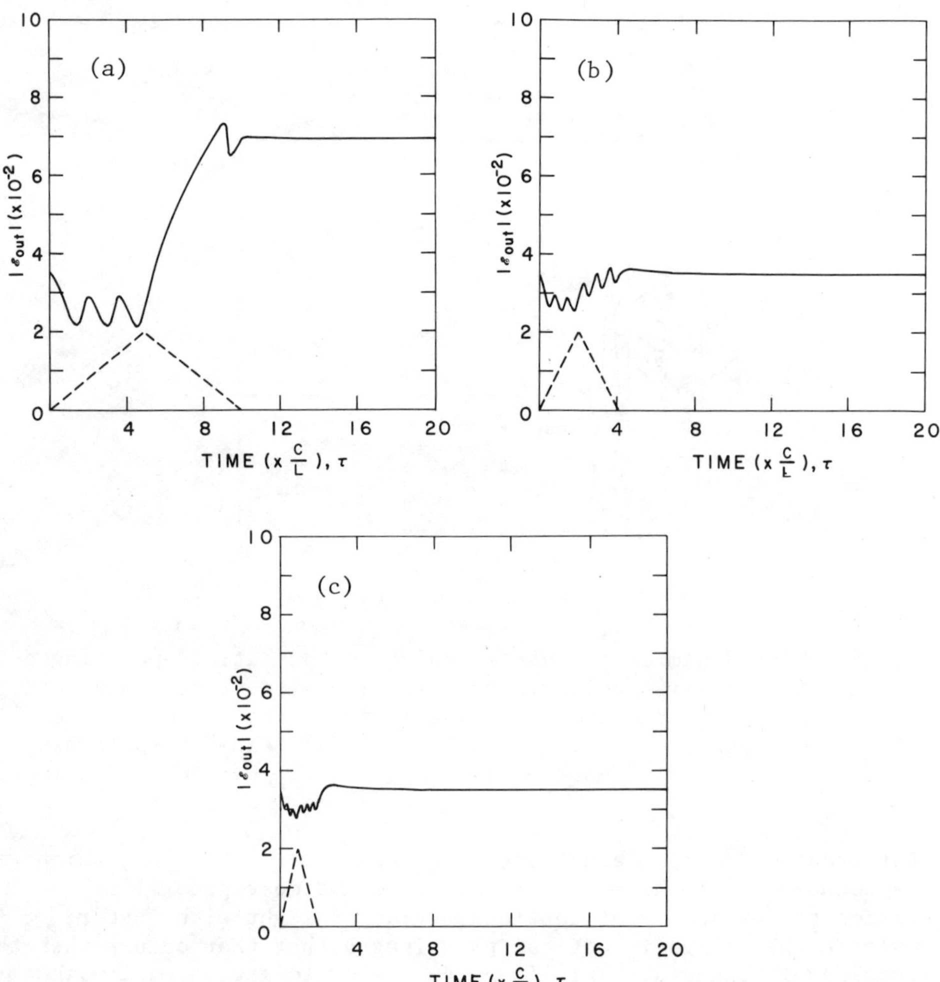

Fig. 5.   (a) "slow" up-switching 1 → 3.   (b) and (c) no switching.

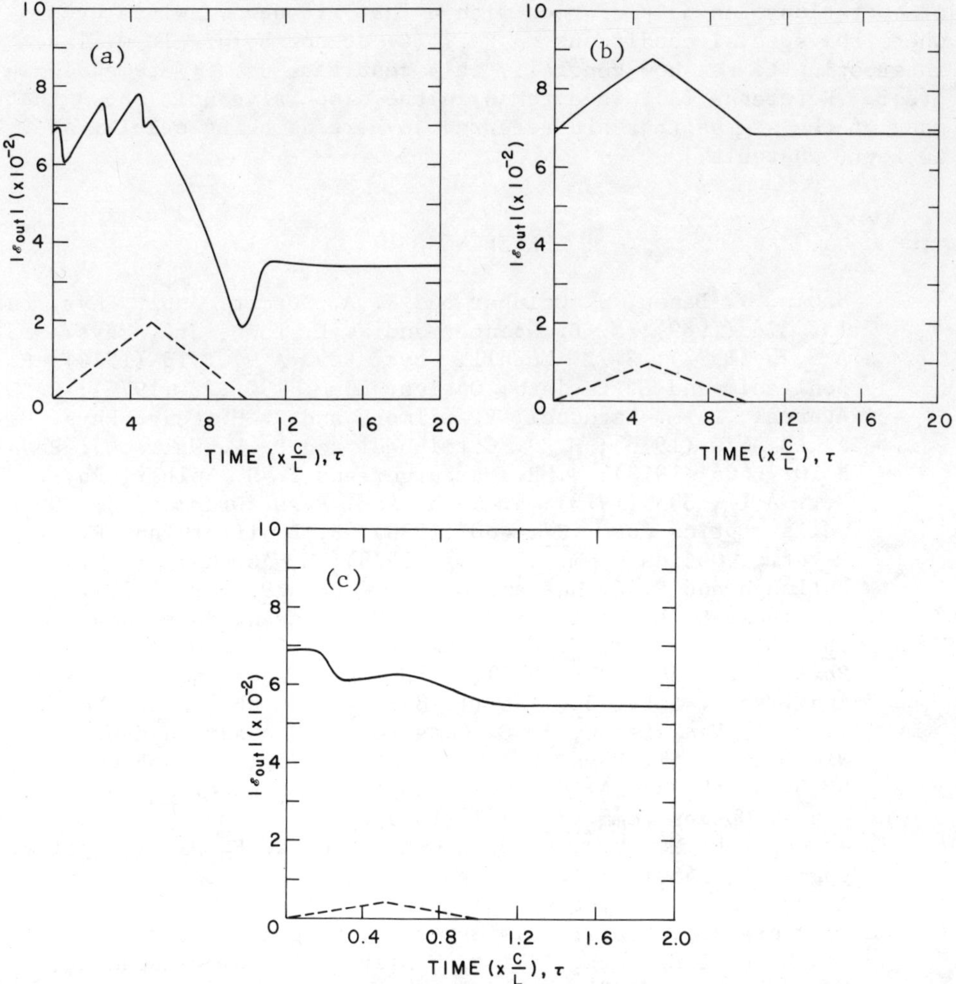

Fig. 6.   (a) "slow" down-switching 3 → 1. (b) no switching.
         (c) "fast down-switching 3 → 2.

We have calculated the response of the system to frequency
switching in the case of a cavity with modest finesse, no special
choice of frequency, and many stable configurations. We find the
switching operations n → n', n' > n, and n → 1 can be done adiabatically. The cases n' → n, n' > n and 1 → 2 can be done quickly. While
only a single example has been computed, the study of phase-switching[5] has shown that there is little difference from one case to the
next within the regime considered here. In the case of very high
finesse, low power, and special tuning conditions that characterized
the earliest study of phase switching[4], both 1 → 2 and 2 → 1 were
achievable by phase changes. We have shown here that this same pair

of operations can be performed with a fast frequency switch even where the special conditions on Ref. (4) do not hold. It will be interesting to see how generally this result holds. (Note added in proof. Reference (3), when taken in the dispersive unit shows that many of the states that are referred to here as being stable, are, in fact, unstable.]

REFERENCES

1. A. Szöke, V. Daneu, J. Goldhar and N. A. Kurnit, Appl. Phys. Lett, 15, 376 (1969); M. B. Spencer and W. E. Lamb, Jr., Phys. Rev. A 5, 5 (1972); S. L. McCall, Phys. Rev. A 9, 1515 (1974); R. Bonifacio and L. Lugiato, Optics. Comm. 19, 172 (1976); G. S. Agarwal, L. M. Narducci, R. Gilmore and D. H. Feng, Phys. Rev. A. 18, 620 (1978); H. J. Carmichael and D. F. Walls, J. Phys. B 10, L685 (1978); J. H. Marburger and F. S. Felber, Phys. Rev. A 17, 335 (1978); S. S. Hassan, P. D. Drummond and D. F. Walls, Optics Comm. 27, 480 (1978); R. Bonifacio and P. Meystre, Optics Comm. 27, 147 (1978); E. Abraham, S. K. Bullough and S. S. Hassan, Optics Comm. 29, 109 (1979).

2. H. M. Gibbs, S. L. McCall and T. N. Venkatesan, Phys. Rev. Lett. 28, 731 (1976); T. N. C. Venkatesan and S. L. McCall, Appl. Phys. Lett. 30, 282 (1977); T. Bischofberger and Y. R. Shen, Appl. Phys. Lett. 32, 156 (1978); H. M. Gibbs, S. L. McCall, T. N. C. Venkatesan, A. C. Gossard, A. Passner, and W. Wiegmann, Appl. Phys. Lett. 35, 451 (1979); D. H. B. Miller, S. D. Smith, and A. Johnston, Appl. Phys. Lett. 35, 658 (1979).

3. K. Ikeda, Optics Comm. 30, 257 (1979).

4. F. A. Hopf, P. Meystre, P. D. Drummond and D. F. Walls, Optics. Comm. 31, 245 (1979).

5. F. A. Hopf and P. Meystre, Optics Comm. to be published.

6. See for example, "Laser Physics" by M. Sargent III, M. O. Scully and W. E. Lamb, Jr., Addison-Wesley Publ. Comp., Reading, Massachusetts (1974).

DISPERSIVE OPTICAL BISTABILITY WITH FLUCTUATIONS

R. Graham and A. Schenzle

University of Essen
Essen, W-Germany

Abstract: We consider the phenomenon of dispersive optical
bistability as an example of a non-equilibrium steady state lacking
detailed balance. The Fokker Planck equation for the transmitted
electromagnetic field is derived. For a special case of parameters
we present an exact solution of the stationary distribution. In the
limit of small fluctuations the Fokker Planck equation is reduced
to the form of a Hamilton Jacobi equation for a "non-equilibrium
thermodynamic potential". This equation is solved approximately by
analytical methods. The "non-equilibrium thermodynamic potential"
acts like a free energy for the first order type phase transitions
far from thermodynamic equilibrium lacking the property of detailed
balance.

I.   INTRODUCTION

The various nonlinear phenomena in Quantum Optics have been
given considerable theoretical attention over the years due to their
interesting statistical properties as representing processes far
from thermal equilibrium. Compared with other problems in this field,
the optical processes seem to be rather elementary examples, because
to a certain extent the effects of spatial propagation in the optical
cavities can be neglected, leaving models with only a few degrees of
freedom. For systems with a small number of relevant macroscopic
variables, however, it may be possible to give a complete stochastic
description taking fluctuations fully into account. This aspect makes
nonlinear optical processes particularly attractive from a statistical
point of view.

Systems in steady states far from thermal equilibrium may exhibit instabilities analogous to the phase transitions in thermodynamic equilibrium. The statistical description of these phenomena would have to include externally imposed fluctuations as well as the intrinsic thermal- and quantum fluctuations. In Quantum Optics such analogies have been discussed first for the single mode laser[1,2] the threshold of which may be compared to the transition point of a second order phase transition of the mean field type.

In recent years analogies to first order phase transitions have been found in optical devices which exhibit regimes of metastable behaviour with a hysteresis cycle analogous to ferromagnetic systems in an external magnetic field. Examples are lasers with saturable absorbers[3] and possibly dye lasers[4]. Beside these examples of incoherently excited systems there exist optical bistable devices based on the nonlinear response of a passive medium contained in a Fabry Perot étalon to externally applied coherent fields[5,6,7,8].

For comparing these instabilities with equilibrium phase transitions the key role is played by the steady state distribution function $P(\{x_i\})$ depending on the relevant macroscopic variables x_i of the model like e.g. the complex amplitude of the electromagnetic field. P can now be used to define a generalized "non-equilibrium potential"

$$\Phi = -\ln P$$

which may be compared with an equilibrium thermodynamic potential like e.g. the free energy. The steady state distribution function P and hence Φ can only be derived in a straightforward and systematic way up to a quadrature when the system under consideration satisfies the condition of detailed balance. For systems in thermodynamic equilibrium this property is always guaranteed as a consequence of microscopic reversibility[9]. In non-equilibrium steady states, however, the condition of detailed balance will in general not be satisfied apart from some fortunate special cases. Important physical examples, when detailed balance is observed even in steady states far from thermal equilibrium, are the single mode laser with its slow amplitude and phase fluctuations near threshold[10], and the fluctuations of the electromagnetic field for purely absorptive optical bistability on time scales long compared to the typical atomic relaxation times[11,12].

Dispersive optical bistability however is an example of a process far from thermal equilibrium that does not satisfy detailed balance. It is the aim of the present paper to develop methods for obtaining the non-equilibrium potential Φ and then demonstrate the practical usefulness of this approach using the example of dispersive optical bistability.

Having determined the potential Φ it becomes possible to make
a real comparison of first order phase transitions in thermal equilibrium and corresponding steady state transitions without detailed
balance. The relative stability of the different branches of the
hysteresis cycle can be obtained by comparing the relative depth of
the minima of Φ. In this way it is possible to predict the point of
equal probability, the point where the two branches exchange global
stability, and generalize the condition of the Maxwell construction
to processes far from thermal equilibrium. It should be emphasized,
that this information is an essential property of the full nonlinear
problem which cannot be derived by any local linearization approximation.

In the deterministic limit the potential Φ acts as a Lyapunoff
function of the deterministic equation of motion. This interpretation of Φ allows one to identify the locally or globally stable
attractors with its local or global minima. It also allows one to
cast the deterministic equations of motion into the standard form
of thermodynamics given by Onsager[13]. The drift can be separated
into a superposition of two parts, one proportional to the gradient
of Φ and the other one having the potential Φ as a constant of
motion. This latter part characterizes essentially the stationary
probability current. In thermodynamic equilibrium this current is
either zero or totally reversible under time reversal transformations, while far from thermal equilibrium the remaining probability
current also contains irreversible contributions.

By constructing the stationary probability distribution
$P = \exp(-\Phi)$ for the case of dispersive optical bistability we will
be able to demonstrate and discuss all the above mentioned properties
explicitly.

II. NON-EQUILIBRIUM THERMODYNAMIC POTENTIAL FOR STEADY STATES
     WITHOUT DETAILED BALANCE

We consider systems in steady states subject to time independent or periodic external fields. We assume that a Fokker Planck
description with a discrete set of variables is valid, i.e. that a
clear separation of time scales between macroscopic variables x_i,
$i = 1,2,\ldots,n$ and microscopic variables exists and the macroscopic
variables form a continuous Markoff process. The Fokker Planck
equation of this process is written in the form:

$$\frac{\partial P}{\partial t} = -\frac{\partial}{\partial x_i} K_i(\{x_j\}) P + \frac{1}{2} \varepsilon Q_{ij} \frac{\partial^2}{\partial x_i \partial x_j} P \qquad (2.1)$$

where $P(\{x_i\}, t)$ is the conditional probability density which reduces
to the n-dimensional δ-function in the limit $t \to 0$, K_i is the drift

vector determined by the deterministic equations of motion $\dot{x}_i = K_i$ and $Q_{ij}\,\varepsilon$ is the positive definite diffusion matrix which characterizes the strength of the fluctuations. εQ_{ij} is assumed to be independent of the variables x_i and will later be considered as small $(\varepsilon \to 0^+)$. Natural boundary conditions are assumed.

The stationary properties of the process will be described by the steady state solution of (2.1)

$$\frac{\partial P_o}{\partial t} = 0, \qquad P_o = \exp(-\varepsilon^{-1}\,\Phi) \quad . \tag{2.2}$$

In the absence of any constants of motion of the process x_i, the steady state distribution is unique under certain general conditions[14]. If the externally applied forces allow the system to reach thermodynamic equilibrium, the potential Φ as described by Eq. (2.2) is proportional to the corresponding thermodynamic potential. The drift vector K_i then can be separated into a dissipative part $d_i(\{x_j\})$ and a reversible part $r_i(\{x_j\})$:

$$K_i(\{x_j\}) = d_i(\{x_j\}) + r_i(\{x_j\}) \quad . \tag{2.3}$$

According to the physical meaning of the variables x_i we can distinguish between even and odd variables under time reversal. Denoting the time reversed of x_i by $\tilde{x}_i$

$$\tilde{x}_i = \varepsilon_i\, x_i \, , \qquad \varepsilon_i = \pm 1 \, , \tag{2.4}$$

we can identify:

$$d_i(\{\tilde{x}_j\}) = \varepsilon_i\, d_i(\{x_j\}) \tag{2.5}$$

and

$$r_i(\{\tilde{x}_j\}) = -\varepsilon_i\, r_i(\{x_j\}) \tag{2.6}$$

In expressions containing ε_i the summation convention is dropped. The dissipative part then is of the form given by Onsager[13]

$$d_i = -\frac{1}{2}\, Q_{ij}\, \frac{\partial \Phi}{\partial x_j} \tag{2.7}$$

The reversible part of r_i is not determined by Φ, but leaves the thermodynamic potential and the volume element in phase space invariant:

$$r_i \frac{\partial \Phi}{\partial x_i} = 0 \quad \text{and} \quad \frac{\partial r_i}{\partial x_i} = 0 \ . \tag{2.8}$$

Defining the reversible drift through (2.3) - (2.6), we can derive from the Fokker Planck equation a more general form of (2.8) which makes use only of the detailed balance condition of the assumed equilibrium case,

$$\frac{1}{\epsilon} r_i \frac{\partial \Phi}{\partial x_i} - \frac{\partial r_i}{\partial x_i} = 0 \ . \tag{2.9}$$

For steady states lacking detailed balance we use (2.3) and (2.7) instead of (2.3) - (2.6) to define r_i by

$$K_i = - \frac{1}{2} Q_{ij} \frac{\partial \Phi}{\partial x_j} + r_i \ . \tag{2.10}$$

Inserting (2.10) into the time independent Fokker Planck equation we obtain the analogy of (2.9)

$$\frac{1}{\epsilon} r_i \frac{\partial \Phi}{\partial x_i} - \frac{\partial r_i}{\partial x_i} = 0 \ . \tag{2.11}$$

Note, however, that here we did not assume, that r_i is the reversible part of K_i; we merely define it through (2.10). Therefore, we can no longer conclude, like in equilibrium, that the stationary probability current j_i

$$j_i = (\frac{1}{2} Q_{ij} \frac{\partial \Phi}{\partial x_j} + K_i) \, P_o = r_i \, P_o \tag{2.12}$$

is reversible under time reversal transformations.

However, the formal structure of the equilibrium theory expressed in terms of a thermodynamic potential carries over to the non-equilibrium theory expressed in terms of a generalized potential Φ. This gives us the key for a direct comparison of equilibrium and non-equilibrium steady states lacking detailed balance.

While for equilibrium states Φ can easily be given in terms of quadratures, in the absence of detailed balance we have to deal with an elliptic nonhermitian boundary value problem for which no general method of solution exists.

Recognizing however that in many relevant physical examples fluctuations are extremely weak we may obtain the leading contribution to Φ in the limit $\varepsilon \to 0^+$ by solving the first order problem

$$\frac{1}{2} Q_{ij} \frac{\partial \Phi}{\partial x_i} \frac{\partial \Phi}{\partial x_j} + K_i \frac{\partial \Phi}{\partial x_i} = 0 \ . \tag{2.13}$$

It should be emphasized, that in this limit Φ becomes independent of ε while P still depends on ε in the form

$$P = \exp\left(-\frac{1}{\varepsilon} \Phi\right) \ .$$

The relation (2.9) between r_i and Φ reduces, for $\varepsilon \to 0^+$, to the orthogonality relation

$$r_i \frac{\partial \Phi}{\partial x_i} = 0 \ . \tag{2.14}$$

Equation (2.13) is a first order partial differential equation of the form of a time independent Hamilton Jacobi equation in classical mechanics for vanishing total energy. From a mathematical point of view we have replaced the second order boundary value problem of the Fokker Planck equation by a first order problem with Φ given on a hypersurface intersecting the field of characteristics. Mathematical literature exists, where these two problems are related in the limit of small ε for the corresponding time dependent equation[15].

As these theorems are restricted to finite time intervals it is not clear in which sense the solution of (2.13) approximates the solution of the full stationary Fokker Planck equation.

In the absence of a well developed mathematical framework for our procedure we have adopted a pragmatic attitude: If from a general solution of Eq. (2.13) we can derive a unique single valued one with the property $\Phi \to \infty$ for $\{x_i\} \to \infty$ necessary for normalizability, it will be taken as an approximation of the corresponding solution of the time independent Fokker Planck Eq. (2.1).

Apart from some trivial examples, unfortunately, a general solution of (2.13) cannot be obtained in a systematic way and we have to resort to further approximations based on the existence of a suitable small parameter λ in the drift vector K_i:

$$K_i = K_i^1 + \lambda K_i^2 \ . \tag{2.15}$$

An approximate solution of Eq. (2.13) can then be derived systematically in the form

$$\Phi = \sum_{n=0}^{\infty} \lambda^n \Phi_n \tag{2.16}$$

satisfying the hierarchy of equations:

$$\sum_{l=0}^{n} Q_{ij} \frac{\partial \Phi_{n-1}}{\partial x_i} \frac{\partial \Phi_1}{\partial x_j} + K_i^1 \frac{\partial \Phi_n}{\partial x_i} + K_i^2 \frac{\partial \Phi_{n-1}}{\partial x_i} = 0 . \tag{2.17}$$

In the following sections, we will derive special exact solutions as well as approximate solutions utilizing the perturbation approach (2.15) - (2.17) for the case of dispersive optical bistability.

The potential Φ, as defined by (2.2), (2.3), (2.7) and (2.14) is a Lyapunoff function of the deterministic equations. This may be seen as follows: Due to Eq. (2.2) Φ is a positive function. Due to Eq. (2.4) Φ can only decrease if it evolves under the deterministic equations of motion

$$\dot{\Phi} = \frac{\partial \Phi}{\partial x_i} \quad \dot{x}_i = \frac{\partial \Phi}{\partial x_i} K_i = - \frac{1}{2} Q_{ij} \frac{\partial \Phi}{\partial x_i} \frac{\partial \Phi}{\partial x_j} \leq 0 . \tag{2.18}$$

Thus the locally (globally) stable attractors of the deterministic equations are identified with the local (global) minima of Φ.

We summarize what actually has been achieved once a single valued solution of (2.13) has been obtained, one way or the other:

   (i)   The steady state probability is known in the limit of
          small fluctuations.

  (ii)   A Lyapunoff function of the deterministic equations of
          motion is available characterizing the locally and
          globally stable attractors.

 (iii)   The condition replacing the Maxwell construction for a
          first order type phase transition far from thermal
          equilibrium can be given.

  (iv)   The deterministic drift K_i can be decomposed into a
          gradient part which stabilizes the attractors and a
          remaining part r_i containing Φ as a constant of motion.
          This separation is a generalization of the separation
          into reversible and irreversible parts in thermodynamic
          equilibrium.

Now we will turn to our special example of optical bistability.

III. EQUATIONS OF MOTION OF A MODEL OF OPTICAL BISTABILITY WITH FLUCTUATIONS

We consider an example of N atoms with a homogeneously broadened absorption line contained in an optical cavity of volume V interacting via dipole interaction g with a single mode of the electromagnetic field proportional to E. The resonance frequency of the two level atoms is given by ν, the eigenfrequency of the empty resonator is ω_0. The coherent external field that drives the system is characterized by the amplitude E_0 and the frequency ω. We introduce the different dimensionless detuning parameters:

$$\delta = \frac{\omega_0 - \omega}{\kappa} \, , \quad \text{the detuning of the cavity mode with respect to the external field}$$

$$\Delta = (\omega - \nu)T_2, \quad \text{the detuning of the polarization with respect to the external field.}$$

Neglecting the spatial variation of the field E, (an approximation which is known in the theory of optical bistability under the somewhat unfortunate name of "mean field approximation" which must not be confused with the term mean field approximation in the theory of phase transitions) we arrive at the following set of coupled Maxwell Bloch equations[7,11,12,22]:

$$\dot{E}^+ = (i(\omega_0 - \omega) - \kappa)\, E^+ + igP^+ + \kappa E_0^+ + F^+ \tag{3.1}$$

$$\dot{P}^+ = (i(\nu - \omega) - \frac{1}{T_2})\, P^+ - igWE^+ + \Gamma^+ \tag{3.2}$$

$$\dot{W} = -\frac{1}{T_1}\, (W - W_0) - 2ig(P^+ E^- - P^- E^+) + \Gamma_0 \, . \tag{3.3}$$

For an extensive list of references on optical bistability c.f. Ref. 16. This model has first been proposed by Bonifacio and Lugiat for the description of optical bistability.

The incoherent processes like spontaneous emission, pressure broadening, finite lifetime of the photons in the cavity and externally imposed fluctuations have been taken into account by coupling the macroscopic variables, i.e. the cavity field E, the medium polarization P, and the inversion W of the two level atoms to external reservoirs by introducing the corresponding relaxation rates n, $1/T_1$, $1/T$. Along with the dissipative terms, the interaction with the reservoir introduces fluctuations which we describe by the fluctuating forces $F^\pm$, $\Gamma^\pm$, Γ_0. We assume, that the fluctuations can be reasonably approximated by Gaussian white noise. The fluctuating part of the external field which may easily be the largest source of noise can be lumped into $F^{\pm}$. It is assumed, that the atomic response to the electromagnetic field is very fast i.e., $\kappa T_2 \ll 1$ allowing the adiabatic elimination of the atomic variables[10] and leaving us with the following equation of motion[17-22]:

$$\dot{E}^+ = (i\delta - 1)\, E^+ + E_o^+ - \Gamma^2 \frac{(1-i\Delta)}{1+|E|^2}\, E^+ + R^+ \tag{3.4}$$

where we used normalized variables[8,22] and

$$\Gamma^2 = -\frac{gW_o}{2\kappa(1+\Delta^2)} \sqrt{\frac{T_1}{T_2}}\;. \tag{3.5}$$

For passive media the equilibrium inversion W_o is negative. $R^{\pm}$ combines all additive fluctuating forces with the property

$$\langle R^{\pm}(t)\rangle = 0 \tag{3.6}$$

$$\langle R^+(t + \tau)\, R^-(t)\rangle = 2Q\delta(\tau)$$

while the multiplicative fluctuations have been neglected. For models where the different contributions to $R^{\pm}$ are considered explicitly, see e.g. Ref. 11, 21. Under the time reversal transformation $t \to -t$, ReE $\to$ ReE, ImE $\to$ -ImE. Equation (3.4) does not satisfy the condition of detailed balance as long as $\Delta,\ \delta \neq 0$.

The stationary Fokker Planck equation corresponding to Eq. (3.4) can be written in the form

$$\frac{\partial P_o}{\partial t} = 0 = \frac{1}{r}\frac{\partial}{\partial r}\, r\Big(\frac{r}{1+r^2}\,(1 + \Gamma^2 + r^2) - r_o\cos\phi + \frac{Q}{2}\frac{\partial}{\partial r}\Big)P_o$$

$$+ \frac{1}{r}\frac{\partial}{\partial\phi}\Big(\delta r + \Delta\Gamma^2\frac{r}{1+r^3} + r_o\sin\phi + \frac{Q}{2}\frac{1}{r}\frac{\partial}{\partial\phi}\Big)P_o \tag{3.7}$$

using polar coordinates,

$$E^{\pm} = x \pm iy = r\exp(\pm i\phi), \qquad E_o^{\pm} = r_o\;.$$

The main task of the paper is now to derive solutions (at least approximate ones) for Eq. (3.7) and to discuss the non-equilibrium properties of the process.

IV.   THE SPECIAL CASE OF EQUAL DETUNING $\delta = \Delta$

We assume in this section, that the mismatch between the cavity mode and the external field measured in units of the resonator damping constant κ equals the detuning of the two level atoms from the external field measured in units of the transverse relaxation constant $1/T_2$ i.e., $\Delta = \delta$. In this case the stationary Fokker Planck Eq. (3.7) can be solved exactly despite the fact that the condition of detailed balance is not satisfied. The solution is given by $P \sim \exp (-\Phi/\varepsilon)$ with

$$\Phi = \frac{1}{Q} (r^2 - 2 \frac{rr_o}{\sqrt{1+\Delta^2}} \cos(\phi - \psi_o) + \Gamma^2 \ln(1 + r^2) + \frac{r_o^2}{1+\Delta^2}) \qquad (4.1)$$

and

$$\tan \psi_o = -\delta = -\Delta ,$$

as may be verified by substituting into the Fokker Planck Eq. (3.7). This distribution resembles the case $\delta = \Delta = 0$ of absorptive bistability[11,12] but is turned away from the real axis by the angle $\psi_o = -((\omega_o-\omega)/\kappa)$. The most probable phase shift between the cavity mode and the external field therefore is neither zero nor $\pm \pi/2$ but intermediate to pure dissipation or pure dispersion.

The attractor of the deterministic limit of Eq. (3.4) can be discussed completely in this special case $\delta = \Delta$ because Φ also serves as a Lyapunoff function. It is obvious that Φ is non negative in the entire plane and one verifies easily that Φ decreases under the evolution of Eq. (3.4):

$$\Phi \geq 0, \quad \frac{d\Phi}{dt} \leq 0 . \qquad (4.2)$$

The only attractors of Eq. (3.4) therefore are the isolated minima of Φ representing fixed points, excluding the possibility of limit cycles.

According to the general discussion of part II, Eq. (2.10), we are now in the position to decompose the drift vector into the gradient part and the residual part r_i which governs the steady state probability current j_i Eq. (2.12). We can write r_i as a superposition of a reversible and an irreversible part:

$$r_i = r_i^{rev} + r_i^{irr} . \qquad (4.3)$$

Using the potential (4.1), we find explicitly in polar coordinates:

$$r^{rev} = \frac{\Delta r_o}{1+\Delta^2} \begin{pmatrix} \sin \phi \\ \cos \phi \end{pmatrix} - \Delta r \begin{pmatrix} 0 \\ \dfrac{1+\Gamma^2+r^2}{1+r^2} \end{pmatrix} , \tag{4.4}$$

$$r^{irr} = - \frac{\Delta^2 r_o}{1+\Delta^2} \begin{pmatrix} - \cos \phi \\ + \sin \phi \end{pmatrix} . \tag{4.5}$$

If no external field r_o is applied, the system relaxes to equilibrium and r_i transforms like a reversible current depending on the sign of Δ in accordance with our general consideration of Eq. (2.3). A non-vanishing external field r_o drives the system away from thermal equilibrium changing the reversible part (4.4) and introducing a non zero irreversible part (4.5). Notice that r_i^{irr} exists only for a detuned system far from thermal equilibrium i.e.,

$$r^{irr} \sim r_o \Delta^2$$

and that r_i^{irr} does not change sign with the frequency mismatch Δ.

In the next section we will find an analogous situation for the more general case $\Delta \neq \delta$ which we will solve in an approximate way. As a guide line for the perturbative approach we will keep in mind: For the derivation of the irreversible residual part r_i it was sufficient to retain only the term up to first order in r_o; For the dependence on the detuning parameter Δ higher than first orders in Δ have to be considered.

V. APPROXIMATE SOLUTION IN THE LIMIT OF WEAK FLUCTUATIONS

We will utilize now the perturbation scheme outlined in section II in Eqs. (2.16) and (2.17) based on the weak fluctuation limit $\varepsilon \to 0^+$ which allows us to approximate the Fokker Planck Eq. (2.1) formally by the Hamilton Jacobi equation (2.13). In accordance with the arguments above we will use the amplitude r_o of the external field as the expansion parameter. Setting $\Phi_n = 1/Q \; \phi_n$ we obtain in zero order

$$\phi_o = (r^2 + \Gamma^2 \ln (1 + r^2)) .$$

For the correction in first order in the field r_o we obtain from (2.16) the following linear inhomogeneous partial differential equation of first order:

$$\frac{\partial \tilde{\phi}_1}{\partial r} \left(- \frac{r}{1+r^2} (1 + \Gamma^2 + r^2)\right) + \frac{1}{r} \frac{\partial \tilde{\phi}_1}{\partial \phi} r(\delta + \frac{\Delta \Gamma^2}{1+r^2}) =$$

$$- \frac{rr_o}{1+r^2} (1 + \Gamma^2 + r^2) \cos \phi \quad . \tag{5.1}$$

This equation can be solved exactly using the ansatz

$$\phi_1 = \tilde{\phi}_1 - 2 \frac{rr_o}{\sqrt{1+\delta^2}} \cos (\phi - \psi_o) \quad , \tag{5.2}$$

with ψ_o defined in (4.1), by the standard method of characteristics. In polar coordinates the differential equation of the characteristics can be written in the form

$$\frac{d\phi}{dr} = - \frac{\delta(1+r^2) + \Delta \Gamma^2}{r(1 + \Gamma^2 + r^2)} \quad . \tag{5.3}$$

Upon integration we find for $\phi = \phi (r)$,

$$\phi = \phi(0) + \frac{(\delta - \Delta)}{2(1+\Gamma^2)} \Gamma^2 \ln \frac{r^2}{1+\Gamma^2+r^2} - \delta \ln r \quad . \tag{5.4}$$

Along these characteristics, the potential $\tilde{\phi}_1$ varies according to

$$\tilde{\phi}_1 = - \int \frac{2(\delta - \Delta) r_o \Gamma^2}{\sqrt{1+\delta^2} (1+\Gamma^2+r^2)} \sin (\phi(r) - \psi_o) \, dr \quad . \tag{5.5}$$

This integral can be evaluated in a straightforward way[23] and we can write the potential Φ up to first order in r_o in the form:

$$\Phi = \frac{1}{Q} \left(r^2 - 2 \frac{rr_o}{\sqrt{1+\delta^2}} \cos (\phi - \psi_o) + \Gamma^2 \ln (1 + r^2)\right)$$

$$- 2 \frac{\Gamma^2}{Q} (\delta - \Delta) \frac{rr_o}{\sqrt{1+\delta^2}} \operatorname{Im} \frac{e^{i(\phi - \psi_o)}}{1+\Gamma^2 - i(\Delta \Gamma^2 + \delta)} F(- \frac{r^2}{1+\Gamma^2}) \tag{5.6}$$

where $F(z)$ is a hypergeometric function in the notation of[23]

$$F(z) = {}_2F_1 \left(1, \frac{1}{2} - \frac{i\delta}{2} \; ; \; \frac{3}{2} - \frac{i}{2} \frac{\delta + \Gamma^2 \Delta}{1 + \Gamma^2} \; ; \; z\right) . \tag{5.7}$$

The potential Φ is a unique solution of the linearized Hamilton Jacobi equation under the requirement of single valuedness and approximates the exact result rigorously to first order in r_o and $1/Q$. This is the most general result we have derived for this problem for finite Δ, δ, $\Delta \neq \delta$, and r_o by analytical methods. It illustrates the scope and the power of the general methods described in section II and allows us to describe the stationary statistical properties of dispersive optical bistability in detail.

Discussion:

1. From the explicit form of (5.6) we can easily derive the various
    limiting cases for which the potential Φ can be understood as
    a generalization. The case of the single mode laser with
    detuning is easily recovered by setting $r_o = 0$, $\delta = 0$, where the
    laser is above threshold for $\Gamma^2 < -1$. Purely absorptive bi-
    stability is contained in the limit $\delta = \Delta = 0$ as well as the case
    of equal detuning described in the previous chapter. From these
    limiting cases only the last one does not satisfy detailed
    balance.

2. As the potential is only an approximate result we cannot make any
    exact statements about the stability of the deterministic equa-
    tions of motion as we could in the previous example by using Φ
    as a Lyapunoff function. Nevertheless, the above result makes
    the existence of a limit cycle highly improbable.

3. In the limit of very small and very large amplitudes the potential
    Φ obtained in (5.6) can be simplified using the asymptotic pro-
    perties of the hypergeometric function. We have found that it
    coincides with the corresponding limit of the Fokker Planck
    equation[22].

4. For a more detailed discussion we restrict ourselves to the case
    $\delta = 0$, $\Delta \neq 0$ because this limit simplifies the results somewhat
    but it is certainly not necessary to do so if one is interested
    in the results in full generality. In this limit we can express
    Φ by

$$\Phi = \frac{1}{Q} \left[r^2 - 2rr_o \cos \phi + \Gamma^2 \ln (1 + r^2) \right.$$

$$\left. + 2 \frac{\Gamma}{1+\Gamma^2} rr_o \ \mathrm{Im} \ \frac{e^{i\phi}}{(1-i\frac{\Delta\Gamma^2}{1+\Gamma^2})} \ {}_2F_1 \left(1, \frac{1}{2}; \frac{3}{2} - \frac{i\Gamma^2\Delta}{2(1+\Gamma^2)}; - \frac{r^2}{1+\Gamma^2}\right) \right]$$

$$(5.8)$$

a) A rather simple approximation of (5.8) in terms of elementary functions can be obtained by expanding this expression
with respect to Δ. As the leading contribution we find:

$$\Phi = \frac{1}{Q} \left(r^2 - 2rr_o \cos \phi + \Gamma^2 \ln (1 + r^2) \right.$$

$$\left. + 2 \frac{\Delta\Gamma^2 r_o}{\sqrt{1+\Gamma^2}} \sin \phi \ \mathrm{arc} \ \tan \frac{r}{\sqrt{1+\Gamma^2}} \right) . \qquad (5.9)$$

This is certainly a crude approximation but it still contain
the main feature of bistability. It is easily seen, that
(5.9) describes a single or double peaked distribution
depending on the parameters Γ^2, Δ and r_o. These peaks no
longer lie at the same phase angle ϕ as was the case for
the previously described distributions. This simple analyti
cal expression allows us to give an intuitive impression of
the stationary distribution in a "three dimensional" plot
making it rather obvious that we have obtained a single
valued, positive definite and normalizable distribution
(Fig. 1). For the picture we have chosen a small value
of Δ as it can not be expected that the simplification
(5.9) is valid over a wide-range of values Δ.

b) At first sight the appearance of (5.8) is rather awkward
and it seems that it may not be simple to deduce some fundamental properties from this expression, keeping in mind that
the hypergeometric function is not an entire function.
Closer inspection however reveals that the functions in Eqs.
(5.6) and (5.8) are of special form and can be related to
Legendre polynomials[24]. The quadratic transformations which
are available for these special cases allow us to transform
our results into hypergeometric functions with an argument
varying only between 0 and 1/2.

$$F\left(a, a + \frac{1}{2}; c; -z\right) = (1+z)^{-a} F(2a, 2c-2a-1; c; \zeta) \qquad (5.10)$$

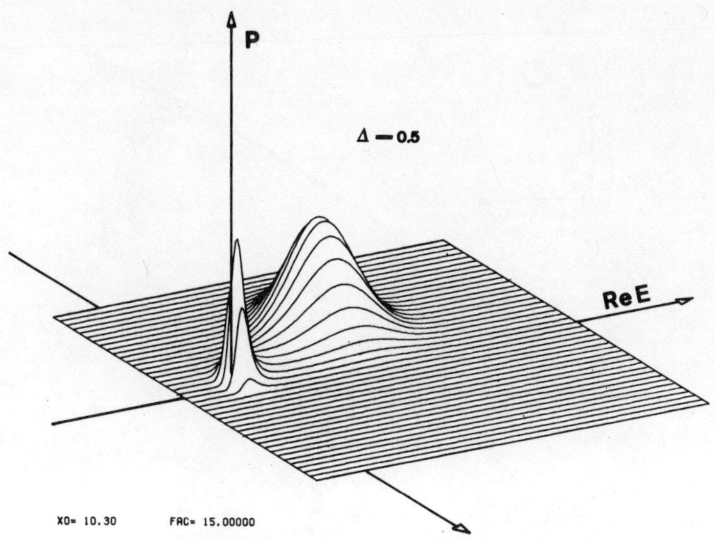

Fig. 1. Probability distribution for the dispersive optical bistability ($\Delta = 0.5$, $\Gamma^2 = 25$, $r_0 = 10.1$).

with

$$\zeta = \frac{1}{2} \left(1 - (1+z)^{-\frac{1}{2}}\right).$$

This transformation creates a rapidly converging power-
series so that the essential content of (5.6) or (5.8)
is already reproduced by the first few terms, making these
expressions easily tractable.

c) According to the weak fluctuations approximation on which
the Hamilton Jacobi formulation has been based the most-
probable value of P and the deterministic steady states are
identical. This is obvious from Eq. (2.13). Utilizing
the transformation properties mentioned in the previous
section we can easily compare the most probable values with
the deterministic steady states. In this way we can check
the validity of the approximation made in these special
physical examples. In the Figures 2a and 2b we have plotted the amplitudes and the phases of the deterministic
stationary points and compared them with the most probable
amplitudes and phases from the Fokker Planck description
in Figures 3a and 3b. Keeping in mind, that we have used
the field amplitude r_0 as an expansion parameter the agreement is unexpectedly good. This can be seen e.g., from the

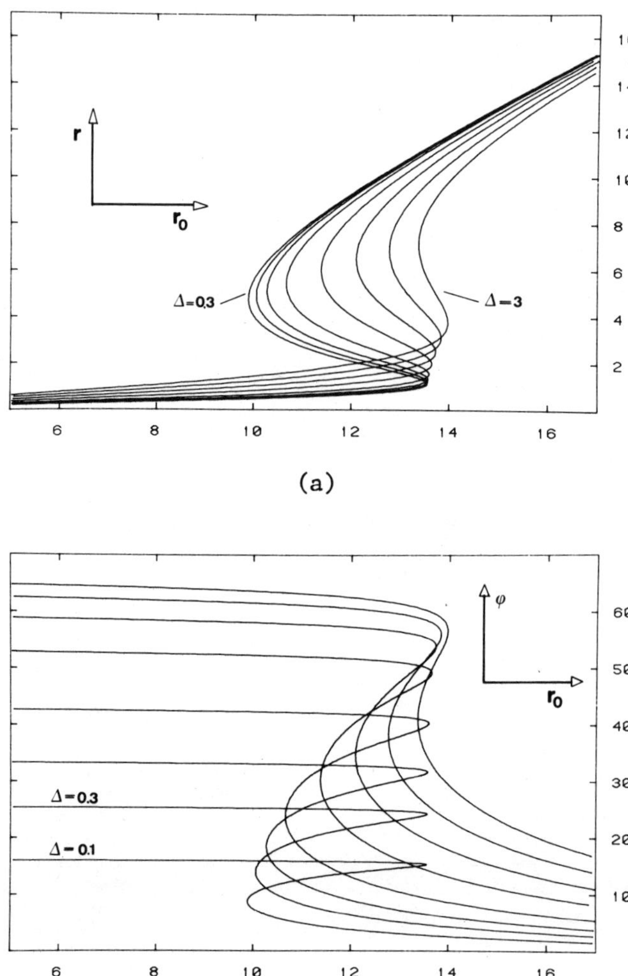

Fig. 2. Deterministic stationary points in polar coordinates for $\Gamma^2 = 25$, $\Delta = 0.3$, 0.5, 0.7, 1.0, 1.5, 2.0, 2.5, 3.0.

plot of the phase ϕ, where, over a wide range of r_0 the values agree better than one percent. While for larger amplitudes e.g., the range of bistability is reproduced rather well, the accuracy for the asymptotic limit $r_0 \to \infty$ becomes continuously worse but is still approaching the correct limiting value:

In the limit $r_0 \to \infty$ we find, allowing $\delta \neq 0$ in this calculation

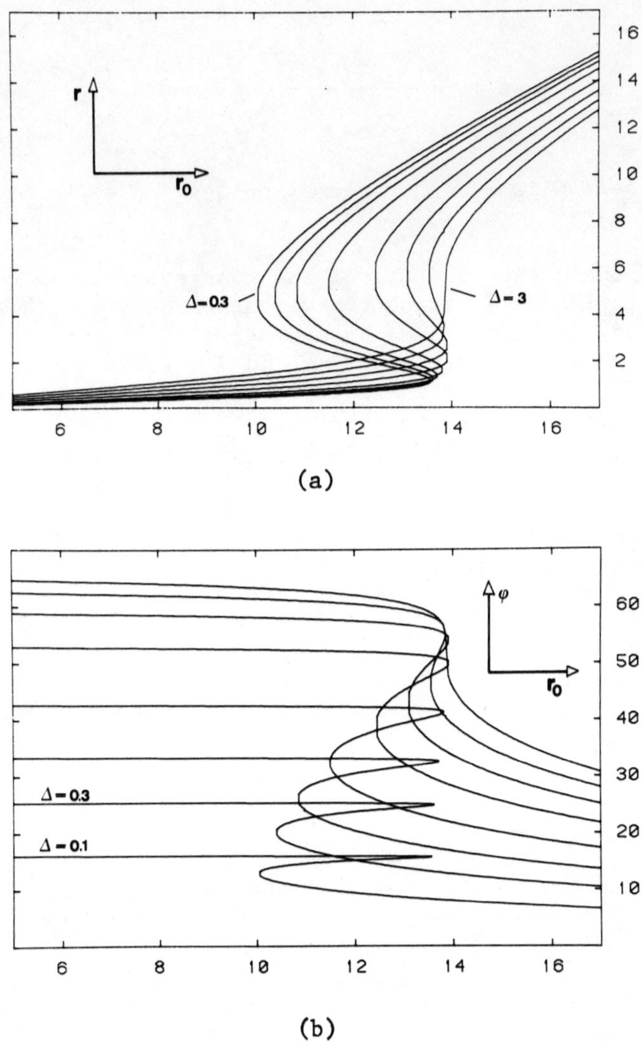

Fig. 3.   Most probable values of the stationary distribution (5.8) in polar coordinates for $\Gamma^2 = 25$, $\Delta = 0.3, 0.5, 0.7, 1.0, 1.5, 2.0, 2.5, 3.0$.

| Deterministic fixed point | Most probable value |

$$\lim_{r_o \to \infty} r = \frac{r_o}{\sqrt{1+\delta^2}} \; , \qquad\qquad\qquad \frac{r_o}{\sqrt{1+\delta^2}}$$

$$\lim_{r_o \to \infty} \phi = -\delta + \frac{(\delta-\Delta)\Gamma^2(1+\delta)^2}{r_o^2} \; , \qquad -\delta + \frac{(1+\delta^2)\,C}{r_o} \; , \qquad (5.12)$$

where C is a rather clumsy collection of constant parameters[22].

d) Knowing the location of the most probable values of the
    stationary distribution we can calculate with the help of
    Eqs. (5.6) or (5.8) the relative stability of the two
    branches. In this way we can determine the field strength
    r_0 for which both branches are simultaneously stable in
    order to generalize the idea of the Maxwell construction to
    processes far from thermal equilibrium. In Fig. 4 the
    relative probability P_1/P_2 of the two branches is plotted
    as a function of r_0 for different parameters Δ. The point
    where the curves attain the value unity is the point of the
    generalized construction.

    In Fig. 5 we have indicated by dashed lines the special
    values of r_0 where global stability is exchanged by the two
    branches. By rotating this plot by 90 degrees we get a
    picture which qualitatively resembles the typical picture

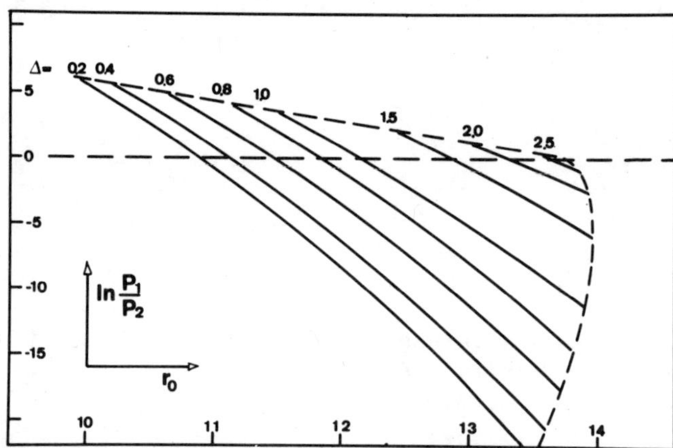

Fig. 4. Ratio of the probabilities of the two branches in the
         bistable domain.

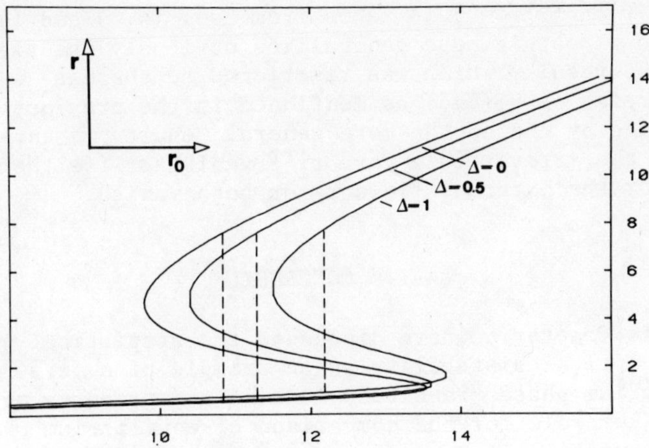

Fig. 5.   Generalization of the Maxwell construction.

of the coexistence curve of a van der Waal gas.   For $\Delta = 0$, $\delta = 0$, it is easy to prove that the Maxwell construction is still valid.

5.   In this last paragraph we want to discuss the interpretation of the potential Eq. (5.6) as a generalized thermodynamic potential for a system far from thermal equilibrium which can be used to define the residual drift r_i by

$$r_i = K_i + \frac{Q}{2} \frac{\partial \Phi}{\partial x_i} \tag{5.13}$$

and the steady state probability current $j_i = r_i \, P_0$.   Abbreviating the potential (5.6) formally by

$$\Phi = \frac{1}{Q} \left[\phi_0 + i r_0 (\delta - \Delta)(Ae^{i\phi} F(r) - c.c.) \right]$$

we can separate r_i into its reversible and irreversible components: In polar coordinates we find

$$r^{rev} = r^{rev}(\Delta=\delta) - \frac{1}{2}(\delta-\Delta)r_0 \begin{pmatrix} (AF' + A^*F^{*'}) \sin \phi \\ (AF + A^*F^*) \cos \phi \end{pmatrix} \tag{5.14}$$

$$r^{irr} = r^{irr}(\Delta=\delta) + \frac{i}{2}(\delta-\Delta)r_0 \begin{pmatrix} (AF' - A^*F^{*'}) \cos \phi \\ -(AF - A^*F^*) \sin \phi \end{pmatrix} \tag{5.15}$$

where $r^{irr,rev}(\delta=\Delta)$ is taken from Eqs. (4.4) and (4.5). The result we obtain here generalizes obviously the previously derived result, which was restricted to the case $\delta = \Delta$. The properties of $r_i(\Delta=\delta)$ as mentioned in the previous chapter carry over one by one to the more general result presented here as, e.g., the irreversible part r_i^{irr} vanishes if either the detuning Δ, δ or the external field r_o or both vanish.

VI.   CONCLUSION

In this Chapter we have discussed the statistical properties of dispersive optical bistability as an example of a first order type non-equilibrium phase transition lacking the property of detailed balance. The key role for the comparison of equilibrium phase transitions and instabilities far from thermal equilibrium is played by the steady state distribution function $P_o = \exp(-\Phi)$. We have shown that the potential Φ can be understood as a generalization of the equilibrium thermodynamic potentials like the free energy. The formal aspects of this analogy have been discussed in Chapter II. A model of dispersive optical bistability with fluctuations has been develope in Chapter III. An exact solution in the special case of equal detun ing has been derived along with an approximate solution for the general case. These solutions then were used to define non-equilibrium thermodynamic potentials.

With the explicit analytical result of the Chapters IV and V it has been possible to illustrate the formal arguments of Chapter II. For the example of dispersive optical bistability we have found:

  i) the stationary probability distribution

 ii) the generalized Maxwell construction

iii) the residual drift vector and the stationary probability current
     with its reversible and irreversible contribution.

ACKNOWLEDGEMENT

We want to thank Mr. M. Dörfle for his help in performing the necessary numerical computations.

REFERENCES

1. R. Graham and H. Haken, Z. Phys. __31__, 237 (1970).
2. De Giorgio and M. O. Scully, Phys. Rev. __A2__, 1170 (1970).

3. J. F. Scott, M. Sargent, III and C. D. Contrell, Optics Comm. 15, 13 (1975), S. T. Dembinsky and A. Kossakowski, Z. Phys. B25, 20 (1976).

4. R. B. Schaefer and C. R. Willis, Phys. Lett. 58A, 53 (1976).

5. A. Szöke, V. Danen, S. Goldhar, and N. A. Kurnit, Appl. Phys. Lett. 15, 376 (1969).

6. S. L. McCall, Phys. Rev. A9, 1515 (1974).

7. H. M. Gibbs, S. L. McCall and T. N. C. Venkatesan, Phys. Rev. Lett. 36, 1135 (1976).

8. R. Bonifacio and L. Lugiato, Optics Comm. 19, 172 (1976).

9. S. R. DeGroot and P. Mazur, Nonequilibrium Thermodynamics, North-Holland, Amsterdam (1962).

10. H. Haken, Handbuch der Physik, Springer, Berlin, Vol. XXV/2c.

11. A. Schenzle and H. Brand, Optics Comm. 27, 485 (1978), Optics Comm. 31, 401 (1979).

12. R. Bonifacio, M. Gronchi and L. A. Lugiato, Phys. Rev. A18, 2266 (1978).

13. L. Onsager, Phys. Rev. 37, 405 (1931); 38, 2265 (1931).

14. J. L. Lebowitz and P. G. Bergman, Ann. Phys. 1, 1 (1957).

15. W. H. Fleming, J. Diff. Equ. 5, 515 (1969).

16. R. Bonifacio, L. A. Lugiato and M. Gronchi, in Laser Spectroscopy IV, ed. H. Walther and K. W. Rothe, Springer (1979).

17. G. P. Agarwal and H. J. Carmichael, Phys. Rev. A19, 2074 (1979).

18. S. S. Hassan, P. D. Drumond and D. F. Walls, Opt. Comm. 27, 480 (1978).

19. R. Bonifacio and L. A. Lugiato, Lett. Nuovo Cimento 21, 517 (1978).

20. G. S. Agarwal, L. M. Narducci, R. Gilmore and D. Hsua Feng, Phys. Rev. A18, 620 (1978).

21. A. Schenzle and H. Brand Phys. Rev. A20, 1628 (1979).

22. R. Graham and A. Schenzle, Phys. Rev. A, to be published.

23. I. S. Gradshteyn and I. M. Ryzhik, Tables of Integrals Series and Products, Academic Press (1965).

24. M. Abramowitz and I. A. Stegun, Handbook of Mathematical Functions, Dover Publ., New York (1965).

FLUCTUATIONS AND TRANSITIONS IN THE ABSORPTIVE OPTICAL BISTABILITY

J. C. Englund, W. C. Schieve, W. Zurek, and R. F. Gragg

Center for Studies in Statistical Mechanics and
Thermodynamics
University of Texas at Austin
Austin, Texas 78712

Abstract: From a previously suggested stochastic differential
equation (S.D.E.), the fluctuations in amplitude, x, *both* in and
out of the steady state are described for the absorptive bistability.
The Stratonovic Fokker-Planck (F.P.) equation obtained is utilized
to develop a theory of switching between locally stable states of
amplitude x. This approach is a generalization of Kramers' early
work and the ideas of nucleation theory to the situation of non-
constant diffusion, an interesting characteristic of the optical
bistability. The importance of fluctuations (noise) is emphasized
in switching. This result is compared to the approximation of
Kramers and Landauer-Swanson. Also, comparisons are made to mean
first passage estimates, and a recent work of Hanggi, Bulsara and
Janda. The important question of the dominance of the low eigenvalue of the F.P. equation is investigated numerically by the development of a variational eigenvalue calculation for the bistability.
The one eigenvalue approximation is found to hold for a wide range
of y values and it is found that the variational treatment and the
above theory agree well. Critical slowing is seen for $c = 4$, $q = 0.01$.
The numerical algorithm may readily be applied to other one-dimensional bistable models.

I. INTRODUCTION

There is a phenomenal growth of interest in physics, chemistry
and even the social sciences in far from equilibrium phenomena, particular attention being focused upon the appearance of bifurcations
and the fluctuations and spatial-temporal behavior associated with
them[1-5]. Far from the bifurcations (critical point) there is a

minimal interest in fluctuation phenomena since then the deterministic equations give the qualitative features. This is a reflection of the theorem of Kurtz[6] which says that in the thermodynamic limit the internal fluctuations are negligible, i.e.,

$$\lim_{v \to \infty} \mathrm{Prob}[\sup_{s \leq t} |x_s^v - x(s)| > \varepsilon] = 0 \qquad\qquad \varepsilon > 0 ,$$

where $x_s^v = x_s/v$ for the stochastic variable x_s in volume v, and x(s) is the deterministic value at time s $\leq$ t. This theorem does not apply at a bifurcation since the stochastic transition rates/volume are no longer analytic in 1/v. The fluctuations no longer scale as 1/v at the critical point but more as $v^{-\frac{1}{2}}$ [5]. The importance of fluctuations is greatly enhanced at the bifurcation point and beyond.

The absorptive optical bistability (AOB) is an interesting system for the discussion of non-linear fluctuations in the region of multiple stationary states and bifurcations. Particularly important will be the discussion of *switching* and time dependent phenomena by means of the Fokker-Planck (F.P.) equation. There is a great interest in this quantum optics example because of the *microscopic* basis for the model developed by Bonifacio, Lugiato and others[7].

We will in this paper develop a *mesoscopic* approach much in the spirit of Risken's work[8] utilizing the Langevin equation for the laser. Here we will treat amplitude fluctuations in the AOB utilizing stochastic differential equations suggested by the deterministic semiclassical theory of Bonifacio and Lugiato[9]. One may obtain a F.P. equation for the probability distribution of field amplitude in the cavity, $P(x,t)$[10,11]. From this the steady state distribution, $P(x,\infty)$, has been examined and compared to the microscopic theory. In the next section (II) we will obtain the F.P. equation for P(x,t) and comment on important aspects of the results in the steady state, particularly the effects of noise on the critical point.

Such a mesoscopic approach utilizing the F.P. equation describing the fluctuations has also been used by Schenzle and Brand[12] who discussed fluctuations introduced by the external driving field, y, in a Langevin theory. Kondo, Mabuchi, Hasegawa[13] have utilized the same steady state solution as in Ref. 10, and comment on the non-validity of the Maxwell construction (see next section) suggesting that it holds in a rescaled coordinate system. Also, F. T. Arecchi and A. Politi[14] have rederived the F.P. equation and obtained the microscopic noise amplitude by the fluctuation dissipation theorem, following Risken.

The main interest here is to consider the *time dependence* and *switching* between locally steady states from the point of view of the F.P. equation. In section III we will develop a simple generalization of Kramers' description of decay from a metastable state[15-17].

This approach is generalized to the case of non-constant diffusion which is an interesting property of the F.P. equation for the AOB. We should comment that the approximations made are the same as in nucleation theory so we might properly call this nucleation in the optical bistability[18-20]. The mean first passage time calculation will be discussed[21,22] and a comparison will be made to the "nucleation theory" approach. Recently work on time dependent transitions in the AOB has begun. Bonifacio and Meystre[23] have examined transient response by numerically solving the deterministic equations. They have obtained a sort of deterministic "critical slowing".

Finally, in section IV we will describe a computer algorithm to approximate the lowest lying eigenvalues of the one dimensional F.P. equation for $P(x,t)$. The important question of the separation of the eigenvalues will be examined. Comparison is then made to the theory of section III and Kramers' approach.

II.   THE F.P. EQUATION: STEADY STATE

Let us write the deterministic equations for the AOB as

$$\dot{x} = -2 \; \partial V/\partial x$$

$$\dot{\phi} = - \frac{2}{x^2} \; \partial V/\partial \phi \; , \tag{2.1}$$

where x is the scaled field amplitude of the cavity field and ϕ is its phase[7]. The deterministic "potential" is given by

$$V(x,\phi) = \frac{1}{4} \; [x^2 + y^2 - 2xy\cos(\phi-\phi_0)+2c\ln(1+x^2)] \; . \tag{2.2}$$

Here y, ϕ_0 are the amplitude and phase of the incident driving field. It is a simplification of the AOB that such a potential may be obtained; there is no such potential for the dispersive case. The "order parameter" analogous to temperature is $c=N\mu^2\omega/4\kappa\gamma_\perp\hbar\epsilon_0 V$[7,9]. To obtain the above equations we have assumed $\kappa<<\gamma_\perp,\gamma_\parallel$ and allowed the polarization phase, θ, and amplitude, P, and the two level population difference, d, as described by the optical Bloch equations to achieve their steady state values. Bifurcation analysis shows $c=4$ is a critical point, bistable states appearing for $c>4$. The deterministic potential is shown in Fig. 1.

The two minima and maximum determining the bistable behavior are clearly present. It is the growth and decline of the two minima as a function of y which characterize the typical S shaped x vs. y curve. The extrema may be readily shown to be the roots of the simple equation

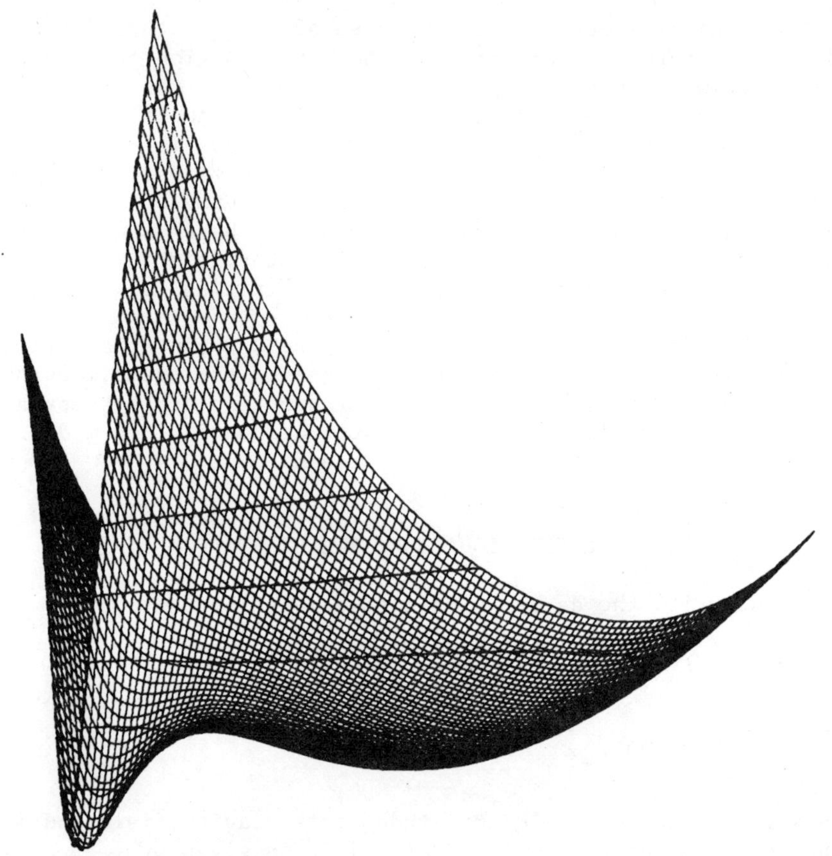

Fig. 1   The deterministic potential of Eq. 2.2 for c=20, y=13.5,
           $0 \leq \mathrm{Re}(x) \leq 15$ and $-4 \leq \mathrm{Im}(x) \leq 4$. The axis of symmetry
           is the Re(x) axis.

$$y - x - \frac{2cx}{1 + x^2} = 0, \tag{2.3}$$

where x follows the "valley' of the potential picture. It is the
simple characteristic of the AOB that the maxima and minima of this
curve are at $\phi = \phi_0$. This simplicity is lost in the dispersive
case[23].

   Fluctuations may distort this picture and more important for
the point of this paper they may cause transitions from the then
locally stable state *over* the potential saddle into the globally
stable state. This is the fluctuation picture of the switching and
the main point of the following discussion. It is important to

emphasize that the fluctuations cause a state which is deterministically stable to become metastable. The locally stable (higher valley) is in fact under fluctuations *unstable*.

In order to introduce the fluctuations from the mesoscopic point of view let us reinterpret the Bloch equation variables as stochastic variables. Then the optical Bloch equations become stochastic differential equations[24-26]. We assume that the phase, ϕ, polarization, P, and inversion, d, are strongly damped about their steady deterministic values. The amplitude x is taken to be the unstable mode. This is strongly suggested by the deterministic potential, Fig. 1. The population inversion is taken to be a *random force* driving the field amplitude, x. We obtain[10]*

$$dx = [y-x- \frac{2cx}{1+x^2}]dt + \sigma \frac{x}{1+x^2} \dot{W} dt, \qquad (2.4)$$

where the driving force is taken to be a Wiener (Brownian) process W,

$$\dot{d} = \hat{\sigma}\dot{W}$$

and,   (2.5)

$$\sigma = 2\kappa c/\gamma_{\|} \hat{\sigma} \quad .$$

Here, we have assumed

a) $\bar{P}(t) = 0$, $\dot{P}$ contributes negligibly to the fluctuations,
b) $\bar{\dot{d}}(t) = 0$,
c) $\dot{\phi}(t) = 0$,

and

$$<\dot{d}(t)\dot{d}(t')> = \hat{\sigma}^2\delta(t-t') \quad . \qquad \text{(white noise)}$$

The $\hat{\sigma}$ in Eq. (2.4) is the phenomenological noise amplitude. The source of fluctuations may be taken to be the microscopic fluctuations, spontaneous emission, in which case we have[8] $\sigma^2=2q\equiv c/N_s$ where $N_s=\gamma_{\perp}\gamma_{\|}/\bar{g}^2$ is the saturation photon number. We may also ascribe other sources to the noise such as the fluctuation in the atomic density in the cavity†.

This is a generalization of the Langevin equation and an example of a *multiplicative stochastic process*. The noise term $\dot{W}dt$ is multiplied by the stochastic saturation term $x/(1+x^2)$. There is now (unlike the Langevin theory where $\dot{W}dt$ is multiplied by a constant) mathematical ambiguity in defining the integral

---

*Here t is the dimensionless time κt, κ the cavity loss constant.

†c itself, then, may be viewed a stochastic variable driven by atom density fluctuations.

$$\int g(x)dW \quad .$$

Two suggestions on how to proceed mathematically have been made by Ito[27] and Stratonovic[28]. It may be shown that the Stratonovic definition allows the classical rules of calculus[27,28]. Uncertainties may thus arise in going to the F.P. equations from Eq. (2.4). Gray and Caughy[29] and Wong and Zakai[30] have discussed this. The latter show that a series of processes $W^n(t)$, $n=1,2,\ldots,\infty$ approach the Wiener process whose Markov process, $x(t)$, is a solution interpreted as a *Stratonovic* integral. Thus, it would seem for continuous processes the Stratonovic point of view and ordinary calculus should be used.

Exactly as in Langevin theory we may obtain the F.P. equation[31] for $P_{S,I}(x,t)$, the *probability* of field amplitude x,

$$\frac{\partial P_{S,I}(x,t)}{\partial t} = -\frac{\partial}{\partial x}\left[a_{S,I}(x)P_{S,I}(x,t)\right] + \frac{1}{2}\frac{\partial^2}{\partial x^2}[b(x)P_{S,I}(x,t)],$$

$$(2.6)$$

where

$$b(x) = \left(\frac{\sigma x}{1+x^2}\right)^2$$

and $$(2.7)$$

$$a_S = a_I + \frac{1}{4}b'(x) , \qquad a_I = y - x\left(1 + \frac{2c}{1+x^2}\right) .$$

Here both the Stratonovic and Ito F.P. equations for P_S and P_I respectively are written together for comparison. Ito and Stratonovic F.P. equations differ slightly usually in having the a_S replaced by a_I in the drift term. Multiplicative stochastic processes have the interesting property of leading to non-constant diffusion, $b(x)$.

Before turning to time dependent phenomena let us briefly make a few comments concerning the steady solutions. We may show in one dimension[2,31]

$$P_S(x,\infty) \propto b^{\frac{1}{2}} P_I(x,\infty) ,$$

$$(2.8)$$

where

$$P_I(x,\infty) = (Nb)^{-1}\exp\left(\frac{-u_I(x)}{q}\right)$$

$$(2.9)$$

and

$$u_I(x) = y/x - 2xy - x^3y/3 + x^4/4 + (1+c)x^2 + (1+2c)\ln x .$$

One characteristic of the AOB for q = 0.01 is the extreme sharpness of the transition in the probability weights of the two steady states[7,11]. This is not physically relevant as we shall see later. The transition region does not obey Maxwell's equal area construction. This is the result of non-constant diffusion. Let us consider the qualitative effect of the noise upon the steady state as a function of $2q = \sigma^2$. If one calculates the extremal x by the F.P. equation (2.6), it may be easily seen that the deterministic x curves are *qualitatively* changed as a function of the noise intensity (see Fig. 2). This has been shown for the AOB at the deterministic critical point, c = 4 [10] (q = 0.01), shown here by curve a.

For larger amplitude of noise (q = 15, curve b) the threshold is shifted to lower c values. The noise shifts the bifurcation point. This has been discussed by Horsthemke et al.[32] for the Schlögl model and a model of tumor immunology. It may further be seen that the noise may qualitatively introduce *new bifurcations* not present at q = 0 (Horsthemke, Ref. 4,5). This can be seen by calculating the extremal x from the extremum of the steady state equation for $P(x,\infty)$. It is found that the order of polynomial equation analogous to Eq. (2.3) may be changed for $q \neq 0$. Noise may introduce new steady states not otherwise present in the deterministic theory.

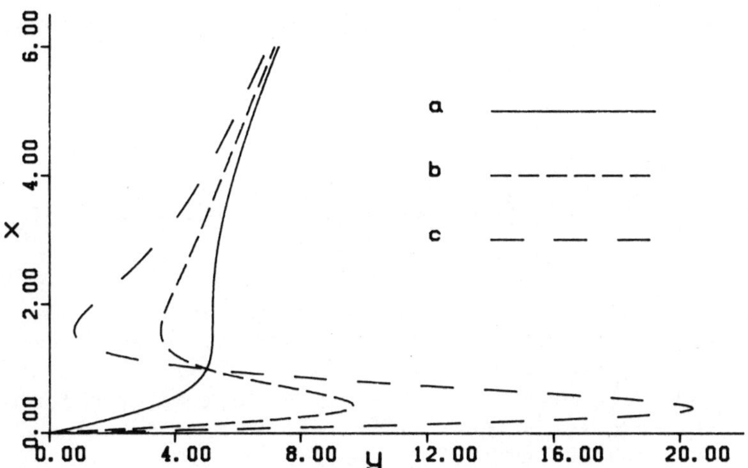

Fig. 2   The locations of the extrema of $P(x,\infty)$ for c = 4 and
           (a) q = .01, (b) q = 15 and (c) q = 40.

The F.P. equation, Eq. (2.6), may be compared to that of Bonifacio, Gronchi and Lugiato[7] obtained from the microscopic Liouville equation and familiar lengthy approximations. After utilizing the Glauber P representation they obtain a F.P. equation for $P(x,\phi,t)$

$$\frac{\partial P(x,\phi,t)}{\partial t} = -\left[\frac{\partial A_x}{\partial x} + \frac{\partial A_\phi}{\partial \phi}\right] P(x,\phi,t)$$

$$+ \frac{1}{2}\left[\frac{\partial^2 B_{xx}}{\partial x^2} + \frac{\partial^2 B_{\phi\phi}}{\partial \phi^2}\right] P(x,\phi,t) \;, \tag{2.10}$$

where

$$B_{xx} = 2q\left(\frac{x}{1+x^2}\right)^2 \tag{2.11}$$

and

$$A_x = A_x^{\text{det}} = y\cos\phi - x - \frac{2cx}{1+x^2} \;.$$

The other coefficients do not interest us here. Comparison with our previous comments indicate that they obtain the equivalent of the Ito result which is curious in the light of Wong and Zakai's arguments. This may be traced[33] to an expansion in $\varepsilon = 1/4N_s$. There Bonifacio, Gronchi and Lugiato obtain a term proportional to $2c\varepsilon \frac{\partial}{\partial x} x \frac{\partial}{\partial x} \frac{x^2}{(1+x^2)^2}$ which they write as $2c\varepsilon \frac{\partial^2}{\partial x^2} \frac{x^3}{(1+x^2)^2}$. With these changes the P representation goes over to the Stratonovic F.P. equation, Eq. (2.6) with

$$A_x = A_x^{\text{det}} + \frac{1}{4} B'_{xx} \;.$$

III. TIME DEPENDENT FLUCTUATIONS IN THE OPTICAL BISTABILITY: THEORY

Let us now consider the time dependent transitions between multiple steady states. We will utilize the F.P. equation, Eq. (2.6). There is much recent interest in time dependent stochastic phenomena[34-36]. The problem can be understood by considering the deterministic F.P. potential, Fig. 1, or more exactly, a plot of $u_I(x)$, Eq. (2.9) shown in Fig. 3.

Here x_1, x_3 are the maxima in $P(x,\infty)$ and x_2 the minimum. The location of these extrema are, of course, given by Eq. (2.8) as previously discussed. As y is increased one minimum at x_1, first globally stable, becomes unstable, the minimum x_3 being then

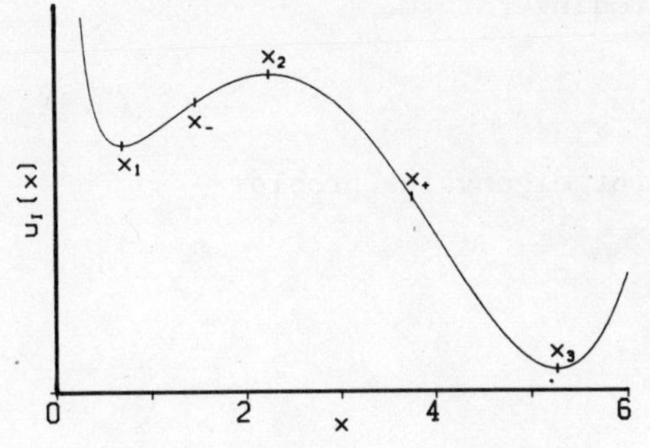

Fig. 3 The deterministic potential $u_I(x)$ for c=8, q=1 and y=8.2. x_1, x_2 and x_3 are the extrema and x_-, x_+ define the coarse-graining of the probabilities P_1 and P_3 in the Kramers theory.

globally stable. This situation is illustrated in the figure. We are interested in the transition rate problem: given the system in the "Potential" well near x_1 what is the time dependence of the switching to the stable state around x_3? The potential picture is only partially correct since this is a time-dependent, not a steady state, problem. We will first adopt the point of view of Kramers and others[15-17]: the potential hump growing at x_2 introduces a long time scale of stochastic diffusion which is longer than the local rapid damping within the well.

Let us briefly outline generally this approach for multiplicative stochastic processes and non-constant diffusion. We write the F.P. equation, Eq. (2.6), as

$$\frac{\partial P(x,t)}{\partial t} = - L(x)P(x,t) ; \qquad (3.1)$$

the probability flux is

$$J(x,t) = a(x)P(x,t) - \frac{1}{2}\frac{\partial}{\partial x}[b(x)P(x,t)] \qquad (3.2)$$

and the "Liouvillian" of the F.P. equation is

$$L(x)P = \frac{\partial}{\partial x} J, \qquad (3.3)$$

with appropriate boundary conditions. Knowing the initial distribution we may obtain $P(x,t)$ as the solution. The steady state solution $P(x,\infty)$ has been written in Eq. (2.8). L is not Hermitian. However, Eq. (3.3) may be readily written in a self-adjoint form. Let

$$H(x) = P^{-\frac{1}{2}}(x,\infty)L(x)P^{\frac{1}{2}}(x,\infty)$$

$$\psi(x,t) = P^{-\frac{1}{2}}(x,\infty)P(x,t) , \qquad (3.4)$$

then the F.P. equation takes a Schrödinger form,

$$\frac{\partial \psi(x,t)}{\psi t} = - H(x)\psi(x,t) \ .$$

(3.5)

We are interested in the self-adjoint eigenvalue problem

$$H\psi_k(x) = -\lambda_k \psi_k(x) \ , \ 0 \le x \le \infty \ .$$

(3.6)

The time dependence is then obviously

$$\psi(x,t) = \sum_{k=1}^{\infty} c_k \psi_k(x) \exp(-\lambda_k t) + P^{\frac{1}{2}}(x,\infty) \ .$$

It would seem a simple one-dimensional Schrödinger like problem but there are few analytically soluble two-well problems[15,16]. Thus, we must resort to approximations or numerical computation. We will do both.

The fundamental assumption is to assume a well separated lowest eigenvalue λ_1, so that the long time decay will be dominated by it on a time scale $t >> (\lambda_2 - \lambda_1)^{-1}$. Define

$$P_1(t) = \int_{o}^{x_- \le x_2} dx \ P(x,t),$$

$$P_3(t) = \int_{x_+ \ge x_2}^{\infty} dx \ P(x,t) \ ,$$

(3.7)

P_1 being the coarse grain probability of being at $x < x_-$ near x_1. Assume initially $P_1(0) = 1$, $P_3(0) = 0$ then for a dominant eigenvalue we have simply

$$P_1(t) = P_1(\infty) + P_3(\infty)\exp(-\lambda_1 t) \ ,$$

$$P_3(t) = P_3(\infty) - P_3(\infty)\exp(-\lambda_1 t) \ .$$

(3.8)

Let us estimate the lowest eigenvalue by the Kramers method, a *generalized nucleation theory*. We write similarly to Eq. (2.8)

$$P(x,\infty) = N^{-1}\exp(-U(x)),$$

(3.9)

where

$$U(x) = \ell n b(x) + u(x),$$

$$u(x) = 2 \int_{o}^{x} dx' a(x')/b(x') \ ,$$

(3.10)

and by integration,

$$\dot{P}_1(t) = J(x_-,t), \quad x_- \le x_2 . \tag{3.11}$$

We may formally integrate the F.P. equation and write

$$P(x,t) = \left[2 \int_0^{x} \frac{x_2}{} dx' \frac{J(x',t)}{b(x')P(x',\infty)} + D(t) \right] P(x,\infty), \tag{3.12}$$

where $D(t)$ is an arbitrary constant. Now, we assume that the diffusion constant $b(x)$ is slowly varying compared to the exponential
behavior of $P(x,\infty)$. The potential maximum of $u(x)$ gives a minimum
to $P(x,\infty)$ near x_2, the bottleneck a "nucleus" leading to the dominant contribution of the integral in this region. We assume $J(x,t)$
is slowly varying near x_+ and write

$$P(x,t) = \left[2\dot{J}(x_2,t) \int_0^{x_+} dx' [(b(x')P(x',\infty))]^{-1} + D(t) \right] P(x,\infty) . \tag{3.13}$$

This is the nucleation approximation. We further assume local
equilibrium near x_1 and x_3. This is the strong damping approximation of Kramers. The system is sufficiently damped so that it
reaches equilibrium locally on a shorter time scale than the diffusion over the barrier. We assume

$$P(x_-,t) = P(x_-,\infty) \, P_1(t)/P_1(\infty) ,$$

$$P(x_+,t) = P(x_+,\infty) \, P_3(t)/P_3(\infty) . \tag{3.14}$$

With these two assumptions and Eq. (3.11) we obtain

$$\dot{P}_1(t) = J(x_2,t) = \lambda[P_1(\infty)P_3(t) - P_3(\infty)P_1(t)] , \tag{3.15}$$

where we identify the lowest eigenvalue λ_1 from Eq. (3.8) as

$$\lambda_1 = [NBP_1(\infty)P_3(\infty)]^{-1}, \tag{3.16}$$

and

$$B = 2 \int_{x_-}^{x_+} dx' \exp u(x'),$$

$$N = \int_0^{\infty} dx' \, \exp(-U(x')) . \tag{3.17}$$

This is slightly more general than Kramers previous results. To
obtain Kramers' result we expand in a Taylor series about x,
$a/b = [Q^2/b(x_2)](x-x_2) + ...$ assuming $b(x)$ is slowly varying near x

compared to $a(x)$. We also approximate N, $P_1(\infty)$, $P_3(\infty)$ by expanding $U(x)$ about the two minima to the quadratic order, $U''(x_{1,3}) \equiv Q^2_{1,3}$ and we obtain the result

$$\lambda_1 = (2\pi)^{-1} \left(\frac{b(x_2)}{2}\right)^{\frac{1}{2}} Q_2 \exp(-U(x_2))$$

$$\cdot [Q_1 \exp(U(x_2)) + Q_3 \exp(U(x_3))] \,. \tag{3.18}$$

We shall shortly compare these results. If $Q_1 \exp U(x_1) >> Q_3 \exp U(x_3)$, then

$$\lambda_1 = (2\pi)^{-1} \left(\frac{b(x_2)}{2}\right)^{\frac{1}{2}} Q_1 Q_2 \exp[U(x_1)-U(x_2)] \,, \tag{3.19}$$

the familiar activation energy formula. Tomita et al.[38] have applied the W.K.B. method to the Schrödinger form of the F.P. equation, Eq. (3.5), and obtained this result, Eq. (3.19), (save for a curious $2/\sqrt{\pi}$ which may be an error). An alternative estimate of switching times may be made from mean first passage times[21,22], as mentioned in the introduction. Let us review this approach which as we shall see at first neglects back diffusion. Introduce the stochastic time variable $T(x_f|x_0)$ defined as the first passage time of the random variable $x(t)$ from x_0 to some value x_f. The distribution of first passage times is

$$F(x_f,t|x_0)dt = \text{Prob}\{t < T(x_f|x_0) < t + dt\} \,. \tag{3.20}$$

It satisfies the "backward" F.P. equation. The moments of the passage time are

$$T_n(x_f|x_0) = \int_0^\infty dt \; t^n \; F(x_f t|x_0) \,. \tag{3.21}$$

In particular $n = 1$ is the mean time of first passage. A hierarchy of equations for the moments may be formed. An important simplification occurs in that the equation for T_1 separates from the other elements and may be integrated. Assuming that $x = 0$ is a reflecting barrier and $x = \infty$ absorbing, we obtain

$$T_1(x_f|x_0) = 2 \int_{x_0}^{x_f} dx' \; \exp u(x') \int_0^{x'} d\xi \; \exp(-U(\xi)), \tag{3.22}$$

where u and U are defined by Eq. (3.10). We choose $x_0 = x_1$ and $x_f = x_3$. $T_1(x_3|x_1)$ is the mean time of first passage from local minimum x_1 to the global minimum x_3.

This may be further simplified[33]. If $\exp u(x')$ is appreciable in $x_- < x' < x_+$ and, $\int_0^\eta d\xi \, \exp(-U(\xi))$ is N_1 (approximately constant), for $x > x_2$, then

$$T_1(x_3|x_1) = 2N \int_{x_-}^{x_+} dx' \, \exp U(x')$$

or

$$T_1^{-1} = \frac{1}{2\pi} \left(\frac{b(x_2)}{2}\right)^{\frac{1}{2}} Q_1 Q_2 \, \exp \, (U(x_1) - U(x_2)) \,. \tag{3.23}$$

This agrees with the W.K.B. result mentioned earlier. A simple estimate of λ_1 including both forward and backward diffusion is obtained from the first passage from x_1 to x_3 and reverse; call them T_1^+ and T_1^- respectively. Take $\lambda_1 = (T_1^+)^{-1} + (T_1^-)^{-1}$ (add in parallel).*

Let us now numerically compare these results for the AOB. (For our choice of parameters, there is only a slight disagreement between the two versions of the F.P. equation (2.6); therefore, we use the (simpler) Ito version.) In Fig. 4 we have a comparison of Kramers' formula, Eq.(3.18)(curve b) with the result Eq.(3.16)(curve a). In Fig. 5a,b, we have Kramers' result superimposed below the deterministic values for c=8, q=1, for comparison. An initial state around x=1/3, y=7.6 is stable. If we rapidly switch y to 9.1, the state becomes unstable and the switching will take place on a deterministic time scale, as described by Bonifacio and Meystre[23]. This is the maximum value of the λ curve shown near y=9.1 $(T=\lambda^{-1}{\approx}\kappa^{-1})$. However, if we switch to y=8 (near λ min) $T=\lambda^{-1}$ is nearly three orders of magnitude longer than the deterministic value. Because of the potential barrier the statistically metastable state like a supersaturated drop may exist for long periods. For sufficiently low q values (q=0.01) the statistically unfavored state has extremely long decay times, $10^{25}\kappa^{-1}$, and the system may follow the deterministic trajectory and only switches at the end points executing a hysteresis cycle. Why is there a minimum in λ? If we switch to $7.5 \leqslant y < 8$ then there is a relaxation by forward and backward diffusion over the barrier back to the initial state of low x. As y approaches 7.5 this nears the first deterministic time scale of relaxation near y = 7.6, x = 0.3. Note in all these cases the system is put in an initial state near x = 0.3. Exactly the same time scaling holds if we initially place the system at x = 6.0, y = 9.20 and switch by reducing y to the lower state. The arrows in Fig. 5a indicate this. Note if diffusion is important λ is very sensitive to the y value switched to. Also note the broad decrease in λ encompassing the entire range of deterministic bistability.

---

*This estimate of the long time decay was also made by J.D.Farina et.al. See their contribution to these proceedings.

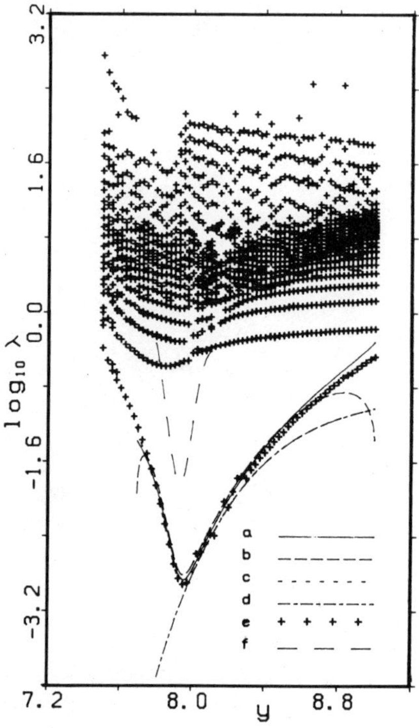

Fig. 4   Eigenvalues of AOB for
         c=8, q=1: (a) "general-
         ized" Kramers theory
         (Eq. (3.16)), (b) Kramers
         theory (Eq. (3.18)),
         (c) combined mean first
         passage times, (d) mean
         first passage time from
         x_1 to x_3 (e) variational
         calculation and (f) HBJ
         theory.

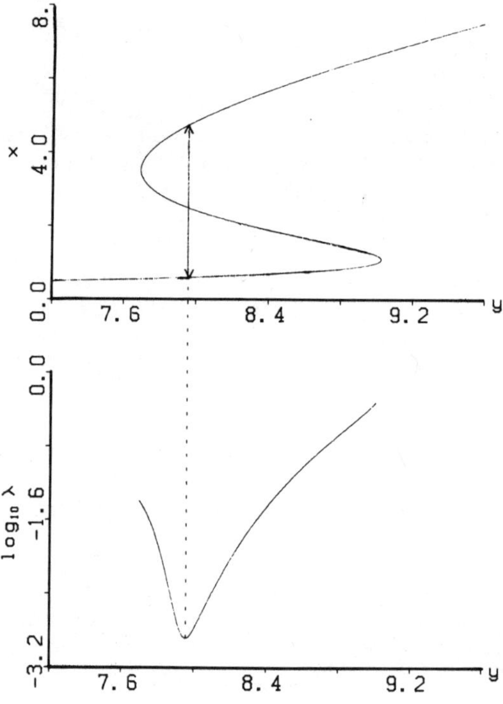

Fig. 5   Comparison of
         (a) the locations
         of the extrema of
         $P(x,\infty)$ with
         (b) result a of
         Fig. 4, for c=8,
         q=1.

The behavior is also very sensitive to q; Fig. 6 shows the result for $q = 0.4$, $c = 8$. a and b are the result of Eqs. (3.16) and (3.18). Curve c is the mean first passage time as in Fig. 5. Here λ decreases by seven orders of magnitude. This very sensitive behavior is analogous to the supersaturation region in nucleation. A word should be said about the first passage estimates. There is no minimum in the first passage calculations since it estimates the first arrival at x from x_1, and as the potential well near x_1 becomes deep the first passage grows long exponentially as can be seen from Eq. (3.23). This is curve a of Fig. 4. We may also calculate the mean first passage time from x_3 to x_1 and add these as discussed previously. This is shown by a in Fig. 4 and Fig. 6. Together we see they estimate λ_1. The formula and numerical integrations are not more simple than Eq. (3.16).

The first conclusion in comparing Fig. 5 and Fig. 6 is that the first passage time estimates are lower than the Kramers result, Eqs. (3.16) and (3.18). The results of Kramers et al. become poorer near the deterministic switching value $\lambda_1=1$, $T=\kappa^{-1}$ and evidence anomalous maxima. More comments will be made in the next section on this comparison.

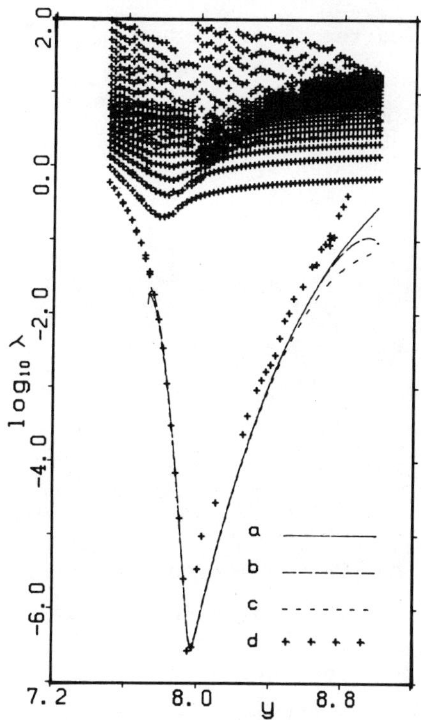

Fig. 6   Eigenvalues of AOB for
c=8, q=.4: (a) "generalized" Kramers theory,
(b) Kramers theory,
(c) combined mean first
passage times and
(d) variational
calculation.

IV. TIME DEPENDENT FLUCTUATIONS IN THE OPTICAL BISTABILITY: VARIATIONAL CALCULATION

To further study the dependence of the λ on y let us introduce a numerical variational scheme. The principle object will be to test the assumptions made in III, particularly the question of well separated eigenvalues. Variational schemes for one dimensional Schrödinger equations for two well problems in chemistry are not immediately adaptable to this problem. Dekker and van Kampen have recently considered the diffusion in a quartic potential. They have not seen well separated eigenvalues at the critical point. Schenzle and Brand[39] have begun variational calculations to investigate the lowest eigenvalue for the optical bistability with the noise introduced by fluctuations in y in a Langevin theory. They have considered the phase in their variational scheme but not systematically. The results are qualitatively similar to Figs. 4 and 6. We must note however, that variational procedures give upper bounds to correct values of λ_1 and it is important to have a theory or experiment to check them against.

First, similarly to Eq. (3.4) we transform the F.P. equation to a self-adjoint equation by letting $P_n(x) = P(x,\infty)\psi_n(x)$. We have

$$\hat{L}\psi_n(x) \equiv a(x)\psi_n'(x) + \frac{1}{2} b(x)\psi_n''(x) = -\lambda_n\psi_n(x) \ , \tag{4.1}$$

and the inner product is,

$$(w,v) = \int_o^\infty dx \ P_o(x)w(x)v(x) \ . \tag{4.2}$$

$L_{mn} \equiv (\psi_m,\hat{L}\psi_n)$ is a real symmetric matrix. In terms of an arbitrary but complete basis ϕ_n,

$$\sum_{j=1}^\infty (L_{ij}c_{jk} + \lambda_k S_{ij}c_{jk} = 0 \ , \tag{4.3}$$

where

$$S_{ij} \equiv \int \phi_i\phi_j \ P_o(x)dx \ ,$$

and

$$L_{ij} = \int \phi_i (\hat{L} \ \phi_j)P_o(x)dx \ .$$

The transformation has the following advantages:

1) The evaluation of (ϕ_i,ϕ_j) may be restricted to domains where $P_o(x)$ is appreciable.
2) The steady state is now $\psi_0(x)=1$ consistent with the use of a polynomial basis, ϕ_n.

The exact ψ_n of the self-adjoint operator minimize the Rayleigh quotient

$$\lambda_n = -(\psi_n, \hat{L}\psi_n)/(\psi_n, \psi_n) . \tag{4.4}$$

Using the Rayleigh-Ritz method[40] we approximate ψ_n on a *finite dimensional* basis: $\psi_n = \sum_{i=1}^{n} c_{ni}\phi_i$; then the condition

$$\partial\lambda_i/\partial c_{ni} = 0 \qquad i = 1,\ldots,N \tag{4.5}$$

is equivalent to solving the eigenvalue equation, Eq. (4.3). We have utilized this for the AOB using the polynomial basis

$$\phi_n = (x-\alpha)^{n-1} \qquad n = 1,\ldots, N; \tag{4.6}$$

with this choice,

$$L_{ij} = -\frac{1}{2}(i-1)(j-1)\int_{\infty}^{\infty} b(x)(x-\alpha)^{i+j-4}P_o(x) ,$$

$$S_{ij} = \int_o^{\infty} (x-\alpha)^{i+j-2}P_o(x)dx . \tag{4.7}$$

All matrix elements may be found from such integrals.

The choice of α is important. Because of finite precision α must be chosen such that basis functions contribute comparably in both wells. The following scheme is adopted; let

$$\alpha = P_1(\infty)x_1 + P_3(\infty)x_3 , \tag{4.8}$$

where as in Eq. (3.8)

$$P_1(\infty) = \int_o^{x_2} P(x,\infty)dx,$$

$$P_3(\infty) = \int_{x_2}^{\infty} P(x,\infty)dx$$

are stationary distributions of probability in the wells around x_1 and x_3 respectively. Further details of the variational calculation will be discussed elsewhere.

The results are shown in Figs. 4 and 6 for c=8, q=1.0, 0.40 respectively. The variational calculations are indicated by +++ both for λ_1 (curves 4e, 6d) and the excited states. Let us first comment upon Fig. 4. The agreement between Eq. (3.16) and the

variational calculation shown is excellent. The approximate formula
of Kramers Eq. (3.18) fails near $y = 8.8$ and $y = 7.7$ resulting in the
anomalous maxima. Within this range, however, it gives excellent
results. The first passage time (with diffusion both ways!) gives
somewhat lower answers without the anomalous maxima. At $y = 8.8$,
$T_{F.P.} = 3\,T_V$. However, the position and depth of the minima in λ
are in agreement. Similar results have been found for other q values
The failure of Kramers can be seen to take place at the limits of
the deterministic bistability curves as can be seen from a comparison of Figs. 4 and 5a,b. Figure 4 emphasizes the asymmetry of the
x vs. y curves. This of course is also noted in Fig. 5a,b. Recall
that the "collective state" exists for $y < 7.75$ and the single atom
state for $y > 9.10$. The collective states seem to enhance the nonlinearity.

Hanggi, Bulsara and Janda[41] using the Mori[42] theory of nonlinear dynamics of fluctuations have by means of continued fractions
calculated λ_1 for AOB. A few points of the H.B.J. data are shown
in Fig. 4(f). Although they have a qualitative asymmetry similar
to the present results and the minimum in λ_1 lies at the same value
of y, the dip is too narrow and the decrease in λ_1 much less (10^{-2}
compared to 10^{-3}). We must conclude that many higher elements in
the continued fraction expansion are required to give good results
for $q = 1$.

Let us now turn to the main point of the variational results,
the calculation of the excited eigenvalues, λ_{ex}, also shown by +++
in Figs. 4 and 6. Again, focus on Fig. 4. Within range of bistability inside the "ends" of the deterministic curve, $7.7 < y < 9.1$
in Fig. 5a, λ_1 is well separated from the excited states. Within
this range where the potential barrier of Fig. 3 is formed there
are well separated time scales. The λ_{ex} qualitatively describe relaxation *within* the potential well on a short time scale compared
to the diffusion over the barrier governed by λ_1. The λ_{ex} show a
dip near $y = 8$, however, less pronounced than λ_1. Figure 6 indicates
that the dip in λ_{ex} is approximately independent of q. This is consistent with the physical interpretation of the separated time
scales. As q becomes smaller the slow diffusion becomes slower, as
in Eq. (3.19). This is the familiar characteristic of such rate
processes. From the variational calculation we see that λ_{ex} are
little changed with q. This is a new result.

Outside the extremes of the bistability range the Kramers
theory fails for a second reason. All eigenvalues λ are comparable
and thus the time dependence *may* be described by the entire spectrum. The mean first passage time does contain contributions from
all λ [21], and in region $y > 9.1$ *may* give an estimate. This is also
the region where diffusion is unimportant so that the deterministic
switching also may be correct. To sum up, Kramers' estimate of

switching time $T = \lambda_1^{-1}$ is very good *within* the bistable region and the deterministic theory should be valid outside.

Near the end points (y = 7.7, 9.1, Fig. 4) of the bistability both approaches are uncertain. The time scales are not well separated, the potential well having not appeared, and yet stochastic diffusion is important having a time scale comparable to deterministic motion. This is particularly the case near the critical point c = 4. The eigenvalues at c = 4, q = 0.01 are shown in Fig. 7. The noise is here taken comparable to the earlier estimates of q[7,10]. This shows *critical slowing* down[2,43] as reflected in the Fokker-Planck eigenvalues. From the previous results we have some confidence in these numerical variational results. All the eigenvalues are comparable. The switching is then very sensitive to *initial conditions* and is not describable by a single parameter. We see a reduction of the lowest λ_1 by an order of magnitude to 1/30 (T = $30\kappa^{-1}$). This suggests a significant critical slowing down.

Bonifacio and Meystre[23] have utilized the deterministic theory to calculate the relaxation from an *unstable* state to the stable deterministic state for c = 20. For low x they have suddenly changed y at t = 0 to values leading to switching to the then stable high x branch. They saw increases in the "delay time" from κ^{-1} to the

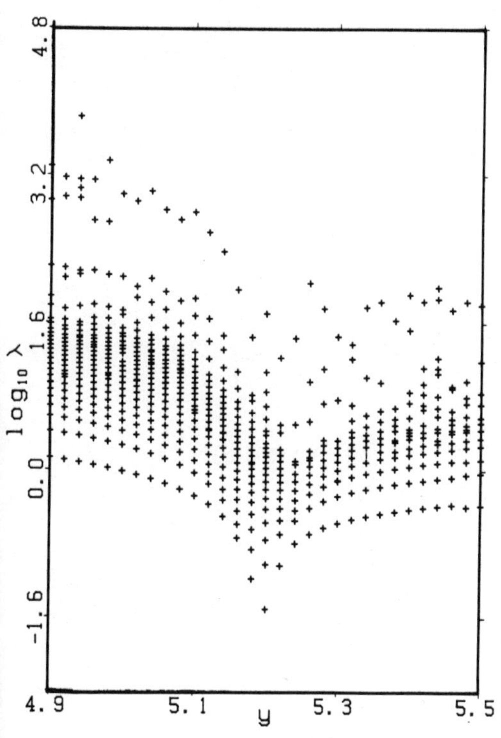

Fig. 7   Variational calculation of eigenvalues at the deterministic AOB critical point c = 4 (q = .01).

order of $10\kappa^{-1}$ (for various initial y). Strictly speaking, they did not observe *critical slowing* down but rather a delay well outside the region of bistability.

To generalize these results to more than two dimensions including the phase, ϕ, is difficult. Few multiple dimension solutions to F.P. equations are known, particularly for multiplicative noise[1-4]. Exact solutions even in the steady state (without detailed balance) are unknown and needed to begin a Kramers-like theory in higher dimensions[16,17].* However, we may expect the enthusiastic interest in optical bistabilities in both theory and the laboratory to help us extend the theoretical description of stochastic time dependent phenomena in physics and chemistry.

REFERENCES

1. G. Nicolis, I. Prigogine, "Self-Organization in Non-Equilibrium Systems," John Wiley, New York (1977).
2. H. Haken, "Synergetics," Springer-Verlag, Berlin (1977).
3. Solvay Conference in Physics, Nov. 1978. (Prooceedings to appear.)
4. "Dissipative Structures in the Social and Physical Sciences," ed. by W. C. Schieve and P. Allen (University of Texas Press, Austin) (to appear).
5. Austin Conference on Dissipative Structures in Chemistry and Physics, March, 1980. (Proceedings in preparation).
6. T. G. Kurtz, J. Chem. Phys. **57**, 2976 (1972); Math. Prog. Study **5**, 67 (1976); Stoch. Proc. Appl. **6**, 223 (1978).
7. R. Bonifacio, M. Gronchi, L. A. Lugiato, Phys. Rev. **A18**, 2266 (1978); F. Casagrande and L. A. Lugiato, Nuovo Cimento **B48**, 287 (1978).
8. H. Risken, Statistical Properties of Laser Light, in "Progress in Optics," Vol. XII, 241, ed. by E. Wolf, North Holland, Amsterdam (1974).
9. R. Bonifacio, L. A. Lugiato, Opt. Comm. **19**, 172 (1976).
10. A. Bulsara, W. C. Schieve, R. F. Gragg, Phys. Lett. **68A**, 294 (1978).
11. R. F. Gragg, W. C. Schieve, A. R. Bulsara, Phys. Rev. **A19**, 2052 (1979); J. C. Englund, W. C. Schieve and R. F. Gragg, Int. J. Q. Chem. Symp. **13**, 695 (1979).
12. A. Schenzle and H. Brand, Opt. Comm. **27**, 485 (1978). (See also A. Schenzle, these proceedings.)
13. K. Kondo, M. Mabuchi, H. Husegawa, Opt. Comm. **32**, 136 (1980).
14. F. T. Arecchi and A. Politi, Opt. Comm. **29**, 361 (1979).
15. H. A. Kramers, Physica **7**, 284 (1940).
16. S. Chandrasekhar, Rev. Mod. Phys. **15**, 1 (1943).
17. R. Landauer and J. A. Swanson, Phys. Rev. **121**, 1668 (1960).

*See the contribution of A. Schenzle these proceedings.

18. F. F. Abraham, "Homogeneous Nucleation Theory," Academic Press, New York (1974) (references therein).
19. K. Binder and D. Stauffer, Adv. Phys. 25, 343 (1976).
20. W. Zurek and W. C. Schieve, "The Nucleation Paradigm," (see Ref. 4).
21. C. H. Weiss, First Passage Time Problems in Chemical Physics, in Adv. Chem. Phys. 13, 1 (1966).
22. A. R. Bulsara and W. C. Schieve, Opt. Comm. 26, 384 (1978).
23. R. Bonifacio and P. Meystre, Opt. Comm. 27, 147 (1978) and Opt. Comm. 29, 131 (1979); F. A. Hopf, P. Meystre, P. D. Drummond and D. F. Walls, Opt. Comm. 31, 245 (1976). (See also the papers of P. Meystre, et al., and F. A. Hopf, et al., in these proceedings.)
24. L. Arnold, "Stochastic Differential Equations," Wiley-Interscience, New York (1974).
25. R. E. Mortensen, J. Stat. Phys. 1, 271 (1969).
26. R. F. Gragg, W. C. Schieve, J. Englund, "Stochastic Differential Equations in the Optical Bistability," (in preparation).
27. J. C. Doob, "Stochastic Processes," John Wiley, New York (1953).
28. R. L. Stratonovic, SIAM J. Control 4, 363 (1966).
29. A. H. Gray, Jr., and T/ K. Caughy, J. Math. and Phys. 44, 288 (1965).
30. E. Wong and M. Zakai, Ann. Math. Stat. 36, 1560 (1965).
31. N. Goel and N. Richter-Dyn, "Stochastic Problems in Biology," John Wiley and Sons, (1973).
32. L. Arnold, W. Horsthemke, R. Lefever, Z. Physik, B29, 367 (1978); W. Horsthemke, and R. Lefever, Phys. Lett. 64A, 19 (1977).
33. R. F. Gragg, Ph.D. Thesis, University of Texas, Austin, August, 1980.
34. M. Suzuki, Proceedings of XVII Conf. on Phys, Nov. 1978.
35. R. C. Desai and R. Zwanzig, J. Stat. Phys. 19, 1 (1978).
36. N. G. van Kampen, J. Stat. Phys. 17, 71 (1977).
37. H. Dekker and N. G. van Kampen, Phys. Lett. 73A, 374 (1979).
38. H. Tomita, A. Ito and H. Kidachi, Prog. Theor. Phys. 56, 786 (1976).
39. A. Schenzle and H. Brand, Opt. Comm. 31, 401 (1979).
40. S. G. Mikhlin, "Variational Methods in Math. Phys.," transl. by Boddington, Macmillan, (1964).
41. P. Hanggi, A. Bulsara, R. Janda, "Spectrum and Dynamic Response Function of Transmitted Light in the Absorptive Optical Bistability," Phys. Rev. (to appear).
42. H. Mori, H. Fujisaka and H. Schigematso, Prog. Theor. Phys. 51, 1209 (1974); see also L. S. Garcia-Colin and J. L. del Rio, J. Stat. Phys. 16, 235 (1978) and references therein.
43. L. van Hove, Phys. Rev. 95, 1374 (1954); S. Ma and G. F. Mazenko, Phys. Rev. B11, 4077 (1975).

SHORT- AND LONG-TIME TRANSIENT EVOLUTION IN ABSORPTIVE OPTICAL

BISTABILITY

J. D. Farina, L. M. Narducci, and J. M. Yuan

Department of Physics and Atmospheric Science
Drexel University
Philadelphia, PA 19104

and

L. A. Lugiato
Istituto di Scienze Fisiche
Università di Milano
via Celoria 16
Milano, Italy

Abstract: We discuss the time evolution of an absorptive bistable device driven by a resonant external field. The dynamics of
a bistable system perturbed from a steady state configuration is
characterized by two widely separated time scales: one, of the
order of a few cavity relaxation times, brings the system to a
metastable state; the other is responsible for the attainment of the
final steady state, and is typically much longer as long as the only
source of fluctuations is the internal quantum noise. Explicit
analytic expressions have been obtained for the rates of decay associated with both relaxation processes.

I. INTRODUCTION

The stationary properties of a bistable ring cavity in the
mean field limit are well known[1,2]. In addition to providing a
simple analytic description of optical bistability, the ring cavity
model has played the role of a useful prototype for the study of
open systems far from thermal equilibrium[2,3]. When studied with the
help of a fully quantum mechanical treatment, an externally driven
absorbing system in a ring cavity reveals the existence of a range
of values of the input field, such that the probability density of

the transmitted field amplitude is bimodal in character. The bimodal
structure of this density function is a consequence of the existence
of two minima in the effective "free energy" of the system[2]; the
widths of the peaks of the density function reflect the existence of
fluctuations of the internal cavity field which are due to spontaneou
emission.

Although external fluctuations of instrumental origin are likely
to be of much more practical significance than the intrinsic quantum
fluctuations[4], for most macroscopic systems, it is of interest to
focus on the quantum mechanical aspects because of their relevance
to the general subject of multiplicative stochastic processes. For
this reason only, we have ignored phase and amplitude fluctuations
of external origin in our discussion.

The mean field limit of the quantum mechanical theory of absorptive optical bistability leads to the one dimensional Fokker-Planck
equation[2]

$$\frac{\partial}{\partial t} P_y(x,t) = - \frac{\partial}{\partial x} (A_y(x) \, P_y) + q \frac{\partial^2}{\partial x^2} (D(x) \, P_y) \qquad (1.1)$$

where y and x denote the dimensionless input and output field amplitudes as in Ref. (1), and $P_y(x,t)$ is the probability density for the
transmitted field x when the system is driven by the external field
y. The drift and diffusion coefficients are defined as

$$A_y(x) = y - x - \frac{2Cx}{1+x^2} \qquad (1.2)$$

$$D(x) = \left(\frac{x}{1+x^2}\right)^2 \qquad (1.3)$$

where C, the cooperation parameter, is proportional to the density
of absorbing atoms in the cavity. More precisely, $2C = \alpha L/T$ is the
ratio between the absorption constant αL of the passive system and
the mirror transmissivity T. The parameter q, which is a measure
of the strength of the quantum fluctuations, is defined as $C/2N_s$,
where N_s is the so-called saturation photon number.

The solution of the Fokker-Planck equation (1.1) yields a
statistical description of the time evolution of a bistable system.
A characteristic feature of this evolution is the existence of two
generally quite different time scales. The first, and most rapid
relaxation mechanism, is one in which the initial probability densit
comes to a metastable state within each well of the double-welled
free energy. We call this process "local relaxation"[6]. One main

feature of this phase of the evolution is that virtually no flow of
probability exists across the local potential maximum, separating
the two wells. This means that if the probability of finding a
system in the low transmission branch (left well of the free energy)
is α_0 at t = 0, the probability will still be α_0 at the end of the
local relaxation, even if the driving field has been switched to a
new bistable operating point. This also means that spontaneous
switching during this time interval is very unlikely. The flow of
probability across the local potential maximum, or "tunneling"[6],
is responsible for the second longer-lived phase of the evolution.
As tunneling proceeds, the probability of occupation of a given
well changes due to noise induced spontaneous switching.

Direct evidence of the existence of two relaxation processes
is provided by the numerical time-dependent solutions of the Fokker-
Planck equation shown in Figs. 1 and 2. The situation can be further clarified with the help of Fig. 3. Here we consider an ensemble of systems initially prepared in a given bimodal configuration

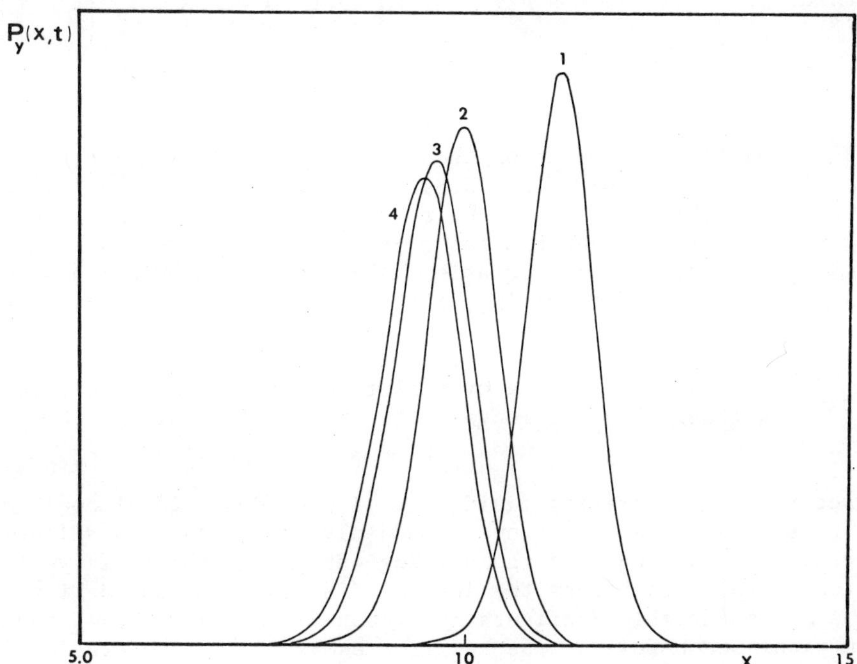

Fig. 1.   The time evolution of the probability distribution $P_y(x,t)$
for y_0 = 13 and y_{op} = 11.5, C = 10 and q = 5. The initial
(curve 1) and final configuration (curve 4) of the system
are monostable along the high transmission branch. The
evolution proceeds at the local relaxation rate.

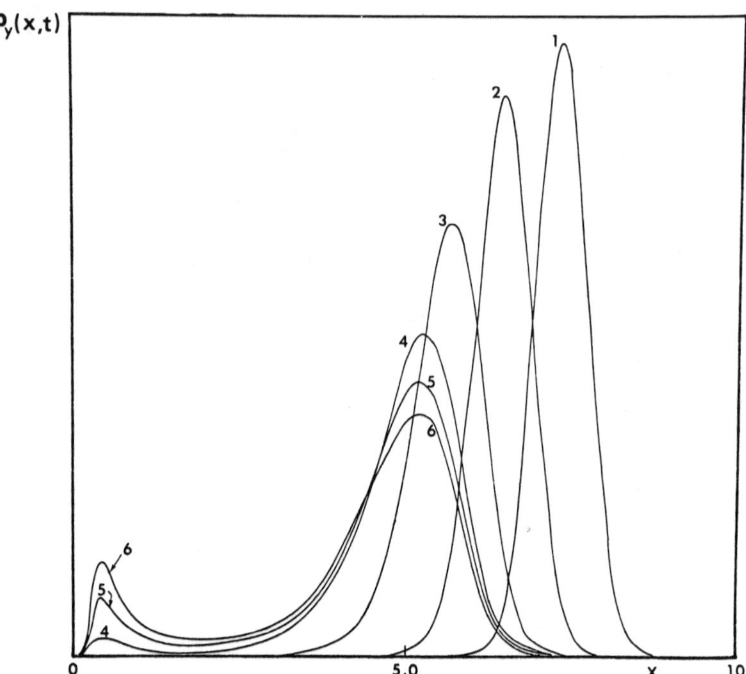

Fig. 2. The time evolution of the probability distribution $P_y(x,t)$
for $y_o = 10$ and $y_{op} = 8.8$, $C = 10$ and $q = 5$. The initial
distribution is centered around a monostable configuration
in the high transmission branch; the final steady state
configuration is bistable. The computer simulation shows
the onset of tunneling, although the final curve (6) shown
in the figure is still removed from the actual steady state.
The large value of q was chosen to enhance the tunneling
process. Still, the overall time scale of the evolution
is about 15 times longer than that of Fig. 1.

characterized by a driving field, y_o. The ensemble is allowed to
come to steady state, and then is suddenly perturbed by a slight
change in the magnitude of the driving external field ($y_o \rightarrow y_{op}$).
The local evolution brings the initial distribution (solid line 1)
into a new metastable configuration (dashed line 2) of local equilibrium within each well of the new free energy (b). The area of
each dashed peak remaining essentially the same as that of the corresponding initial peak, i.e., the occupation probability of each
state of transmission remains constant up to this point. However,
the metastable distribution is quite different from the actual steady
state configuration corresponding to the new driving field y_{op}.
After a sufficiently long time, usually much longer than the local

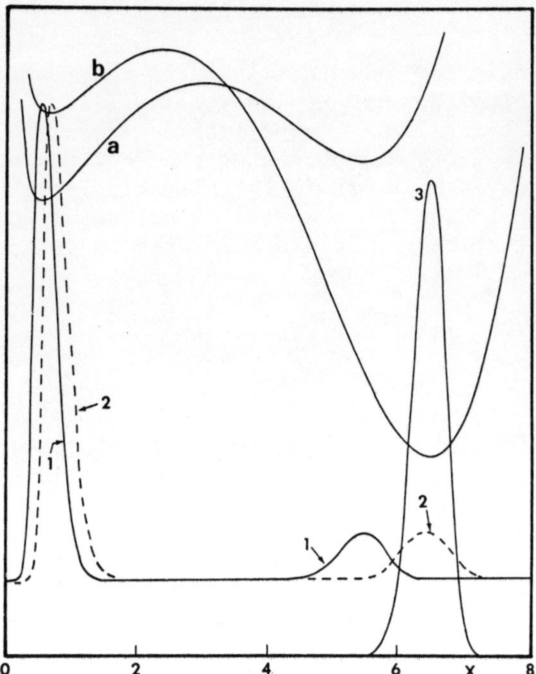

Fig. 3.   Schematic representation of the local and global tunneling
          processes.   Curves (a) and (b) represent the initial and
          final free energies.   Curves 1, 2 and 3 show qualitatively
          the initial P-function, the P-function of the metastable
          state and the final distribution.

relaxation time, the tunneling process brings the system to its
global steady state (solid line 3).

     It is this vast difference in time scales that is responsible
for the existence of a quasi steady state configuration and, ultimately, for the occurence of hysteresis.   In the following development we shall focus on the transient behavior of an absorptive
optically bistable device.   The previous qualitative discussion of
the time scales plays a central role in that it allows consideration of the local relaxation and of the tunneling process as entirely
separate events.   We shall describe the local relaxation by developing a linearization procedure that is capable of handling bimodal
configurations.   The tunneling process, instead, will be studied
with the help of the first passage time technique.[7]   Both phases of
the transient evolution can be described by simple analytic formulas,
which will be compared for accuracy with exact numerical solutions
of the Fokker-Planck Eq. (1.1) and of the associated first passage
time equation.

II.   LOCAL RELAXATION

The starting point of our analysis is the Bonifacio-Lugiato ring cavity model of absorptive bistability (Fig. 4) in the good cavity and mean field limits. The intrinsic nonlinearity of the absorbing two level atoms, coupled to the feedback provided by the mirrors of the cavity results in the bistable behavior. It is well known[1] that in the semiclassical limit the input and output scaled field variables, y and x are related by the cubic state equation

$$0 = y - x - \frac{2Cx}{1+x^2} . \tag{2.1}$$

The existence of multiple real roots x_i (i = 1,2,3) for C > 4 and y bounded between the lower and upper threshold values y_1 and y_2 (Fig. 5), provides the first indication of bistability. In fact, a simple linear stability analysis shows that only two of the three roots correspond to stable steady states. The third root is unstable against fluctuations, and is therefore physically unrealizable. The result of this situation is the existence of a hysteresis cycle (Fig. 5) whenever fluctuations can be neglected over the time scale of variation of the input field y from a value, say, smaller than y_1 to a value larger than y_2 and back.

The first problem of interest in our discussion can be stated as follows: given an initial steady state corresponding to the external field y_o, one wants to describe the evolution that results from a sudden change in the strength of the driving field to a new operating value y_{op}. The semi-classical Maxwell-Bloch equations[8],

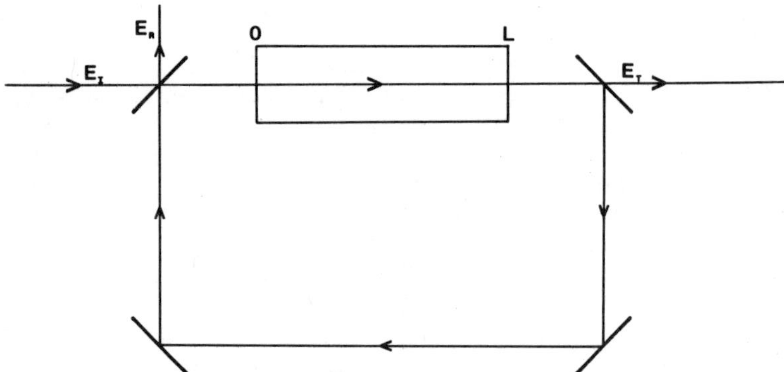

Fig. 4.   Schematic representation of a ring cavity containing the sample of absorbing atoms; E_I and E_T are the incident and transmitted field amplitudes.

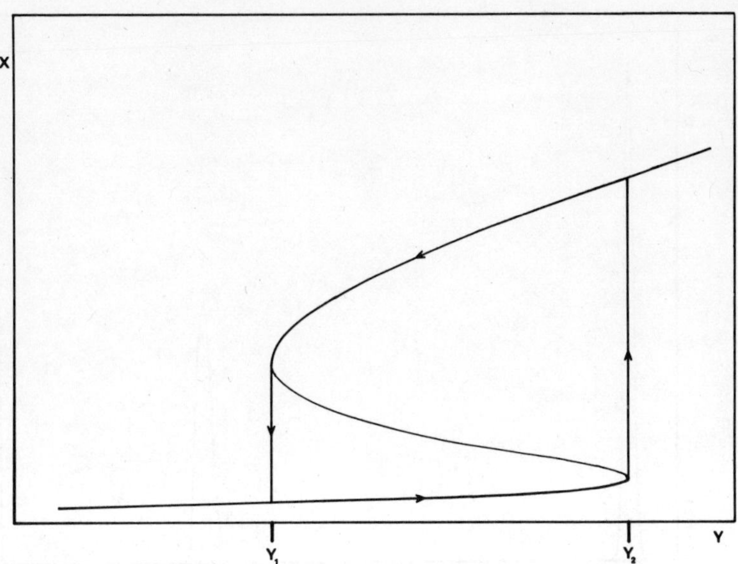

Fig. 5.   State equation of a bistable system for C > 4 and hysteresis
          loop. The values of y_1 and y_2 of the incident field ampli-
          tude correspond to the switching points.

of which Eq. (2.1) represents the steady state solution, are not very
useful for this purpose. A hint as to the best procedure is provided
by the structure of the stationary solution of the Fokker-Planck
equation. This is given by

$$P_y(x,\infty) = N \exp(-U_y(x)/q) \tag{2.2}$$

where N is a normalization constant and the "free energy" $U_y(x)$ is
defined by

$$U_y(x) = -\int \frac{A_y(x)}{D(x)} \, dx + q \, \ln D(x) \; .$$

In fact, as shown in Fig. 6, for sufficiently small values of the
fluctuation parameter q, the stationary solution (2.2) is very
sharply peaked around the stable semiclassical roots[6]. Moreover,
to excellent accuracy, the individual peaks of the bimodal distribution can be well represented by properly weighted Gaussian functions
$G(x)$ whose widths are given by

$$\sigma_1 = \left[\frac{1}{q} \left(\frac{\partial^2 U_y(x)}{\partial x^2} \right)_{x_1} \right]^{-1/2} \tag{2.3}$$

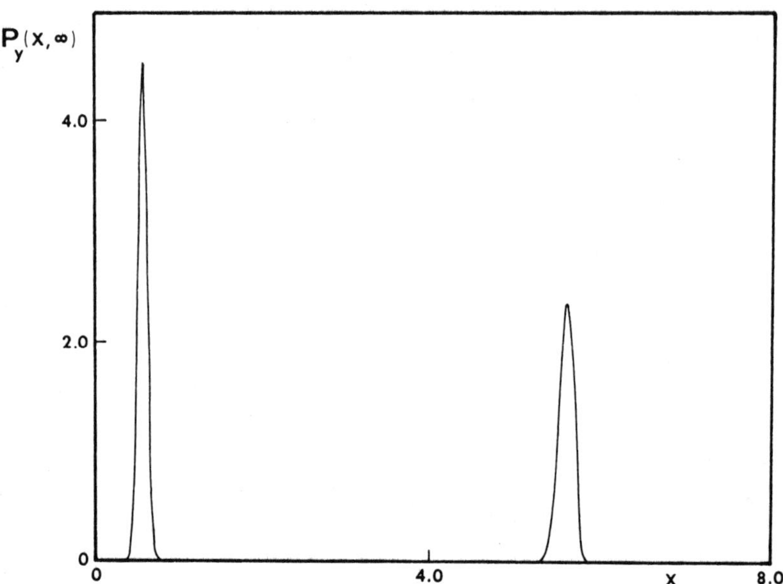

Fig. 6. Stationary solution of the Fokker-Planck equation $P_y(x,\infty)$
         for $C = 10$, $q = 0.1$ and $y = 8.06$.

and where x_i is one of the stable roots of the state equation (Fig.
7). Thus, a good quantitative fit of the steady state solution is
given by

$$P_y(x,\infty) = \omega_1\, G_1(x) + \omega_2\, G_2(x) \tag{2.4}$$

where the weight factors ω_1 and ω_2 represent the areas of the left
and right peaks of the density function (2.2), or the ensemble
averaged probability of the low and high transmission states, respectively.

At $t = 0_+$ the external field is suddenly switched to a new
operating value y_{op} which for the sake of simplicity will be assumed
to be not too different from y_o. For sufficiently small values of
the fluctuation parameter q (typically $q \lesssim 1$), the occupation probability of either state of transmission stays nearly constant
during the initial relaxation to local equilibrium. Thus, the weight
factors can be assumed to remain equal to their initial values $\omega_1^{(0)}$,
$\omega_2^{(0)}$ corresponding to the driving field y_o.

Because of the sharply peaked nature of the stationary solution
of the Fokker-Planck equations and of the absence of probability flow
from one well to the other, the evolution of each peak of the initial

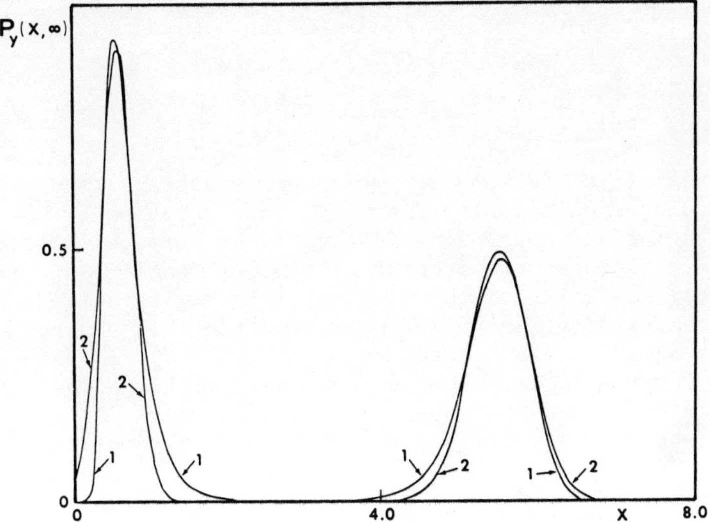

Fig. 7. Comparison between the steady state solution of the Fokker-Planck equation (curve 1) and the Gaussian approximation (curve 2) for C = 10, q = 2 and y = 8.02. For small values of q (e.g. q = 0.1 as in Fig. 6) the agreement is so close that the exact and the Gaussian curves are indistinguishable in this scale.

distribution can be treated as a separate Brownian process. Moreover, if $|y_o - y_{op}| \ll y_o$, y_{op} the position of each peak undergoes a simple exponential relaxation of the type

$$x_i(t) = (x_i^{(o)} - x_i^{(op)})\, e^{-\lambda_i t} + x_i^{(op)} \tag{2.5}$$

where the relaxation rates are given by

$$\lambda_i = \frac{dA_{y_{op}}(x)}{dx}\Bigg|_{x = x_i^{(op)}} . \tag{2.6}$$

In conclusion, the time dependent solution of the Fokker-Planck Eq. (1.1) is well represented by the linear superposition of two time-dependent Gaussian functions

$$P_{y_{op}}(x,t) = \sum_{i=1}^{2} \omega_i^{(o)}\, G_i(x - x_i(t)) \tag{2.7}$$

and the ensemble average transmitted field takes the form

$$<x(t)> = \sum_{i=1}^{2} \omega_i^{(o)} \cdot \{(x_i^{(o)} - x_i^{(op)}) \, e^{-\lambda_i t} + x_i^{(op)}\} \,. \tag{2.8}$$

A comparison of Eq. (2.8) with the ensemble average transmitted field calculated numerically from the exact solution of the Fokker-Planck equation has been shown in Fig. 10 of Ref. 6a. The agreement is very satisfactory even after several local relaxation times, λ_i^{-1}. If the new value of the driving field y_{op} is sufficiently removed from y_o, the linearized approximation (2.5) is no longer adequate. Equation (2.7) needs to be modified by replacing $x_i(t)$ (Eq. (2.5)) with the solution of the nonlinear Langevin equation

$$\dot{x}(t) = A_{y_{op}}(x) \tag{2.9}$$

III.   LONG TIME EVOLUTION - TUNNELING

The description of the time dependence of a bistable system proposed in the previous section is limited to the local relaxation phase as a result of the requirement that the weighting factors ω_i be constant. However, it has been argued in the Introduction that the overall evolution consists of two widely separated phases, the latter one being associated with transfer of probability from one well of the bistable potential to the other. Obviously, relevant information on the long-time behavior of the system is contained in the time dependence of the weight factors. This, in fact is not easy to extract from the Fokker-Planck equation. It is possible, however, to obtain an estimate of the tunneling time scales and, in the process, to arrive at a confirmation of our qualitative discussion using the following argument.

We define

$$\alpha(t) = \int_{0}^{\tilde{x}} dx \, P_y(x,t) \tag{3.1}$$

as the time dependent probability of occupation of the left well ($\tilde{x}$ is the unstable root of the state equation and the position of the local maximum of the potential). The function $\alpha(t)$ satisfies the exact equation of motion

$$\frac{d}{dt} \alpha(t) = \frac{\partial}{\partial x} D(x) \, P_{y_{op}}(x,t) \Big|_{x = \tilde{x}_{op}} \tag{3.2}$$

as one can readily verify from Eq. (3.1) and the Fokker–Planck Eq.
(1.1). Of course, the solution of Eq. (3.2) requires knowledge of
the density function $P_{y_{op}}(x,t)$ for all times. Nevertheless it is
clear that in general, $\alpha(t)$ evolves at an exceedingly slow rate
because its time derivative is directly related to the value of the
density function at the local maximum $\tilde{x}$ (see Fig. 6). A more quantitative estimate of this rate of change can be obtained by evaluating
the right hand side of Eq. (3.2) at $t = 0$. In this case, a few
simple algebraic steps lead to

$$\frac{d}{dt}\alpha(t)\bigg|_{t=0} = \frac{y_o - y_{op}}{q}\, P_{y_{op}}(x,0)\bigg|_{\tilde{x}_{op}} = \frac{y_o - y_{op}}{q}\, P_{y_o}(\tilde{x}_{op},\infty)\ . \qquad (3.3)$$

Several methods have been brought to bear on the problem of
estimating the magnitude of the tunneling time[7,9,10]. Perhaps the
treatment that offers the greatest insight into the tunneling mechanism is the so-called first passage time approach. This can be
illustrated as follows: one considers a fictitious Brownian particle
in an external field of force and under the action of random collisions. In the absence of random perturbations, the particle would
naturally seek the nearest stable equilibrium position. The addition
of collisions will result in a diffusion process which makes it
possible for the particle, at least in principle, to move away from
a position of local equilibrium and to become trapped in another
local minimum of the external potential. After identification of
an arbitrary domain bounded by, for instance, x_A and x_B, the first
passage time technique seeks to answer the following question:
given that the "particle" is initially placed at some starting point,
how long, on the average, will it take for it to escape the chosen
domain (x_A, x_B). It is understood that, once the "particle" reaches
one of the boundaries, the "clock" is stopped and the next measurement is started with a new "particle" in the same initial position.

This viewpoint is ideally suited to the tunneling problem on
hand. Here we may imagine a Brownian particle trapped, for example,
in the left well of the potential. Spontaneous switching from one
state of transmission to the other is equivalent to the escape of
the particle from the chosen domain $(x_A < x < x_B)$. Thus we look for
the average time required for a particle to hop across the local
potential maximum, e.g., from the left well into the adjacent minimum. The boundaries of interest in this case are $x = 0$ and $x = \tilde{x}_{op}$.
We note that $x = 0$ is an impenetrable boundary. On the contrary
$\tilde{x}_{op}$ is an absorbing boundary, in the sense that a particle placed
at $x = \tilde{x}_{op}$ will hop out of the domain instantaneously. If $M(x)$
denotes the first passage time for a "particle" placed at a position
x at $t = 0$, the boundary conditions corresponding to the escape process out of the left well are

$$\frac{d}{dx} M(x = 0) = 0; \qquad M(x = \tilde{x}_{op}) = 0 \tag{3.4}$$

The first passage time function satisfies the ordinary differential equation[7]

$$qD(x) \frac{d^2 M}{dx^2} + A_{y_{op}}(x) \frac{dM}{dx} = -1 . \tag{3.5}$$

The solution of Eq. (3.5) is very easy to obtain by direct quadratures. Corresponding to the boundary conditions (3.4) the solution takes the form

$$M_L(x) = \int_x^{\tilde{x}_{op}} dx' \int_0^{x'} dx'' \frac{1}{qD(x'')} \exp\left(- \frac{U(x') - U(x'')}{q}\right) \tag{3.6}$$

where the subscript L denotes explicitly the escape time out of the left well. The boundary conditions corresponding to the escape process out of the right well are

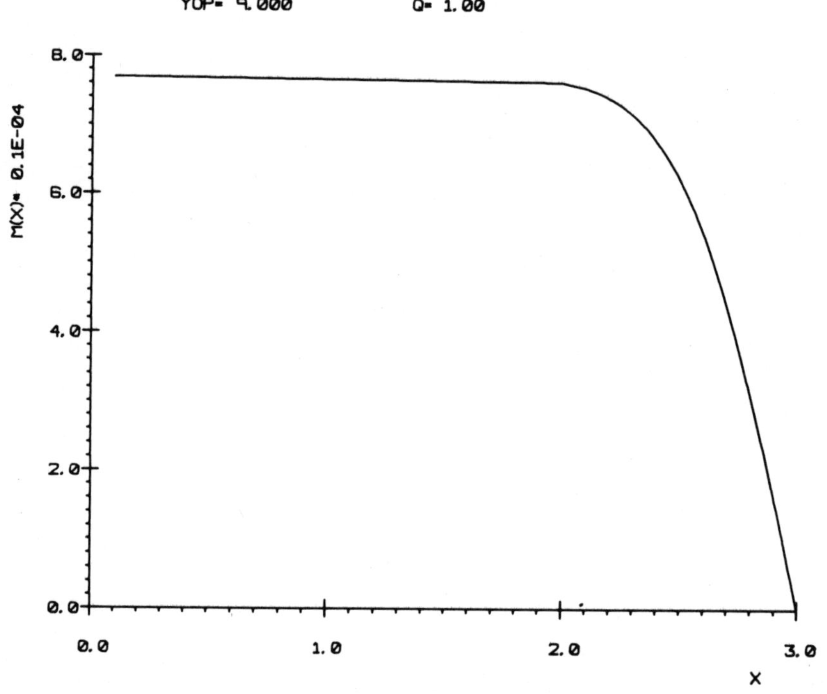

Fig. 8(a)

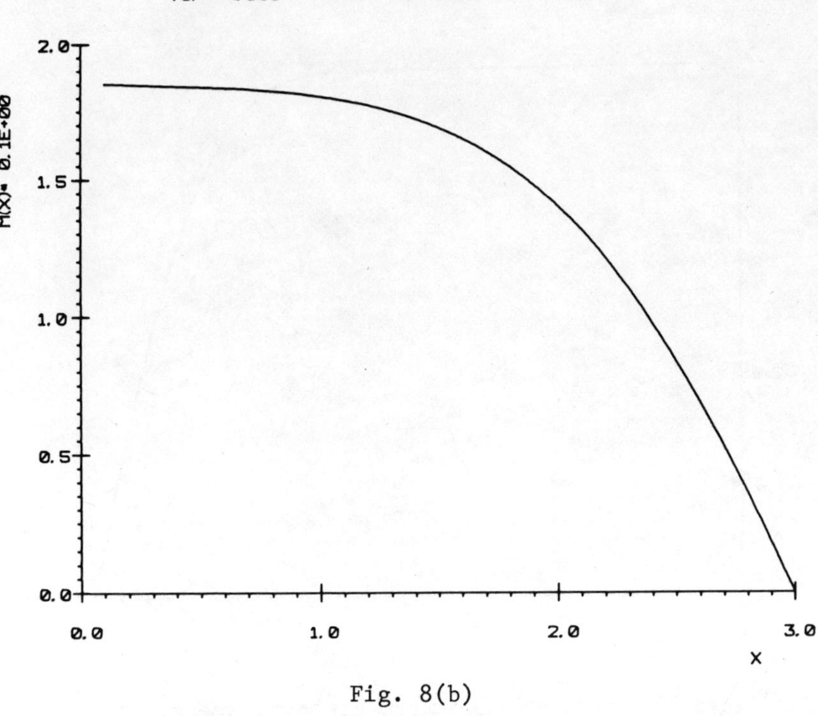

Fig. 8(b)

Fig. 8. Exact solution of the first passage time equation for the
escape time M_L out of the left well for q = 1 (a) and
q = 5 (b). In both cases the operating field is y_{op} = 9.

$$M_R(x = \tilde{x}_{op}) = 0; \qquad \frac{d}{dx} M(x = \infty) = 0 \tag{3.7}$$

and the appropriate first passage time is given by

$$M_R(x) = \int_{\tilde{x}_{op}}^{x} dx' \int_{x'}^{\infty} \frac{1}{qD(x'')} \exp\left(- \frac{U(x') - U(x'')}{q}\right) . \tag{3.8}$$

Not surprisingly, the first passage times $M_L(x)$ and $M_R(x)$ are very
sensitive functions of the fluctuation parameter q and of the value
of the applied external field (Figs. 8, 9). It is surprising,
instead, that for sufficiently small values of q, both M_L and M_R
are practically independent of x over most of the range of interest
except in the vicinity of the top of the potential barrier $\tilde{x}_{op}$ where
the escape times become vanishingly small. As a consequence, each

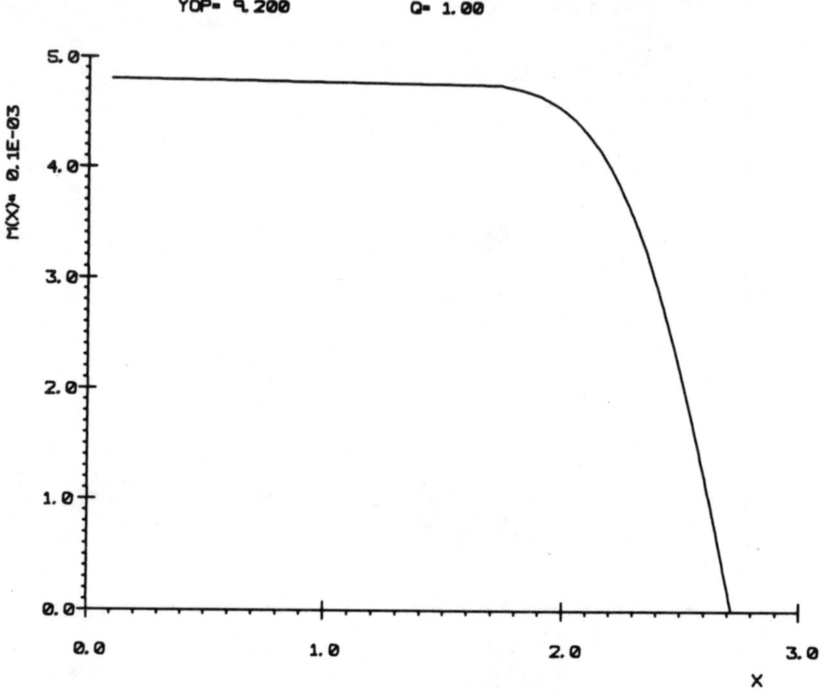

Fig. 9. Exact solution of the first passage time equation for the escape time M_L out of the left well for y_{op} = 9.2 and q = 1.0.

sharply peaked distribution corresponding to values of $q \gtrsim 1$ is characterized by two well defined escape times M_L and M_R. It is natural to identify the global relaxation time, M i.e., the time scale over which steady state is achieved as

$$\lambda \equiv \frac{1}{M} = \frac{1}{M_L} + \frac{1}{M_R} \; . \tag{3.9}$$

This, in fact, turns out to be a reasonable identification not only on intuitive grounds, but also on the strength of more precise mathematical arguments. For a more detailed discussion of the finer points of the long time relaxation process, the reader can consult Ref. 11. An interesting point concerning the dependence of λ on the applied external field is the existence of a minimum in the vicinity of the operating value y_{op} for which the steady state probabilities of the two states of transmission are equal to one another. This value of y_{op} insures the greatest stability of the bistable system against spontaneous switching due to internal noise and appears to be optimum

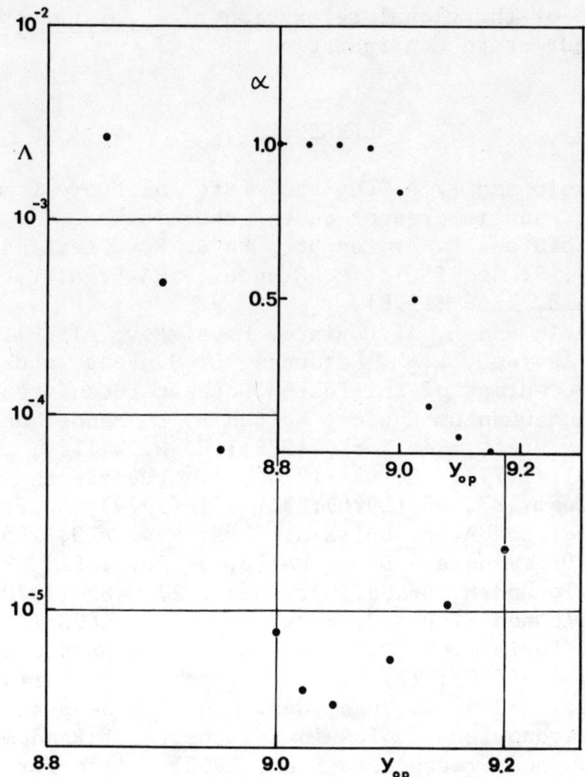

Fig. 10. Behavior of the global tunneling rate as a function of y_{op}.
The inset shows the behavior of the area of the left peak
in the same range of operating fields.

as a bias for logical operations. The behavior of λ as a function
of y_{op} is illustrated in Fig. 10.

IV.   CONCLUSIONS

We have analyzed the time evolution of a bistable system following a sudden perturbation from steady state. Two distinct stages
of the time development can be recognized. At first, the perturbed
system relaxes into a metastable quasi steady state through a process
that is very reminiscent of the approach to equilibrium of a Brownian
particle. Over a much longer time scale, spontaneous switching can
occur as a result of "tunneling" from one metastable state to the
other. The escape times out of each state have been calculated using
the first passage time method. Finally, we have suggested a natural

identification of the global relaxation time for the approach to
the final steady state configuration.

REFERENCES

1. R. Bonifacio and L. A. Lugiato, Lett. al Nuovo Cimento 21, 505
    (1978), and references quoted therein.
2. R. Bonifacio and L. A. Lugiato, Phys. Rev. Lett. 40, 1023, 1538
    (1978); R. Bonifacio, M. Gronchi, and L. A. Lugiato, Phys.
    Rev. A18, 2266 (1978).
3. R. Bonifacio and L. A. Lugiato, Phys. Rev. A18, 1129 (1978);
    G. S. Agarwal, L. M. Narducci, D. H. Feng, and R. Gilmore
    in Proceedings of the Fourth Rochester Conference on Coher-
    ence and Quantum Optics, Edited by L. Mandel and E. Wolf
    (Plenum Press, New York, 1978); C. R. Willis, Opt. Comm.
    23, 151 (1977); 26, 62 (1978); A. Schenzle and H. Brand,
    Opt. Comm. 27, 85 (1978); 31, 401 (1979); R. F. Gragg, W. C.
    Schieve, and A. R. Bulsara, Phys. Rev. A19, 2052 (1979);
    P. D. Drummond and D. F. Walls, J. Phys. B13, 725 (1980).
4. A. Schenzle and H. Brand, Opt. Comm. 27, 485 (1978).
5. A. Schenzle and H. Brand, Phys. Rev. A20, 1628 (1979).
6. (a) L. A. Lugiato, J. D. Farina and L. M. Narducci, Phys. Rev.
    A22, 253 (1980); (b) L. A. Lugiato, J. D. Farina, L. M.
    Narducci and J. M. Yuan, Opt. Eng. (to be published).
7. R. L. Stratonovich, Topics in the Theory of Random Noise,
    (Gordon and Breach, New York, 1963). Also for a review of
    the first passage time problem in the context of the theory
    of multistate relaxation processes, see: I Oppenheim, K. E.
    Shuler, and G. H. Weiss, Adv. Mol. Rel. Proc. 1, 13 (1967-
    68); G. H. Weiss, Adv. Chem. Phys. 13, 1 (1966).
8. R. Bonifacio and L. A. Lugiato, Opt. Comm. 19, 172 (1976).
9. H. A. Kramers, Physics (Utrecht), 7, 284 (1940).
10. R. Landauer, J. Appl. Phys. 33, 2209 (1962); N. G. van Kampen,
    J. Stat. Phys. 17, 71 (1977); A. Schenzle and H. Brand, Phys.
    Lett. A68, 427 (1978); M. Morsch, H. Risken and H. D.
    Vollmer, Z. Phys. B32, 245 (1979).
11. R. Bonifacio, L. A. Lugiato, J. D. Farina and L. M. Narducci,
    IEEE J. Quant. El. (to be published).

COOPERATION IN AN "OPTICAL-BISTABILITY" SYSTEM*

I. R. Senitzky and Jan Genossar

Department of Physics
Technion - Israel Institute of Technology
Haifa, Israel

Abstract: The behavior of an "optical-bistability" system in
which the atoms are described by a single angular-momentum oscillator of constant (large) total angular momentum is investigated.
It is shown that, as the input field is increased through a critical
value, the steady state output field turns into a modulated field,
the modulation frequency increasing from zero to the Rabi frequency
with increasing input. The reason for qualitatively different predictions by other authors based on the same model is attributed to
differences in the interpretation of quantum mechanics applied to a
single macroscopic system.

A number of models have been used to analyze optical bistability.
The active atomic medium responsible for this phenomenon has been
described by a nonlinear susceptibility,[1] by a collection of two-
level systems satisfying the Bloch equations,[2] and by a single angular momentum oscillator (AMO) of constant total angular momentum.[3-5]
The last model, while not being necessarily the best for conditions
which produce the most useful type of optical bistability, focuses
on cooperative atomic behavior - a subject of central interest in
quantum optics - to a greater extent than any other model. One
cannot help noting that despite the existence of this model for over
a quarter of a century,[6] and its widespread use,[7] the literature
associated with it continues to display conceptual problems and con-
troversy.[8-12] It is the purpose of the present paper to present an
analysis of an "optical-bistability" type of system based on this
model, with conclusions and interpretations qualitatively different

---

*Work supported, in part, by the U.S. Army through its European
Research Office.

from those found in recent literature.

   We consider a number of identical two-level systems, to be
referred to as "atoms," contained in a cavity of which one mode is
in resonance with the atoms, at frequency ω. This cavity mode is
coupled not only to the atoms, but also to a prescribed input mechanism and to an output mode, the latter being a travelling wave that
may be regarded as emanating from the cavity into free space. The
atoms, for their part, are assumed to be coupled not only to the
cavity mode, but also to free space. In a discussion of optical
bistability, one is interested in the relationship between the field
of the output mode and the prescribed input field. In the case of
resonance fluorescence, one is also interested in the free space
radiation by the atoms.

   The validity and limitations of the AMO model has been discussed in detail by a number of authors,[4,5,13] and need not be discussed further here. The notation associated with this model is
that of several previous articles,[9,13] as is the formalism, which
has the important feature of being interpretable and valid in a
classical analysis as well as in a quantum mechanical analysis. Let
the Hamiltonian of the entire system be given by

$$H = \hbar\omega\ell_3 + \hbar\omega(a_c^\dagger a_c + \tfrac{1}{2}) + \tfrac{1}{2}\hbar[(\gamma a_c^\dagger + \Sigma_k \gamma_k a_k^\dagger)\ell_-$$

$$+ \text{h.c.}] + H_{\text{input}} + H_{\text{cavity-loss}} + H_{\text{radiation}}.$$

$$(1)$$

The variables ℓ_3, ℓ_+ are dimensionless angular momentum components
of the AMO that represents the collection of atoms, and obey the
commutation rules $[\ell_+,\ell_-] = \ell_3$, $[\ell_3,\ell_+] = \pm \ell_+$. (Classically, commutators are to be interpreted as Poisson brackets multipled by
i^{13}). The variables a_c and $a_c^\dagger$ are the annihilation and creation
operators for the resonant cavity mode, with $[a_c, a_c^\dagger] = 1$, while the
a_k's and $a_k^\dagger$'s refer to the modes of the free-space radiation field.
H_{input} accounts for the effect of the input mechanism on the cavity
field, $H_{\text{cavity-loss}}$ accounts for both the effect of cavity losses
and the transmission into the output mode, and $H_{\text{radiation}}$ refers to
the modes of the free-space radiation field designated by the index
k. (None of the last three terms contain atomic variables.) The
coupling between atoms and field has been approximated by the
rotating-wave approximation. Although the field of the output mode
is an important part of our system, we need not refer to it explicitly in the Hamiltonian, but can assume,instead, that this mode
is driven by transmission from the cavity, and is, thus, characterized by an amplitude proportional to that of the cavity-mode field.
It is therefore sufficient to describe the field of the cavity mode.

   The dynamical variables of interest are a_c, a_k, $\ell_\pm$ and ℓ_3.

As in previous analyses,[9,13] it is convenient to use the "reduced" variables A_c, $L_\pm$, L_3 specified by

$$a_c = A_c e^{-i\omega t}, \quad \ell_\pm = L_\pm e^{\pm i\omega t}, \quad \ell_3 = L_3,$$
(2)

which vary slowly compared to $\exp(i\omega t)$ (due to the fact that the coupling constants are assumed to be much smaller than ω), and to introduce the variables

$$a \equiv \frac{1}{2} i\Sigma_k \gamma_k^* a_k e^{i\omega t}, \quad A \equiv \frac{1}{2} i\gamma^* A_c .$$
(3)

The Heisenberg equations of motion for the reduced atomic variables, as obtained directly from the Hamiltonian, are

$$\dot{L}_+ = (A^\dagger + a^\dagger)L_3 , \quad \dot{L}_- = (\dot{L}_+)^\dagger ,$$

$$\dot{L}_3 = - [L_+(A + a) + (A^\dagger + a^\dagger)L_-] .$$
(4)

The derivation of the equations of motion for a_c and a_k, or, more conveniently, for A and a, involves the consideration of the cavity-loss mechanism and the free radiation field, both thermal-reservoir type systems. The coupling to such systems has been analysed previously.[13,14] With sufficient accuracy for present purposes, the results of this analysis can be expressed by the approximate relationships

$$a \simeq a_o + \alpha L_-$$
(5)

$$A \simeq A_o + \frac{1}{4} |\gamma|^2 \int_o^t dt_1 L_-(t_1) e^{-\xi(t-t_1)}$$
(6)

where a_o and A_o are defined formally as a and A [in Eq. (3)], but with $a_k{}^o$ and $A_c{}^o$ that appear in the definition representing, instead, the field in absence of the atoms; α is the spontaneous emission rate of an excited atom into the free-space modes, and 2ξ is the decay rate of cavity energy in the absence of atoms. It is assumed that the coupling between the atoms and the resonant cavity mode begins at $t = 0$. If we assume, furthermore, that the cavity is sufficiently lossy so that the part of the cavity field due to the atoms follows the atomic polarization adiabatically (that is, ξ is much larger than the rate of change of L_+), which we do henceforth, then one can write, instead of Eq. (6),

$$A \simeq A_o + (|\gamma|^2/4\xi)L_-(t) \; . \tag{7}$$

Substituting the above expressions for a and A into the atomic equations of motion, we obtain

$$\dot{L}_+ = (A_o^\dagger + a_o^\dagger + \beta L_+)L_3 \; ,$$

$$\dot{L}_3 = - [L_+(A_o + a_o + \beta L_-) + h.c.] \; , \tag{8}$$

where

$$\beta \equiv \alpha + |\gamma|^2/4\xi \; .$$

One sees that β is the total spontaneous emission rate of an excited atom, α being the emission rate into free space, and $|\gamma|^2/4\xi$ being the emission rate into the cavity.

These equations are valid both quantum mechanically and classically, the dynamical variables being either operators or c-numbers, respectively. The state of the free-space radiation field is taken to be the ground state, so that we have $a_o|> = <|a_o^\dagger = 0$, quantum mechanically, and $a_o = 0$, classically. As for A_o, it consists of the sum of two parts, the first being due to the prescribed (c-number) input field, and the second being an operator that operates on the ground state of the cavity mode. The latter vanishes classically, of course.

Our interest lies in the cooperative behavior of a large number of atoms, or a large value of L_o, the total angular momentum quantum number. Except for the case of complete initial inversion of the two-level systems *and* the absence of an external field (which yields unstable equilibrium classically and spontaneous emission quantum mechanically), the intrinsic statistical aspects of quantum mechanics may be neglected for most purposes (that is, in the answer to most questions of interest), and the system may be described classically.[13] We proceed to do so, and will discuss quantum mechanical aspects of the problem later.

The equations of motion can be put into simpler form by the

following notational changes:

$$\Omega^2 = 2|A_o|^2, \qquad x = (A_o L_+ + A_o^* L_-)/\Omega L_o,$$

$$y = -i(A_o L_+ - A_o^* L_-)/\Omega L_o, \qquad z = L_3/L_o, \qquad (9)$$

$$c = L_o \beta/\Omega, \qquad \tau = \Omega t .$$

The classical equations of motion can now be converted into

$$x' = z(1+cx), \quad y' = cyz, \quad z' = -x - c(1-z^2) , \qquad (10)$$

where the prime indicates differentiation with respect to τ. The
variable z is the energy per atom (in units of $\tfrac{1}{2}\hbar\omega$), x is proportional to the component of dipole moment that interacts with the
input field to produce atomic energy changes, and y is the component
in quadrature with x.

It is seen that the pair of equations for x and z can be solved
independently of y. If y is zero initially, it remains zero. It is
also zero in the steady state, if $z \neq 0$. Otherwise, y can be obtained simply from the relationship

$$x^2 + y^2 + z^2 = 1. \qquad (11)$$

This relationship also shows that, in the xz plane, the trajectory
that describes the solution remains within the unit circle.

The solution to the classical equations of motion obtained
from Eqs. (8) has been studied previously. Approximate solutions
for a weak and a strong input field, respectively, are given in
Ref. (13). Exact solutions, using a complex transformation of coordinates due to Glauber and Haake[15], are given by Drummond and
Carmichael[4]. Here, we present an exact solution using the coordinates x and z.

It is instructive to display, first, the trajectories of the
solution in the (x,z) plane, for which the differential equation is

$$\frac{dz}{dx} = - \frac{x + c(1-z^2)}{z(1+cx)} . \qquad (12)$$

The solutions fall into two categories, one for $c > 1$, and another for $c < 1$. Families of trajectories for several values of c are shown in Fig. 1. For $c > 1$ there exist two steady states, or singular points, at $x = -1/c$, $z = \pm[1 - (1/c^2)^{\frac{1}{2}}$. The one with $z > 0$ is unstable and the one with $z < 0$ is stable. For $c < 1$, there exists only one steady state, at $x = -c$, $z = 0$. We can visualize the trajectories for all c as follows: For $c = \infty$, that is, in the absence of an input field (or $\Omega = 0$), the trajectories are ellipses with the major axis along the z axis and the minor axis along the x axis. All trajectories have the same major axis, namely, unity, and minor axes that range from 0 to 1. They are all tangent to each other at $z = \pm 1$, the two singular points. Limiting curves are the straight line (of minor axis zero) and the circle (of minor axis equal to major axis). Let c decrease, now. The two singular points move left along the circumference of the unit circle [see Fig. 1] until they join for $c = 1$. The trajectories may be considered as distorted continuously with the motion of the singular points, the straight verticle trajectory between them remaining straight and vertical, the circular trajectory remaining "stuck" to the circle, and all other trajectories remaining smooth and tangent at the singular points. As c decreases below unity (the bifurcation point[16]) the trajectories separate and become closed curves surrounding the only singular point at $x = -c$, $z = 0$, assuming a more circular shape as c approaches zero.

The method of solution of the equations of motion [Eqs. (10)] is given in the Appendix. The explicit solutions are as follows: For $c > 1$, with $x(0) = x_0$, $z(0) = 0$, and $\eta = (c^2 - 1)^{\frac{1}{2}}\tau$, we have

$$x(\tau) = -\frac{(x_0+c)\cosh\eta - c(x_0c+1)}{c(x_0+c)\cosh\eta - (x_0c+1)},$$

$$z(\tau) = -\frac{(x_0+c)(c^2-1)^{\frac{1}{2}}\sinh\eta}{c(x_0+c)\cosh\eta - (x_0c+1)}. \tag{13a}$$

For $c < 1$, with $x(0) = -c$, $z(0) = z_0(1-c^2)^{\frac{1}{2}}$, and $\eta = (1-c^2)^{\frac{1}{2}}\tau$, we have

$$x(\tau) = \frac{z_0\sin\eta - c}{1 - cz_0\sin\eta}$$

$$y(\tau) = \frac{\pm(1-c^2)^{\frac{1}{2}}(1-z_0^2)^{\frac{1}{2}}}{1 - cz_0\sin\eta}$$

$$z(\tau) = \frac{z_0(1-c^2)^{\frac{1}{2}}\cos\eta}{1 - cz_0\sin\eta} \tag{13b}$$

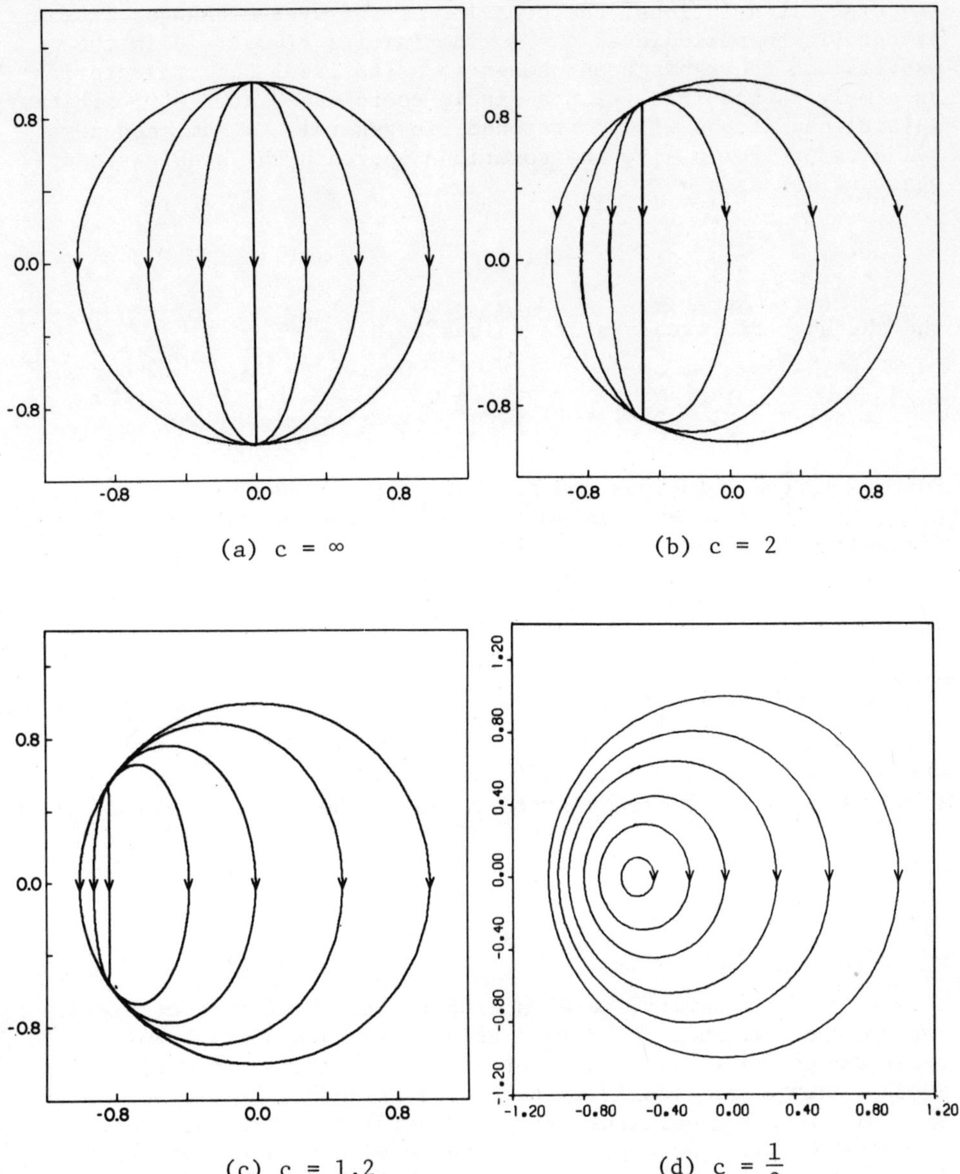

(a) $c = \infty$

(b) $c = 2$

(c) $c = 1.2$

(d) $c = \frac{1}{2}$

Fig. (1). Trajectories in the x, z plane of solutions of Eqs. (10) for four values of c. The point representing the solution travels in the sense indicated by the arrows. Different curves for a given c correspond to different initial conditions.

As demonstrated both by the discussion of the trajectories and the explicit solutions, the behavior of the system changes from asymptotic to periodic at $c = 1$. The initial time $\tau = 0$ in the expressions above has been chosen so as to label the trajectories as simply as possible, with a single coordinate, at $\eta = 0$; arbitrary initial conditions will correspond, in general, to some non-zero value of η. For $c > 1$, the solutions approach the stable steady-state values as $\eta \to \infty$,

$$x(\infty) = -\frac{1}{c}, \quad z(\infty) = -[1 - c^{-2}]^{\frac{1}{2}}, \tag{14}$$

and the unstable steady-state values as $\eta \to -\infty$,

$$x(-\infty) = -\frac{1}{c}, \quad z(\infty) = [1 - c^{-2}]^{\frac{1}{2}} . \tag{15}$$

For $c < 1$, the solutions are periodic with frequency $(1 - c^2)^{\frac{1}{2}}\Omega$. As the input field becomes sufficiently strong so that $c \ll 1$, the frequency approaches the Rabi frequency Ω.

Our main interest lies in the cavity field A_c. From Eqs. (7), (9) and (3), we obtain

$$A_c = A_c^{(o)}[1 + c_1(x-iy)], \tag{16}$$

where $c_1 = |\gamma|^2 L_o/4\xi\Omega$, and $A_c^{(o)}$ is the cavity input field. For $c > 1$, the (stable) steady-state value of the cavity field is given by

$$A_c = A_c^{(o)}(1 - \frac{c_1}{c}) = A_c^{(o)} \frac{4\alpha\xi}{4\alpha\xi + |\gamma|^2} = A_c^{(o)} \frac{\alpha}{\beta} . \tag{17}$$

We see that the cavity field is proportional to the input field, the constant of proportionality being less than unity. For $c < 1$, there exists no steady state for arbitrary initial conditions, since x and y vary periodically in time with frequency $(1-c^2)^{\frac{1}{2}}\Omega$. For the special initial conditions, $z(0) = 0$, $x(0) = -c^{-1}$, there does exist the steady state corresponding to these values, but it is not stable in the sense of asymptotic orbital stability[16], that is, in the sense of exponential damping of a perturbtion. Any disturbance of the steady state will remain undamped, and produce a periodic variation in x and y. Explicitly, we have for this "semi-stable" steady state

$$A_c = A_c^{(o)}[1 - c_1 c \pm ic_1(1-c^2)^{\frac{1}{2}}] \tag{18a}$$

and

$$|A_c|^2 = |A_c^{(o)}|^2 (1 - 2c_1 c + c_1^2) \ . \tag{18b}$$

For the solution in which x and z oscillate with the largest amplitude, that is, the circular trajectory (of unit-radius) in the x,
z plane, we have (noting that y = 0)

$$A_c = A_c^{(o)} (1 + c_1 \frac{\sin \eta - c}{1 - c \sin \eta}) \ . \tag{19}$$

This is also the solution that corresponds to initial conditions in
which the atoms are in the ground state. [Note that, for these
initial conditions, z_o = 1, and the initial η is given by η_o =
$\sin^{-1}c$.] In general, for arbitrary initial conditions, the cavity
field will oscillate both in amplitude and phase. As $A_c(o)$, or Ω,
increases, this oscillation will become relatively small, but of
constant amplitude in absolute terms.

We come now to the question that is of greatest interest from
a conceptual viewpoint, as far as the present model is concerned.
Are there any significant differences between the results of a
quantum mechanical analysis and those of a classical analysis for
systems with large L_o?   A case in which such a difference does
exist was mentioned earlier, namely, that for Ω = 0 and the AMO
initially in its highest energy state. Our present result yields
unstable equilibrium, while quantum mechanically, as is well known,
there exists spontaneous emission. In order to investigate further
the problem raised by this question, we look at an equation that
can be obtained quantum mechanically for the *operator* $z(\equiv L_3/L_o)$,

$$<z''> + \frac{3}{2} c(\lambda/L_o)<z'> - \frac{3}{2} c <zz' + z'z> + <z>$$
$$\tag{20}$$
$$= c^2 <[1 - z^2 - (\lambda/L_o)(1 + z)][z - (\lambda/L_o)]>.$$

This equation has been derived in Ref. (13) for the case of free-
space radiative loss only (in which case c = $\alpha L_o/\Omega$), but it is
generalized immediately to apply to the present case by using the
present definition of c. It is written in a form that can be read
both quantum mechanically and classically, due to the use of λ,
which is defined by

$$\lambda = \begin{cases} 1 \text{ quantum mechanically,} \\ 0 \text{ classically.} \end{cases}$$

Classically, the expectation-value brackets are unnecessary. We note
that the classical form of this equation is identical to the second
order differential equation for z that can be derived from Eqs. (10):

$$z'' - 3czz' + z = c^2 z(1 - z^2) \ . \tag{21}$$

For the sake of simplicity, we will consider the strong field
case only, that is, the case $c \ll 1$. We can, therefore, drop the
c^2 term in Eq. (20), which becomes

$$\langle z'' \rangle + \frac{3}{2} c(\lambda/L_o) \langle z' \rangle - \frac{3}{2} c \langle zz' + z'z \rangle + \langle z \rangle \simeq 0, \tag{22}$$

and which can also be written as

$$\langle \ddot{z} \rangle + \frac{3}{2} \lambda\beta\langle \dot{z} \rangle + \Omega^2 \langle z \rangle = \frac{3}{2} L_o \beta\langle \dot{z}z + z\dot{z} \rangle \ , \tag{23}$$

where, it is recalled, $\beta = \alpha + |\gamma|^2/4\xi$. Although no explicit exact
solution has been found for this equation, it has been analyzed both
by perturbation theory and by the approximation $\langle \dot{z}z + z\dot{z} \rangle \simeq 2\langle z \rangle\langle \dot{z} \rangle$,
in Ref. (13); it is concluded there that $\langle z \rangle$ is damped approximately
exponentially with the damping factor $\exp(- 3/4 \ \lambda\beta t)$. Now, one
should note that this factor is unaffected by the value of L_o, so
that even for L_O large, that is, for a macroscopic AMO, the quantum
mechanical result is qualitatively different from the classical
result. It is true that, in order to maintain the condition $c \ll 1$
(where, it is recalled, $c = \beta L_O/\Omega$), Ω must be large for L_O large,
which means that the decay rate in terms of Rabi cycles will decrease
for increasing L_O (and c fixed). The decay rate in terms of absolute
time, however, is unaffected. The qualitative difference between
the classical and quantum mechanical results appears, at first glance
paradoxical.

This apparent paradox is discussed in Ref. (13). It is pointed
out that the quantity which is damped in the quantum mechanical
analysis is an *average* over an *ensemble* of (macroscopic - for large
L_O) AMO's. Now, any uncertainty in the frequency of oscillation,
that is, a random spread about a center frequency among members of
the ensemble, will produce a damping of the oscillation of the
average. (For instance, a Lorentzian frequency distribution will
produce exponential damping.) The most reasonable explanation of
the qualitative difference in behavior between the quantum mechanical
expectation value and the classical value is, therefore, offered by
the existence of an uncertainty, or randomness, in the frequency of
oscillation of z about the Rabi frequency. (A randomness in amplitudes is qualitatively irrelevant as far as damping is concerned.)
Both uncertainty-principle arguments and higher-moment considerations are shown in Ref. (13) to support this explanation. We

conclude, therefore, that (for the AMO model) the dynamical variable
z oscillates without damping in any given experiment, even though
$\langle z \rangle$ approaches a steady state. This conclusion is in contradiction
with that of Drummond and Carmichael[4], as well as with that of Walls
et $al.$[5], who identify the quantum mechanical expectation value with
the result of an experiment, and claim that the present model leads
to a steady state. Although they also obtain the undamped oscillation by a limiting procedure in which L_0, $\Omega \to \infty$ while $c < 1$, they
then use a stochastic averaging method, that simulates the averaging
procedure inherent in taking quantum mechanical expectation values,
to obtain the "exact" result. They interpret the qualitative difference between the oscillating result and its steady-state ensemble
average as due to an "incorrect" factorization implicit in the classical calculation, and regard the averaging procedure as a "quantum
correction" to this calculation.

This brings us to the question of the validity of "classical
factorization"[11], "semiclassical decorrelation"[5] or the "decoupling
approximation"[12]. These expressions are all meant to denote the
replacement of $\langle AB \rangle$ by $\langle A \rangle \langle B \rangle$ in a quantum mechanical calculation.
Now, it is well known that

$$\langle AB \rangle_{av} \neq \langle A \rangle_{av} \langle B \rangle_{av} \, ,$$

in general, even in a classical calculation, when a statistical
description is used. For instance, for a harmonic oscillator,
$\langle q^2 \rangle \neq \langle q \rangle^2$, not only when the oscillator is described quantum
mechanically by an energy state, but also when it is described
classically by a microcanonical ensemble. Aside from properties
of noncommutativity, which may usually be neglected when the variables have macroscopic magnitudes, the problem associated with
"classical factorization" (for variables of this magnitude) is one
of $statistics$. A statistical spread does not necessarily disappear
in the classical limit. (The harmonic oscillator in an energy state
of high quantum number is an example of such a limit.) However,
when we are dealing with a macroscopic system of a few degrees of
freedom, we need not introduce this statistical spread at the beginning of the calculation and enquire about averages at every step.
Even though the initial values of the problem may not be specified
precisely, one may proceed with the calculation as though it were
deterministic, in which case no "classical factorization" approximation is involved. If one is interested, at the end of a calculation, in an average over a distribution of initial values, the final
result can be averaged. We have seen, however, that this average
may have no qualitative resemblance to the result of a particular
experiment.

Helpful suggestions by Dr. Ady Mann are acknowledged with
appreciation.

REFERENCES

1. S. L. McCall, Phys. Rev. A9, 1515 (1974).
2. R. Bonifacio and L. A. Lugiato, Opt. Commun. 19, 172 (1976).
3. L. M. Narducci, D. H. Feng, R. Gilmore and G. S. Agarwal, Phys. Rev. A18, 1571 (1978).
4. P. D. Drummond and H. J. Carmichael, Opt. Commun. 27, 160 (1978).
5. D. F. Walls, P. D. Drummond, S. S. Hassan and H. J. Carmichael, Prog. Theor. Phys. (Japan) Supp. 64, 307 (1978).
6. R. H. Dicke, Phys. Rev. 93, 99 (1954).
7. See, for instance, "Cooperative Effects in Matter and Radiation, C. M. Bowden, D. W. Howgate and H. R. Robl, eds., Plenum, New York (1977).
8. G. S. Agarwal, A. C. Brown, L. M. Narducci and G. Vetri, Phys. Rev. A15, 1613 (1977).
9. I. R. Senitzky, Phys. Rev. Lett. 40, 1334 (1978).
10. G. S. Agarwal. D. H. Feng, L. M. Narducci, R. Gilmore and R. Tuft, Phys. Rev. A20, 2040 (1979).
11. H. J. Carmichael, Phys. Rev. Lett. 43, 1106 (1979).
12. G. S. Agarwal, R. Saxena, L. M. Narducci, D. H. Feng and R. Gilmore, Phys. Rev. A21, 257 (1980).
13. I. R. Senitzky, Phys. Rev. A6, 1175 (1972).
14. I. R. Senitzky, Phys. Rev. 155, 1387 (1967), Sec. I.
15. R. J. Glauber and F. Haake, in "Cooperative Phenomena," H. Haken, ed., North Holland, Amsterdam (1979).
16. See, for instance, H. Haken, "Synergetics," Springer, Berlin (1977), Chap. 5.

APPENDIX

We exhibit here the method of solution of the equations

$$x' = z(1 + cx), \quad z' = - x - c(1 - z^2),$$

which are two of the three equations given in Eqs. (10) of the text. The third equation, for y, can be obtained from Eq. (11). In terms of the variables X, z, where $X \equiv x + c$, the equations become

$$X' = z[1 - c^2 + cX], \quad z' = - X + cz^2. \tag{A1}$$

Introducing polar coordinates, $z = r \cos \theta$, $X = r \sin \theta$, we obtain

$$r' = cr(r - c \sin \theta) \cos \theta, \quad \theta' = 1 - c^2 \cos^2 \theta. \tag{A2}$$

The equation for θ can be integrated immediately. For $c < 1$, we have

$$\tan \theta = (1 - c^2)^{\frac{1}{2}} \tan (1 - c^2)^{\frac{1}{2}}\tau, \tag{A3}$$

and for $c > 1$, we have

$$\tan \theta = - (c^2 - 1)^{\frac{1}{2}} \coth (c^2 - 1)^{\frac{1}{2}}\tau. \tag{A4}$$

The equation for z' becomes

$$z' = cz^2 - z \tan \theta, \tag{A5}$$

which is now a nonlinear equation for z only. It can be linearized by the change of variable $z = u^{-1}$, so that we have

$$u' = u \tan \theta - c. \tag{A6}$$

By standard methods for a linear first order equation, one obtains for $c < 1$,

$$u = \frac{K - k \sin \eta}{\cos \eta} \tag{A7}$$

and for $c > 1$,

$$u = \frac{K - k \cosh \eta}{\sinh \eta} \tag{A8}$$

where, in Eq. (A7),

$$k = c/(1 - c^2)^{\frac{1}{2}}, \quad \eta = (1 - c^2)^{\frac{1}{2}}\tau ,$$

and, in Eq. (A8),

$$k = c/(c^2 - 1)^{\frac{1}{2}}, \quad \eta = (c^2 -1)^{\frac{1}{2}}\tau,$$

and K is a constant of integration. Setting $K = z_0^{-1}$ in Eq. (A7), and setting $K = [(x_0 c+1)/(x_0+c)(c^2-1)]$ in Eq. (A8), yields the expressions for z given in Eqs. (13). The expressions for x are obtained by noting that $X = z \tan \theta$.

It is interesting to note that, for all c, solutions of the differential equation (12), that is, the trajectories of Fig. 1, are a set of ellipses given by

$$\frac{z^2}{a^2} + \frac{(x-x_c)^2}{b^2} = 1,$$

where

$$\left[-c, \ -\frac{1}{c}\right]_{max} \leqslant x_c \leqslant 0 \ ,$$

$$a^2 = 1 + \frac{x_c}{c} \ , \quad b^2 = \left(1 + \frac{x_c}{c}\right)(1 + cx_c) \ ,$$

$[\ , \ ]_{max}$ indicating the larger of the two quantities inside the bracket, and x_c being a constant of integration.

THE DRIVEN DICKE MODEL AND ITS MACROSCOPIC EXTENSION:

BISTABILITY OR BIFURCATION?

S. S. Hassan* and R. K. Bullough

Department of Mathematics
U.M.I.S.T., P.O. Box 88
Manchester M60 1QD, U.K.

Abstract: We are concerned with the relation between recent
work on the "driven Dicke model" of N two-level atoms, on the same
site, driven by a c.w. laser field Ω, and a corresponding theory for
the more realistic macroscopically extended system. We review the
results on the driven Dicke model: two different decorrelation
schemes yield different results; in the steady state at resonance
a semiclassical approximation without damping best approximates the
exact solution of the quantum model also described. The exact solution of the quantum model does not display normal optical bistability
(OB):calculation of $g^{(2)}(0) = G^{(2)}(0)/\{G^{(1)}(0)^2\}$ (where $G^{(n)}(0) =$
$\langle(S_+)^n(S_-)^n\rangle$ and $S_\pm$ are collective spin operators) shows $g^{(2)}(0)$
$\rightarrow 1.2$ and there is a simple bifurcation point at $\theta\{\propto \lim\Omega N^{-1}, N\rightarrow\infty\}$
$= 1$.   The inversion r_3 plays the role of the order parameter: $r_3 =$
$\pm \frac{1}{2}(1-\theta^2)^{\frac{1}{2}}, \theta < 1; = 0, \theta > 1$.   There is a second-order type phase
transition, and by moving off-resonance and relating to the decor-
related model, we are able to identify one set of equivalent thermodynamic parameters for the model. We find "critical exponents"
$\alpha = \frac{1}{2}, \beta = \frac{1}{2}, \gamma = 1.5$ and $\alpha + 2\beta + \gamma > 2$ in this manner. Results
are compared with the operator theory for the extended system also
presented (unlike the Dicke model this model does not have total
spin as a constant of the motion). Decorrelation of operator products with self-correlation (radiation damping) leads of course to
the c-number theory of cusp catastrophe OB. An operator theory involving a natural power dependent refractive index is sketched and
we believe that it is this which should appear as the parameter in
the usual treatment of the Fabry-Perot interferometer. But, alternatively, by extracting a single mode theory in the "mean-field"

____

*On sabbatical leave from: Ain Shams University, Faculty of Science,
Applied Mathematics Department, Cairo, Egypt.

approximation, we regain both the Bloch equations and the master
equation of the driven Dicke model. The spectra and correlation
functions shown in Figs. 1-4 are calculated from these in a decorrelation approximation which retains single-particle damping and
which differs from the exact solution of the master equation. The
hierarchy of different models relates to the realistic extended
system model in ways very similar to those of a similar hierarchy
in the theory of superfluorescence. It is concluded that mean field
theory maltreats the analysis. However, it is expected that the
decorrelation scheme adopted for the spectra we have calculated
will be adequate to describe their essential features.

I.   INTRODUCTION

Two problems of recent interest in both theoretical and experimental quantum optics have been: the interaction of an intense
single mode resonant c.w. laser field with a single two-level atom
(the problem of resonance fluorescence); that of cooperative behavior in the spontaneous emission from an inverted sample of many
two-level atoms (the problem of super-radiance or super-fluores-
cence). Two early theoretical papers on the resonant single atom
problem are Mollow's quantum mechanical treatment[1] of 1969 and
Stroud's and Jaynes's neoclassical treatment[2] of 1970. Both the
interpretations and to some extent the predictions differ in the
two cases. The quantal treatment[1,3-5] shows that: (i) the atomic
population inversion oscillates at the Rabi frequency but is exponentially damped so that the atom saturates in the intense field
limit with equal probability of occupying its upper or lower level;
(ii) the power spectrum of the resonantly scattered light in the
intense field limit is incoherent and three-peaked Stark split,
separated by the Rabi frequency; (iii) the exponential damping is
a consequence of the quantal treatment of spontaneous emission and
this plays a fundamental role in forming the Stark spectrum.

The neoclassical treatment[2] shows that the atom oscillates much
as in the quantal case; but precisely on resonance at least the oscillations are not damped and the atomic inversion does not reach a
saturated non-oscillatory steady state. The neoclassical theory of
a single atom provided a stimulating episode in the recent history
of quantum optics[6-9]; but experiments (see the first ref. of Ref.6)
tend to support the quantum theory rather than the neoclassical
theory, and the first and subsequent observations[10] of the spectrum,
the so called dynamical Stark spectrum, of single-atom resonance
fluorescence in particular conformed very well to the predictions
of the quantum theory.

The cooperative spontaneous emission from a group of inverted
atoms was first discussed quantum mechanically by Dicke[11] whose

treatment of the extended many-atom system was, however, very obscure to us. Considerable understanding of this difficult aspect
of the problem has been achieved recently[12] stimulated by the successful experiment of Skribanowitz et al.[13a] in 1973 and the several
other experiments[13b] which closely followed it.

The Dicke model of N atoms on a single site led to a soluble
quantum mechanical model (cf. e.g. refs. in Ref. 12f). The natural
theoretical generalization of single atom resonance fluorescence is
therefore the same single site Dicke model with the atoms driven by
a c.w. laser field[14-18]. For finite but very small N ($\lesssim 5$) some of
these authors have given an exact solution of the master equation of
the atomic system and have studied the atomic correlation functions,
searching for the additional sidebands in the spectrum first predicted semiclassically[19,20]. In the large N limit other work[21,22]
showed that the atomic behavior is of second order critical phase
transition type. The possibility of optical bistability was also
discussed[16,17] (and cf. Ref. 18). The master equation was solved[16,17]
subject to a factorization procedure which allows self correlation
of the atoms but decorrelates operator products involving different
atoms: in the intense field limit the emission spectrum[16] and the
absorption spectrum[17] were the same as for a single atom and this
independent behavior of the atoms was reflected in the normalized
intensity-intensity correlation function $g^{(2)}(0)$ which had the value
2^{17}.

Subsequently Puri and Lawande[23] found an exact steady state
solution of the master equation valid at exact resonance for arbi-
trary N and normalized Rabi frequency $|g|$. In the 'thermodynamic
limit' $N \rightarrow \infty$ with $2|g|N^{-1} \equiv \theta$ = constant, the derivatives of the
atomic observables with respect to θ showed a discontinuity at $\theta=1$,
although their fluctuations were finite[20-21]. Hassan et al.[24] calculated the atomic correlation functions and found in the thermodynamic limit $g^{(2)}(0)$ = 1.2; the intensity fluctuations were reduced compared with the earlier work[17] by cooperative interactions.
A simple bifurcation point occurs at $\theta = 1$ [24] and this can be viewed
as a phase transition at the critical 'temperature' $\theta^2 = 1$ [24].
Recently the exact solution of Ref. 23 has been extended to the
off-resonance situation[25]. In the large N limit dispersive effects
destroy the critical bifurcation behavior.

This paper is concerned with the relationship of these various
results for the point, driven Dicke model of optical bistability
to the more realistic but much more intractable theory which describes the behavior of N two-level atoms irradiated by a c.w. laser
field in a macroscopically extended region V. The main point is to
discover whether the results achieved so far on the point system
can apply at all to the extended system, and particularly whether
the exact quantum solution[23,25] is relevant to the extended system

for which no exact results are available.

Our main conclusion is essentially that it is not, although surprisingly perhaps we believe that the approximate solution of Refs. 16,17 (which illustrates the conventional picture of optical bistability) probably is. A pointer to this conclusion is that the squared total angular momentum $\underset{\sim}{S}^2$ is a constant of the motion of the N atom driven Dicke model; the exact solution[23,25] preserves this condition, but the approximate one[16,17,18] does not; $\underset{\sim}{S}^2$ is not a constant of the motion for the extended system. The result displays itself in the need to look rather carefully at propagational effects in the extended system.

The paper is organized roughly as follows: we review the different results achieved for the driven Dicke model and discuss their physical consequences: it is here, in the semiclassical approximation (§II.B) to the N atom problem that the neoclassical work[2] proves to be directly relevant. We extend a little more the second-order phase transition interpretation of the exact quantum theory before going on, in §III, to develop the theory of the extended system. Within a single- or two-mode ansatz for the collective dipole operators and a spatial average of "mean field type," both the collective Bloch equations and the corresponding master equation take on essentially the same form as the corresponding equations for the driven Dicke model. Consequently the best solution would seem to be the exact quantum solution[23,24] which does not show the usual features of optical bistability. This suggests that on the one hand the theory should not be spatially averaged; and on the other that decorrelated theories like that of Refs. 16, 17 are a better guide to the calculation of the correlation functions. Spectra shown in Figs. 2 and 3 are calculated on this basis; and the conventional bistable features also found this way are shown in Fig. 1. The intensity-intensity correlation function $g^{(2)}(0)$ is shown in Fig. 4.

Although our conclusion is that spatial averaging is inappropriate, the details of the theory which do not include this step are by no means complete, and the work must be presented elsewhere. At the time of writing we cannot say whether the spatially averaged correlation functions and bistable behavior shown in the figures are good as coarse grained results. However, we do not expect qualitatively different behavior from the more complete theory.

II.   THE DRIVEN DICKE MODEL

A.   Master Equation and Operator Bloch Equations

We consider a system of N identical two-level atoms, each with transition frequency ω_o, occupying a volume of dimension less than

a wavelength and driven by an imposed single mode c.w. laser field
of frequency ω. We actually place all the atoms on the same single
site, so this is the atomic model of Dicke[11] driven by the field imposed at that atomic site. For this system the master equation for
the reduced atomic density operator $\hat{\rho}_A$ in the rotating frame reads
(cf. Refs. 14, 23-25)

$$\frac{d\hat{\rho}_A}{dt} = -i\Omega[S_+ + S_-, \hat{\rho}_A] + i\delta_o[S_z, \hat{\rho}_A]$$
$$+ \gamma_o[2S_-\hat{\rho}_A S_+ - S_+ S_-\hat{\rho}_A - \hat{\rho}_A S_+ S_-] \tag{1}$$

where Ω is the laser Rabi frequency, γ_o is one half of Einstein's
A-coefficient $\Gamma_o \equiv 4\hbar^{-1}p^2\omega_o^3/3c^3$, p is the atomic dipole matrix element and $\delta_o \equiv (\omega-\omega_o)$ is the frequency mismatch (detuning) between
the atoms and laser. The collective atomic dipole operators $S_\pm$ together with the atomic inversion operator S_z satisfy the angular
momentum commutation relations

$$[S_+, S_-] = 2S_z, \qquad [S_z, S_\pm] = \pm S_\pm \quad . \tag{2}$$

The master equation (1) assumes dipole and rotating wave approximations.

Within the same approximations Heisenberg's equations of motion
yield the Bloch equations

$$\frac{dS_+}{dt} = i\delta_o S_+ + 2\gamma_o S_+ S_z - 2i\Omega S_z$$

$$= \left(\frac{dS_-}{dt}\right)^* \tag{3}$$

$$\frac{dS_z}{dt} = -2\gamma_o S_+ S_- - i\Omega(S_+ - S_-) \quad .$$

The master equation (1) and the Bloch equations (3) are equivalent,
for although the details of the treatment of spontaneous emission actually used in each case are different, the results are exactly the
same and indeed (3) can be derived from (1). In §III we show briefly
how to derive the master equation (1) by using reaction field theory
(cf. Ref. 7) and this can also be used directly for (3). An important property of both Eqs. (1) and (3) is that the total angular
momentum operator $\underset{\sim}{S}^2$ is a constant of the motion.

We now summarize three different situations, two approximate,
one exact, in which Eqs. (1) or (3) have been solved and describe
the different behavior of the solutions.

B.   Exact Semiclassical Results

Drummond and Hassan[20] gave the complete time dependent solution of Eqs. (3) at resonance ($\delta_o = 0$) by adopting the direct semi-classical factorization (decorrelation) $\langle S_+ S_- \rangle = \langle S_+ \rangle \langle S_- \rangle$, etc. The solution depends on the parameter $\theta \equiv 2\Omega(\gamma_o N)^{-1}$: for $\theta < 1$ the atomic system reaches a stable saturated steady state monotonically; at the threshold $\theta = 1$ the atomic 'observables' $\langle S_+ \rangle$ and $\langle S_z \rangle$ have a discontinuous derivative with respect to θ (a type of critical phase transition); for $\theta > 1$ the system is unstable and has cyclic trajectories of Lotka-Volterra type on the Bloch sphere, whilst, because there is no damping, these oscillations continue indefinitely. Similar results were found by S. Ja. Kilin[26] whilst Senitzky[19b] also discussed the limiting cases of weak and strong fields. The same set of equations (3), including the detuning δ_o, was actually assumed semiclassically and solved by Stroud and Jaynes[2] who predicted the same behavior, but their results were intended to describe the behavior of a single two-level atom within the neoclassical theory[6], and no approximation to an operator theory was intended. Irrespective of the merits of the neoclassical theory as a physical theory their mathematical results apply directly to this many-atom case!

It was shown also in Ref. 20 that consideration of cyclic path diffusions induced by spontaneous decay will bring the system to a non-oscillating steady state very similar to the result found from the exact quantum mechanical solution[23] discussed below. The spectrum in this semiclassical factorization[20] shows additional sharp peak (δ-function) sidebands of small weight situated at all multiples of $\pm 2\Omega$ in addition to the usual Stark spectrum. These sidebands arise (cf. Senitzky (1978), Ref. 19 who examined the case $\Omega \gg \gamma_o$) because the semiclassical cyclic polarization solution contains non-sinusoidal oscillations; the spectrum must be averaged over all cycles and is not oscillatory, but the correlation is, and the Fourier transform shows sidebands at all harmonics.

If the driving field is detuned ($\delta_o \neq 0$) no threshold behavior occurs at $\theta = 1$ because of dispersive non-resonance effects[25]. The atomic observables take stable steady state values through the whole range $0 < \theta < \infty$. In fact in Ref. 25 we found that the exact semiclassical steady state solutions for $\delta_o \neq 0$ were increasingly close to the exact quantum results for increasing N as described below. The exact time-dependent semiclassical solutions have been obtained by a Fokker-Planck equation analysis and will be reported elsewhere (cf. Hassan et al.[27]).

Note that within the semiclassical decorrelation Eq. (3) shows that the *factorized* angular momentum $\langle \underset{\sim}{S} \rangle^2$ is a constant of the motion: $\langle \underset{\sim}{S} \rangle^2 = (N/2)^2$.

C. Semiclassical Decorrelation and Self-correlations

If, in addition to the semiclassical decorrelation, the self-correlations of the atoms are retained[16,17], for example

$$<S_+ S_-> = \sum_{i=1}^{N} <S_+^{(i)} S_-^{(i)}> + \sum_{j \neq i}^{N} <S_+^{(i)} S_-^{(j)}>$$

$$\to <S_z + \frac{1}{2} N> + (1 - N^{-1}) <S_+><S_-> , \tag{4}$$

etc., in an obvious notation, then this plainly allows individual atomic decay to take place through spontaneous emission. The atomic observables governed by (3) now exhibit bistable behavior of the usual sort[28-31] in the steady state. In particular the relation between the input field Ω and an internal field defined as the input field minus the induced polarization takes the usual cubic form. This internal field for point systems is the natural one to choose for a comparison with the mean field theory of extended systems[28]. Both the emission[16] and absorption spectra[17] for large values of the driving field (corresponding to the upper branch on the bistability curve) have the same structure as in the resonance fluorescence from a single atom[1,4,5,32]. These results can be expected because single-atom self-correlations are retained in the theory whilst single atom contributions could be expected to be additive because the second order normalized intensity-intensity correlation function $g^{(2)}(0)$ ~ 2, and the scattered field is incoherent[17]. Note that decorrelations like (4) mean that $<\underset{\sim}{S}>^2$ is not now a constant of the motion and $<\underset{\sim}{S}^2>$, now necessarily interpreted via (4), is not a constant either.

Bowden and Sung[18] have also treated cooperative bistable behavior in a small volume (that is the Dicke model) as a first order phase transition. They used thermodynamic Green's functions introducing an effective spin temperature. At lower temperatures the state equation at resonance is similar to those in Refs. 16,17 and to that obtained for a macroscopic ring cavity in the spatially averaged "mean field" theory[27,29]. The model of Ref. 18 nevertheless differs significantly from that of Refs. 16, 17: a decoupling scheme was adopted equivalent to ignoring correlation between different atoms very much as in (4) above; the phenomenological damping was included so that the photon escape rate was much faster than the spontaneous decay, i.e., the cavity line width was much greater than $\gamma_0 = \frac{1}{2} \Gamma_0 \equiv T_2^{-1}$. Thus, the effective Hamiltonians in the two cases are not directly comparable.

Note the important point that the work of Refs. 16–18 shows that bistable behavior is possible in a small volume without appeal to cavity boundary conditions and mirrored surfaces (even though

these could be incorporated in Ref. 18 through the cavity width).
Optical bistability has been observed in a Fabry-Perot interferometer[30], but also in InSb without mirrored surfaces[33].

D.   Exact Quantum Mechanical Results

   The different approximations made in §II.B and §II.C to solve
the driven Dicke model were introduced before any exact quantum
mechanical solution was available. Since then an exact steady state
solution for resonance ($\delta_0 = 0$) of the master equation (1) has been
obtained by Puri and Lawande[23]. It is valid for any N and any value
of the driving field Ω. Their solution was

$$(\hat{\rho})_{ss} = \frac{1}{N_F} \sum_{m,n=0}^{N} (g^*)^{N-m} g^{N-n} (S_-)^m (S_+)^n \tag{5}$$

with

$$N_F \equiv \sum_{p=0}^{N} \sum_{k=0}^{p} \frac{(N-p+k)!p!}{(p-k)!(N-p)!} |g|^{2(N-k)} \tag{6}$$

and $g \equiv i\Omega\gamma_0^{-1}$ is a normalized Rabi frequency. The steady state
solutions for the atomic inversion and polarization for any finite
N depend smoothly on $|g|$. But it is interesting and relevant to a
comparison with the previous approximated calculations to consider
also a formal thermodynamic limit in which both N and $|g| \to \infty$ so that
$\theta \equiv 2|g|N^{-1}$ remains finite. In this case there is a bifurcation
point at $\theta = 1$ (at exact resonance) and we shall scrutinize this interesting feature carefully below. At $\theta = 1$ the quantum fluctuations
were found to be finite[23], and previous numerical work[21] and asymptotic arguments for large N[22] gave similar results.

   The exact solution (5) can be used[24] for the calculation of the
atomic correlation functions. One of our results is[24] that for a
weak driving field ($|g| \ll 1$)

$$g^{(n)}(0) = 1, \qquad\qquad n = 1,2,\ldots \quad .$$

This characterizes a coherent state field. For N, $|g| \to \infty$, $G^{(1)}(0)$
$= \frac{1}{6} N (N+2)$ (as speculated by Amin and Cordes[15]) and $g^{(2)}(0) = 1.2$.
This last result means that the cooperative interactions reduce the
intensity fluctuations compared with those in the approximate solution of Ref. 17 described in §II.C (here the possibility of independent atom decay caused $g^{(2)}(0) \sim 2$). The value $g^{(2)}(0) = 1.2$ is
close to the numerical value $g^{(2)}(0) = 1.18$ found by Walls et al.[22]
and differs from that of Agarwal et al.[34] who found $g^{(2)}(0) \sim 1.65$
for $\Omega \gg \gamma_0$. We find this value incorrect, however[24]. The correlation functions, like the atomic observables, have discontinuous
derivatives at $\theta = 1$.

The exact solution (5) has been extended by Puri, Lawande and Hassan[25] to the case of finite detuning. The solution is then

$$(\hat{\rho}_A)_{ss} = D^{-1} \sum_{m,n=0}^{N} (g^*)^{-m}(g)^{-n} \; a_{mn} \; (S_-)^m (S_+)^n \tag{7}$$

where

$$a_{mn} = \frac{\Gamma(m-\Delta+1)\Gamma(n+\Delta+1)}{m!\,n!\,\Gamma(1+\Delta)\Gamma(1-\Delta)} \; , \tag{8}$$

$$D = \frac{\sinh(\pi|\Delta|)}{\pi|\Delta|} \sum_{m=0}^{N} \frac{\Gamma(m-\Delta+1)\Gamma(m+\Delta+1)(N+m+1)!}{(N-m)!\,(2m+1)!} \; |g|^{-2m} \; , \tag{9}$$

and $\Delta = i\,\delta_0 \gamma_0^{-1}$ ($\Gamma(x)$ is the usual Γ-function). The solution (7) reduced to (5) for $\delta_0 = i^{-1}\gamma_0\Delta = 0$. From the exact result (7) exact expressions are obtained for the atomic operator expectation values (the 'observables'), the atomic correlation functions and the fluctuations: these fluctuations reduce when $\Delta \neq 0$ and numerical plots indicate that the sharp bifurcation at $\theta = 1$ occurs only at resonance, $\Delta = 0$.

In Ref. 25 a comparison was made between the exact results and the semiclassical solutions of §II.B. For fixed large N (~ 500) and increasing Δ, agreement between the exact quantum mechanical results and the semiclassical ones becomes very close indeed, and for $\Delta \neq 0$, however small, the semiclassical results may be considered to be the asymptotic limit of the quantum results as $N \to \infty$, a satisfying conclusion. When $\Delta \neq 0$ the semiclassical solutions gain a damped contribution from dispersion so there is a steady state solution, particularly above threshold where $\theta > 1$. This compares with the undamped Lotka-Volterra cycles occurring in the semiclassical solutions as described in §II.B at exact resonance, $\Delta = 0$.

For large N ($\gtrsim 100$) and increasing Δ, the exact $g^{(2)}(0)$ falls below 1.2[25]: dispersion, that is detuning, thus reduces the intensity fluctuations as one might expect. The atomic observables and their correlations have continuous derivatives with respect to θ for the whole range $0 < \theta < \infty$ for any value of $\Delta \neq 0$.

E.   An Analogy With a Magnetic Phase Transition

A fruitful analogy has been drawn between the lasing transition and critical phenomena in a ferromagnet[35] and is of particular interest because the laser is pumped and so is in a far-from-equilibrium state. The results obtained for optical bistability in the approximation of §II.C can be interpreted as a ferromagnetic type phase transition taken in mean field approximation[29]. The essential point for present purposes is that phase transitions can display certain universal features, notably the critical exponents.

Whilst there is no certain guide to the proper identification of analogous thermodynamic variables, we have at least been able to make one possible such identification for the exact quantum mechanical solution described in §II.D. We have noted[24,25] that the atomic inversion $N^{-1}\langle S_z\rangle \equiv r_3$ has *two* solutions

$$r_3 = \mp \frac{1}{2}\sqrt{\xi} \tag{10}$$

where

$$\xi = \frac{1}{2}\left[+\sqrt{\left(\frac{1}{4}\phi^2 + \theta^2 - 1\right)^2 + \phi^2} - \left(\frac{1}{4}\phi^2 + \theta^2 - 1\right)\right] \tag{11}$$

with $\phi \equiv 2N^{-1}|\Delta|$ the scaled detuning parameter. At resonance $\phi = 0$,

$$\begin{aligned} r_3 &= \mp \frac{1}{2}\sqrt{1 - \theta^2} & \theta &< 1, \\ &= 0 & \theta &\geq 1, \end{aligned} \tag{12}$$

a result reached in the exact quantum theory in the thermodynamic limit $|g|$, $N \rightarrow \infty$ [24] and after cyclic averaging in the semiclassical theory[20] (compare also Refs. 21–23). Evidently r_3 has two branches for $\theta < 1$ and a discontinuous derivative with respect to θ^2 at $\theta = 1$. This suggests that r_3 is the natural order parameter of a phase transition theory analogous to the magnetization M in the ferromagnet and that θ^2 plays the role of the reduced temperature (T/T_c). There is a simple bifurcation point at $\theta^2 = 1$, i.e., where $T = T_c$; for $\theta^2 < 1$ the system is ordered (and cooperative) for $\theta^2 > 1$ the system is disordered (and chaotic). Thus $r_3 \propto (1 - \theta^2)^\beta$ where the 'critical exponent' β has the value $\beta = \frac{1}{2}$. This is the exact universal result for a thermodynamic phase transition in $d = 4$ dimensions and the result in mean field approximation for $d = 3$, although there is no reason to suppose either condition applies here. Notice the rather special feature of (12) for $\theta^2 < 1$: $r_3 \rightarrow \mp \frac{1}{2}$ as $\theta^2 \rightarrow 0$ so the branch tending to $r_3 = +\frac{1}{2}$ must be unstable through spontaneous emission, a complication which does not arise in the ferromagnetic case. There is here a point we have not properly understood since the semiclassical approximation §II.B eliminates spontaneous emission and the exact quantum solution §II.D whilst necessarily containing it does not explicitly display it in the argument. Moreover the exact quantum argument certainly preserves the constant of the motion $\langle \underset{\sim}{S}^2\rangle$ which spontaneous emission processes tend to break.

At resonance in the thermodynamic limit r_3 and θ^2 exhaust the available parameters. But off resonance the detuning adds the extra parameter $\phi^2 \equiv N^{-2}(2|\Delta|)^2$ which we choose to identify as the analogue of the magnetic field H: $\phi^2 \cdot \operatorname{sgn} r_3$ is thus the symmetry breaking parameter for the driven Dicke model, and is the variable conjugate to the order parameter; note however that an effective

Hamiltonian might be said to be proportional to $\Delta \cdot r_3$, and this is proportional to ϕ but not to ϕ^2.

In the laser transition the analogies with the ferromagnet are[33]: $M = E$ (laser field), $T = \sigma$ (population inversion) and $H = S$ (injected signal) and the critical exponents are the mean field exponents[33] $\beta = \frac{1}{2}$, $\gamma = 1$, $\alpha = \frac{1}{2}$. The different analogy we are obliged to draw for the exact quantum theory of the driven Dicke model is $M = r_3$ (population inversion), $T = \theta^2 = (2\Omega\gamma_0^{-1}N^{-1})^2$ and $H = (2|\Delta|.N^{-1})^2$. The susceptibility $\chi_\theta \equiv (\partial r_3/\partial\phi^2)_\theta \sim (1-\theta^2)^{-\gamma}$ and the specific heat $C_\phi \equiv (\partial r_3/\partial\theta^2) \sim (1-\theta^2)^{-\alpha}$ defining the critical exponents α, γ. One easily finds that, as $\phi \to 0$,

$$r_3 \sim (1 - \theta^2)^{\frac{1}{2}} \quad , \quad C_\phi \sim (1 - \theta^2)^{-\frac{1}{2}}$$
$$\chi_\theta \sim \theta^2(1 - \theta^2)^{-3/2}, \quad \theta < 1 \tag{13}$$

so $\beta = 1/2$, $\alpha = 1/2$, $\gamma = 3/2$ and $\alpha + 2\beta + \gamma = 3 > 2$. These results are not critical exponent mean field results therefore, and this is consistent with the exact quantum character of the solution of the driven Dicke model.

The corresponding Gibbs free energy $G(r_3, \theta^2, \phi^2)$ appears to be

$$G = G_0 + r_3^2 \mp \sqrt{\xi}\, r_3 \tag{14}$$

so that the minimum condition $\partial G/\partial r_3 = 0$ for each branch is the state equation (10). The free energy (14) is not of Landau's fourth power type in the order parameter. At resonance $\phi^2(\equiv H) = 0$, $\sqrt{\xi} = \sqrt{1 - \theta^2}$, $\theta^2 < 1$; $= 0$, $\theta^2 > 1$ so that

$$G - G_0 = r_3^2 \mp \sqrt{1 - \theta^2}\, r_3 \, , \qquad 0 < \theta^2 < 1;$$
$$= r_3^2 \qquad , \qquad \theta^2 > 1. \tag{15}$$

The free energy is least for the negative sign for $r_3 > 0$ and for the positive sign for $r_3 < 0$. The curve minimizing G for each r_3 is therefore not differentiable at $r_3 = 0$ where there is a cusp. The absolute minima occur symmetrically at the roots of the state equation (10) as noted. The symmetry in r_3 from this point of view is puzzling because there is no sign of the instability from spontaneous emission expected for the branch of the state equation (10) which has $r_3 > 0$. It is possible that a more fruitful analogy with a second order phase transition is still to be drawn (the choice we have made does at least emphasize that the exact solution of the driven Dicke model is not a critical mean field type result). In our view the most satisfying result of the analogy we have drawn is the way each 'thermodynamic parameter' approaches the semiclassical values of §II.B as $N \to \infty$ in the thermodynamic limit (our

numerical results which show this result very clearly must be presented elsewhere).

III. MACROSCOPIC EXTENSION OF THE DRIVEN DICKE MODEL

A.   Coarse-grained Operator Bloch Equations

In order to describe an extended system we shall place N 2-level
atoms labelled by i at distinct sites $\underline{x}_i$ (subsequently we average
the positions $\underline{x}_i$ over a macroscopic region V). The usual second
quantized Hamiltonian in dipole approximation is

$$H = \sum_{\underline{k},\lambda} \hbar\omega_k \, a^{\dagger}_{\underline{k},\lambda}(t) a_{\underline{k},\lambda}(t) + \frac{1}{2} \hbar\omega_o \int \sigma_z(\underline{x},t)\,d\underline{x}$$

$$- \int \underline{p}(\underline{x},t)\cdot\underline{e}(\underline{x},t)\,d\underline{x}, \qquad (16)$$

in which the operator densities are

$$\sigma_z(\underline{x},t) = \sum_{i=1}^{N} \sigma_z^{(i)}(t)\delta(\underline{x}-\underline{x}_i) \ ,$$

$$\underline{p}(\underline{x},t) = \underline{p}\sigma_x(\underline{x},t) = \underline{p} \sum_{i=1}^{N} \sigma_x^{(i)}(t)\delta(\underline{x}-\underline{x}_i),$$

etc.

The commutation relations for the Pauli spin operators $\sigma^{(i)} = (\sigma_x^{(i)}, \sigma_y^{(i)}, \sigma_z^{(i)})$ for each atom i mean that

$$[\sigma_x(\underline{x},t), \sigma_y(\underline{x}',t)] = 2i\sigma_z(\underline{x},t)\delta(\underline{x}-\underline{x}') \ , \qquad (17)$$

etc.

The field operator $\underline{e}(\underline{x},t)$ is

$$\underline{e}(\underline{x},t) = i \sum_{\underline{k},\lambda} \underline{C}(\underline{k},\lambda) \left\{ a_{\underline{k},\lambda}(t)e^{i\underline{k}\cdot\underline{x}} - a^{\dagger}_{\underline{k},\lambda}(t)e^{-i\underline{k}\cdot\underline{x}} \right\} \qquad (18)$$

and

$$\underline{C}(k,\lambda) = \sqrt{\frac{2\pi\hbar\omega_k}{V_o}} \ \hat{\varepsilon}_{\underline{k},\lambda} \ .$$

The index $\lambda = (1,2)$ labels two distinct unit polarization vectors
$\hat{\varepsilon}_{\underline{k},\lambda}$ for each $\underline{k}$: $\underline{k}\cdot\hat{\varepsilon}_{\underline{k},\lambda} = 0$. The volume V_o is the quantization
volume ($V_o \to \infty$ and contains V). It is a significant feature of the
extended system with Hamiltonian (16) that the square $\underline{S}^2$ of the

total angular momentum

$$\underline{S} = \left(\sum_{i=1}^{N} \sigma_x^{(i)}, \sum_{i=1}^{N} \sigma_y^{(i)}, \sum_{i=1}^{N} \sigma_z^{(i)} \right)$$

is not now a constant of the motion because of the phase factors $e^{\pm i\underline{k}\cdot\underline{x}_j}$ in the interaction term of the Hamiltonian.

From (18) the total field operator $\underline{e}(\underline{x},t)$ is transverse, div $\underline{e} = 0$, whilst the Hamiltonian (16) essentially means that it satisfies Maxwell's wave equation, as an operator equation, driven by $4\pi c^{-2} (\partial^2 \underline{p}/\partial t^2)^{6,7,36,37}$. Thus its integral form is

$$\underline{e}(\underline{x},t) = \underline{e}_o(\underline{x},t) + \int_o^t dt' \int_V d\underline{x}' \ \underline{\underline{F}}(\underline{x},\underline{x}';t-t')\cdot\underline{p}(\underline{x}',t') \tag{19}$$

in which $\underline{e}_o$ is the free field operator and takes the form of (18) with however $a_{\underline{k},\lambda}(t)$, $a_{\underline{k},\lambda}^\dagger(t)$ replaced by $a_{\underline{k},\lambda}(0)e^{-i\omega_k t}$, $a_{\underline{k},\lambda}^\dagger(0)e^{i\omega_k t}$ respectively.

We assume that the photon propagator $\underline{\underline{F}}$, a second rank tensor, is causal, i.e., retarded, whilst it is certainly transverse, div $\underline{\underline{F}} = 0$. Evaluated at two distinct atomic coordinates it is

$$\underline{\underline{F}}(\underline{x}_i,\underline{x}_j;t-t') = (\nabla\nabla - \nabla^2\underline{\underline{U}})\delta(t-t'-r_{ij}c^{-1})r_{ij}^{-1} \tag{20}$$

and $r_{ij} \equiv |\underline{x}_i - \underline{x}_j| \neq 0$ since $i \neq j$; $\underline{\underline{U}}$ is the unit tensor and the derivatives can be taken either both at $\underline{x}_i$ or both at $\underline{x}_j$. The time Fourier transform of $\underline{\underline{F}}$ is

$$\underline{\underline{F}}(\underline{x}_i,\underline{x}_j;\omega) = (\nabla\nabla - \nabla^2\underline{\underline{U}})(e^{i\omega c^{-1}r_{ij}} r_{ij}^{-1}) \ . \tag{21}$$

The field operator $\underline{e}(\underline{x},t)$ drives each distinct atom i at its particular site $\underline{x} = \underline{x}_i$. Since the source operator $\underline{p}(\underline{x}',t')$ in (19) contains the dipole operator density at each atomic site we must also consider the case $r_{ij} = 0$ for $i = j$. This can be done (with however some problems![7,8,36]) by expressing the total field operator (19) at $\underline{x} = \underline{x}_i$ as the sum of three parts:

$$\underline{e}(\underline{x}_i,t) = \underline{e}_o(\underline{x}_i,t) + \underline{e}_{self}(\underline{x}_i,t) + \underline{e}_{int.}(\underline{x}_i,t) \ . \tag{22}$$

The field operator $\underline{e}_{self}(\underline{x}_i,t)$ describes the interaction of atom i on itself and can be used successfully to obtain the usual descriptions of spontaneous emission and radiative level shifts in low energy q.e.d.[7,8,36]. The field $\underline{e}_{int}(\underline{x}_i,t)$ is the interatomic field due to all the atoms $j \neq i$, excluding i, acting on the atom i at $\underline{x}_i$. This field operator carries the propagating field through the

macroscopic system and of course describes the cooperative interaction between the atoms; apparently both $\underline{e}_{self}$, describing spontaneous emission, and $\underline{e}_{int}$ describing the cooperation of the atoms
are essential for a theory of optical bistability. The contribu-
tions of neither $\underline{e}_{int}$ nor $\underline{e}_{self} + \underline{e}_o$ to the interaction term in the
Hamiltonian (16) commutes with $\underline{S}^2$, thus both terms contribute to
the result noted by (17) that $\underline{S}^2$ is not a constant of the motion.

From (19) we have

$$\underline{e}_{int}(\underline{x}_i,t) = \int_o^t dt' \sum_{\substack{j=1 \\ j \neq i}}^N \underline{\underline{F}}(\underline{x}_i,\underline{x}_j;t-t') \cdot \underline{p}^{(j)}(t') \tag{23}$$

where
$$\underline{p}^{(j)}(t) = p \, \underline{\hat{u}} \, \sigma_x^{(j)}(t)$$
$$= p \, \underline{\hat{u}} \left[\sigma_+^{(j)}(t) + \sigma_-^{(j)}(t) \right] \tag{24}$$

in terms of raising and lowering operators $\sigma_\pm^{(j)}(t)$ for each atom;
p is the dipole matrix element for the 2-level atoms as before and
$\underline{\hat{u}}$ is the direction of the vector matrix element. In a vapour, for
example, $\underline{\hat{u}}$ takes on all orientations but we shall be a bit cavalier
with this aspect of the argument.

The simplest treatment of self-interactions via the self-field
$\underline{e}_{self}$ splits this and the free field operator $\underline{e}_o(\underline{x}_i,t)$ into positive and negative frequency part fields[7,8,36]. It is therefore
convenient (but certainly not essential[36]) to split $\underline{e}_{int}$:

$$\underline{e}_{int}^\pm(\underline{x},t) = \int_o^t dt' \sum_{j=1}^N p \, \underline{\hat{u}} \cdot \underline{\underline{F}}(\underline{x}_i,\underline{x}_j;t-t') \sigma_+^{(j)}(t') \; . \tag{25}$$

The driving field in the problem will be a coherent state field of
frequency ω_k close to ω_o. Thus we can expect to write the operators $\sigma_\pm^{(j)}(t)$ in the form

$$\sigma_\pm^{(j)}(t) = R_\pm^{(j)}(t)e^{\pm i\omega_k t} , \tag{26}$$

where the operators $R_\pm^{(j)}(t)$ vary slowly on the time scale ω_k^{-1}. We
can then perform the time integration in (25) to get

$$\underline{e}_{int}^\pm(\underline{x}_i,t) = \sum_{\substack{j=1 \\ j \neq i}}^N \left[(\nabla\nabla - \nabla^2 \underline{\underline{U}}) \cdot p\hat{u} (R_\mp^{(j)}(t-r_{ij}c^{-1})e^{\mp i\omega_k(t-r_{ij}c^{-1})} r_{ij}^{-1}) \right] . \tag{27}$$

For a sample as big as 1 cm. $r_{ij}c^{-1} < 10^{-10}$ sec. and we suppose the $R_{\pm}^{(j)}(t)$ change negligibly in this time. Thus we take

$$R_{\pm}^{(j)}(t - r_{ij}c^{-1}) = R_{\pm}^{(j)}(t) \quad . \tag{28}$$

Contributions to the fields (27) do not arise from atoms j so far from atom i that $r_{ij} > ct$. We shall take no account of this in the theory so our result would be in doubt for time $\sim 10^{-10}$ sec. from the time the driving field is switched on.

Because of (28) we can extract the whole atomic operators $\sigma_{\pm}^{(j)}(t)$ given by (26) from (27) so that

$$e_{-int}^{\pm}(\underline{x}_i,t) = \sum_{\substack{j=1 \\ j \neq i}}^{N} p\,\hat{u}\,\sigma_{\mp}^{(j)}(t) \cdot \begin{cases} \underline{\underline{F}}(\underline{x}_i,\underline{x}_j;\omega_k) \\ \underline{\underline{F}}^*(\underline{x}_i,\underline{x}_j;\omega_k) \end{cases} ; \tag{29a}$$

$\underline{\underline{F}}$ and $\underline{\underline{F}}^*$ are defined by (21) and the factor $p\,\hat{u}$ is conveniently moved to the left by appeal to the symmetry of $\underline{\underline{F}}$. Later we need to integrate smoothly through $r_{ij} = |\underline{x}_i - \underline{x}_j| = 0$. This can be done by operating with the ∇^2 in $\underline{\underline{F}}$ and removing the δ-function at $r_{ij} = 0$ as a transverse 'contact' term[37]. At the same time one can conveniently add the Coulomb interactions not originally in the Hamiltonian (16). The result is[37]

$$\underline{\underline{F}}(\underline{x},\underline{x}';\omega_k) = (\nabla\nabla + \omega_k^2 c^{-2}\underline{\underline{U}})(e^{i\omega_k c^{-1} r} r^{-1}) \tag{29b}$$

(for $r \equiv |\underline{x}-\underline{x}'|$) and is not transverse.

The fields $e_{-int}^{\pm}(\underline{x}_i,t)$ depend only on the atomic operators $\sigma_{\mp}^{(j)}(t)$ for $j \neq i$ and so commute at all equal times with all the atomic operators of atom i, and this is why the ordering of e_{-int}^{+} and e_{-int}^{-} is immaterial (compare Ref. 36). They do not commute with all the atomic operators of any other atom, and this is why the contribution to H of e_{-int} does not commute with S^2. At arbitrary points $\underline{x}_i = \underline{x}$ the interatomic fields $e_{-int}^{\pm}(\underline{x},t)$ fail to commute with the atomic operators of atom j at $\underline{x} = \underline{x}_j$ only through self-interactions but the value of $e_{-int}^{\pm}(\underline{x},t)$ is only of interest at $\underline{x} = \underline{x}_i$.

With the Hamiltonian (16) Heisenberg's equations of motion imply the operator Bloch equations

$$\dot{\underline{\sigma}}(\underline{x},t) = \underline{\omega}(\underline{x},t) \underset{\wedge}{\,}\underline{\sigma}(\underline{x},t) \quad , \tag{30}$$

where

$$\underline{\sigma}(\underline{x},t) = \sum_{i=1}^{N} \underline{\sigma}^{(i)}(t)\delta(\underline{x}-\underline{x}_i) \ ,$$

$$\underline{\omega}(\underline{x},t) \equiv (-2\,p\,\hat{\underline{u}} \cdot \underline{e}(\underline{x},t),0,\omega_o)$$

and 'dot' means $\partial/\partial t$. In detail Eqs. (30) are

$$\dot{\sigma}_x(\underline{x},t) = -\omega_o \sigma_y(\underline{x},t)$$

$$\dot{\sigma}_y(\underline{x},t) = \omega_o \sigma_x(\underline{x},t) + 2\,p\,\hbar^{-1}\,\hat{\underline{u}} \cdot \underline{e}(\underline{x},t)\sigma_z(\underline{x},t)$$

$$\dot{\sigma}_z(\underline{x},t) = -2\,p\,\hbar^{-1}\,\hat{\underline{u}} \cdot \underline{e}(\underline{x},t)\sigma_y(\underline{x},t) \ . \tag{31}$$

We introduce raising and lowering operator densities

$$\sigma_\pm(\underline{x},t) = \tfrac{1}{2}\,(\sigma_x \pm i\sigma_y) = \sum_{i=1}^{N} \sigma_\pm^{(i)}(t)\delta(\underline{x}-\underline{x}_i)$$

and normally order the positive and negative frequency parts of the fields[6],[7],[36]. Then (31) becomes

$$\dot{\sigma}_+(\underline{x},t) - i\omega_o\sigma_+(\underline{x},t) = i\,p\,\hbar^{-1}\,\hat{\underline{u}} \cdot \underline{e}^-(\underline{x},t)\sigma_z(\underline{x},t)$$

$$\dot{\sigma}_-(\underline{x},t) + i\omega_o\sigma_+(\underline{x},t) = -i\,p\,\hbar^{-1}\,\hat{\underline{u}}\cdot\sigma_z(\underline{x},t)\underline{e}^+(\underline{x},t)$$

$$\dot{\sigma}_z(\underline{x},t) = -2\,p\,\hbar^{-1}\,\hat{\underline{u}} \cdot \left\{\underline{e}^-(\underline{x},t)\sigma_y(\underline{x},t) \right.$$

$$\left. + \sigma_y(\underline{x},t)\underline{e}^+(\underline{x},t)\right\} \tag{32}$$

to an approximation of rotating wave type in which we discard terms in $\sigma_z e^+$ and $e^-\sigma_z$ in the equations for $\sigma_\pm$ respectively.

The positive and negative frequency parts of the fields are

$$\underline{e}^\pm(\underline{x},t) = \underline{e}_o^\pm(\underline{x},t) + \underline{e}_{self}^\pm(\underline{x},t) + \underline{e}_{int}^\pm(\underline{x},t) \ . \tag{33}$$

In the single atom coupled to vacuum problem the normal ordering prescription was introduced as a convenience to eliminate the free field operator from all (or almost all) relevant expectation values[7],[8]. The position is similar here but the many atom system is coupled to a driving field which will be taken to be a coherent field state in the single mode $(\underline{k},\lambda)$. This means that in relevant expectation values we need consider only the contributions to the free field operators

$$\underline{e}_o^\pm(\underline{x},t) = \pm i\,\underline{C}(\underline{k},\lambda) \begin{cases} a_{\underline{k},\lambda}(0)e^{-i(\omega_k t - \underline{k}\cdot\underline{x})} \\ \\ a_{\underline{k},\lambda}^\dagger(0(e^{+i(\omega_k t - \underline{k}\cdot\underline{x})} \end{cases} \ . \tag{34}$$

For the slowly varying envelope operators constructed like (26)

$$\underline{e}^{\pm}_{self}(\underline{x}_i,t) = \pm\, i\, \frac{2}{3}\left(\frac{\omega_o}{c}\right)^3 \underline{p}\, \sigma^{(i)}_{\mp}(t)$$

$$\pm\, \frac{2}{3\pi}\left(\frac{\omega_o}{c}\right)^3 \log\left(\frac{\omega_c}{\omega_o}\right)\left[\sigma^{(i)}_{+}(t) - \sigma^{(i)}_{-}(t)\right] \tag{35}$$

in which ω_c is the Compton frequency[7,8,36]; the fields $\underline{e}^{\pm}_{int}(\underline{x}_i,t)$ are given by Eq. (29).

The equation for $\sigma_{+}(\underline{x},t)$ in (32) becomes

$$\dot{\sigma}_{+}(\underline{x},t) - i\omega_o\sigma_{+}(\underline{x},t) = i\,p\,\hbar^{-1}\,\hat{\underline{u}}\cdot[\underline{e}^{-}_{self}(\underline{x},t)\sigma_z(\underline{x},t)$$

$$+ \underline{e}^{-}_{int}(\underline{x},t)\sigma_z(\underline{x},t) + \underline{e}^{-}_{o}(\underline{x},t)\sigma_z(\underline{x},t)] \tag{36}$$

and the properties of the atomic operators $\sigma^{(i)}_{\pm}(t)$, $\sigma^{(i)}_{z}(t)$ mean that

$$i\,p\,\hbar^{-1}\,\hat{\underline{u}}\cdot\underline{e}^{-}_{self}(\underline{x},t)\sigma_z(\underline{x},t) = (-\tfrac{1}{2}\Gamma_o - i\Delta_o)\sigma_{+}(\underline{x},t) \tag{37}$$

where $\Delta_o = \pi^{-1}\Gamma_o\ln(\omega_c\omega_o^{-1})$ is the vacuum radiative level shift for each atom: $\Gamma_o = (4/3)p^2\hbar^{-1}\omega_o^3c^{-3}$, the A-coefficient. From (34)

$$i\,p\,\hbar^{-1}\,\hat{\underline{u}}\cdot\underline{e}^{-}_{o}(\underline{x},t)\sigma_z(\underline{x},t) = p\hbar^{-1}\hat{\underline{u}}\cdot\underline{C}(\underline{k},\lambda)a^{+}_{\underline{k},\lambda}(0)\sigma_z(\underline{x},t)e^{i(\omega_k t-\underline{k}\cdot\underline{x})} \tag{38}$$

As a matter of convenience we introduce the reduced operators

$$\bar{\underline{\sigma}}(\underline{x},t) = \langle\alpha_{\underline{k},\lambda}|\underline{\sigma}(\underline{x},t)|\alpha_{\underline{k},\lambda}\rangle$$

where $|\alpha_{\underline{k},\lambda}\rangle$ is the single mode coherent state for the mode $(\underline{k},\lambda)$ driving the system. Then (38) gives

$$\langle\alpha_{\underline{k},\lambda}|i\,p\,\hbar^{-1}\hat{\underline{u}}\cdot\underline{e}^{-}_{o}(\underline{x},t)\sigma_z(\underline{x},t)|\alpha_{\underline{k},\lambda}\rangle = E^*\sigma_z(\underline{x},t)e^{i(\omega_k t-\underline{k}\cdot\underline{x})} \tag{39}$$

where

$$E \equiv p\hbar^{-1}\hat{\underline{u}}\cdot\underline{\varepsilon}_{\underline{k},\lambda}(2\pi\hbar\omega_k V_o^{-1})^{1/2}\alpha_{\underline{k},\lambda} \equiv i\Omega \tag{40}$$

is the laser Rabi frequency.

Notice that the use of these reduced operators in no way approximates the operator Bloch Eqs. (32) since the only bilinear terms are the products of field and matter operators. But the situation is apparently different in the correlation functions: for example, for the spectrum we need collective matter operator products at two different times and this involves the commutation relations of the matter operators and the *free* field operators at two different times (cf. Mollow (1975) in Ref. 1). We have already solved[38] the full quantized problem for a single two-level atom interacting with a field in a combination of coherent and chaotic field states by using reaction field theory[7,38], that is, the theory of the self-field described in this paper, and evaluating the free field matter commutators at different times. The analysis[38] shows that reduced operators can be used *exactly* in the single atom case. But in the present problem the situation is complicated by the presence of $\underline{e}_{int}$. The effect of this is still to be determined. Thus, the use of reduced operators does not approximate the Bloch equations (32) but the possibility of their more general application is still open. Henceforth we drop the bar on the reduced operators and σ is a reduced operator throughout the rest of this paper. The coherent state field in state $|\alpha_{k,\lambda}\rangle$ is thus equivalent to the classical field E defined by (39) and (40) in the analysis based on the Bloch equations.

Using (37) and (39) in (36) we get

$$\sigma_+(\underline{x},t) - i\omega_o'\sigma(\underline{x},t) + \frac{1}{2}\Gamma_o\sigma_+(\underline{x},t)$$
$$= E^*\sigma_z(\underline{x},t)e^{i(\omega_k t - \underline{k}\cdot\underline{x})}$$
$$+ i p \hbar^{-1}\,\underline{\hat{u}}\cdot\underline{e}_{int}^-(\underline{x},t)\sigma_z(\underline{x},t) \tag{41}$$

where $\omega_o' \equiv \omega_o - \Delta_o$ includes the radiative level shift and replaces ω_o subsequently. The interpretation of $\underline{e}_{int}^-(\underline{x},t)$ in (41) is that as $\underline{x}$ takes on the values $\underline{x}_i$ in $\sigma_z(\underline{x},t)$, $\underline{e}_{int}^-(\underline{x},t)$ takes on the values $\underline{e}_{int}^-(\underline{x}_i,t)$ as defined by (29).

There are no specific atomic sites in a vapour and we know only that each $\underline{x}_i$ always lies somewhere inside V. We therefore ensemble average the operators over the sites $\underline{x}_i$ and at the same time move into the rotating frame of the driving field by setting

$$S_\pm(\underline{x},t) = n^{-1}\langle\sigma_\pm(\underline{x},t)\rangle_{Avr.}\,e^{\mp i\omega_k t}$$

$$2S_z(\underline{x},t) = n^{-1}\langle\sigma_z(\underline{x},t)\rangle_{Avr.} \tag{42}$$

where

$$n \equiv \langle\sum_{i=1}^{N}\delta(\underline{x}-\underline{x}_i)\rangle_{Avr.}$$

The factor 2 on S_z is introduced for a direct comparison with the driven Dicke model of §II. For the same reason we reintroduce $\gamma_o \equiv \frac{1}{2}\Gamma_o$ and $\Omega \equiv -i\mathcal{E}$. We shall arbitrarily assume that the commutation relations of the ensemble averaged operators are the same as those of the microscopic operators (17), but it is by no means clear that this should be so. We make no appeal to this assumption for manipulation of the Bloch equations; but for the spectrum calculated in §III.B and for the master equation derived in §III.C the assumption is necessarily implicit.

From (42) we readily arrive at the following Bloch equations in the rotating frame

$$\dot{S}_+(\underline{x},t) + (\gamma_o + i\delta_o)S_+(\underline{x},t) = 2i(E^+(\underline{x},t)-\Omega e^{-i\underline{k}\cdot\underline{x}})S_z(\underline{x},t) \ , \quad (43a)$$

The adjoint of this equation in $S_-(x,t)$, and

$$\dot{S}_z(\underline{x},t) + 2\gamma_o\left[\frac{1}{2}+S_z(\underline{x},t)\right] = -i\left\{E^+(\underline{x},t)S_-(\underline{x},t) - S_+(\underline{x},t)E^-(\underline{x},t)\right\}$$

$$+ i\Omega\left\{e^{-i\underline{k}\cdot\underline{x}} S_-(\underline{x},t) - e^{i\underline{k}\cdot\underline{x}} S_+(\underline{x},t)\right\} \quad (43b)$$

Again δ_o is the detuning, $\delta_o \equiv (\omega_k - \omega_o')$, whilst

$$E^+(\underline{x},t) = np^2\hbar^{-1} \int_V d\underline{x}' \ g(r)S_+(\underline{x}',t)\underline{\hat{u}}\,\underline{\hat{u}} : \underline{\underline{F}}^*(\underline{x},\underline{x}';\omega_k)$$

$$= \{E^-(\underline{x},t)\}^* \quad (44)$$

and g(r) is the two-body correlation function defined by

$$g(r) = n^{-2} \left< \sum_{\substack{i,j \\ i\neq j}}^{N} \delta(\underline{x}-\underline{x}_i)\delta(\underline{x}'-\underline{x}_j) \right>_{\text{Avr.}} \quad (45)$$

(with $r = |\underline{x}-\underline{x}'|$). An approximation is involved in reducing the problem to one in terms of the two-body correlation function only: correlation functions of all orders appear even in the linearized problem[37] and the situation seems still more complicated here. The approximation may be adequate in the present context at low enough densities n and we shall not pursue the problem further here.

In c-number form, with Maxwell's equations in differential form regained from the propagator $\underline{\underline{F}}$, Eqs. (43) can be traced back to those we have used for a c-number theory of optical bistability in a Fabry-Perot cavity of finite size[39]. Despite the use of rotating wave and slowly varying envelope approximations and of ensemble averaged operators, a comprehensive operator solution of Eqs. (43) still presents some unresolved problems. A physically based two-mode ($\pm$ k,λ) ansatz seems to contain the essentials of the solution but at the time of writing this theory is incomplete. Instead, by adopting a deliberately rough "mean-field"[28,29] approach to the problem we are able to show in §III.B next that an averaged two-mode ansatz does predict bistable behavior.

B.   The Mean Field Two-mode Ansatz Theory of Optical Bistability

In the "mean field" theories developed, for example in Ref. 28, spatially dependent quantities are averaged so that a time dependent but spatially independent theory emerges. The prescription for the product of two spatially dependent operators is to replace them by the product of their spatial averages. It works well in the *limit* of small transmissivity in the c-number theory[39], but it is not good at high transmissivity, and it is not at all good in the corresponding operator theory of superfluorescence[12d,e,f]. We indicate below the problems a similar point of view raises here.

We shall here simply set[36,12e,12f]

$$2S_{\pm}(\underline{x},t) = R_{\pm}^{(+)}(t)e^{\mp i\underline{k}'\cdot\underline{x}} + R_{\pm}^{(-)}(t)e^{i\underline{k}'\cdot\underline{x}}$$

$$S_z(\underline{x},t) = R_3(t)$$

$$(46)$$

where $\underline{k}' \neq \underline{k}$. This ansatz raises problems about the commutation relations assumed for $S_{\pm}(\underline{x},t)$ and $S_z(\underline{x},t)$. This apart, it appears to avoid the problem of spatial averages if it can be adopted consistently. We believe it can but the details have still to be worked out: the crucial steps are to associate a refractive index $m = |\underline{k}'|/|\underline{k}| \neq 1$ with the wave vectors $\underline{k}'$, invoke the optical extinction theorem[37] through the interaction fields $E^{\pm}(\underline{x},t)$, show that consequently, if V is a slab with, say, $\underline{k}$ (the wave vector of the driving field) normally incident, the extinction theorem generates all the theory of the Fabry-Perot interferometer in terms of m[37], but finally show that m depends on Ω so that the system is necessarily bistable. We hope to report the details of such an analysis elsewhere: a brief note follows §IV.

Here we take the much less elegant point of view that within linear theory $m \approx 1$ at resonance so that $\underline{k}'$ in (46) can be approximated by $\underline{k}$, the wave vector of the incident field. In this case it is known from the corresponding theory of superfluorescence[12e,f] that consistency can only be achieved by averaging over all points $\underline{x}$, and that whilst consistency is then achieved 'on the average' it is not achieved in detail. However, the condition $m \approx 1$ is a high transmissivity condition. The justification for averaging the theory here is therefore that the slab V may be supposed to have

silvered surfaces so that the transmissivity is small. Of course
this silvering does not appear in the theory otherwise since it was
never put into the Hamiltonian (16)!

In these terms the result is that there are no reflected waves
$R_{\pm}^{(-)} e^{\pm i\underline{k}\cdot\underline{x}}$ so that we can drop the superscript (+) on $R_{\pm}^{(+)} e^{\mp i\underline{k}\cdot\underline{x}}$,
whilst the operators $R_{\pm}$, R_3 satisfy

$$\dot{R}_+(t) + (\gamma_o + i\delta_o)R_+(t) = I_o\, R_+\, R_3 - 2i\Omega R_3$$

$$\dot{R}_-(t) + (\gamma_o - i\delta_o)R_-(t) = I_o^*\, R_3\, R_- + 2i\Omega R_3$$

$$\dot{R}_3(t) + 2\gamma_o\left[\tfrac{1}{2} + R_3(t)\right] = \tfrac{1}{4}\,(I_o + I_o^*)R_+R_- - i\Omega(R_+ - R_-) \qquad (47)$$

in which

$$I_o = 2iNp^2\hbar^{-1}V^{-2}\int_V d\underline{x}\int_{V'} d\underline{x}'\,\hat{\underline{u}}\hat{\underline{u}}:\underline{\underline{F}}^*(\underline{x},\underline{x}';\omega_o)g(r)e^{i\underline{k}\cdot(\underline{x}-\underline{x}')} \ . \qquad (48)$$

The quantity I_o is simply a complex number. It is finite because
by assumption $g(r) \to 0$ sufficiently rapidly as $r \to 0$ to ensure convergence. It depends on the shape of V and particularly the direction of $\underline{k}$ in relation to that shape. It is here, of course, that
the cavity defined by V obtrudes itself into the theory and it is
here also that for $|\underline{k}'| \neq |\underline{k}|$ the optical extinction theorem emerges[37].
Sufficiently far from the atomic resonance at $\omega_k = \omega_o$ the whole of
the operator theory goes over to a linear boson-coupled-to-boson
theory (see Refs. 37 and especially the last reference there) and
from this the conventional theory of refractive index can be developed. In that theory the integral in $\underline{x}'$ in I_o (before the
average over $\underline{x}$ is taken) is used to construct, via the extinction
theorem, all the features of the Fabry-Perot interferometer once V
is chosen to be a slab[37]. This feature must survive in the nonlinear theory near resonance since Maxwell's equations remain lin-
ear and its Green's function $\underline{\underline{F}}$ is unchanged. But $m \neq 1$ since it
will now be field dependent, and these remarks reinforce those made
at the beginning of this section. They are illustrated in the note
at the end of §IV.

Nevertheless, in this paper we really confine ourselves to
studying the spatially averaged theory. Taking I_o defined by (48)
on its merits we find near the atomic resonance $\omega_k \approx \omega_o$ when $\omega_k = ck$
that for a slab of width L and k normal to its surface

$$I_o = I_o^r + i\, I_o^i + 2J \qquad (49)$$

with

$$I_o^r = 4\pi n p^2 L\omega_o \, c^{-1} \hbar^{-1} \equiv \frac{1}{2} \tau_R^{-1}$$

$$I_o^i = \frac{8}{3} \pi p^2 \hbar^{-1} n \equiv 2\delta_c$$

$$J = i p^2 \hbar^{-1} n \int (g(r)-1)\hat{\underline{u}}\hat{\underline{u}} : \underline{\underline{F}}(\underline{r};\omega_o) e^{i\underline{k}\cdot\underline{r}} d\underline{r} \quad .$$

The small complex quantity J appears in the linear refractive index theory: it essentially describes the scattering and changes the
effective value of Γ_o[37]. We shall simply ignore it here (notice however that the integration is over all space, not V, because of the
property $g(r) \to 1$ in the order of an atomic correlation length). The
quantity τ_R defined through I_o^r is exactly the 'super-radiance' time
we introduced in the theory of superfluorescence[12e,f]. It is also
exactly that assumed in the earlier heuristic single-mode theory of
super-radiance from a rod-like dielectric of length L[12a]. The number δ_c is the 'cooperative shift' due to the Lorentz cavity field.
It arises through the longitudinal part of the field, namely the
Coulomb dipole-dipole interaction (see Eq. (29b)) and is actually
very small at vapour densities [40] ($\sim$100 Hz for $n \sim 10^{11}$ but 10^{14} Hz
for $n \sim 10^{23}$).

The important result of the spatially averaged theory is that
Eqs. (47) which have now emerged are operator equations with exactly
the structure of those analyzed in §II derived there from the driven
Dicke model. Indeed they are the Bloch equations (3) except that
the linear damping describing spontaneous emission appears explicitly. Decorrelated they will yield exactly the results described
in §II.C. In particular they will show conventional optical bistability since this was the result found in Refs. 16,17, as described
in §II.C. We now indicate the details of the analysis.

First we note that Eqs. (47) are not quite the same as (3) because by (42) the operators are scaled to a single atom. They can
be scaled to N atoms by rewriting (47) in terms of the total operators (rather than operator densities)

$$\int_V d\underline{x} \, n \, R_{\pm}(t) = N \, R_{\pm}(t)$$

$$\tag{50}$$

$$\int_V d\underline{x} \, n \, R_3(t) = N \, R_3(t) \, ,$$

and taking up the extra factors N^{-1} arising in the bilinear products
into the number I_o. Notice that since there is no reflected wave

$$N\,R_\pm(t) = \int_V d\underline{x}\; n\, R_\pm(t) = \int_V d\underline{x}\; S_\pm(\underline{x},t)\, e^{\pm i\underline{k}\cdot\underline{x}} =$$

$$= \left\langle \sum_{i=1}^{N} \sigma_\pm^{(i)}(t)\, e^{\pm i\underline{k}\cdot\underline{x}_i}\, e^{\mp i\omega_k t} \right\rangle_{\text{Avr.}} . \tag{51}$$

It follows from this that the total operators (5) actually satisfy
the commutation relations (2) providing the ensemble average of the
commutator is taken and not the commutator of the ensemble averaged
operators (this is a second problem of the theory referred to later).
However, that the operators (50) satisfy the commutation relations
(2) even in this limited sense is an accident of the single mode
theory – since R_+, R_3 actually arise in (47) as operator *densities*.
Evidently (46) means $R_\pm(t)\, e^{\mp i\underline{k}\cdot\underline{x}}$ and $R_3(t)$ should satisfy

$$n^2[S_+(\underline{x},t),\, S_-(\underline{x}',t)] = 2n\, S_z(\underline{x},t)\delta(\underline{x}-\underline{x}')$$

$$n^2[S_z(\underline{x},t),\, S_\pm(\underline{x}',t)] = \pm n\, S_\pm(\underline{x},t)\delta(\underline{x}-\underline{x}')$$

the relations assumed after (42) for the densities $S_\pm$ and S_z.   The
total operators $\int_V d\underline{x}\, n\, S_\pm(\underline{x},t)$ and $\int_V d\underline{x}\, n\, S_z(\underline{x},t)$ satisfy (2).

One can see that the operators in (47) can be restroed as densities by restoring the phase factors and integrating over V, replacing operators by total operators. But to do this consistently
in the first two equations of (47) means that Ω must be replaced by
the average fields $\int_V \Omega\, e^{\pm i\underline{k}\cdot\underline{x}}d\underline{x}$, whilst in the last equation the term
$N \int_V d\underline{x}\, R_+(t)R_-(t)$ becomes $N \int_V d\underline{x}\, \left(R_+(t)e^{-i\underline{k}\cdot\underline{x}}R_-(t)e^{+i\underline{k}\cdot\underline{x}}\right)$ which
can be expressed in terms of total field operators only via the
"mean field" prescription

$$N \int_V d\underline{x}\left(R_+(t)e^{-i\underline{k}\cdot\underline{x}}\right) \cdot V^{-1}\int_V d\underline{x}'\left(R_-(t)e^{+i\underline{k}\cdot\underline{x}'}\right) .$$

We shall not actually need the commutation relations and the
"mean field" prescription for the decorrelated solutions of the
Bloch equation (47), although this equation is already spatially
averaged, as we have noted. We do need the commutation relations
(2) for the operators (50) in the calculation of the correlation
functions and spectra which follow; and we need them for the exact
solution of the master equation given later in §III.C. These results are therefore really restricted by the spatial averages like
that for (47) *and* by the restriction of the single mode theory.

We now proceed to the different solutions of (47).

(a) **Steady-state decorrelated solution in terms of the internal field**

Following the previous work we define the quantities

$$X_1 = Y - 2\gamma_o^{-1} \, \text{Im}\{\Gamma R_+\} \, , \qquad X_2 = 2\gamma_o^{-1} \, \text{Re}\{\Gamma R_+\} \, ,$$

$$\Gamma \equiv I_o^r - i I_o^i (\approx I_o^*) \, ,$$

$$X = X_1 + i X_2 \qquad , \qquad Y = 2\sqrt{2} \, \gamma_o^{-1} \Omega \, ,$$

$$R_x = \frac{1}{2}(R_+ + R_-) \qquad , \qquad R_y = \frac{1}{2} i^{-1}(R_+ - R_-) \, . \tag{52}$$

In the steady state the *decorrelated* solutions of (47) are

$$N\langle R_3\rangle_o = -\frac{N}{2} \, \frac{1 + \Delta^2}{1 + \Delta^2 + |X|^2}$$

$$N\langle R_y\rangle_o = \frac{N}{\sqrt{2}} \, \frac{X_1 + X_2\Delta}{1 + \Delta^2 + |X|^2}$$

$$N\langle R_x\rangle_o = \frac{N}{\sqrt{2}} \, \frac{X_1\Delta - X_2}{1 + \Delta^2 + |X|^2} \tag{53a}$$

where X now denotes the steady state expectation value. The number N comes in from the equilibrium value of $N\langle R_3\rangle + \frac{1}{2} N$. Consistently with these results we find the 'state equation' connecting the (real) driving field Y with the magnitude of the internal field X

$$Y^2 = |X|^2 \, \frac{[(1 + \Delta^2 + |X|^2 + N_1)^2 + N_2^2]}{(1 + \Delta^2 + |X|^2)^2} \, . \tag{53b}$$

The definitions are (Δ is $i^{-1} \times \Delta$ used previously)

$$\Delta \equiv \delta_o \gamma_o^{-1} , \quad N_1 \equiv 2(F_1 - F_2\Delta), \quad N_2 \equiv 2(F_1\Delta + F_2),$$

$$F_1 \equiv \frac{1}{2} \gamma_o^{-1} N^{-1} I_o^r = \frac{1}{4} \gamma_o^{-1} N^{-1} \tau_R^{-1} \, ,$$

$$F_2 \equiv \frac{1}{2} \gamma_o^{-1} N^{-1} I_o^i = \gamma_o^{-1} N^{-1} \delta_c \, . \tag{53c}$$

Notice that the inphase component X_1 of X is

$$X_1 = Y + 2\gamma_o^{-1}[\delta_c \langle R_x\rangle_o - \frac{1}{2} \tau_R^{-1} \langle R_y\rangle_o] \tag{54}$$

so that this is the usual Lorentz internal field due to the in-phase dipole density $\langle R_x \rangle_0$ together with the contribution from the dipole density $\langle R_y \rangle_0$ in phase quadrature essentially due to the resonance condition $\omega_k \approx \omega_0$ and $\omega_k = ck$ used in computing I_0. For $F_2 = 0$ ($\delta_c = 0$) the theory depends only on the bistability parameter $C \equiv F_1 = \tfrac{1}{4}(N\gamma_0\tau_R)^{-1}$ and we regain the familiar state equation for a ring cavity in mean field approximation[30]. At resonance ($\Delta = 0$) with $F_2 = 0$ and $F_1 = N$ ($\tau_R^{-1} = 4N^2\gamma_0$) Eq. (53b) coincides with that found in Refs. 16, 17, for the point system of the driven Dicke model.

From (53b) the condition for bistability is

$$27(1+\Delta^2)^4(N_1^2 + N_2^2) \geqslant [2N_1(1+\Delta^2) - (N_1^2 + N_2^2)]^3 . \tag{55}$$

This reduces to that of Ref. 30 when $F_2 = 0$ and $F_1 = C$. It is of interest to see the effect of the cooperative shift δ_c upon the characteristic curves. In Fig. 1, $\delta_c \approx \tau_R^{-1}$, fairly large. Figure 1a shows that at resonance ($\Delta = 0$) and for fixed F_1 the shift δ_c in F_2 moves the threshold to larger values of the driving field Y whilst the bistable region becomes larger. For fixed F_2 increasing F_1 corresponds to increasing N: Fig. 1b shows the effect of this when $\Delta \neq 0$.

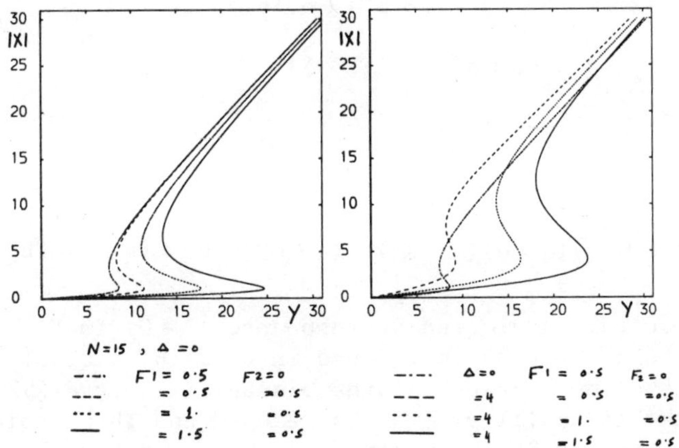

Fig. 1.   Characteristic curves ($|X|$ against Y) according to Eq. (53b). The parameters F_1, F_2 are defined in Eq. (53c).

(b)   Emission and absorption spectra and intensity correlation

In order to calculate these quantum mechanical correlation functions, Eqs. (47) are linearized in the neighborhood of the steady state. The resulting equations are

$$\langle \dot{R}_+ \rangle + b \ \langle R_+ \rangle = -i \ b_1 \langle R_3 \rangle$$

$$\langle \dot{R}_- \rangle + b^* \ \langle R_- \rangle = i \ b_1^* \langle R_3 \rangle$$

$$\langle \dot{R}_3 \rangle + 2\gamma_o \langle R_3 \rangle = i \ b_2 \langle R_+ \rangle + i \ b_2^* \langle R_- \rangle \qquad (56)$$

where

$$b = (\gamma_o - I_o^r)\langle R_3 \rangle_o + i(\delta_o + I_o^i \langle R_3 \rangle_o)$$

$$b_1 = \tfrac{1}{2} \sqrt{2} \ \gamma_o X, \qquad b_2 = \Omega - i4I_o^r \langle R_- \rangle_o \ . \qquad (57)$$

Since we have already executed a spatial average we can only define
the emission spectrum as the Fourier transform (for the steady state)
of the correlation function $\langle R_+(\tau)R_-(0)\rangle$. One way to calculate this
is by assuming the quantum regression theorem[41] and the factorization
scheme of Eq. (4)[16,17]. After some manipulation (cf. Refs. 17,4) we
find the incoherent part of the spectrum in the form

$$g_{inc.}(\omega-\omega_k) \propto \frac{N}{(1+\Delta^2+|X|^2)} \ \mathrm{Re}\left\{\frac{M_1(q = i(\omega-\omega_k))}{Q(q = i(\omega-\omega_k))}\right\} \qquad (58)$$

where

$$M_1(q) = |X|^4 \ ((q + 2\gamma_o)(q + b^*) + b_1^* \ b_2^*)$$

$$+ \ i\sqrt{2} \ b_1(q + b^*)(\Delta - i)X^*|X|^2$$

$$- \ b_1 b_2^*(\Delta - i)(X^*)^2,$$

and

$$Q(q) = (q + \Gamma_o)(q + b)(q + b^*) + (q + b^*)b_1 b_2 + (q + b)b_1^* b_2^* \ . \qquad (59)$$

This spectrum is plotted at resonance ($\Delta = 0$) in Fig. 2a. It
has the same character as that found in Ref. 16: a broad Lorentzian
develops on the lower branch of the bistability curve (53b) which
narrows as the instability point is reached and then rapidly changes
to the usual triplet Stark spectrum on the upper branch as Y increases. For $\Delta \neq 0$ the behavior is shown in Fig. 2b. The effect of
the cooperative shift δ_c, which makes the Lorentzian on the lower
branch asymmetric, is also shown. Note the reduction in the central
peak above threshold also.

In the presence of a weak probe field of amplitude ε and frequency ν, the absorption line scaled to a single atom is given by[32]

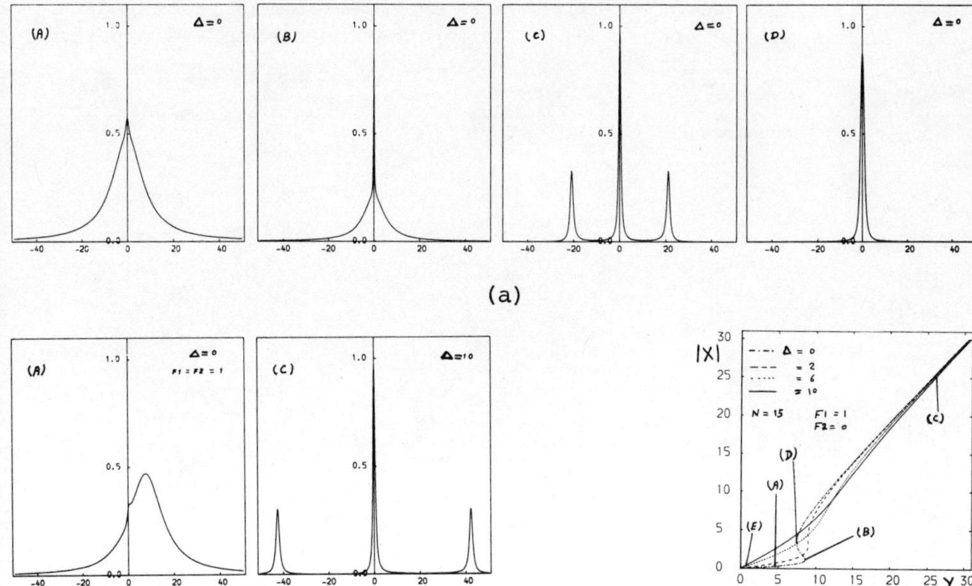

Fig. 2.   The (incoherent) emission spectrum, $g_{inc.}(\omega-\omega_k)$, plotted against $(\omega-\omega_k)$ at various points (A)...(E) on the bi-stable curve. Note the asymmetry in the spectrum at point (A) shown in (2b), which is due to the cooperative shift ($F_2 = 1$).

$$g_{abs.}(\Delta\nu) = 4\epsilon^2 \int e^{i\nu\tau} <[R_-(0), R_+(\tau)]>d\tau \tag{60}$$

in which the commutator is calculated in the presence of the driving field Ω alone[32]. The result scaled to N proves to be

$$g_{abs.} \propto \frac{4\epsilon^2 N}{(1 + \Delta^2 + |X|^2)} \text{ Re} \left\{\frac{M_2(q = i\Delta\nu)}{Q(q = i\Delta\nu)}\right\} \tag{61}$$

in which $\Delta\nu \equiv \nu-\dot\omega_k$ and

$$M_2(q) \equiv (1+\Delta^2)((q+2\gamma_o)(q+b^*)+b_1^* b_2^*)-\gamma_o(1+i\Delta)(q+b^*)|X|^2 . \tag{62}$$

This absorption spectrum is plotted at resonance in Fig. 3a. It has the characteristics found (cf. Ref. 17) for the one-atom case: above threshold, for values of $|\Delta\nu| < 2\Omega$, the profile is two peaked and negative (so it is amplified), and it changes sign (signifying absorption) for $|\Delta\nu| > 2\Omega$. Below threshold the spectrum is broad Lorentzian narrowing near the instability point. The off-resonance case, $\Delta \neq 0$, is shown in Fig. 3b and above threshold it again follows the one-atom line shape. This profile has been observed experiment-ally[32].

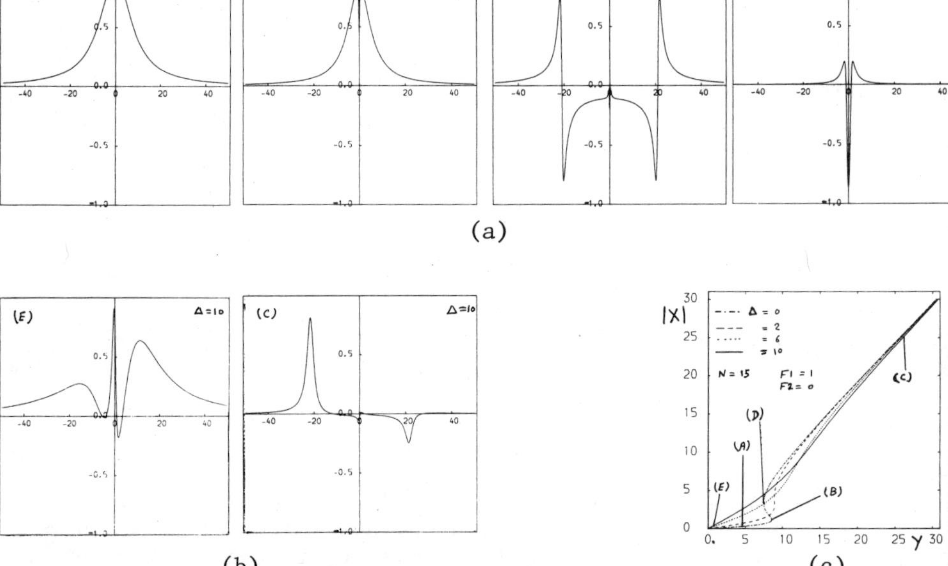

Fig. 3.   The absorption spectrum $g_{abs.}(\Delta\nu)$ plotted against $\Delta\nu$, at resonance (3a) and off-resonance (3b).

Agarwal et al.[42] have calculated both the emission and absorp-tion spectrum for a bistable system using a Gaussian decoupling scheme instead of the decorrelation (4). Their results are quanti-tatively the same as those derived here both above and below threshold.

In order to calculate the normalized intensity-intensity cor-relation function, $g^{(2)}(0)$, we adopt the same decorrelation as (4). We find (and cf. Ref. 17)

$$g^{(2)}(0) = \frac{[N^2(1+\Delta^2)^2+4N(1+\Delta^2)(1+\Delta^2+|X|^2)+2(1+\Delta^2+|X|^2)^2]}{[|X|^2+N(1+\Delta^2)]^2} \qquad (63)$$

At resonance $(\Delta=0)$, $g^{(2)}(0)$ reduces to the result obtained in Ref. 17, namely $g^{(2)}(0) = 1$, $|X| < 1$; $= 2$, $|X| \sim N_1$ for which values it shows bistable behavior. For $\Delta \neq 0$ the bistable behavior tends to disappear but $g^{(2)}(0) \to 2$ for, however, larger values of the driving field Y (Fig. 4).

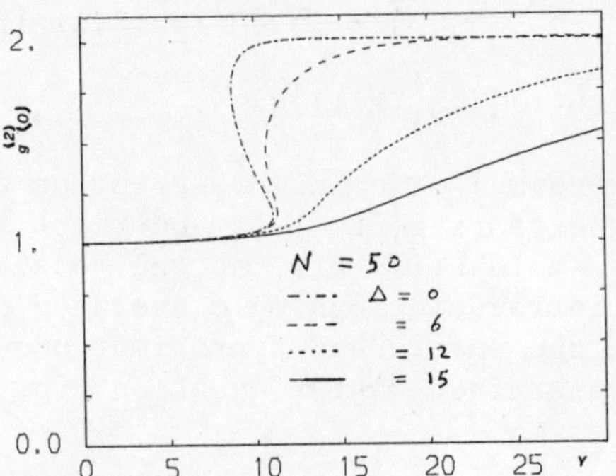

Fig. 4.   Normalized intensity-intensity correlation function
$g^{(2)}(0)$ plotted agains the input field Y, for various
detuning parameters.

C.   Macroscopic Master Equation and Bifurcation

In this section we use reaction field theory[7,34] to derive the
master equation for the ensemble averaged operators of the extended
but spatially averaged driven two-level atom model. We show it takes
the same form as the master equation (1) for the point system of the
driven Dicke model. It follows that up to ensemble averaging and
within the assumptions of the two-mode theory, we shall find for the
averaged extended system the same bifurcation behavior as for the
point system described in §II.D.

The Hamiltonian (16) can be taken in the form

$$H = H_0 + H_1 ;$$

$$H_0 = \sum_{\underline{k},\lambda} \hbar\omega_k\, a^+_{\underline{k},\lambda} a_{\underline{k},\lambda} + \frac{1}{2} \hbar\omega_o \int \sigma_z(\underline{x},t)\,d\underline{x} - \int \underline{p}(\underline{x},t)\cdot\underline{e}_o(\underline{x},t)\,d\underline{x},$$

$$H_1 = - \int \underline{p}(\underline{x},t)\cdot\underline{e}_1(\underline{x},t)\,d\underline{x} \tag{64}$$

where the free field $\underline{e}_o$ is given by the free field form of (18) and
$\underline{e}_1$ is the remaining part of the total field operator

$$\underline{e}_1 = \underline{e}_{self} + \underline{e}_{int}$$

given by (35) and (27). For any matter operator $\rho(t)$ Heisenberg's
equations yield

$$\dot{\rho} = i\hbar^{-1}[H,\rho] = i\hbar^{-1}[H_0,\rho] - i\hbar^{-1} \int \{\underline{e}_1^-(\underline{x},t)\cdot[\underline{p}(\underline{x},t),\rho(t)]$$

$$+ [\underline{p}(\underline{x},t),\rho(t)]\cdot\underline{e}_1^+(\underline{x},t)\}d\underline{x} \quad . \tag{65}$$

We have adopted the normal ordering prescription described in §III.A placing the field operators $\underline{e}_1^{\mp}$ to left and right respectively (cf. Refs. 7, 36 also). We identify $\rho(t)$ as the total reduced density operator. Then by introducing ensemble averaged operators as in §III.A and adopting the ansatz and approximations of §III.B we arrive at the coarse grained master equation for ρ

$$\dot{\rho} = -i\Omega[R_+ + R_-, \; \rho] + i\delta_o[R_3,\rho]$$

$$+I_1[2R_-\rho R_+ - R_+R_-\rho - \rho R_+R_-] - iI_2[R_+R_-,\rho] + \Lambda\rho \; . \tag{66}$$

The quantities I_1 and I_2 are defined as before by (48) and (49), and more precisely should be corrected by Re J and Im J as before. The self-interactions have appeared in the term $\Lambda\rho$ which is defined to be

$$\Lambda\rho = \gamma_o n^{-1}\langle\sum_i \delta(\underline{x}-\underline{x}_i)\{2\sigma_-^{(i)}(t)\rho\sigma_+^{(i)}(t)-\sigma_-^{(i)}\sigma_+^{(i)}\rho-\rho\sigma_+^{(i)}\sigma_-^{(i)}\}\rangle_{\text{Avr.}} \tag{67}$$

Since e.g. $\sigma_+^{(i)}\sigma_-^{(i)} = \sigma_z^{(i)}+ \tfrac{1}{2}$, $\sigma_-^{(i)}\sigma_+^{(i)} = \sigma_z^{(i)} - \tfrac{1}{2}$ the two terms with negative sign in the curly bracket can be expressed in terms of the ensemble averaged operator $R_3(t)$. However, the remaining term offers difficulties since $\rho(t)$ is supposed coarse grained, i.e. ensemble averaged and its commutation relations with $\sigma_\pm^{(i)}(t)$ are not defined. In fact as we noted by (42) and (51) there is a difficulty throughout the whole ensemble averaging process: it is arbitrarily *assumed* that the commutation relations of the ensemble averaged operators R_+, R_3 as well as those of the more fundamental operators $S_\pm(\underline{x},t)$, $S_z(\underline{x},t)$ defined by (42) follow the commutation relations of $\sigma_\pm(\underline{x},t)$ and $\sigma_z(\underline{x},t)$ which are derived from those of $\sigma_\pm^{(i)}(t)$ and $\sigma_z^{(i)}(t)$. As we have noted by (51) it is not at all clear that e.g.

$$\langle[\sigma_+(\underline{x},t),\sigma_-(\underline{x},t)]\rangle_{\text{Avr}} \equiv \langle\sigma_+(\underline{x},t)\sigma_-(\underline{x},t)\rangle_{\text{Avr}}$$

$$- \langle\sigma_-(\underline{x},t)\sigma_+(\underline{x},t)\rangle_{\text{Avr}} = [\langle\sigma_+(\underline{x},t)\rangle_{\text{Avr}}, \; \langle\sigma_-(\underline{x},t)\rangle_{\text{Avr}}].$$

At least (66) becomes consistent with, and alternative to, the Bloch equations (47) if the microscopic commutation relations are assumed to survive the ensemble averaging process - so this is what we assume.

Because of the damping term $\Lambda\rho$ in (66) the total angular momentum scaled to a single atom, namely

$$\underline{R}^2 = R_3^2 + \frac{1}{2} (R_+ R_- + R_- R_+),\tag{68}$$

is not a constant of the motion. But since in many cases of interest $I_1 = \frac{1}{2} \tau_R^{-1} \gg \gamma_0$ and one can see that the term in I_1 in (66) involves both self-interactions and interparticle interactions, we can approximate (66) by dropping the spontaneous damping (of course we could have done the same in (47), but then, in order to produce the damping needed for conventional bistability we should need to have extracted it from the bilinear products $R_+ R_-$, etc. writing these as $(<R_3> + \frac{1}{2})n^{-1} + <R_+><R_->$, etc.). If we define

$$g_1^* = -i\Omega I_1^{-1}, \quad \Delta_1 = i\delta_0 I_1^{-1}, \quad \Delta_c = iI_2 I_1^{-1}, \quad \tau = I_1 t\tag{69}$$

then (66) takes the form

$$\frac{\partial \rho}{\partial \tau} = g_1^*[R_+ + R_-, \rho] + \Delta_1 [R_3, \rho] + \Delta_c [R_+ R_-, \rho] - [2R_- \rho R_+ - R_+ R_- \rho - \rho R_+ R_-].\tag{70}$$

Of course these operators only satisfy the commutation relations (2) in the sense discussed in connection with Eqs. (50) and (51). But in this sense the total angular momentum $\underline{R}^2$ defined in (68) is now a constant of the motion. And, apart from the cooperative shift Δ_c, the master equation (70) has exactly the same form as the master equation (1) for the driven point system. The method of solution in the steady state is, therefore, now immediately available.

We proceed as in Ref. 25 and set

$$\rho_{ss} = \sum_{m,n=0}^{N} C_{mn} R_-^m R_+^n .\tag{71}$$

Then we reach the following difference equation for the coefficients C_{mn} $(= C_{nm}^*)$

$$(m+1)g_1^* C_{m+1n} - [(m+1)(1+\Delta_c) - \Delta_1]C_{mn} = 0 .\tag{72}$$

The solution is thus.

$$\rho_{ss} = D_1^{-1} \sum_{m,n=0}^{N} b_{mn} (G_1^*)^{-m} (G_1)^{-n} (R_-)^m (R_+)^n\tag{73}$$

in which

$$D_1 = \frac{\sinh(\pi|B|)}{(\pi|B|)} \sum_{m=0}^{N} \frac{\Gamma(m+1-B)\Gamma(m+1+B)(N+m+1)!\,|G_1|^{-2m}}{(N-m)!\,(2m+1)!} ,$$

and

$$G_1 \equiv g_1 [1 + \Delta_c]^{-1}, \quad B = \Delta_1 [1 + \Delta_c]^{-1} .$$

The conclusion is then that the exact solution of the ensemble and spatially averaged quantum theory of the extended system, within the two-mode ansatz and neglect of the spontaneous emission has all the features of the exact solution of the point driven Dicke model; and in particular, there is a second order type phase transition with a bifurcation point, at resonance ($\Delta_1 = 0$), where $\theta (\equiv \lim\{2|G_1|/N\})=1$. Thus within these approximations the model of an extended system of two-level atoms with Hamiltonian (16) does *not* display conventional optical bistability!

IV.   CONCLUSION

The results of §III.C naturally raise the question of how far the approximations of the argument leading to (70) have damaged the theory. First of all notice that in the absence of the driving field ($\Omega=0$) and with $I_2 = \delta_0 = 0 (\Delta_c = \Delta_1 = 0)$ the master equation (66), together with the commutation relations assumed for it, are exactly the master equation and commutation relations long since written down heuristically as a single-mode theory by Bonifacio et al[12a] to describe super-radiance from a rod-shaped region V (Bonifacio et al.[12a] include an empirical damping term analogous to $\Lambda\rho$). This single mode theory does not describe the super-radiant emission (that is superfluorescence) actually observed by Vrehen et al.[43] from Cs vapour.

It might be said that the spontaneous emission must not be neglected since the work of §II.C suggests that this is vital to a theory leading to conventional bistability. But this is not really the case since sufficient damping can be extracted from the self-interactions locked-up in the last term in (70) (or in the term in I_1 in (66)). What seems to be essential is to extract these damping terms explicitly. And the reason for this is that the spatial average obscures the fact that operator products on the same site describing self-interactions (and which linearize through operator algebra) are different from operators on different sites. After this the theory is not sensitive to the replacement of operators $\sum_{j \neq 1} \sigma_{\pm,z}^{(j)}(t)\delta(\underline{x}'-\underline{x}_j)$ by some collective and ensemble averaged operator at $\underline{x}'$ which includes j=i in the interaction field $\underline{e}_{int}(\underline{x},t)$, providing $\underline{x}'$ and $\underline{x}$ are separated and the fields carried by $\underline{\underline{F}}(\underline{x},\underline{x}';\omega_k)$ are made to propagate. But this means that the spatial average should *not* be performed. The evidence from the preliminary work on the operator theory of the extended system which searches for the intensity dependent refractive index m as sketched at the beginning of §III.B is that just as in the problem of superfluorescence the most damaging approximation in the argument leading to (70) is the

spatial average. This conclusion is accentuated by the fact that
the refractive index theory is not a single mode theory. Thus the
commutation relations (2) apply to the operators of (70) only if
these are total operators interpreted in "mean field" approximation.
Evidently the "mean field" approximation is once again the damaging
aspect of the theory, and this conclusion is consistent with the observation that $\underline{S}^2$ is not a constant of the motion in the extended
case.

We conclude from this that the c-number theory of the extended
system obtained by decorrelating all operator products in the expectation value of Eqs. (43) is already likely to give a good description of the real physical situation obtaining in a resonantly
driven Fabry-Perot cavity. And this suggests, but certainly does
not prove, that both the spectrum and the intensity-intensity correlation functions for such a real system will be better described
by the decorrelated results of §III.B rather than the results (e.g.
$g^{(2)}(0) = 1.2$) which follow from (70) and which, given that equation,
are exact results. Of course all the correlation functions for the
real extended system have still to be calculated: the calculations
in §III.B are approximated by the spatial average and assumptions
about the commutation relations, but even then are the only results
nominally applicable to the extended system so far available with
the exception of the calculations of the spectrum by Agarwal et al.[42]
and of Lugiato[42] based on a Fokker-Planck equation approach.

The proof that the c-number equations are a good approximation
to (43) would rest on the fact that from (42), in which S_+, S_z are
normalized to a single atom, the commutators are $O(n^{-1})$: for example

$$[S_+(\underline{x},t), S_-(\underline{x}',t)] = n^{-1}S_z(\underline{x},t)\delta(\underline{x}-\underline{x}') \tag{74}$$

This would seem to be an adequate description in the cooperative region, and, providing the self-interactions are extracted, could be
good still in the chaotic state associated with the upper branch of
the bistability curve (although this seems quantal, the quantal
character is contained in the quantum spontaneous emission). However, all of this has still to be investigated in the operator
theory without spatial average sketched at the beginning of §III.B.

We have been impressed by the fact that the semiclassical decorrelation of §II.B reproduces so well the results of the exact
quantum treatment of §II.D for very large N (as all our numerical
work has shown). Because of (74) we believe that the same will be
true for the extended system provided decorrelations like (4) are
adopted. We remain puzzled by the fact that the work of §II.B does
not include spontaneous emission whilst that of §II.D, which must do
so, keeps $\underline{S}^2$ a constant of the motion and despite the exact character of the solution appears to conceal it. There appears to be no

part of the exact results of §II.D which exhibits effects of spontaneous emission, and even the two branches of the solution for r_3
for $\theta < 1$ show no instability.

Insofar as a main conclusion of the work is that decorrelations
like (4) are a good approximation to the physical situation, it
would be helpful to have confirmation of this from experimental data
on the spectra and on the intensity-intensity correlation functions
for all parts of the observed bistability curves.

ACKNOWLEDGMENTS

We are glad to acknowledge a helpful correspondence with R. R.
Puri, Bhabha Institute. We are grateful to the SRC for their finan-
cial support.

APPENDIX

This note concerns the theory of the refractive index m sketched
at the beginning of §III.B. In the c-number expectation value theory
in which all operators are decorrelated in the fashion of §II.B we
find that the two-mode ansatz (46) when put into the Eqs. (43) yields
expressions which, after some deliberate simplifications take the
forms

$$\delta_o - i\gamma_o = \frac{-4|E|^2}{\gamma_o|m+1|^2}\left[\frac{i(m^{*2} - m^2)}{m^2 - 1}\right] - \frac{4\pi n p^2}{\hbar(m^2 - 1)} \tag{A.1}$$

$$\langle R_3\rangle_0 = \frac{-2\hbar|E|^2}{\gamma_o p^2|m+1|^2}\left[\frac{i(m^{*2} - m^2)}{4\pi n}\right] - \frac{1}{2} \tag{A.2}$$

$$p\langle R_+\rangle_0 = \frac{2}{(m+1)}\left(\frac{m^2-1}{4\pi n}\right) E\hbar p^{-1} . \tag{A.3}$$

As in (53a) the quantities $\langle R_3\rangle_0$ and $\langle R_+\rangle_0$ are evaluated in the
steady state. Additional standing wave contributions arise (com-
pare our work in Refs. 12e,f) but these are neglected here. The
expression (A.1) determines m as a function of δ_o and $|E|$ in the
same steady state. The cooperative shift δ_c has been dropped, but
if it is included its effect is to replace each quantity m^2-1 by
$3(m^2 - 1)\cdot(m^2 + 2)^{-1}$ and similarly for $m^{*2}-1$. The expression (A.2)
determines the inversion as a function of m and $|E|$. The optical
extinction theorem has been used in the form for a slab with the
wave vector $\underline{k}'$ normally incident[37]; and (A.3) is the result up to
neglect of formal phase factors depending on m in which however, by

(A.1), m is in general complex valued. The term in the reflection coefficient $(m-1) \cdot (m+1)^{-1}$ which arises naturally in the analysis[37] and which describes the usual theory of the Fabry–Perot interferometer in terms of m[37] is also neglected here for simplicity.

There are two obvious limiting cases:

(a) Weak fields, $|E| \ll \gamma_o$: $R_3 \to -\frac{1}{2}$; the relation (A.3) is then the usual linear relation between ε and the dipole density $n\,p\,R_+$, whilst (A.1) after correction for the cooperative shift and for the complex number J defined by (49) takes the usual form in the rotating wave approximation[37]

$$\frac{m^2-1}{m^2+2} = \frac{4\pi}{3} \frac{n\alpha(\omega_k)}{1 + n\alpha(\omega_k)\hbar p^{-2} iJ}$$

$$\alpha(\omega_k) = p^2 h^{-1}(\omega_o' - \omega_k + i\gamma_o)^{-1} \ .$$

(b) Strong fields, $|E| \to \infty$ with $m \to 1$ and $R_3 \to 0$ with $p\,R_+ \to 0$; this result is consistent with the three–peaked Stark spectrum computed in §III.B.

For intermediate values of $|E|$ one can see from (A.1) that m is complex and satisfies a quartic equation. We still tentatively infer from this that for some range of input field amplitude $|E|$ the transmitted field amplitude $|2m(m+1)^{-2}E|$ follows the usual S–shaped portion of a conventional bistability curve complicated now by the fact that in the physically significant region close to resonance m is certainly complex valued.

At the time of writing we have not explored the theory further.

REFERENCES

1.   B. R. Mollow, Phys. Rev. <u>188</u>, 1969 (1969); Phys. Rev. A<u>12</u>, 1919 (1975).
2.   C. R. Stroud, Jr., and E. T. Jaynes, Phys. Rev. A<u>1</u>, 106 (1970).
3.   C. R. Stroud, Jr., Phys. Rev. A<u>3</u>, 1044 (1971).
4.   S. S. Hassan and R. K. Bullough, J. Phys. B<u>8</u>, L147 (1975).
5.   G. S. Agarwal, Springer Tracts in Modern Physics, <u>70</u>, (Springer–Verlag: Heidelberg, N.Y., 1974); M. E. Smithers and H. S. Freedhoff, J. Phys. B<u>7</u>, L432 (1974); S. Swain, J. Phys. B<u>8</u>, L437 (1975); H. J. Carmichael and D. F. Walls, J. Phys. B<u>9</u>, 1199 (1976); H. J. Kimble and L. Mandel, Phys. Rev. A<u>13</u>, 2123 (1976); C. Cohen–Tannoudji and S. Reynaud, J. Phys. B<u>10</u>, 345 (1977).

6.   Proc. 3rd Conf. on Coherence and Quantum Optics, Rochester, (Plenum, N.Y., 1972), eds: L. Mandel and E. Wolf - see articles by: R. K. Bullough, pp. 121-56; J. R. Ackerhalt, J. H. Eberly and P. L. Knight, pp. 635-44.

7.   R. Saunders, R. K. Bullough and F. Ahmad, J. Phys. A8, 759 (1975).

8.   J. R. Ackerhalt and J. H. Eberly, Phys. Rev. D10, 3350 (1974).

9.   "IV Rochester Conf. on Coherence and Quantum Optics," (Plenum, N. Y., 1977), eds: L. Mandel and E. Wolf.

10. F. Schuda, C. R. Stroud, Jr. and M. Hercher, J. Phys. B7, L198 (1974); also, W. Hartig, W. Rasmurren, R. Schieder and H. Walther, Z. Physik A278, 205 (1976); F. Y. Wu, R. E. Grove and S. Ezekiel, Phys. Rev. Lett. 35, 1426 (1975).

11. R. H. Dicke, Phys. Rev. 93, 99 (1954).

12. (a) R. Bonifacio, P. Schwendimann and F. Haake, Phys. Rev. A4, 302, 854 (1971); (b) F. Haake and R. Glauber, Phys. Rev. A5, 1457 (1972); 13, 357 (1976); 20, 2047 (1979); "Cooperative Phenomena," (North-Holland, Amsterdam, 1974), ed. H. Haken, p. 71; M. F. Schuurmans, D. Polder and Q. Vrehen, Phys. Rev. A19, 1192 (1979); (c) G. Banfi and R. Bonifacio, Phys. Rev. Lett. 33, 1259 (1974); Phys. Rev. A12, 2068 (1975); (d) R. Bonifacio and L. Lugiato, Phys. Rev. A11, 1507 (1975); 12, 587 (1975); (e) R. Saunders, S. S. Hassan and R. K. Bullough, J. Phys. A9, 1725 (1976); R. K. Bullough et al., in Ref. 9, p. 263; J. MacGillivray and M. S. Feld, Phys. Rev. A14, 1169 (1976); (f) "Cooperative effects in matter and radiation," (Plenum, N. Y., 1977), eds. C. M. Bowden, D. W. Howgate and H. R. Robl - see especially articles by: J. MacGillivray and M. S. Feld, pp. 1-14; R. Bonifacio et al., pp. 193-208; R. Saunders and R. K. Bullough, pp. 209-256.

13. (a) N. Skribanowitz, I. P. Herman, J. C. MacGillivray and M.S. Feld, Phys. Rev. Lett. 30, 309 (1973); (b) M. Gross, C. Fabre, P. Pillet and S. Haroche, Phys. Rev. Lett. 36, 1035 (1976); H. M. Gibbs, in Ref. 12f, pp. 61-78; Q.H.F. Vrehen, in Ref. 12f, pp. 79-100.

14. G. S. Agarwal, A. C. Brown, L. M. Narducci and G. Vetri, Phys. Rev. A15, 1613 (1977); G. S. Agarwal, D. H. Feng, L. M. Narducci, R. Gilmore and R. A. Tuft, Phys. Rev. A20, 2040 (1979); G. S. Agarwal, R. Saxena, L. M. Narducci, D. H. Feng and R. Gilmore, Phys. Rev. A21, 257 (1980).

15. A. S. Amin and J. G. Cordes, Phys. Rev. A18, 1298 (1978); C. Mavroyannis, Phys. Rev. A18, 185 (1978); Opt. Comm. 33, 42 (1980); H. J. Carmichael, Phys. Rev. Lett. 43, 1106 (1979).

16. H. J. Carmichael and D. F. Walls, J. Phys. B10, L685 (1977).

17. S. S. Hassan and D. F. Walls, J. Phys. A11, L87 (1978).

18. C. M. Bowden and C. C. Sung, Phys. Rev. A19, 2392 (1979); also C. M. Bowden in this volume.

19. I. R. Senitzky, Phys. Rev. Lett. 40, 1334 (1978); Phys. Rev. A6, 1171, 1175 (1972).

20. P. D. Drummond and S. S. Hassan, Phys. Rev. A22, 662, (1980).

21. L. M. Narducci, D. H. Feng, R. Gilmore and G. S. Agarwal, Phys. Rev. A18, 1571 (1978).

22. P. D. Drummond and H. J. Carmichael, Opt. Comm. 27, 160 (1978); D. F. Walls, P. D. Drummond, S. S. Hassan and H. J. Carmichael, Prog. Theoret. Phys. Suppl. 64, 307 (1978).

23. R. R. Puri and S. V. Lawande, Phys. Lett. 72A, 200 (1979); Physica A 101, 599 (1980).

24. S. S. Hassan, R. K. Bullough, R. R. Puri and S. V. Lawande, Physica A (in press); P. D. Drummond also obtained the value $g^{(2)}(0) = 1.2$ using the solution in Ref. 23 (preprint).

25. R. R. Puri, S. V. Lawande and S. S. Hassan, Opt. Comm. (to appear).

26. S. Ja. Kilin, J. Applied Spect. (USSR) 28, 255 (1978).

27. S. S. Hassan et al., (in preparation).

28. R. Bonifacio and L. Lugiato, Opt. Comm. 19, 172 (1976); Phys. Rev. A18, 1129 (1978).

29. G. S. Agarwal, L. M. Narducci, D. H. Feng and R. Gilmore in Ref. 9, pp. 281-292; Phys. Rev. A18, 620 (1978). Notice that "mean field" arises in two distinct contexts in this paper: one is the "mean field" theory of Ref. 28 in which all atomic operator densities are averaged on the space variable (in the context of §III operate with $V^{-1} \int dx$) with the prescription that an operator or c-number product is replaced by the product of averages (on $A(x)$ $B(x)$ form $V^{-2} \int dx\, dx'\, A(x)\, B(x')$). The other usage arises in the theory of phase transitions and is essentially a decorrelation approximation of Hartree type. It is the first usage which is referred to in the discussion of the work of Ref. 18 (§II.C); but it is the second one which is referred to at the beginning of §II.E. We shall use double quotes to indicate the first usage throughout the paper.

30. S. S. Hassan, P. D. Drummond and D. F. Walls, Opt. Comm. 27, 480 (1978); R. Bonifacio and L. Lugiato, Lett. Nuovo. Cim. 21, 517 (1978).

31. H. M. Gibbs, S. L. McCall and T. N. Venkatesan, Phys. Rev. Lett. 36, 1135 (1976); also see S. L. McCall, Phys. Rev. A9, 1515 (1974).

32. B. R. Mollow, Phys. Rev. A5, 2217 (1972); F. Y. Wu, S. Ezekiel, M. Ducloy and B. R. Mollow, Phys. Rev. Lett. 38, 1077 (1977).

33. D. A. Miller and S. D. Smith, Opt. Comm. 31, 101 (1979).

34. G. S. Agarwal, L. M. Narducci, D. H. Feng and R. Gilmore, Phys. Rev. Lett. 42, 1260 (1979); 43, 238 (1979).

35. V. Degiorgio, Physics Today, 29, (October, 1976); N. Corti and V. Degiorgio, Phys. Rev. Lett. 36, 1173 (1976).

36. S. S. Hassan, Ph.D. Thesis (U. of Manchester, 1976); R. Saunders, Ph.D. Thesis (U. of Manchester, 1973).

37. R. K. Bullough, J. Phys. A1, 409 (1968); 2, 477 (1969); 3, 708, 726, 751 (1970); R. K. Bullough, A. S. F. Obada, B. V.

Thompson and F. Hynne, Chem. Phys. Lett. $\underline{2}$, 293 (1968); R. K. Bullough and F. Hynne, Chem. Phys. Lett. $\underline{2}$, 307 (1968); F. Hynne and R. K. Bullough, J. Phys. $A\underline{5}$, 1272 (1972); and "The Scattering of Light" (to be published, 1980).

38. S. S. Hassan and R. K. Bullough (to be published); also see Phil. Trans. Roy. Soc. London $A\underline{293}$, 232 (1979).

39. E. Abraham, R. K. Bullough and S. S. Hassan, Opt. Comm. $\underline{28}$, 109 (1979); $\underline{33}$, 93 (1980); E. Abraham and S. S. Hassan, Opt. Comm. (to appear); P. Meystre, Opt. Comm. $\underline{26}$, 277 (1978); R. Bonifacio and L. Lugiato, Lett. Nuovo Cim. $\underline{21}$, 505 (1978); J. A. Herman, Opt. Acta $\underline{27}$, 159 (1980).

40. R. Saunders and R. K. Bullough, J. Phys. $A\underline{6}$, 1348 (1973).

41. M. Lax, Phys. Rev. $\underline{157}$, 213 (1967).

42. G. S. Agarwal, L. M. Narducci, R. Gilmore and D. H. Feng, Phys. Rev. $A\underline{18}$, 620 (1978); $\underline{20}$, 545 (1979); L. A. Lugiato, Nuovo Cim. $B\underline{50}$, 89 (1978), and references therein; G. S. Agarwal and S. P. Tewari, Phys. Rev. $A\underline{21}$, 1638 (1980); S. P. Tewari, Opt. Comm. $\underline{34}$, 273 (1980).

43. H. M. Gibbs, Q. H. F. Vrehen and H. M. Hikspoors, Phys. Rev. Lett. $\underline{39}$, 547 (1977); and in Ref. 12f.

OPTICAL BISTABILITY BASED UPON ATOMIC CORRELATION IN A SMALL VOLUME

C. M. Bowden

Research Directorate, U.S. Army Missile Laboratory
U. S. Army Missile Command
Redstone Arsenal, Alabama 35809

Abstract: A model for optical bistability which emphasizes
the effects of atomic pair correlation in a small volume is presented. A unitary transformation is introduced to remove the explicit time dependence of the original Hamiltonian and an ensemble
representation for the system is imposed by introducing a ficticious, ("spin") temperature in the rotating frame. The ensemble
averages have the effect, in the results, of breaking the J^2-sym-
metry intrinsic to the Hamiltonian. Adiabatic elimination of the
field variables in the mean-field limit leads to a mean-field,
atomic-level Stark-shift-dependent interatomic interaction. We use
this retarded dipole-dipole interaction in a small volume, to derive the properties of the quasi-thermodynamic ensemble representing the system in the rotating frame, and obtain the equation of
state relating the externally-applied field y to the internal field
x and the inverse of the effective temperature β_s. The equation
shows hysteresis and bistability among the three quantities x, y,
and β_s for suitable values of the parameters in the model. Stability
conditions in the hysteresis zone in the limit of a single radiation field mode are analyzed and the optical spectrum is derived.
The results predict intrinsic mirrorless optical bistability in a
small volume.

I.   INTRODUCTION

We analyze the effects of atomic correlation as a cause of
optical bistability[1] (hereafter referred to as OB). For the present, we shall confine our interests to qualitative aspects, so we
consider here the simplest possible model which offers physical

insight. The model consists of a collection of identical two-level
atoms coupled to the electromagnetic field in the dipole approximation and to an externally-applied coherent driving field. The atoms
are considered uncoupled from one another except, possibly, via their
mutual radiation field (Dicke model)[2]. We are interested primarily
in the role of interatomic interaction via the radiation field[3,4,5]
(virtual photon exchange); therefore the atoms are assumed to be
contained within a volume small compared to an atomic bare transition wavelength. This restriction allows us to analyze the effects
of atomic correlation in an unencumbered fashion, i.e., without
having to deal explicitly with retardation effects over a distributed volume.

It is well known that a first-order phase transition cannot
occur due to atomic correlation for a system described by a J^2 conserving Dicke model Hamiltonian[6,7]. Therefore, a crucial step in
the derivation is to break the effects of J^2 conservation in the
results. For this purpose, we represent the system as a canonical
ensemble in a unitary representation that removes the explicit time
dependence of the Hamiltonian which is contained in the coupling of
the atoms to the externally-applied field (rotating frame). This
requires the introduction of an effective or "spin temperature" in
the transformed representation, which is equivalent to making an
explicit statement about the form of the density operator in the
transformed frame. The results are obtained as ensemble averages,
which do not preserve J^2 as a quantum number.

The second major step in the calculation is to introduce the
mean field dressed atomic frequency in the adiabatic elimination of
the field variables. This amounts to introducing the mean field
atomic Stark shift as a variational parameter, to be determined
self-consistently later on in the calculation. This gives rise to
an atomic pair interaction which is nonlinearly dependent upon the
mean field Stark shift. Ordinarily, the Stark shift correction in
the adiabatic elimination provides only negligible contributions in
the results; however, in this case, it is absolutely necessary, together with J^2 symmetry breaking, for the existence of a first-
order phase transition.

It is shown that OB can be generated entirely by atomic correlation in a small volume. The results show hysteresis and bistability between the externally-applied field y, the internal field
of the atoms x, and the inverse of the effective temperature β_s.
The equation of state is formally similar (in fact identical for
perfect tuning and absolute zero of effective temperature) to that
obtained for absorptive OB from the single-mode, Maxwell-Bloch model
in the mean field limit[8,9]. The physics, however, as already discussed earlier[1], is quite different in this case.

The optical spectrum is derived and discussed and it is shown that the spectral line shape for the scattered as well as transmitted light undergoes an abrupt change at the transition between lower and upper bistable states. For the scattered light, the lower state corresponds to a single-line spectrum, whereas the upper branch gives rise to a double-peaked spectrum for lower values of the effective temperature, T_s, whereas for high values of T_s a triple-peaked spectrum is obtained resembling the cases for resonance fluorescence, and absorptive OB[9].

The next section is used to present the explicit description of the model and to obtain the main results of the calculation in the form of the equation of state relating the externally-applied field y to the internal field x and the inverse of the effective temperature β_s. Section III is used to discuss the implications of the equation of state in terms of bistability and hysteresis among the three quantities x, y and β_s. The "free energy" for the system in the transformed representation is derived in Section IV and is used to discuss stability conditions in the hysteresis zone. In Section V the optical spectrum is derived and discussed. A summary of the results and conclusions is presented in the last section.

II.   MODEL HAMILTONIAN AND DERIVATION OF THE EQUATION OF STATE

The model Hamiltonian and derivation of the equation of state is presented in detail in reference 1. For purposes of clarity and continuity, we restate the Hamiltonian here and merely outline the derivation presented earlier. In addition, we include here atomic relaxation in the derivation, which was ignored in the earlier treatment.

The Hamiltonian for the system is given by

$$H = H_o + H' \tag{2.1}$$

where, in units such that

$$\hbar = c = 1$$

$$H_o = \sum_{k=0}^{\infty} \omega_k a_k^{\dagger} a_k + \sum_{\substack{n=1,2 \\ j=1}}^{N} \varepsilon_n c_{nj}^{\dagger} c_{nj} \quad , \tag{2.2a}$$

and

$$H' = \sum_{j=1}^{N} \sum_{k=0}^{\infty} g_k a_k^{\dagger} c_{1j}^{\dagger} c_{2j} + h \cdot c + \alpha^* e^{i\omega_o t} \sum_{j=1}^{N} c_{1j}^{\dagger} c_{2j} + h \cdot c. \tag{2.2b}$$

The first and second terms in (2.2a) represent the free-field and the free-atomic system respectively, whereas the first two terms in (2.2b) describe the interaction of the atoms with the internal field and the last two terms describe their interaction with the externally-applied field of normalized amplitude α and carrier frequency ω_0. The field creation and annihilation operators conform to Bose commutation relations.

$$\left[a_k, a_\ell^\dagger\right] = \delta_{k\ell} \ , \qquad \left[a_k, a_\ell\right] = 0 \ , \tag{2.3a}$$

and the atomic operators are Bose operators[10] which satisfy

$$\left[c_{j\ell}, c_{ik}^\dagger\right] = \delta_{ji}\, \delta_{k\ell} \ , \qquad \left[c_{j\ell}, c_{ik}\right] = 0 \ . \tag{2.3b}$$

The Hamiltonian (2.2) is J^2-conserving, and will not, as it stands, lead to a first-order phase transition in the atomic-radiation field interaction due to atomic pair correlation[6]. To eliminate the effects of the intrinsic J^2-symmetry associated with (2.2), we consider the system in a canonical ensemble representation in which the results are computed in terms of ensemble averages. This has the effect of averaging over the full Hilbert space associated with (2.2) resulting in a breaking of the J^2 symmetry.

For the purpose of constructing a canonical ensemble representing the system based upon (2.2), we introduce the unitary transformation, U, that renders (2.2) explicitly time-independent. The transformation desired is

$$U = \exp\left\{\frac{i}{2}\,\omega_0 t \sum_{j=1}^{N}\left[c_{2j}^\dagger c_{2j} - c_{1j}^\dagger c_{1j}\right] + i\omega_0 t \sum_{k=0}^{\infty} a_k^\dagger a_k\right\} , \tag{2.4}$$

which is seen to be simply a transformation to a doubly rotating frame where each orthogonal axis rotates at the rate ω_0, the laser carrier frequency.

The transformed canonical Hamiltonian H_T is

$$H_T = H_{oT} + H_T' \tag{2.5}$$

where

$$H_{oT} = \sum_{k=0}^{\infty} \Omega_k a_k^\dagger a_k + \frac{1}{2}\sum_{\substack{n=1,2 \\ j=1}}^{N} \Omega_n c_{nj}^\dagger c_{nj} \ , \tag{2.6a}$$

$$H_T' = \sum_{j=1}^{N} \sum_{k=0}^{\infty} g_k a_k^\dagger c_{1j}^\dagger c_{2j} + h \cdot c \cdot$$

$$+ \alpha^* \sum_{j=1}^{N} c_{1j}^\dagger c_{2j} + h \cdot c. \tag{2.6b}$$

Here,

$$\Omega_k = \omega_k - \omega_o \quad , \quad \Omega_2 = \varepsilon - \omega_o \quad , \quad \Omega_1 = \omega_o - \varepsilon \tag{2.7}$$

and

$$\varepsilon = \varepsilon_2 - \varepsilon_1 \quad , \quad \varepsilon_2 > \varepsilon_1 \quad . \tag{2.8}$$

The canonical ensemble representing the system in steady state in the transformed representation is defined if we assume that the density operator, ρ, has the form

$$\rho \sim e^{-\beta_s H_T} \tag{2.9}$$

in the transformed frame. This assumption is valid if an effective or "spin temperature," β_s, exists[11]. The criterion for its existence is that the dispersion for the total energy for the system be well defined and non-zero. This condition will be satisfied in the present model if we introduce explicit linear damping γ_k, of the field modes. Results can then be calculated as "thermal averages," i.e., for an operator A,

$$<A> = tr \; A \; e^{-\beta_s \hat{H}_T}/Z \quad ,$$
$$Z = tr \; e^{-\beta_s \hat{H}_T} \tag{2.10}$$

where

$$\hat{H}_T = H_T - u\,N \tag{2.11}$$

and u is the chemical potential, and N is the total number of atoms. The quantity β_s is thus a parameter in the model and must be determined from experiment.

To perform the trace indicated in (2.10) for any arbitrary operator A, it is necessary to eliminate one set of variables from H_T, either field or atomic, or to linearize the Hamiltonian. A formal integral of Heisenberg's equation of motion generated from (2.6) gives for $a_k^\dagger(t)$

$$a_k^\dagger(t) = i\, g_k \sum_{j=1}^{N} \int_{-\infty}^{t} dt'\; c_{2j}^\dagger(t') c_{1j}(t') e^{i\tilde{\Omega}_k(t-t')} , \qquad (2.12)$$

where

$$\tilde{\Omega}_k = \Omega_k + i\,\gamma_k \qquad (2.13)$$

and γ_k^{-1} is added phenomenologically and is the photon lifetime in the field mode k.

Here, we take a very crucial step. In order to evaluate the integral in (2.12), we replace the atomic operators in the integrand using the relation,

$$c_{2j}^\dagger(t) = c_{2j}^\dagger(t') e^{\frac{i}{2}\,\bar{\Omega}_{\frac{1}{2}}(t-t')} , \qquad t > t' . \qquad (2.14)$$

Here

$$\bar{\Omega}_{\frac{1}{2}} = \pm\,\sigma + i\,\bar{\gamma}_{\frac{1}{2}} \qquad (2.15)$$

where σ is the mean field "dressed" atomic frequency,

$$\sigma = + \sqrt{(\varepsilon - \omega_o)^2 + 4|\Delta|^2} . \qquad (2.16)$$

In (2.16), $|\Delta|$ is the mean field Stark shift, here treated as a variational parameter to be determined self-consistently later on.

The damping term in (2.15), $\bar{\gamma}_{\frac{1}{2}}$, is the natural atomic relaxation and gives rise to the natural atomic spectral width. In order to discuss the effects of atomic correlation, unencumbered, as a cause of OB, it is assumed that $\gamma_k \gg \bar{\gamma}_{\frac{1}{2}}$ for some principle mode k.

The ansatz (2.14) in (2.12) means that the atomic variables propagate within time τ, $\tau = (t-t')$, $\tau \lesssim \gamma_k^{-1}$ determined by mean field interaction with the electromagnetic field, i.e., each atom experiences essentially the same field as all the others, which is composed of the externally-applied field amplitude α together with the reaction field due to all the other atoms. That is, within a correlation time τ on the order of or less than the photon escape time γ_k^{-1}, the atoms experience a mean field Stark shift due to the presence of the externally-applied field and mutual photon exchange.

Using (2.14) in (2.12), we get

$$a_k^\dagger(t) = -\frac{g_k^*}{\lambda_k} \sum_{j=1}^{N} c_{2j}^\dagger(t)\, c_{1j}(t) \quad , \tag{2.17}$$

where

$$\lambda_k = \Omega_k + i(\gamma_k + \bar{\gamma}_1 + \bar{\gamma}_2) - \sigma \quad . \tag{2.18}$$

If (2.17) is used in (2.6b) to eliminate the field variables, the interaction Hamiltonian becomes

$$H_T' = -\sum_{\substack{k=0 \\ }}^{\infty} \sum_{\substack{j,\ell=1 \\ j\neq\ell}}^{N} |g_k|^2 \left(\frac{1}{\lambda_k} + \frac{1}{\lambda_k^*}\right) c_{2\ell}^\dagger c_{1\ell} c_{1j}^\dagger c_{2j}$$

$$+ \left(\alpha* \sum_{j=1}^{N} c_{1j}^\dagger c_{2j} + h\cdot c.\right) \quad , \tag{2.19}$$

where it is understood that the self-energy terms, i.e., those corresponding to $j=\ell$, are included in the diagonal part of the Hamiltonian H_{OT} and give rise to single-atom frequency renormalization. These self-energy terms contribute on the order $1/N$ times the cooperative frequency shifts accounted for by the atomic pair interaction terms in (2.19), and are therefore neglected. It is to be noted that the first term on the right-hand side of (2.19) describes interatomic pair interaction, and the interaction, by (2.18) and (2.16) is Stark-shift dependent, i.e., a function $|\Delta|$ (which is yet to be determined self-consistently).

For simplicity, we consider only one principle or "cavity" mode $k=k_c$, nearly resonant at the atomic bare frequency, and taking the sum over modes in (2.19) to an integral, we obtain[1],

$$H_T' = -\bar{g}/N \sum_{\substack{j,\ell=1 \\ j\neq\ell}}^{N} c_{2\ell}^\dagger c_{1\ell} c_{1j}^\dagger c_{2j} + \left(\alpha* \sum_{j=1}^{N} c_{1j}^\dagger c_{2j} + h\cdot c.\right), \tag{2.20}$$

$$\bar{g} = g_o^2 \frac{(\omega_c - \omega_o - \sigma)}{[(\omega_c - \omega_o - \sigma)^2 + \gamma^2]} \quad . \tag{2.21}$$

Here, ω_c is the principal or "cavity" frequency and

$$\gamma = \gamma_1 + \gamma_2 + \gamma_{k_c} \quad , \tag{2.22}$$

and g_o^2 is given in terms of the Hamiltonian parameters in reference 1.

Equation (2.20) is our working, or effective, interaction Hamiltonian. A number of calculational methods can be used to determine steady-state results using (2.20) consistent with (2.9). We have, for convenience, used the thermodynamic Green's function method[1]. We simply outline the procedure here.

The following set of single particle Green's functions are defined,

$$G_{n,i}(t-t') = -i\theta(t-t') < \left[C_{n,i}(t), C_{n,i}^{\dagger}(t') \right] > , \qquad (2.23a)$$

$$F_{1,i}(t-t') = -i\theta(t-t') < \left[C_{2,i}(t), C_{1,i}^{\dagger}(t') \right] > , \qquad (2.23b)$$

$$F_{2,i}(t-t') = -i\theta(t-t') < \left[C_{1,i}(t), C_{2,i}^{\dagger}(t') \right] > , \qquad (2.23c)$$

where

$$\begin{aligned}
\theta(\tau) &= 1 \qquad & \tau > 0 \;, \\
\theta(\tau) &= 0 \qquad & \tau < 1 \;.
\end{aligned} \qquad (2.24)$$

If (2.20) is used together with (2.6a) to calculate the equations of motion for (2.23), and if the mean field factorization[1]

$$< \left[C_{1,j}^{\dagger}(t) C_{2,j}(t) C_{1,i}(t), C_{1,i}^{\dagger}(t') \right] > = <C_{1,j}^{\dagger}(t) C_{2,j}(t)>$$

$$\times < \left[C_{1,i}(t), C_{1,i}^{\dagger}(t') \right] > \qquad (2.25)$$

is used to terminate the resulting hierarchy of coupled, first-order differential equations, the following equations of motion for the identical single particle Green's functions are obtained in the frequency domain,

$$(\tilde{\omega}_1 - \tfrac{1}{2}\Omega_1) \, G_1(\omega) = \frac{1}{2\pi} - \Delta' F_1(\omega) \;, \qquad (2.26a)$$

$$(\tilde{\omega}_2 - \tfrac{1}{2}\Omega_2) \, F_1(\omega) = -\Delta'^* \, G_1(\omega) \;, \qquad (2.26b)$$

where

$$\Delta' \equiv \bar{g} \, F^* - \alpha^* \;, \qquad (2.27)$$

and

$$F \equiv <C_1^{\dagger} C_2> \qquad (2.28)$$

is proportional to the mean field macroscopic atomic polarization.
Here,

$$\tilde{\omega}_n = \omega + i \, \bar{\gamma}_n (Sg\omega) \ . \tag{2.29}$$

A set of equations similar to (2.26) are obtained for G_2 and F_2.

If we take

$$\bar{\gamma}_1 = \bar{\gamma}_2 = \bar{\gamma} \ , \tag{2.30}$$

the eigenfrequencies, $\Gamma_\pm$, of (2.26) are easily determined as,

$$\Gamma_\pm = - i\bar{\gamma}(Sg\omega) \pm \frac{1}{2} \left[\Omega^2 + 4|\Delta'|^2 \right]^{\frac{1}{2}} \tag{2.31}$$

where

$$\Omega = \varepsilon - \omega_o \ . \tag{2.32}$$

Comparing (2.31) with (2.16) and (2.15) we see that the variational
parameter Δ introduced in (2.14) is now determined self-consistently,
i.e.,

$$\Delta = \Delta' = \bar{g} \, F^* - \alpha^* \ . \tag{2.33}$$

Using the spectral theorem for thermodynamic Green's functions[12],
and normalization from atomic number conservation[1], and taking the
limit of classical statistics[12,1] the atomic polarization F, (2.28),
is calculated,

$$F = \Delta^* \ \frac{\tanh \frac{1}{2} \beta_s \sigma}{\sigma} \ . \tag{2.34}$$

It is useful to make the identification in (2.33),

$$E_I \equiv \bar{g} \, F - \alpha \tag{2.35}$$

for the internal field, and

$$E_A \equiv - \alpha \tag{2.36}$$

for the externally-applied field amplitude. Then, after multiply-
ing both sides of (2.34) by $\bar{g}$ and using (2.35) and (2.36), we arrive at the equation of state[1],

$$E_I - E_A = \bar{g} \, E_I \ \frac{\tanh \frac{1}{2} \beta_s \sigma}{\sigma} \ . \tag{2.37}$$

It is expedient to define dimensionless quantities,

$$x = E_I/\gamma \quad , \quad y = E_A/\gamma \tag{2.38a}$$

$$\delta = (\omega_c - \omega_o)/\gamma \ , \quad \nu = (\epsilon - \omega_o)/\gamma \tag{2.38b}$$

$$C = \tfrac{1}{2} g_o^2/\gamma^2 \quad . \tag{2.38c}$$

If (2.38) is used in (2.37), the result is

$$y = x - 2C \ \frac{(\delta-\sigma')x}{(\delta-\sigma')^2 + 1} \ \frac{\tanh \tfrac{1}{2} \beta_s \gamma \sigma'}{\sigma'} \quad , \tag{2.39}$$

where

$$\sigma' = (\nu^2 + x^2)^{\tfrac{1}{2}} \quad . \tag{2.40}$$

Equation (2.39) is formally identical to Eq. (3.31) of reference 1 obtained by neglecting atomic relaxation. The only difference in the results given here in (2.37) and (2.39) and those obtained earlier[1] is that γ, given by (2.22) now contains the atomic relaxation rates as well as the field damping. That is, the atomic relaxation has no essential qualitative effect on the equation of state for the mean field, small volume model as long as the field linear damping is dominant. If the atomic relaxation or dephasing becomes dominant, the mean field Stark shift approximation, (2.14) and (2.15), becomes invalid.

Equation (2.39) was shown earlier[1] to give rise to a first-order phase transition corresponding to hysteresis and bistability between the input field amplitude y and the local field amplitude x for various values for the detunings ν and δ in the limit of low effective temperature β_s^{-1}.

III. EFFECTIVE TEMPERATURE-INDUCED BISTABILITY

This section is used to discuss the effective temperature dependence associated with the first-order phase transition described by (2.39). For conditions of perfect tuning, (2.39) becomes

$$y = x + 2C \ \frac{x}{1+x^2} \ \tanh \tfrac{1}{2} \Gamma x \quad , \tag{3.1}$$

where

$$\Gamma \equiv \gamma/kT_s \quad . \tag{3.2}$$

This equation is shown plotted in Fig. 1 for y vs. x for $C = 60$ and various values of $0 < \Gamma < 1$. In the limit that $T_s \to 0$, Eq. (3.1) is identical in form to the results of models which describe

absorptive OB from the standpoint of single atom response to the
radiation field[9,13]. As described previously[1], the physics in the
case treated here is, however, quite different. The OB discussed
in reference 1 and depicted in Fig. 1 is the result of catastrophic
breakdown in the interatomic pair correlation due to the large induced collective Stark shift at sufficiently large applied field
amplitudes. In previous models[9,13], OB is caused by the breakdown
of single-atom correlation to the coherent driving field, due to
nonlinear absorption. Our model does not take into account the effects of nonlinear absorption, since the mean field Stark shift approximation (2.14), (2.15), and (2.16) is invalid when the atomic
relaxation rates are dominant over the linear field damping.

It was pointed out in reference 1 that the instability responsible for the first-order phase transition described by (2.39) is
the result of essentially two important properties of our model:
(a) the internal field dependent effective atom-atom interaction,
(2.21), which provides self-induced feedback between the atomic
system and the internal field; and (b) the internal mean field induced Stark shift per atom (2.15), which determines the functional
dependence of the interatomic interaction (2.21), and hence the
functional form for (2.39).

In the lower internal field branch in the hysteresis zone,
Fig. 1, the interatomic interaction (2.21), is large and the reaction field due to the presence of the atoms (first term in (2.35))
opposes the externally-applied field y causing the total field x
to be small. The interaction, (2.21), is interpreted as due to
virtual photon exchange between pairs of atoms. The behavior of
the atomic system corresponding to the lower branch in the hysteresis zone, Fig. 1, is analogous to the response of a diamagnetic
material to an externally-applied magnetic field.

Correspondingly, in the upper (larger internal field) branch
of the hysteresis zone, the atomic interaction, (2.21), is small
due to the relatively large Stark shift Δ, whereas the atomic polarization is large (2.34). Thus, the reaction field due to the atoms
is small, but the macroscopic polarization is large. So, in the
upper branch, the atomic system becomes transparent to the externally-applied driving field with the dipole moment of each atom in-
dividually responding to the driving field (i.e., Rabi cycling).

The results of this model are in strong contrast to those of
models for OB corresponding to single-atom response to the driving
field[9,13] (saturable absorption). In those cases the upper bistable
state represents bleaching (saturation) of the atomic system and
vanishing macroscopic dipole moment. If spatial distribution and retardation effects are taken into account, we believe the results of
our model will reduce to the results of previous models[9,13] when

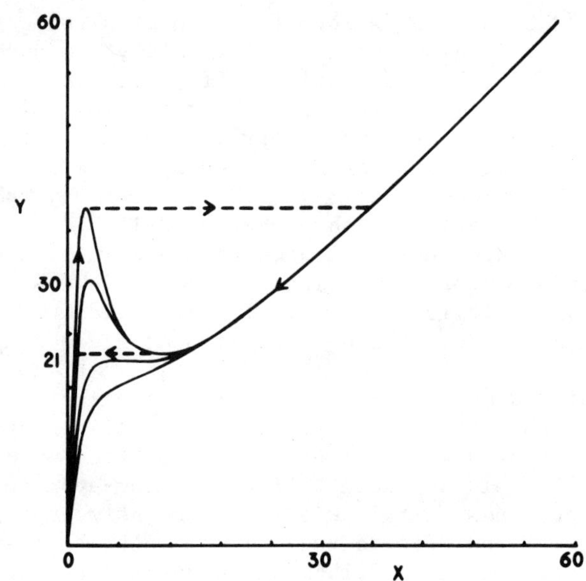

Fig. 1.   Normalized externally-applied field y vs. normalized in-
           ternal field x according to Eq. (3.1) with C = 60.   Each
           curve corresponds to a different value for Γ, and in
           order of decreasing values for the critical value for
           y, Γ = 1.00, 0.65, 0.35, 0.25.   The threshold value of Γ
           for bistability in this case is Γ = 0.35.   The dotted
           lines and arrows indicate a particular hysteresis cycle.

the atomic dephasing rate becomes comparable to or greater than the
open cavity photon escape rate.

It is interesting to invert (3.1) to obtain Γ as a function of
x and y,

$$\Gamma = \frac{1}{x} \ln \left\{ \frac{2Cx + (y-x)(x^2 + 1)}{2Cx - (y-x)(x^2 + 1)} \right\} . \qquad (3.3)$$

The above equation is plotted in Fig. 2 showing the variation of Γ
as a function of the internal field x for various values of the externally-applied field y, and C = 60 as in Fig. 1.   Figure 2 clearly
demonstrates that Eq. (3.3) possesses a first-order phase transition
with two critical temperatures corresponding to the upper and lower
turning points shown in Fig. 2, for the externally-applied field y
fixed at a value greater than a certain critical value.   In the case
presented, for C=60, the threshold value for y is y_c=21.0.   This is
seen to be consistent with the corresponding threshold value of Γ,
Γ = 0.35 shown in Fig. 1.

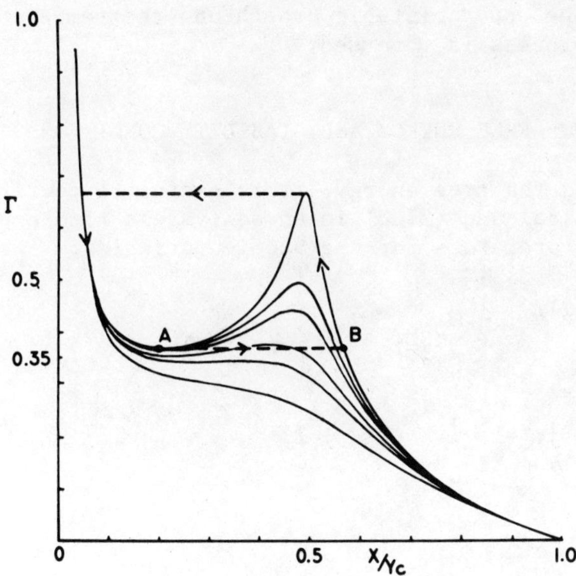

Fig. 2. Γ vs. x/y_c according to Eq. (3.3) for C=60 and various
           values for the fixed externally-applied field y_c. In
           order of decreasing values for the critical value for
           Γ, y_c = 21.8, 21.7, 21.6, 21,3, 21.0, 20.0. The threshold
           value of y_c for bistability is y_c = 21.0 which is seen
           to be consistent with the threshold value of Γ for bi-
           stability in Fig. 1. The dotted lines and arrows indi-
           cate a particular hysteresis cycle.

On the lower bistable branch (small internal field x) the interatomic interaction is large and the reaction field due to the
atoms opposes the applied field. Raising the effective temperature
T_S (i.e., lowering Γ), tends to destroy the atom-atom pair correlation, thus reducing the reaction field (first term on the r.h.s.
of Eq. (2.35)). If the fixed applied field is larger than the
critical value, as the temperature T_S is increased, then at a certain temperature (i.e., point A in Fig. 2), the internal field
abruptly increases due to catastrophic reduction in the atomic pair
interaction and hence the reaction field. The system is then on
the upper branch in the hysteresis zone (i.e., point B in Fig. 2),
corresponding to larger macroscopic polarization and smaller atom-
atom interaction (both due to sudden increase in the Stark shift
Δ). As the temperature T_S is further increased, the internal field
x approaches the applied field y as the macroscopic polarization
approaches zero and the atomic system becomes "bleached." If the

system is in the upper bistable branch and the temperature T_s is lowered, the process is reversed.

IV. FREE ENERGY AND STABILITY CONDITIONS

To discuss the free energy, it is useful to rewrite the Hamiltonian (2.6a) and (2.20) in an equivalent form explicitly in terms of Pauli operators for the atomic variables,

$$H_{oT} = \Omega_f \, a^\dagger a + \frac{1}{2} \Omega \sum_{j=1}^{N} \sigma_j^z \, , \tag{4.1a}$$

$$H_T' = -\frac{\bar{g}}{N} \sum_{\substack{j,\ell=1 \\ j \neq \ell}}^{N} \sigma_j^+ \sigma_\ell^- + \alpha^* \sum_{j=1}^{N} \sigma_j^- + \alpha \sum_{j=1}^{N} \sigma_j^+ \, , \tag{4.1b}$$

where

$$\Omega_f = \omega_c - \omega_o \, . \tag{4.2}$$

We shall now proceed to calculate the free energy F in the rotating frame,

$$e^{-\beta_s F} = \mathrm{tr}\, e^{-\beta_s H_T} = Z \, . \tag{4.3}$$

In keeping with the mean field approximation already made in (2.14) in the adiabatic elimination of the field variables, we linearize (4.1b) to obtain a "thermodynamically equivalent" mean field Hamiltonian. We use the linearization procedure introduced earlier[14,15,16] by making the substitution in (4.1b),

$$\sigma_j^- = (\sigma_j^- - \nu_j) + \nu_j \, , \tag{4.4}$$

where the ν_j are treated as variational parameters. The result is that (4.1b) becomes

$$H_T' = \alpha^* \sum_{j=1}^{N} \sigma_j^- + \alpha \sum_{j=1}^{N} \sigma_j^+ + \frac{\bar{g}}{N} \sum_{\substack{j,\ell=1 \\ j \neq \ell}}^{N} \nu_j^* \nu_\ell$$

$$- \frac{\bar{g}}{N} \sum_{\substack{j,\ell=1 \\ j \neq \ell}}^{N} \sigma_j^+ \nu_\ell - \frac{\bar{g}}{N} \sum_{\substack{j,\ell=1 \\ j \neq \ell}}^{N} \sigma_j^- \nu_\ell^*$$

$$- \frac{\bar{g}}{N} \sum_{\substack{j,\ell=1 \\ j \neq \ell}}^{N} (\sigma_j^+ - \nu_j^*)(\sigma_\ell^- - \nu_\ell) \, . \tag{4.5}$$

The last term on the r.h.s. of the above expression contributes to
the results only corrections due to mean-square deviations, if we
make the ansatz

$$\nu_j \equiv \langle \bar{\sigma}_j \rangle \; . \tag{4.6}$$

If, however, the ν's are determined variationally so as to minimize
the free energy, then the above definition is expected to be actually a result. Thus, we neglect the last term in the above
Hamiltonian.

Consistent with the mean-field approximation made in Section
II, we take all the particles as having identical statistical averages, and in the limit of large N, the Hamiltonian thermodynamically
equivalent to (4.1), $\tilde{H}_T$, is neglecting fluctuations,

$$\tilde{H}_T = H_F + H_A + H_o \; , \tag{4.7}$$

where

$$H_F = \Omega_f \, a^\dagger a \tag{4.8a}$$

$$H_A = \tfrac{1}{2} \Omega \sum_{j=1}^{N} \sigma_j^z - \Delta^* \sum_{j=1}^{N} \sigma_j^- - \Delta \sum_{j=1}^{N} \sigma_j^+ \tag{4.8b}$$

$$H_o = \frac{N}{g} \left[|\Delta|^2 + |\alpha|^2 + \alpha\, \Delta^* + \alpha^*\Delta \right] , \tag{4.8c}$$

and

$$\Delta \equiv \bar{g}\, \nu^* - \alpha^* \; , \tag{4.9}$$

where $\nu = \nu_j$, $j = 1, 2, \ldots, N$.

The Hamiltonian (4.8b) has an immediate interpretation as a mean
field interaction and is consistent with the mean field assumption
in Section II.

The required trace is now easily performed, using (4.7) and
(4.8), to give

$$e^{-\beta_s \tilde{F}} = \mathrm{tr}\; e^{-\beta_s \tilde{H}_T} = \left(\mathrm{tr}\; e^{-\beta_s H_F} \right)\left(\mathrm{tr}\; e^{-\beta_s H_A} \right)\left(\mathrm{tr}\; e^{-\beta_s H_o} \right). \tag{4.10}$$

Thus, from (4.9) and (4.8),

$$\tilde{F} = \frac{N}{g}\left[|\Delta|^2 + |\alpha|^2 + \alpha\Delta^* + \alpha^*\Delta\right] - \frac{N}{\beta_s}\ell n\left[2\,\cosh\frac{\beta_s}{2}\sigma\right]$$

$$+ \frac{1}{\beta_s}\ell n\left(1 - e^{-\beta_s\Omega_f}\right) \quad . \tag{4.11}$$

If we let

$$x = \Delta^* \quad ; \quad y = -\alpha \tag{4.12}$$

then (4.11) becomes

$$\tilde{F} = \frac{N}{g}|x-y|^2 - \frac{N}{\beta_s}\ell n\left[2\,\cosh\frac{\beta_s}{2}\sigma\right] + \frac{1}{\beta_s}\ell n\left(1 - e^{-\beta_s\Omega_f}\right) \tag{4.13}$$

where

$$\sigma = +\sqrt{\Omega^2 + 4|\Delta|^2} \quad . \tag{4.14}$$

From (4.9) and (4.12) we have

$$x = \bar{g}\nu + y \quad . \tag{4.15}$$

From (4.8b) and (4.12), Eq. (4.15) is interpreted as the mean field
acting on a single atom due to the normalized reaction field from
all the other atoms, $\bar{g}\nu$, superimposed on the externally-applied
field y. Thus, (4.15) is equivalent to (2.35) and the linearization
(4.4) and (4.8) is consistent with the mean field approximation in
Section II. Furthermore, it is seen that (4.15) and (2.35) are,
after all, just Maxwell's equation in the rotating frame.

Using (4.15), the free energy (4.13) can be written in terms
of the variational parameter ν,

$$\tilde{F} = N\bar{g}|\nu|^2 - \frac{N}{\beta_s}\ell n\left|2\,\cosh\frac{\beta_s}{2}\sigma\right| + \frac{1}{\beta_s}\ell n\left(1 - e^{-\beta_s\Omega_f}\right) \quad . \tag{4.16}$$

The normal equation

$$\frac{\partial F}{\partial\nu} = 0 \tag{4.17}$$

yields

$$\frac{\partial F}{\partial\nu} = N\bar{g}\nu^* - \frac{N}{\beta_s}\left(\tanh\frac{\beta_s}{2}\sigma\right)\left(\frac{\beta_s}{2}\,\frac{\partial\sigma}{\partial\nu}\right) = 0 \tag{4.18}$$

where, using (4.14) and (4.15),

$$\frac{\partial\sigma}{\partial\nu} = 2\bar{g}\,\frac{x^*}{\sigma} \quad . \tag{4.19}$$

From (4.18) and (4.19) we get

$$\nu = x \; \frac{\tanh \frac{\beta_s}{2} \sigma}{\sigma} \qquad (4.20)$$

which is identical to Eq. (2.34) and is seen to be the macroscopic polarization as anticipated by (4.6). Equations (4.15) and (4.20) constitute the main result of this section and are identical with the corresponding results of Section II, Eqs. (2.27) and (2.34), respectively. Thus, the linearization (4.4) and (4.5) and the resulting free energy (4.13) or (4.16) emphasize the mean field character of the model.

If we multiply both sides of (4.20) by $\bar{g}$ we get the equation of state (2.37),

$$x - y = \bar{g} \; x \; \frac{\tanh \frac{\beta_s}{2} \sigma}{\sigma} \quad . \qquad (4.21)$$

The equation of state, (4.21) can be used to eliminage $\bar{g}$ from the free energy (4.13), under the constraint (4.17), and the result is

$$\tilde{F} = N \; x(x-y) \; \frac{\tanh \frac{\beta_s}{2} \sigma}{\sigma} \; - \; \frac{N}{\beta_s} \; \ln \left| 2 \cosh \frac{\beta_s}{2} \sigma \right| + \frac{1}{\beta_s} \; \ln \left(1 - e^{-\beta_s \Omega_f} \right),$$
$$(4.22)$$

and from this we have, using (4.20)

$$\frac{\partial \tilde{f}}{\partial y} = - \nu \qquad (4.23)$$

where $\tilde{f}$ is the free energy per particle. A fundamental property for thermodynamic stability[17] is

$$\frac{\partial^2 \tilde{f}^2}{\partial y^2} = - \frac{\partial \nu}{\partial y} = - k \qquad (4.24)$$

where k is the polarizability, and it is required that

$$k > 0 \quad . \qquad (4.25)$$

Again, using (4.20) we have

$$\frac{\partial \nu}{\partial y} = \frac{\partial x}{\partial y} \left\{ \frac{\tanh \frac{\beta_s}{2} \sigma}{\sigma} \left[1 - \frac{x^2}{\sigma^2} \right] + \frac{\beta_s}{2} \frac{x^2}{\sigma^2} \, \mathrm{sech}^2 \frac{\beta_s}{2} \sigma \right\} > 0 \quad . \qquad (4.26)$$

Since the term in brackets on the r.h.s. of (4.26) is positive
definite, a condition of stability is that

$$\frac{\partial x}{\partial y} > 0 \quad . \tag{4.27}$$

So, the negative derivative region in the hysteresis zone of Fig. 1
is unstable whereas the lower and upper branches are stable.

V.   THE LIGHT SPECTRUM

Here, we calculate the power spectrum of the transmitted and
scattered light. Results of a photon counting experiment yield a
power spectrum[18]

$$I(\omega) = \int_{-\infty}^{\infty} \langle E^-(t) \, E^+(o) \rangle \, e^{i\omega t} \, dt. \tag{5.1}$$

For simplicity, we take the normalized negative frequency part of
the E-field operator as

$$E^- = \sum_{k=o}^{\infty} a_k^{\dagger} \quad , \tag{5.2}$$

and ignoring contributions from cross-spectral components,

$$I(\omega) = \sum_k I_k(\omega) \quad . \tag{5.3}$$

Thus, the quantities we wish to compute are

$$I_k(\omega) = \int_{-\infty}^{\infty} \langle a_k^{\dagger}(t) \, a_k(o) \rangle \, e^{i\omega t} \, dt \quad . \tag{5.4}$$

Making use of Eqs. (2.17) and (2.28),

$$\langle a_k(t) a_k(o) \rangle = |\alpha|^2 - \frac{g_k^*}{\lambda_k} \, N \, \alpha \, F^* - \frac{g_k}{\lambda_k^*} \, N \, \alpha^* F$$

$$+ \frac{|g_k|^2}{|\lambda_k|^2} \, \sum_j \sum_\ell \langle C_{2j}^{\dagger}(t) C_{1j}(t) C_{2\ell}(o) C_{1\ell}^{\dagger}(o) \rangle \quad . \tag{5.5}$$

The last term in (5.5) can be written as

$$\sum_j \sum_\ell <C_{2j}^\dagger(t)C_{1j}(t)C_{2\ell}(o)C_{1\ell}^\dagger(o)>$$

$$= \sum_{j=1}^{N} <C_{2j}^\dagger(t)C_{1j}(t)C_{2j}(o)C_{1j}^\dagger(o)>$$

$$+ \sum_{\substack{j,\ell=1 \\ j \neq \ell}}^{N} <C_{2j}^\dagger(t)C_{1j}(t)C_{2\ell}(o)C_{1\ell}^\dagger(o)> \quad . \tag{5.6}$$

The Fourier transform of the first term on the r.h.s. of (5.6) gives the single atom light spectrum and is the power spectrum of the scattered light, whereas the second term yields the collective atomic response to the electromagnetic field and contributes an N^2 dependency to the intensity of the transmitted light. The other terms in (5.5) constitute the transmitted light comprised of the externally-applied field and cross terms which arise because of the single atom coherent response to the externally-applied driving field.

If the mean field factorization is used, the second term on the r.h.s. of (5.6) gives contributions,

$$<C_{2j}^\dagger(t)C_{1j}(t)C_{2\ell}(o)C_{1\ell}^\dagger(o)> \rightarrow <C_{2j}^\dagger(t)C_{1j}(t)>$$

$$\times <C_{2\ell}(o)C_{1\ell}^\dagger(o)> = |F|^2 \quad , \tag{5.7}$$

and the first term yields contributions,

$$<C_{2j}^\dagger(t)C_{1j}(t)C_{2j}(o)C_{1j}^\dagger(o)> \rightarrow <C_{2j}^\dagger(t)C_{2j}(o)>$$

$$\times <C_{1j}(t)C_{1j}^\dagger(o)> \quad . \tag{5.8}$$

So, for N large, and taking into account that the atoms are identical, (5.5) can be written as

$$<a_k^\dagger(t)a_k(o)> = \left| \alpha - N \frac{g_k}{\lambda_k^*} F \right|^2$$

$$+ N \frac{|g_k|^2}{|\lambda_k|^2} <C_2^\dagger(t)C_2(o)><C_1(t)C_1^\dagger(o)> \quad . \tag{5.9}$$

The first term on the r.h.s. of (5.9), in the spectral domain, constitutes the coherent part of the light spectrum and is the transmitted component. The second term is the incoherent part and constitutes the scattered light spectrum.

To evaluate (5.9) we must calculate the two-time correlation functions which appear in the last term. This is easily accomplished from the Green's functions (2.23) and the equations of motion (2.26). From the spectral theorem for thermodynamic Green's functions[12],

$$<C_n^\dagger(o)C_n(t)> = -2 \int_{-\infty}^{\infty} \frac{d\omega e^{-i\omega t}\ \mathrm{Im}\ G_n(\omega)}{\left[e^{\beta_s(\omega-u)} - 1\right]} , \qquad (5.10)$$

and $<C_n(t)C_n^\dagger(o)>$ is related to (5.10) by the relation[12]

$$<C_n(t)C_n^\dagger(o)> = -2 \int_{-\infty}^{\infty} d\omega e^{-i\omega t}\ \mathrm{Im}\ G_n(\omega) + <C_n^\dagger(o)C_n(t)> . \qquad (5.11)$$

Equations (5.10) and (5.11) are sufficient to calculate the correlation functions in the last term of (5.9).

From (2.26) and a similar set of equations for G_2 and F_2, we get

$$G_1(\omega) = \frac{1}{2\pi}\ \frac{(\omega + i\bar\gamma + \tfrac{1}{2}\Omega)}{(\omega + i\bar\gamma - \tfrac{1}{2}\sigma)(\omega + i\bar\gamma + \tfrac{1}{2}\sigma)} \qquad (5.12a)$$

and

$$G_2(\omega) = \frac{1}{2\pi}\ \frac{(\omega + i\bar\gamma - \tfrac{1}{2}\Omega)}{(\omega + i\bar\gamma - \tfrac{1}{2}\sigma)(\omega + i\bar\gamma + \tfrac{1}{2}\sigma)} . \qquad (5.12b)$$

Using (5.12) in (5.10),

$$<C_1^\dagger(o)C_1(t)> = \frac{e^{-\bar\gamma t}}{2\sigma\left(e^{\beta_s/2^\sigma} + e^{-\beta_s/2^\sigma}\right)} \left\{(\sigma+\Omega)e^{-\beta_s/2^\sigma}e^{-i\sigma/2t}\right.$$

$$\left. + (\sigma-\Omega)\ e^{\beta_s/2^\sigma}e^{i\sigma/2t}\right\} , \qquad (5.13a)$$

$$<C_2^\dagger(o)C_2(t)> = \frac{e^{-\bar\gamma t}}{2\sigma\left(e^{\beta_s/2^\sigma} + e^{-\beta_s/2^\sigma}\right)} \left\{(\sigma-\Omega)e^{-\beta_s/2^\sigma}e^{-i/2\sigma t}\right.$$

$$\left. + (\sigma+\Omega)\ e^{\beta_s/2^\sigma}e^{i/2\sigma t}\right\} \qquad (5.13b)$$

As a simple check on the results of (5.13), the population difference W in the rotating frame, at equal times,

$$W \equiv \left[<c_2^\dagger c_2> - <c_1^\dagger c_1> \right] \qquad (5.14a)$$

from (5.13) is

$$W = - \frac{|\Omega|}{\sigma} \tanh \frac{\beta_s}{2} \sigma \quad . \qquad (5.14b)$$

Let

$$S_A(t) \equiv <c_2^\dagger(t)c_2(o)> <c_1(t)c_1^\dagger(o)> \quad , \qquad (5.15a)$$

$$S_A(\omega) = \frac{1}{2\pi} \int_{-\infty}^{\infty} S_A(t)e^{-i\omega t} \, dt \quad . \qquad (5.15b)$$

Thus, using the Hermitian conjugate of (5.13b) and using (5.13a) together with (5.11) to obtain the second factor in (5.15a), we get for (5.15b),

$$S_A(\omega) = \frac{\bar{\gamma}}{4\pi} \frac{\left(1 - \frac{\Omega^2}{\sigma^2}\right)}{\left[\omega^2 + 4\bar{\gamma}^2\right]} \left\{3 + \tanh^2 \frac{\beta_s}{2} \sigma\right\}$$

$$+ \frac{\bar{\gamma}}{4\pi} \frac{\left(1 - \frac{\Omega}{\sigma}\right)^2}{\left[(\omega+\sigma)^2 + 4\bar{\gamma}^2\right]} \left\{\frac{1 + \frac{1}{2} e^{-\beta_s \sigma}}{\cosh^2 \beta_s/2\sigma}\right\}$$

$$+ \frac{\bar{\gamma}}{4\pi} \frac{\left(1 + \frac{\Omega}{\sigma}\right)^2}{\left[(\omega-\sigma)^2 + 4\bar{\gamma}^2\right]} \left\{\frac{1 + \frac{1}{2} e^{+\beta_s \sigma}}{\cosh^2 \beta_s/2\sigma}\right\} \quad , \qquad (5.16)$$

where

$$\omega = \omega' - \varepsilon. \qquad (5.17)$$

It is seen that (5.16) predicts in general a three component spectrum for the light scattered by single atoms. The spectrum is composed of a central component centered at $\omega = 0$, and two sidebands of different peak heights, centered on either side of the central component, and separated from it by the Stark shift, i.e., at $\omega = \pm \sigma$.

For perfect tuning, i.e., for $\Omega = 0$, at low effective temperature, $\beta_s \to \infty$, (5.16) exhibits a strong component centered at $\omega = 0$
and a weaker component centered at $\omega = \sigma$, i.e., a two-component
spectrum. The peak intensities are in the ratio 2:1.

For the high temperature limit, $\beta_s \to 0$, (5.16) shows a three-
component spectrum which is symmetric for $\Omega = 0$. It consists of
a strong central component centered at $\omega = 0$ and two weaker sidebands at $\omega = \pm \sigma$, with peak intensities in the ratio of 2:1. This
case is reminiscent of the fluorescence spectrum of absorptive OB[9]
and resonance fluorescence[19].

The first term in (5.9) is delta-correlated in the spectral
domain. Thus, the light spectrum consists of a strong central component for the transmitted spectrum corresponding to the first term
in (5.9) with a contribution which is proportional to N^2. The
scattered light stems from the second term in (5.9) and its spectrum
is given by (5.16) and discussed above. Along the lower bistable
branch, Fig. 1, the scattered spectrum consists of a single component centered at the atomic resonance, i.e., the side-bands are
contained within the line width $\bar{\gamma}$ in the lower bistable branch since
the stark-shift $\sigma < \bar{\gamma}$ along this branch. As the first tuning point
is approached along the lower branch with increased externally-
applied field, α, the scattered light spectrum becomes broader, due
to increased Stark shift, and asymmetric except at high effective
temperatures (note that the effect of higher effective temperature
is to destroy bistability as shown in Fig. 1).

At the phase transition, the scattered light spectrum undergoes a discontinuity; a side-band appears at the high frequency
side, for the case $\beta_s \to \infty$, and the "central" component becomes
narrower as seen from (5.16).

VI.   SUMMARY AND CONCLUSIONS

We have treated atomic pair correlation in a small volume in
the mean field limit and have shown in Section II, and reference 1,
that this leads to a first-order phase transition in the internal
field x as a function of the externally-applied field y, for fixed
effective temperature β_s, (2.39). In addition, we have determined
the effects of the effective temperature on this phase transition
from Eq. (3.1) as shown by the family of curves in Fig. 1, each
curve corresponding to a different value for the inverse effective
temperature β_s. It is seen that the overall effect of increasing
temperature, T_s, is equivalent to decreasing C, (3.1).

Equation (3.1) was inverted, (3.3), to give the internal field
x as a function of the inverse of the effective temperature T_s (in

the form of Γ, (3.2)), for fixed values of the applied external field y. These results are shown by the family of curves in Fig. 2, where each curve corresponds to a particular value for y. The Figs. 1 and 2 show hysteresis and bistability among the three quantities x, y, and β_s, which are related by the equation of state (2.39).

There are two essential assumptions in our calculation to arrive at (2.39): (i) the effects of the J^2 conservation intrinsic to the Hamiltonian (2.2) are destroyed in the calculation of the results by replacing the system by a canonical ensemble. This is done by introducing a ficticious or "spin temperature," (2.9), in the unitary representation which removes the explicit time dependency of (2.2), i.e., if we define a "cooperation number" r by

$r(r+1) \equiv \langle J^2 \rangle$, where $J_i = \sum_{j=1}^{N} \sigma_{ij}$, i = x,y,z, are the usual collective

operators, then by (2.10) and (4.8), $r(r+1) = \frac{3N}{4} + \frac{N}{4} (N-1) \tanh^2 \frac{\beta_s}{2} \sigma$.

Thus the collection of atoms behaves as N independent spin ½ particles for $T_s \rightarrow \infty$, whereas for $T \rightarrow 0$ the system behaves as a giant spin moment r = N/2. The phenomenon has been termed condensation of "cooperation number" in the ground state[14]. (ii) The mean-field Stark shift per atom, Δ, is introduced as a variational parameter in the "dressed" atom frequency in the adiabatic elimination of the field variables (2.12), and determined self-consistently (2.33). Both of these assumptions, ensemble averages (2.10) and mean-field Stark shift (2.15) are essential to the existence of a first-order phase transition in this model for atomic pair correlation in a small volume and to the functional form of the equation of state (2.39).

As cited earlier[1], the atomic pair interaction (2.20) is formally similar to the interaction between Cooper pairs in the BCS theory of superconductivity as well as the interatomic interaction which gives rise to collective effects in the theory of superfluorescence. Equation (2.39), represented in Figs. 1 and 2, shows a first-order phase transition with critical effective temperature T_s and critical input field y_c. Thus, Eq. (2.39) may represent the closest optical analogue to the Meisner effect in superconductivity.

Using the mean-field Hamiltonian (4.8), which is thermodynamically equivalent to (2.6a) and (2.20), the free energy $\tilde{F}$ (4.13) in the transformed representation was derived. From the minimization of the free energy $\tilde{F}$ with respect to the variational parameter ν, the equation of state (4.21) was obtained, which is identical to (2.39). This corroborates the mean-field interpretation of the results. From the free energy (4.22), the stability conditions with respect to the equation of state were analyzed and it was found that all states defined by (2.39) are stable except those between the turning points, Fig. 1, i.e., the states corresponding to $dx/dy < 0$ are unstable.

In Section V, the transmitted and scattered light spectra, (5.1) were calculated (5.9) and (5.16). The transmitted spectrum corresponds to a sharp central peak and has a component which varies as N^2, which is large in the upper (larger internal field) OB branch. The scattered light spectrum (5.16) possesses a strong central peak in the lower OB branch which splits into a three-peaked spectrum in the upper OB branch. The side bands, each separated from the central component by the Stark shift σ, exhibit unequal peak intensities which depend strongly upon the effective temperature T_s. For the lower effective temperature regime (where OB is most probable), the scattered light spectrum in the upper OB branch consists mainly of a strong central peak and a single weaker side band separated from the main peak by the atomic Stark shift and displaced to the higher frequency side.

It should be noted that due to the fact that the interatomic interaction (2.21) is Stark-shift-dependent, the interaction (2.20) provides for nonlinear internal feedback, which depends upon the applied field independent of any cavity. Here ω_c, (2.21) is a preferred axial mode and γ_{k_c}, (2.22), is the photon escape rate from the volume. In the absence of mirrors ω_c is just the atomic frequency ε. The results thus suggest the existence of mirrorless OB in a small volume.

Finally, it has been shown[1] that (2.39) predicts the existence of a second-order phase transition in the limit of zero applied field, i.e., for $y = 0$, (2.39) yields the gap equation

$$\tanh \tfrac{1}{2} \beta_c \varepsilon = \frac{\varepsilon}{2g_o^2} \; \frac{(\omega_c - \varepsilon)^2 + \gamma^2}{\omega_c - \varepsilon} \; . \tag{6.1}$$

If the above equation can be satisfied for explicit values for the Hamiltonian parameters, then (6.1) determines a true critical temperature β_c^{-1} below which the system possesses a non-zero macroscopic dipole moment in the absence of applied field and above which the macroscopic dipole moment vanishes. In this case it is necessary that ω_c be a cavity frequency different from ε. This phenomenon is the so-called superradiance phase transition in thermodynamic equilibrium.[20]

ACKNOWLEDGEMENT

The author gratefully acknowledges useful discussions with C. C. Sung.

REFERENCES

1. C. M. Bowden and C. C. Sung, Phys. Rev. A19, 2392 (1979).
2. R. H. Dicke, Phys. Rev. 93, 99 (1954).
3. P. W. Milonni and P. S. Knight, Phys. Rev. A10, 1096 (1974).
4. G. Banfi and R. Bonifacio, Phys. Rev. A12, 2068 (1975).
5. L. M. Narducci and C. M. Bowden, J. Phys. A9, L75 (1976).
6. P. D. Drummond and H. J. Carmichael, Opt. Comm. 27, 160 (1978).
7. H. J. Carmichael, "Analytical and Numerical Results for the Steady State in Cooperative Resonance Fluorescence," to be published.
8. R. Bonifacio and L. A. Lugiato, Lett. al Nuovo Cim. 21, 505 (1978).
9. R. Bonifacio and L. A. Lugiato, Phys. Rev. A18, 1129 (1978).
10. R. Gilmore, C. M. Bowden and L. M. Narducci, Phys. Rev. A12, 1019 (1975); in "Quantum Statistics and the Many-Body Problem," edited by S. B. Trickey, W. P. Kirk, and J. W. Dufty, Plenum, New York, 1975, p. 249.
11. A. Abragam, "The Principles of Nuclear Magnetism," Oxford University, London, 1961, Chapter V, XII B.
12. D. N. Zubarev, Usp. Fiz. Nauk. 71, 71 (1960) [Sov. Phys. Usp. 3, 320 (1960)].
13. H. J. Carmichael and D. F. Walls, J. Phys. B10, L685 (1977).
14. R. Gilmore and C. M. Bowden, Phys. Rev. A13, 1898 (1976); in "Cooperative Effects in Matter and Radiation," edited by C. M. Bowden, D. W. Howgate and H. R. Robl, Plenum, New York, 1977, p. 335.
15. R. Gilmore and C. M. Bowden, J. Math Phys. (NY) 17, 1617 (1976).
16. C. M. Bowden and C. C. Sung, J. Phys. A11, 151 (1978).
17. L. D. Landau and E. M. Lifshitz, "Electrodynamics of Continuous Media," Addison-Wesley, Reading, MA (1966), p. 54.
18. R. J. Glauber, Phys. Rev. 130, 2529 (1963).
19. B. R. Mollow, Phys. Rev. 188, 1969 (1969).
20. C. C. Sung and C. M. Bowden, J. Phys. A12, 2273 (1979) and references contained therein.

COMPLEX ORDER PARAMETERS IN QUANTUM OPTICS FIRST ORDER PHASE

TRANSITION ANALOGIES

Charles R. Willis

Physics Department
Boston University
Boston, Massachusetts 02215

Abstract: Our purpose in this paper is to present the thermodynamic treatment of quantum optical systems which have first order
phase transition analogies and whose order parameters are two-
dimensional or complex valued such as the electric field variable in
optical bistability. In so doing we show that it is advantageous to
analyze phase transitions by exactly the same procedure as one uses
in equilibrium thermodynamics by means of the appropriate free
energies which are related to each other by Legendre transformations.
We find there exist two classes of quantum optical systems which
differ in the role played by the phase variable in the complex order
parameter. In the first class, consisting of systems such as saturable absorbers, the mean field first order transition is completely
independent of the phase variable as in the analogy between the laser
threshold region and the second order phase transition. In the
second class of systems, such as optical bistability, we find the
first order phase transition depends on the relative phase between
the incident field and the internal field. We show how the necessary
metastable states arise without the addition of physical mechanisms
to break gauge invariance when the external field is zero. We conclude with an analysis of the effect fluctuations in the phase variable have on the mean field phase transition.

I. INTRODUCTION

Our purpose in this paper is twofold. First we analyze mean-
field quantum optical first order phase transitions when the relevent
thermodynamic variable or order parameter is two dimensional or
complex valued. Examples of such transitions are optical bistability
and saturable absorbers. Second, we show that it is advantageous to

analyze phase transitions by exactly the same procedure as one uses
in equilibrium thermodynamics, i.e., we start with a free energy for
the system, obtain the equation of state and use the equation of
state to Legendre transform the original free energy to a new free
energy which is a function of the intensive variable which appears
in the equation of state. The above procedure for the thermodynamic
case of a simple single component system is: we start with the
Helmholtz free energy $F(T,V)$ where T is the temperature and V is the
volume, we obtain the equation of state $p = -\partial F/\partial V$ and then performing the Legendre transformation we obtain the Gibbs free energy
$G(T,p) = F + pV = U - TS + pV$ where p is the pressure and S is the
entropy. For a simple system the Gibbs free energy per particle,
$(G(T,p)/N)$, is just the chemical potential $\mu(T,p)$ use to analyze
phase transitions at constant intensive variables p and T. We show
below that the two intensive variables in the optical bistability
case are the dimensionless complex external electric field, y, and
the dimensionless parameter $(N/N_T) \equiv N$ which is denoted by $2c$ in the
conventional notation.[1] The variables y and N are defined in Sec.
III. The fundamental free energy of the system that contains all
the macroscopic thermodynamic information about the system is independent of the phase of the local field variable. As a consequence,
the free energy analogue of the Gibbs free energy is also independent
of the phase of the external field, and the only phase that can
appear is the relative phase between the local field and the external
field. We show that in a mean field theory the only allowed values
of the relative phase between the local field and the external field
are 0 and π. The term mean field is used throughout this paper in
the conventional statistical mechanical sense that the theory is
obtainable from some underlying microscopic theory by neglecting
correlations and fluctuations between microscopic variables. Example
of mean field theories are Curie-Weiss, Bragg-Williams, and Van der
Waals equation of state. One of the advantages of the thermodynamic
procedure is that it enables us to see clearly how to describe metastable as well as stable states when the relevant thermodynamic variable is complex valued, and reduces the possibility of making incorrect inferences which occur when one treats the external field
variable y and the local field variable x as independent variables
rather than as Legendre transform variables. More fundamentally
once we have the full energy for the quantum optical system then all
the general consequences of thermodynamics are available to us such
as, inequalities involving second derivatives of the free energy and
the Clapeyron-Clausius equation. It is especially important in
optical bistability to have a full thermodynamic treatment of the
free energy rather than just an analogy limited to a small neighborhood around the critical point where we have shown[2] that optical
bistability can be mapped on to the Curie-Weiss first order phase
transition, because far from the critical point the analogy does not
hold. However, in the practically important region far from the
critical point, the full thermodynamic treatment of optical bistability proceeds in exactly the same manner as the analysis of the

Van der Waals equation for a phase transition and not just by
analogy.

     In Sec. II we review the thermodynamics of the two dimensional
Curie-Weiss theory of ferromagnetism in order to introduce a complex
order parameter and to obtain the limitations a thermodynamic treatment of a mean field theory places on the phase variable of a complex
order parameter. Also we show that we can reformulate the thermodynamics of self-consistent field theories such as optical bistability
so that we can use the local field rather than the "magnetization"
as the thermodynamic order parameter. We obtain the free energy in
Sec. III in terms of the local field variable from which we obtain
the optical bistability equation of state for the external field
as function of the local field. Then we obtain the appropriate free
energy in terms of the external field and analyze the first order
phase transition. In Sec. IV we compare the behavior of the phase
variable in saturable absorbers with the phase variable in coherently
driven optical bistability systems. We discuss in Sec. V some of
the effects fluctuations and dynamics have on the different kinds of
phase variable behavior in quantum optical systems which have first
order phase transitions.

II.   REVIEW OF CURIE-WEISS THEORY

     We first review the two dimensional equilibrium mean field
Curie-Weiss theory of magnetism in the presence of an external field,
where there is a first order phase transition in order to find the
consequence a complex order parameter has for mean field theories in
gneral and to provide motivation for the choice of the optical bistability free energy. Furthermore, we find the modifications
required in a free energy formulation when the thermodynamic magnetization variable is replaced by the local field variable as is convenient both in Curie-Weiss theory and in optical bistability. The
order parameter in optical bistability has two degrees of freedom
because the wave equation for the electric field is second order
in the electric field. (The vector character of the electric field
occurs in the dot product between field and atomic dipole, which is
conventionally[1] absorbed in the coupling constant, and is not the
determiner of the number of degrees of freedom of the order parameter in the mean field theory of obtical bistability.) In order
to make the connection with optical bistability more transparent,
instead of using M_x and M_y to describe the two dimensional magnet,
we introduce the complex magnetization $M = M_x + iM_y$ and its complex
conjugate M^*. The complete thermodynamic characterization of the
magnetic part of the two-dimensional Curie-Weiss ferromagnet is
given by the dimensionless free energy per particle, (F_m/NkT),

$$\frac{F_m}{NkT} \equiv -amm^* + \left[\frac{1 + (mm^*)^{\frac{1}{2}}}{2}\right] \ln\left[\frac{1 + (mm^*)^{\frac{1}{2}}}{2}\right] +$$

$$\left[\frac{1 - (mm^*)^{\frac{1}{2}}}{2}\right] \ln\left[\frac{1 - (mm^*)^{\frac{1}{2}}}{2}\right] \tag{2.1}$$

where kT is Boltzmann's constant times the temperature, $a \equiv (v/2kT)$ where $v > 0$, is the dipole interaction energy, and m is the dimensionless magnetic moment per particle. For a complete discussion of Curie-Weiss theory see Brout[3]. In thermodynamics the complex magnetic field B is defined[4] as the derivative of the free energy with respect to the magnetization which in dimensionless units is

$$b \equiv \frac{B}{kT} = \frac{\partial (F_m/NkT)}{\partial m^*} = -am + \frac{m}{2(mm^*)^{\frac{1}{2}}} \ln\left[\frac{1+(mm^*)^{\frac{1}{2}}}{1-(mm^*)^{\frac{1}{2}}}\right] \equiv mK(mm^*). \tag{2.2}$$

Similarly, the definition of b* yields

$$b^* \equiv \frac{B^*}{kT} = \frac{\partial (F_m/NkT)}{\partial m} = m^*K(mm^*). \tag{2.3}$$

We can see that the definition of the magnetic field is equivalent to the self-consistent field approximation representation of the Curie-Weiss theory by writing Eq. (2.2) in the following form

$$|m| = \tanh \left[a|m| + |b|\right] \tag{2.4}$$

where $|m| \equiv (mm^*)^{\frac{1}{2}}$ and $|b| \equiv (bb^*)^{\frac{1}{2}}$. Observe that when $|b| = 0$ we have the typical mean field second order or order - disorder phase transition. The well known laser threshold second order phase transition analogy[5,6] is obtained by identifying $|m|$ with the dimensionless electric field of the laser and by retaining the linear and cubic terms in the expansion of $\tanh a|m|$. In a similar manner I have shown[2] that the first order phase transition as a function of b which occurs at $b = 0$ for the Curie-Weiss theory can be identified with the optical bistability instability in the neighborhood of the critical point of the optical bistability transition. Paranthetically, Eq. (2.4) demonstrates one of the serious problems of the mean field theory of magnetism, that is, one obtains the same equation for $|m|$ independent of the dimension of $|m|$. Consequently the critical indices of the Curie-Weiss theory of magnetism are independent of dimension contrary to experiment. An analysis of dimensionality and instability in quantum optics has been given by Graham.[7]

From Eqs. (2.2) and (2.3) we can obtain an important relationship between complex thermodynamic variables and their conjugates, when the free energy is independent of phase. Consider a complex thermodynamic variable x, a free energy $F(xx^*)$ which depends on xx^* only, i.e., independent of the phase of x, and the conjugate thermodynamic variable $y \equiv \partial F/\partial x^*$. The condition on F implies $y = x\,F'(xx^*)$, where the prime means differentiation with respect to the argument xx^*, and $y^* = x^*\,F'(xx^*)$ which leads to the relation $(y/y^*) = (x/x^*)$. In terms of the phases ϕ and θ where $x = |x|\,\exp(-i\phi)$ and $y = |y|\,\exp(-i\theta)$ we have $\exp[-2i(\theta-\phi)] = 1$, which requires either $\theta = \phi$ or $\theta - \phi = \pi$. Consequently a mean field free energy which is independent of the phase of the thermodynamic variable, has the property that the conjugate thermodynamic variable, y, must either be in phase with x or 180° out of phase with x. As an example, consider the Curie-Weiss free energy, Eq. (2.1), which is independent of the phase of the magnetization; then either the magnetic field is in phase with the magnetization as it is in the stable equilibrium states of the first order Curie-Weiss phase transition or 180° out of phase as it is in the metastable states of the same transition. The second consequence I want to draw from the discussion of the Curie-Weiss theory is the fact that we can reformulate theories with a self-consistent field in terms of a local field variable instead of a magnetization variable. Consider Eq. (2.4) which we can write in the form,

$$(a|m| + |b|) - |b| = a \tanh\,[a|m| + |b|] \,. \tag{2.5}$$

The form of Eq. (2.5) suggests we introduce the dimensionless local field variable $h_\ell = a|m| + |b|$ instead of $|m|$ i.e., we write Eq. (2.5) as

$$|b| = h_\ell - a \tanh h_\ell = \partial g_m/\partial h_\ell \tag{2.6}$$

where $g_m \equiv (1/2)\,h_\ell{}^2 - a \ln \cosh h_\ell$

or to return to a complex variable formulation we have by a similar derivation

$$g_m\,(h_\ell h_\ell{}^*) \equiv \tfrac{1}{2}\,h_\ell h_\ell{}^* - a\ln \cosh\,(h_\ell h_\ell{}^*)^{\frac{1}{2}} \tag{2.7}$$

where now h_ℓ represents the complex local field $h_\ell \equiv am + b$. The important result for us is that we can formulate the theory directly in terms of the external field and the local field instead of the magnetization which we shall do when we treat optical bistability. However, it is important to realize the free energy function in the local field representation, g_m, is a different free energy function than the function, (F/NkT) of Eq. (2.1) in the magnetization representation, although both contain the same physical information. In the next section we apply the results of this section to the mean field theory of optical bistability.

III. THERMODYNAMIC POTENTIALS FOR OPTICAL BISTABILITY

The steady state solution for the Maxwell-Bloch equations for
N two-level systems interacting with a single cavity mode and with
an incident driving field whose frequency is resonant with the two-
level atomic frequency in dimensionless complex variables is[1]

$$y = x + \frac{Nx}{1 + xx^*} \qquad\qquad (3.1)$$

where $x \equiv y + z$, $y \equiv 2\,|\mu|\,(\gamma_\perp\gamma_\parallel)^{-\frac{1}{2}}\,E_i/\hbar$, $z = 2\,|\mu|\,(\gamma_\perp\gamma_\parallel)^{-\frac{1}{2}}\,E_{int}/\hbar$
and $x \equiv 2\,|\mu|\,(\gamma_\perp\gamma_\parallel)^{-\frac{1}{2}}\,E_\ell/\hbar$. E_i is the incident electric field, E_{int}
is the internal electric field, and E_ℓ is the local electric field.
The atomic electric dipole is μ, $\gamma_\parallel^{-1}$ $(\gamma_\perp^{-1})$ is the logitudinal
(transverse) relaxation time, $N \equiv (N/N_T)$ where N is the number of
two-level systems, $N_T \equiv (\kappa\gamma_\perp/|\mu|^2)$ is the laser threshold in the
cavity and κ^{-1} is the photon lifetime in the cavity. The constant
N is equivalent to the constant 2c of Ref. (1). Equation (3.1) is
plotted below in Fig. (1) for $N = 40$. To compare Eq. (3.1) for
optical bistability with the Curie-Weiss theory of magnetism, Eq.
(2.2), we replace b, m, and h_ℓ by y, z, and x respectively. We could
have used y and z which are the conjugate thermodynamic variables
instead of y and x to formulate the equation of state, Eq. (3.1).
We chose y and x for two reasons. First, historically the variables
y and x appear natural for reasons unconnected with thermodynamic
descriptions. Second, although in the neighborhood of the critical
point $|y_c| = 3(3)^{\frac{1}{2}}$, $|x_c| = (3)^{\frac{1}{2}}$ and $N_c = 8$, we can make a one to one
correspondence with the Curie-Weiss first order phase transition[2]
we find the analogy no longer holds even approximately for large
values of $N(\gg 8)$ and consequently for $N \gg 8$ the choice of y and z
no longer has any particular intuitive appeal.

Given the equation of state, Eq. (3.1), we find for the appropriate free energy

$$f(xx^*) = xx^* + N\ln(1 + xx^*) + \text{additional terms} \qquad (3.2)$$

where the additional terms contain the remaining macroscopic variables of the system but do not depend on x or on y and which will
not concern us in this paper. Now the essential point that makes
Eq. (3.1) an equation of state is that the thermodynamical definition
of the variable conjugate to x, namely y, is $y \equiv \partial f/\partial x^*$ and $y^* \equiv
\partial f/\partial x$. In the rest of this paper we show that we can treat $f(xx^*)$
as a free energy in the thermodynamical sense and Eq. (3.1) as the
equation of state obtained from that free energy in exactly the same
sense as obtaining $p = F(V)$ (where p is pressure and V is volume) fo
a matter system or $B = F(M)$ (where B is the magnetic field and M is
the magnetization) from the appropriate free energy. Since $f(xx^*)$
is independent of the phase of x it follows from Sec. II that either
y must be in phase or $180°$ out of phase with x. As we have mentione

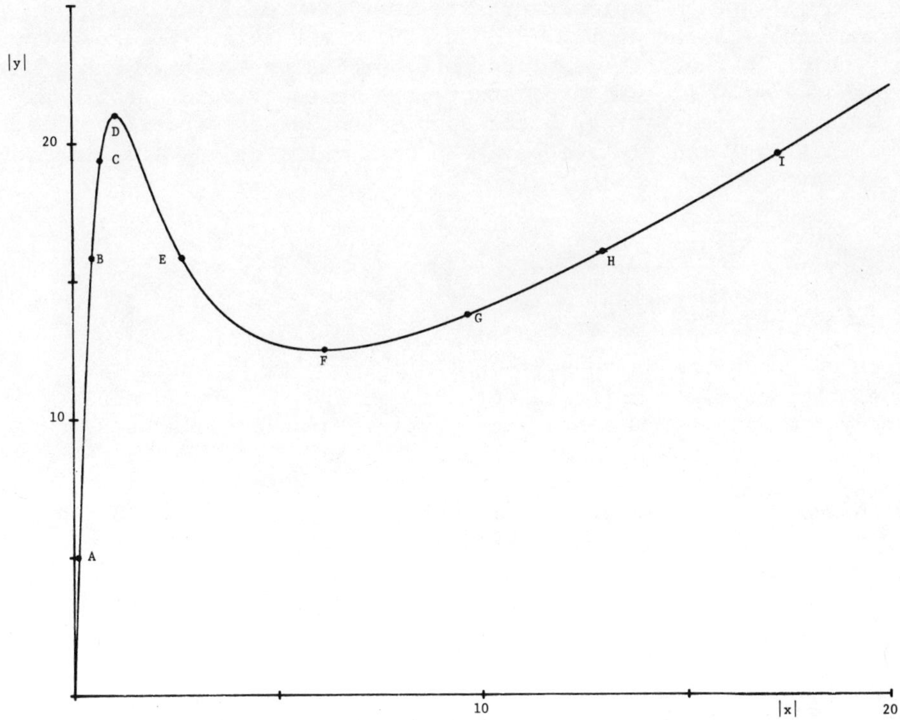

Fig. 1.   Equation of state $|y|$ as a function of $|x|$ for optical
           bistability Eq. (3.1) of the text for $N = 40$. The corre-
           sponding lettered points on all three figures refer to the
           same physical state.

in the Curie-Weiss case the stable states have m in phase with b
while the unstable states have m 180° out of phase with b. However,
in the case of optical bistability we have z 180° out of phase and
x in phase with y for both stable and unstable states. The differences in the Curie-Weiss and optical bistability cases arise from
the fact that the critical point of the Curie-Weiss transition is
at b = 0 while the optical bistability critical point occurs at
$|y| = 3(3)^{\frac{1}{2}}$. At this point we study the graphical representations
fo the equation of state Eq. (3.1) and the "Helmholtz" free energy
before we derive the Gibbs free energy.

     In Fig. 1 we had to plot $|y|$ as a function of $|x|$ because of a
shortage of four dimensional graph paper which we need because both
y and x are complex. However in the four dimensional (x,y) space
the figure is relatively simple since x and y are in phase, so when
x is multiplied by a phase factor y is multiplied by the same phase
factor. Fig. 1 has been thoroughly studied[1] so we will concentrate
on the free energy representations. Since f(xx*) is real and depends

only on $|x|$ we can represent $f(xx^*)$ as a figure of revolution in 3 dimensions by rotating Fig. 2 about the $x = 0$ axis. The most impor-tant point is that all the information relating to the x and y vari-ables at $N = 40$ (modulo a constant) are in the figure. For example we have indicated in Fig. 2 the double tangent construction i.e., at points $|x_B|$ and $|x_H|$ we have the same value of the slope $|y|$, that is we have

$$|y| = \frac{f(|x_H|) - f(|x_B|)}{|x_H| - |x_B|}$$

which we can write as $f_H - |y||x_H| = f_B - |y||x_B|$, which is just the usual thermodynamic criteria for the equilibrium of two phases, namely the Gibbs free energy per particle of the two phases must be equal at constant $T(N)$ and constant $p(y)$ as we discuss in the next paragraph. The points D and F in Fig. 1 correspond to the two points of inflection that occur between B and H in Fig. 2. In fact one could plot all the points A through I of Fig. 1 on the corresponding

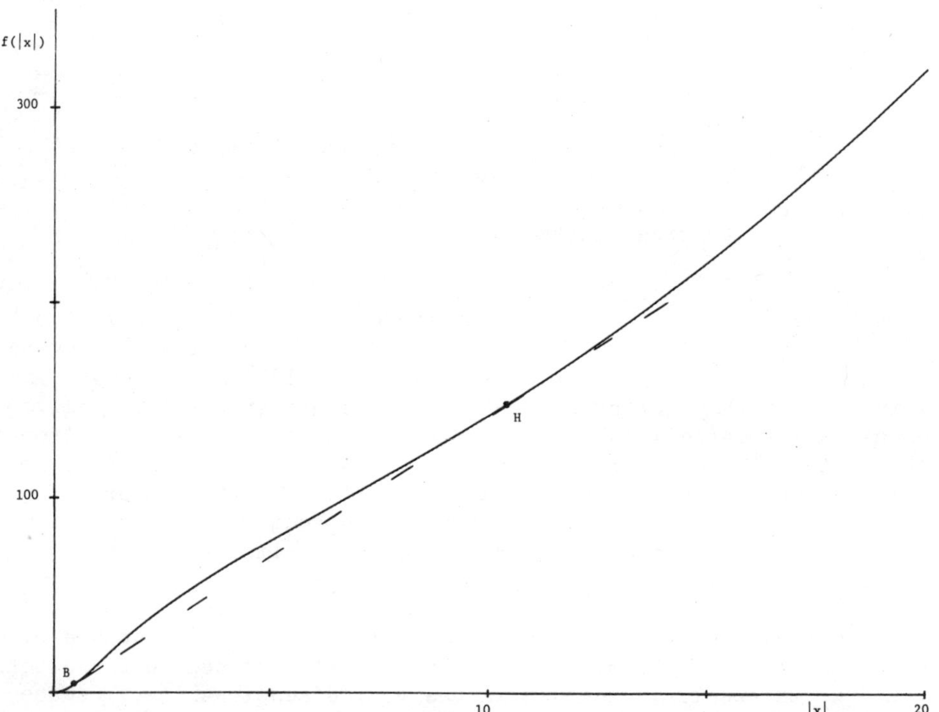

Fig. 2.   "Helmholtz" free energy, $f(xx^*)$, for optical bistability
         Eq. (3.2) of the text, for $N = 40$, as a function of $|x|$.

points of $f(xx^*)$ in Fig. 2. We omitted them to avoid crowding the
figure. We can see from Fig. 2 how in optical bistability we can
have invariance under phase rotations and a first order phase transition with hysteresis involving metastable states. In order to see
this more clearly we proceed as in thermodynamics to obtain the
"Gibbs" free energy. (See Chap. 9 of Ref. 4).

The Gibbs free energy for a simple single component system
$G(T,p) = U - TS + pV$ is just the chemical potential $\mu(T,p) = (G/N)$.
Thus $G(T,p)$ is just the Legendre transformation which replaces the
independent variable S in U by $\partial U/\partial S$ which is just T and V in U by
$\partial G/\partial V$ which is just $(-p)$. Since we have already shown[2] that N is
related to T/T_c i.e., in the neighborhood of the critical point $N - 8$
$\leftrightarrow (T - T_c)/T_c$, we need only to Legendre transform the x dependence
of $f(xx^*)$ replacing the dependence of f on x by $\partial f/\partial x^* = y$. The
Legendre transformation $g(yy^*)$ is

$$g(yy^*) \equiv f[x(yy^*), x^*(yy^*)] - y^*x(yy^*) - yx^*(yy^*) \tag{3.3}$$

where $x(yy^*)$ is obtained by solving the equation of state Eq. (3.1)
for x in terms of y and y* and inserting the result in the right
hand side of Eq. (3.3). g is a function of y and y* and not a
function of x and x*. Futhermore, since f depends only on $|x|$ it
follows that g depends only on $|y|$ and is independent of the phase
of the external field. Consequently $g(|y|)$ is a figure of revolu-
tion and we need only plot $g(|y|)$ as a function of $|y|$ as we do in
Fig. 3 and revolve the figure about the y = 0 axis.

The multivalued structure of Fig. 3 arises because we took a
value of $N > 8$. If we had taken a value $N < 8$ the curve would have
been monotonic and at $N = 8$ we would have a critical point. A series
of figures as a function of N would be qualitatively similar to the
curves of a simple single component system chemical potential $\mu(T,p)$
as plotted in Fig. 9.5 of Ref. 4. We now show the usefulness of the
"Gibbs" free energy for describing the optical bistability phase
transition. The g surface consists of three parts, the first part,
generated by the curve O A B C D represents the cooperative phase,
the second part generated by the curve F G H I to ∞, represents the
"single particle" phase, and the third part generated by the curve
F E D represents unstable states. Each of the phase surfaces are
divided further into stable states from 0 to B in the "cooperative
phase" and H to ∞ in the "single particle" phase and metastable
states from B to D in the cooperative phase and F to H in the single
particle phase. The division into stable, metastable and unstable
states is a deduction from the curve, i.e., for a given $|y|$ the low-
est value of g is the stable state if the curvature of g is positive,
the higher point for a given $|y|$ is metastable if the curvature of g
is positive and the unstable states correspond to negative curvature.
The above remarks apply very closely to the Van der Waal treatment of
a gas - liquid transition with "cooperative" replaced by liquid and

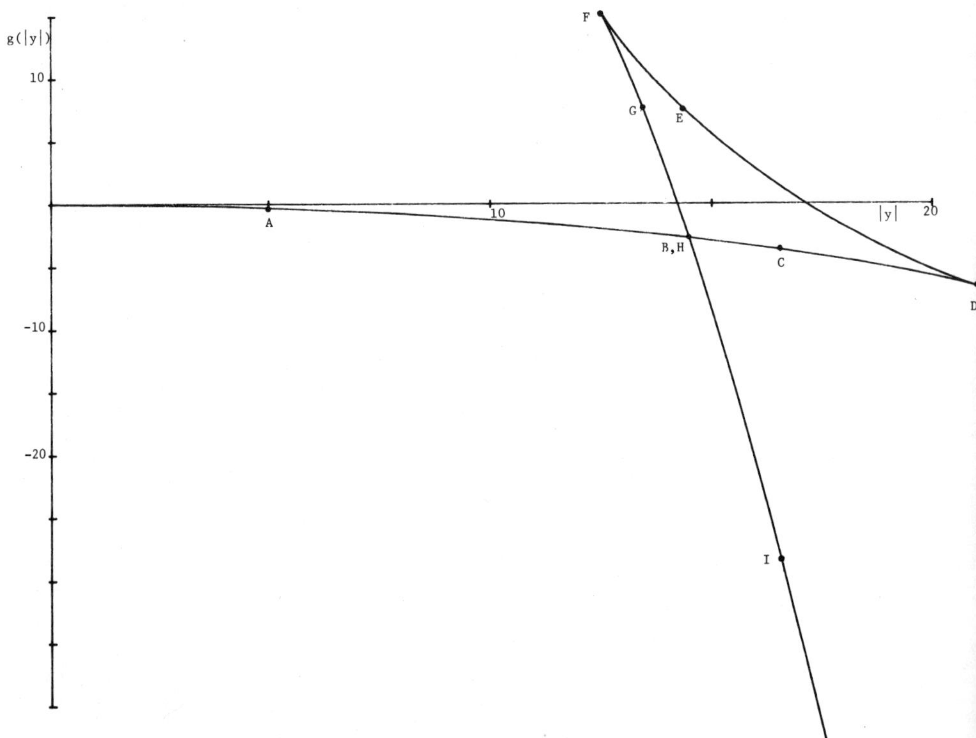

Fig. 3.   "Gibbs" free energy for optical bistability g(yy*), Eq. (3.3
of the text plotted for N = 40 as a function of $|y|$.

"single" particle replaced by gas. The only real difference is that
we have a surface of revolution (because y is complex) instead of a
curve. From the properties of the Legendre transformation we have
$x \equiv -\partial g/\partial y*$ and thus x is in phase with y everywhere in both thermodynamic phases because $-g'(yy*)$ is positive in both phases. The dimensionless "suscrptibility" χ which measures the stability defined by
$\chi \equiv \partial x/\partial y* = -g''(yy*)$ (where prime means differentiation with respect
to the argument xx*) is positive everywhere in the two thermodynamic
phases and negative in the unstable states as we have already indi-
cated. The point B, H in Fig. 3 occurs at the value of $|y|$ when the
two phases have the same value of $g(|y|,N)$ for the value N = 40. In
thermodynamics this is just the criteria for the phase equilibrium
of two phases A and B i.e. $\mu_A(T,p) = \mu_B(T,p)$ where μ_A is the chemical
potential for phase A and μ_B is the chemical potential for phase B
at fixed T and p. Thus the intersection of the two optical bistabil-
ity phases at B, H determine the value of $|y|$ which yields the
Maxwell construction in Fig. 1. (Recall the letters A – I represent
the same physical points on all three diagrams).

The important conclusion from this section is that the properties and treatment of g is essentially the same as the properties
and treatment of the chemical potential μ for a simple single component equilibrium thermodynamic system undergoing a first order
phase transition. The second point of this section is, that it is
clear from Fig. 3 that it is straightforward to have both a metastable
state such as point C and a stable state I present for the same com-
plex variable y and still have phase invariance in $g(|y|)$ and $f(|x|)$
at the same time. Thus there is no problem having metastable states
in a mean field first order phase transition theory with complex
order parameters and at the same time the theory is completely independent of the absolute phase of x and the absolute phase of y. In
the next section we will compare the optical bistability first order
phase transition of this section with the first order phase transition behavior in saturable absorbers.

IV.   LASER FIRST ORDER PHASE TRANSITIONS IN ZERO EXTERNAL FIELD

It is possible to have first order phase transitions in quantum
optical systems with complex order parameters in addition to those
which occur in a nonvanishing external field such as dye lasers[8] or
equivalently saturable absorbers.[9] In Ref. 8, I have shown that
under certain conditions the triplet levels of a dye laser behave
like a saturable absorber and in the mean field approximation the
dye laser can undergo a first order phase transition in zero external
field. The mean field free energy of the saturable absorber in the
neighborhood of the instability is

$$\Phi(zz^*) = \bar{a}zz^* + (\bar{b}/2)(zz^*)^2 + (\bar{c}/3)(zz^*)^3 \tag{4.1}$$

where $\bar{a} \equiv N^{-1} - 1$, $\bar{b} \equiv 1 - \alpha(1 + r)$, $\tilde{c} \equiv (1/2)(\alpha r^2 - \bar{b})$, and where z
is the same z that appears in Eq. (3.1). The two dimensionless para-
meters α and r charaterise the dye molecule, α is a measure of the
triplet losses where $\alpha < 1$ is a necessary condition for laser action
and r is the relative saturability of the singlet to the triplets.
(From the thermodynamic point of view the saturable absorber is no
longer a simple single component system specified by two intensive
variables T and p but requires three intensive variables, the new
"intensive" variable is $\bar{b}$). In all operating cases $\bar{c} > 0$; if $\bar{b}$ also
is greater than zero we have the normal laser second order phase
transition. However, for $\bar{b} < 0$, $\bar{a} > 0$ and $\bar{b}^2 > 4\bar{a}\bar{c}$ we have a maximum at $(zz^*)_-$ and a minimum at $(zz^*)_+$ where

$$[-\bar{b} \pm (\bar{b}^2 - 4\bar{a}\bar{c})^{\frac{1}{2}}](2\bar{c})^{-1} \equiv (zz^*)_\pm \; .$$

We have two minima, one at the origin and one at $(zz^*)_+$ until $\bar{a}$
becomes negative when there is a single minimum at $(zz^*)_+$. For

large positive $\bar{a}$ the origin is the absolute minimum and $(zz*)_+$ is the
relative minimum or metastable state. As $\bar{a}$ decreases eventually
$(zz*)_+$ becomes the absolute minimum, and the origin the relative
minimum, until finally when $\bar{a}$ becomes negative there is a single
minimum at $(zz*)_+$. For a full analysis including fluctuations see
Ref. 8.

For our purposes in this paper the fundamental point is that
Eq. (4.1) contains a first order phase transition in the absence of
an external field. Since $\Phi(zz*)$ is independent of the phase of z,
Φ is a figure of revolution about the vertical axis passing through
the origin. The equation of state for y is obtained in the usual
manner i.e., $y \equiv \partial\Phi/\partial z*$. For y = 0 we have

$$y \equiv \frac{\partial\Phi}{\partial z*} = 0 = z\ \{\bar{a} + \bar{b}(zz*) + \bar{c}(zz*)^2\} \tag{4.2}$$

which describes a first order phase transition. What is different
in the saturable absorber case is that not only the free energy,
Eq. (4.1), but the equation of state itself, Eq. (4.2), for y = 0 is
invariant under change in phase of x. Consequently, the double
tangent construction also occurs at slope y = 0 where the values of
$\Phi[0]$ equals $\Phi[(zz*)_+]$, i.e., when the two minima have the same value
of Φ. In the saturable absorber case Φ itself is already the "Gibbs"
free energy for the transition because it is a function of the intensive variables N and $\bar{b}$ explicitly. One phase is z = 0 and the other
phase is $(zz*)_+$. As we have seen the mean field theory of a phase
transition of a phase invariant free energy requires that y and z are
either in phase or out of phase by 180°. In the saturable absorber
case, Eq. (4.2), where the external field is zero, we find the mean
field places no restriction on the phase of z. In fact, as we see
in the next section when we discuss fluctuations the behavior of the
phase variable in saturable absorbers is the same as the behavior of
the phase variable in conventional laser theory.

V.   FLUCTUATIONS OF COMPLEX ORDER PARAMETERS

So far we have anlyzed the phase variable of complex order parameters undergoing first order phase transitions in mean field theory
and found different kinds of behavior. In this section we investigate the dynamical and fluctuation behavior of the phase variable
behavior obtained from master equations to see if the fluctuation
behavior is consistent with the mean field behavior. Fluctuations
in laser second order transitions have been carefully analyzed in
Ref. 7. Fluctuations in first order phase transitions have also
been thoroughly studied by many authors.[10-13] In Sec. V of Ref. 8
there is a discussion of fluctuations, hysteresis, and bistability
in the saturable absorber problem much of which is relevant to

optical bistability in an external field. We first briefly review
some of the main points about fluctuations in quantum optical phase
transitions.

The mean field Van der Waals theory of the gas-liquid first order
phase transition is incomplete in the sense that it does not pre-
scribe whether one will observe metastable states and hysteresis or
a Maxwell construction. In order to resolve the problem of what
behavior will be observed in any mean field theory and in particular
the mean field of optical bistability we would have to be able to
solve the full time dependent problem to see if the time scale of
fluctuations would be sufficiently fast that a Maxwell construction
would be observed in the time scale of the experiment or whether
the time scale of the fluctuations is so long (except in the immediate neighborhood of the turning points) that in any reasonable
experimental time rate of change of y, the fluctuations are so slow
that they have no effect on the experiment and we observe meta-
stability and hysteresis. What we are faced with here is actually
a form of the ergodic problem. If we have a steady state distribution function describing optical bistability, $P_S(x)$, and if we
assert the time average behavior of the system is given by $\langle x \rangle_S \equiv
\int x P_S(x) dx$, then we will not observe hysteresis but will observe
something like a Maxwell construction. The reason is we expect $P_S(x)$
to have the character of a two peaked distribution function in the
two phase region, the larger peak representing the stable state and
the smaller peak representing the metastable state whose heights
change relative to each other as a function of y. However $\langle x \rangle_S$
calculated from such a distribution function is a smooth (analytic)
but rapidly changing function of the external field y in the neighborhood of the y that gives the Maxwell construction. In fact, the
fluctuations, although relatively small, round off the edges of the
Maxwell construction so that strictly speaking $\langle x \rangle_S$ does not undergo
a phase transition at all but just a rapid variation of $\langle x \rangle_S$ as a
function of y. Furthermore the susceptibility $\chi = \partial \langle x \rangle_S / \partial y^*$ becomes
finite everywhere and is small at the turning points but large and
peaked in the neighborhood of the Maxwell construction where $\langle x \rangle_S$
changes rapidly.

Consequently as we see from the discussion in the preceeding
paragraph, the full understanding of the role played by fluctuations
in experiment requires an analysis of the dynamical solutions of the
Fokker-Planck equation for optical bistability which is beyond the
scope of this paper. However, we can say something about the role
of fluctuations and dynamics in the behavior of the phase of the
complex order parameter in systems with mean field first order phase
transitions. The statistical mechanical master equations for the
density matrix of saturable absorber systems are separable in the
amplitude and phase variables and satisfy detailed balance. The
stationary state is independent of phase, i.e. all phases are equally
probable. The time dependent behavior of the phase variable is

described to a high degree of accuracy by the phase diffusion model exactly as in conventional laser theory. The diffusion constant $D \sim \kappa <zz^*>^{-1}$ where κ^{-1} is the photon lifetime in the cavity. Thus in the time scale D^{-1} the phase distribution of a saturable absorber approaches a uniform phase distribution. The behavior of the phase variable in the coherently driven optical bistability case is fundamentally different. First, driven optical bistability does not satisfy detailed balance and consequently neither are the amplitude and phase variables separable nor do we have a potential for the stationary state Fokker-Planck equation. However, even if we assume the amplitude and phase are approximately separable, as is the case, we find the equation of motion for the distribution of the phase variable satisfies not the phase diffusion equation but a Fokker-Planck equation with the drift term proportional to $\sin(\phi-\theta)$ where $\phi-\theta$ is the relative phase of y and x. When $(\phi-\theta)$ is small the Fokker-Planck equation in the variable $\phi-\theta$ is fundamentally the same as the equation for Brownian motion which is solvable analytically. Consequently, the phase of the local field, ϕ, approaches the phase of the external field, θ, in a time κ^{-1} (typically $\sim 10^{-6}$ seconds) and the fluctuations $<(\theta-\phi)^2>_s$ in the steady state are of the order of the inverse saturation number of photons n_s^{-1} which is typically $\leq 10^{-6}$. Thus instead of the phase diffusing slowly in time, the phase locks on to the phase of the external field in κ^{-1} seconds, even if the external field itself varies in time, as long as the external field phase varies more slowly than a time scale $\sim \kappa^{-1}$. Thus the mean field behavior of the relative phase $\phi-\theta$ in optical bistability, namely that it is zero, is thus compatible with the time dependence and steady state fluctuation behavior of the Fokker-Planck equation for the distribution function of the phase variable.

VI.   CONCLUSIONS

We have shown that we can treat the mean field theory of optical bistability by conventional thermodynamic means. The free energies for optical bistability Eqs. (3.2) and (3.3) are independent of the phase of x and y respectively and consequently they are figures of revolution about the respective vertical axes through the origin. With conventional thermodynamic phase transition analysis we showed g(yy*) could be analyzed into stable phases, metastable phases and unstable states which were simply compatible with the invariance of g under phase rotations. However, if we had considered Eq. (3.3) as an equation for a potential for x and x* with y and y* given external parameters instead of Legendre transform variables we would consider Eq. (3.3) as no longer invariant under rotations about the vertical axis through the origin. This is the "tipped sombero" argument applied to optical bistability which has been used to infer there can be no metastable states, and the lack of metastable states for x complex forbids the occurrence of a first order phase transition unless additional terms are added to the free energy to break

the gauge invariance. The resolution of the apparent difficulty is
now clear; that is, Eq. (3.3) is a function of the single variable
yy^* alone, not a function of x and x^* with y and y^* parameters
because the $x(y,y^*)$ and $x^*(y,y^*)$ in Eq. (3.3) are functions of y and
y^* through the equation of state, Eq. (3.1), and the definition of
the Legendre transformation. Consequently the thermodynamic potentials with complex order parameters are invariant under rotations
of phase and still describe first order phase transitions with
metastable states and hysteresis. (If one insists on considering
the right hand side of Eq. (3.3) as a function of two variable x and
x^* and two parameters y and y^* then there is no invariance under
phase rotations for mean field theories because x and x^* must be
exactly in phase or 180° out of phase wity y and y^*. Thus x and x^*
can have only two phases θ_0 (where $y = |y| \, e^{-i\theta}0$) or $\theta_0 - \pi$, and
the phase symmetry is broken and again there is no difficulty with
having metastable states).

We have shown that if in a steady state nonequilibrium problem
one can treat the steady state equation as a thermodynamic equation
of state relating the thermodynamic variable or order parameter, the
internal field in optical bistability, to its thermodynamic conjugate
variable, the external field in optical bistability, then we can
obtain the thermodynamic potential for the problem at least up to an
undetermined function of the remaining thermodynamic variables. With
the thermodynamic potential we can deduce the same kind of consequences we obtain in equilibrium thermodynamics. By having the
full thermodynamic potentials we are not limited to analogies in the
neighborhood of the critical point (which is important in lasers
because the critical point is the threshold) but we can examine the
full range of the thermodynamic variables (which is important in
optical bistability because one operates far from the critical point).
Furthermore, the "analogous" first order phase transition near the
critical point is the Curie-Weiss first order phase transition in
an external magnetic field, while the full thermodynamic approach is
more versatile and allows us to consider one phase as the "free
particle" phase (gas) and the "cooperative phase" (liquid) in a
systematic fashion rather than just as an appealing analogy. In a
future publication we will carry out the thermodynamic program further
by determining the fundamental relation for the entropy rather than
the free energy (with the undetermined function) and deduce the
standard thermodynamic results such as the Clayperon-Clausius relation. Also we will carry out the full thermodynamic program for the
laser with injected signal.

ACKNOWLEDGEMENTS

The author would like to thank R. A. Bradbury for work on the
computer. The present work was done in part while the author was

a visiting scientist at RADC Hanscom Field as a visiting scientist in the AFOSR sponsored University Resident Research Program.

REFERENCES

1. R. Bonifacio and L. A. Lugiato, Opt. Comm. <u>19</u>, 172 (1976).
2. C. R. Willis, Opt. Comm. <u>23</u>, 151 (1977).
3. R. H. Brout, <u>Phase Transitions</u> (Benjamin, New York, 1965).
4. H. B. Callen, <u>Thermodynamics</u> (J. Wiley, 1960).
5. V. Degiorgio and M. O. Scully, Phys. Rev. <u>A2</u>, 1170 (1970).
6. R. Graham and H. Haken, Z. Physik <u>237</u>, 31 (1970).
7. R. Graham, in: <u>Progress in Optics, Vol. XII</u>, ed. E. Wolf (North Holland, Amsterdam) 1974, p. 234, and the references therein.
8. C. R. Willis, in: <u>Coherence and Quantum Optics IV (1978)</u> eds. L. Mandel and E. Wolf (Plenum, N.Y.) p. 63 and the references contained therein.
9. R. Roy, Phys. Rev. <u>20A</u>, 2093 (1979) and the references contained therein.
10. R. Bonifacio, M. Gronchi, and L. A. Lugiato, Phys. Rev. <u>18A</u>. 2266 (1978) and references contained therein.
11. G. S. Agarwal, L. M. Narducci, R. Gilmore and Da Hsuan Feng, Phys. Rev. <u>18A</u>, 620 (1978) and references contained therein.
12. D. F. Walls, P. D. Drummond, S. S. Hassan, and H. J. Carmichael Prog. of Theo. Phys. Supp. <u>64</u>, 307 (1978) and references contained therein.
13. C. R. Willis and J. Day, Opt. Comm. <u>28</u>, 137 (1979) and references contained therein.

THEORY OF PLANE WAVE REFLECTION AND REFRACTION BY THE NONLINEAR

INTERFACE[†]

A. E. Kaplan

Francis Bitter National Magnet Laboratory*
Massachusetts Institute of Technology
Cambridge, Massachusetts 02139

Abstract: This report includes a review of published as well
as recent results. In previous papers a new kind of optical bistability was proposed which is connected with reflection and refraction of light by a single surface on a nonlinear medium without
a resonator or feedback; its first experimental observations were
made recently by Smith et al. The main conditions required for its
existence are very exact matching of the optical densities of both
media and almost grazing incidence of light. These effects are
available for positive nonlinearity as well as for negative nonlinearity. In the last case, it is possible to excite a new kind of
nonlinear wave (longitudinally inhomogeneous travelling waves) which
could provide a phenomenon of strong nonlinear parallax of refracted
rays along the interface. For more simple observation and some applications of reflection bistability, the use of an electro-optic
element as an "artificial" nonlinearity can be proposed; this light-
feedback method is analogous to that used in hybrid devices. Our
last result is connected with a proposition for a new way to realize
reflection bistability which consists of application of single-mode
optical waveguides (one of which must be nonlinear) rather than
using two semi-infinite media. This allows us to avoid the secondary effects of self-focusing and self-bending of bounded refracted
beams of light in a nonlinear medium. At the same time it con-
serves all features of the main phenomenon of reflection bistability.

[†]Work supported by the U.S. Air Force Office of Scientific Research
under Grant No. AFOSR-80-0188.
*Supported by the National Science Foundation.

I. INTRODUCTION

In a number of articles[1-15] published in recent years a new
class of effects in nonlinear optics was investigated, which arose
under special conditions[1,2] (1976) of almost grazing incidence of
strong light on the interface between a linear and nonlinear medium
whose susceptibilities are very close to each other. All possible
effects, in particular bistability, in such cases are caused by competition between linear mismatch of susceptibilities of these media
and a nonlinear component; under the above conditions this competition can lead to a strong change of interface reflection even if
the nonlinear component is small (as it usually is in nonlinear
optics).

This mechanism provides the main difference between such
phenomena and known optical bistability. At the present time,
the known bistable optical devices are comprised of a Fabry-Perot
interferometer filled with a nonlinear medium, first proposed by
Seidel[16] and Szöke et al.[17] in 1969 and first observed by Gibbs,
McCall and Venkatesan in 1974[18]. (One can see a detailed survey of
this development in Ref. 19.) In these systems, the bistability is
due to the presence of a resonator (interferometer) which provides
a feedback. The media used might have resonant saturating absorption[16,18] or Kerr-nonlinearity[20]; it was also proposed to use a
phase transition in the resonant system of two-level atoms[21]. Use
of resonators or resonance causes these devices to be strongly
selective to the frequency of the incident light.

In contrast to these, the present paper is concerned with
phenomena not involving resonance and therefore might use a broad
spectrum of light. Interaction of light with matter in this nonresonant arrangement can give rise to a number of new effects (besides bistability) which are of physical interest as well as of
importance to applications and, moreover are not possible in
resonators.

Effects due to nonlinear interface reflection predicted by the
plane wave theory include the following:

1)   multistability and hysteresis jumps in reflection coefficient[1-10]
2)   change and scanning of the refraction angle and reflection co-
      efficient by varying the intensity of incident light[1,2,7-9]
3)   total bleaching of interface by incident light with definite
      intensity[1,2]
4)   change of penetration depth of the field into reflecting medium
      in the regime of total internal reflection (TIR)[1,2]
5)   excitation of nonlinear waves of a new kind (so called longi-
      tudinally inhomogeneous travelling waves - LITW[7-9,13] and
      effects connected with LITW, namely
6)   strong nonlinear self-parallax of refracted rays along the

interface[8,9] and
7)   self-limitation of the energy flux of the refracted light[7-10].

The conditions mentioned above (proximity of susceptibilities
and grazing incidence [1,2]) are important for the demonstration of
the existence of all these effects in nonlinear optics. The lack
of these conditions leads to missing the phenomena; this may be the
reason why such effects were not found earlier (beginning from 1962)
either theoretically or experimentally in studies by Bloembergen
et al.[22] devoted to harmonic generation by reflection of light from
nonlinear dielectrics.

The first experimental observation of the new phenomenon was
made recently by Smith, Herman, Tomlinson and Maloney[3], who demonstrated hysteresis behavior of reflection by "positive" nonlinear
(Kerr-effect) medium. Their first evaluations apparently show good
agreement with the plane wave theory[1,2] modified for the bounded
beam case.

The most detailed and analytically exhaustive theory has been
developed for the case of *plane* incident waves[1,2,5-10] (which implies also the assumption concerning one-dimensional behavior of
all waves in the system). There are numerical calculations only
for "bounded beam" cases [11-14], obtained recently which demonstrate
contradictory results for the main issues of the problem and it has
not been possible to check them analytically.

Therefore, this paper is devoted to a brief survey of the published plane-wave theory results as well as to a brief consideration of some new results in this direction. There are several
reasons why the plane-wave theory remains to be of essential interest: 1) it is of general physical interest by itself because it introduces new kinds of waves (such as LITW) and a new problem into
nonlinear electrodynamics (for instance, a "continuum problem" for
LITW); 2) it can be applied to other nonlinear wave problems, for
example, in a plasma; 3) it can point out the new interesting direction of research for real bounded beams and 4) there are at least
several possible experimental situations which can be precisely
described by the equations of plane wave theory (see below).

II.   THE MAIN EQUATIONS AND CONDITIONS

Let a plane wave with amplitude E_{in} be incident from a linear
medium with susceptibility ε_o at the glancing angle ψ (Fig. 1) on
the boundary of a nonlinear medium whose susceptibility depends on
the field amplitude E in the medium:

$$\varepsilon_{NL} = \varepsilon_o + \Delta\varepsilon_L + \Delta\varepsilon_{NL} \ (|E|^2) \tag{1}$$

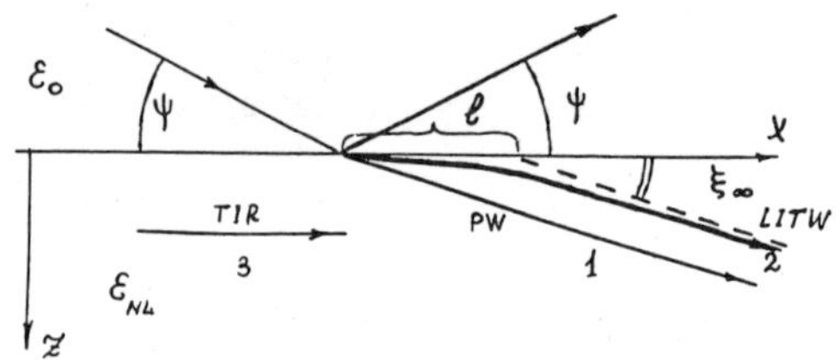

Fig. 1.   Wave diagram of the nonlinear interface. Ray traces in
           nonlinear media: 1) travelling plane wave (PW), 2) longi-
           tudinally inhomogeneous travelling wave (LITW), 3) total
           internal reflection (TIR).

where in the simplest case (for instance, Kerr-nonlinearity)

$$\Delta\varepsilon_{NL} = \varepsilon_2 \left|E\right|^2 \tag{2}$$

and $\Delta\varepsilon_L$ does not depend on the field.

To expect the phenomena concerned, the above mentioned conditions should be satisfied[1,2]

$$1 \gg \left|\Delta\varepsilon_L\right| \sim \left|\Delta\varepsilon_{NL}\right| ; \quad 1 \gg \psi \sim \left(\frac{\Delta\varepsilon}{\varepsilon_o}\right)^{1/2} \tag{3}$$

which amount to $\psi \sim 1°$ at $\left|\Delta\varepsilon_{NL}\right| \sim 10^{-4}$. It can be pointed out that
these conditions simplify the theory and make the phenomena inde-
pendent of the polarization of incident light.

As usual, the field in a linear medium can be represented in
the form of two homogeneous plane waves, one of which is the incident wave and the other the reflected wave, which should be characterized by an unknown complex reflection coefficient r. The wave
equation for the complex amplitude of the field E in a nonlinear
medium in the one-dimensional case can be written as

$$\frac{d^2E}{dz^2} + k_o^2 E\left[\frac{\varepsilon_{NL}(|E|^2)}{\varepsilon_o} - \cos^2\psi\right] = 0; \quad k_o = \frac{\omega\sqrt{\varepsilon_o}}{c} \tag{4}$$

where the z axis is perpendicular to the boundary (the total field
is $\frac{1}{2}E \exp(-i\omega t) + c.c.$). By comparing the tangential components of
the fields on both sides of the boundary, the generalized boundary
condition for the refracted wave E can be obtained[2]

$$i \frac{d\,E(0)}{dz} + k_o \sin\psi\,[2E_{in} - E(0)] = 0 \quad . \tag{5}$$

An expression for the reflection coefficient r is,

$$r = \frac{E(0)}{E_{in}} - 1 \ . \tag{6}$$

The field E in a nonlinear medium can be represented in the form:

$$E = u(z)\exp\left[ik_o\int_o^z \xi(z)dz + i\phi + ik_o x\sin\psi\right] \quad (\phi=\text{const}) \tag{7}$$

where $u(z)$ and $\xi(z)$ are real; ξ is the angle formed by the rays and the x axis at a given point z. For plane waves (PW) u and ξ are constant, and for surface waves (SW), when TIR occurs, $\xi=0$. In general, u and ξ are not constant.

As is usual in the theory of reflection from a semi-infinite medium, the condition at infinity is of great importance. Since there are no sources inside the nonlinear medium, only travelling waves should propagate toward the interior of the medium (or SW which do not carry any energy along the z axis at all). Therefore[2] in the expression (7) for $z \rightarrow \infty$

$$u \rightarrow \text{const} \equiv u_\infty \gtrsim 0, \quad \xi \rightarrow \text{const} \equiv \xi_\infty \gtrsim 0 \ , \tag{8}$$

in essence the Sommerfeld radiation condition (the absence of the reverse wave).

Substituting the field in the form (7) into the wave equation (4), we obtain the first integral of it,

$$I = \xi u^2 = \text{const} \equiv \xi_\infty u_\infty^2 \gtrsim 0 \tag{9}$$

which expresses the conservation of energy flux, and also the equation for the real amplitude u,

$$\frac{d^2u}{dz^2} + k_o^2 u \left(\frac{\varepsilon_{NL}(u^2)}{\varepsilon_o} - \cos^2\psi - \frac{I^2}{u^4}\right) = 0 \ . \tag{10}$$

The first integral of this equation, which satisfies the radiation condition (8), can be written in the form[7],

$$\left(\frac{du}{dz}\right)^2 = k_o^2 \int_{u_\infty}^u [F(u) - F(u_\infty)]\ d\left(\frac{1}{u^2}\right) \tag{11}$$

where the nonlinear "characteristic" function F(u) is introduced,

$$F(u) = u^4\left[\frac{\varepsilon_{NL}(u^2)}{\varepsilon_o} - \cos^2\psi\right] \ . \tag{12}$$

By integrating Eq. (11) and taking into account boundary condition
(5) one can obtain all possible wave solutions for a given problem.

II. EFFECTS AT "POSITIVE" NONLINEARITY

It can be proven[2-7] that in the simplest case of "cubic" nonlinearity (2) with $\varepsilon_2 > 0$ (i.e., Kerr-nonlinearity) only two kinds
of one-dimensional wave regimes can exist:

1) homogeneous plane wave (PW), $u(z) = \mathrm{const}$, which corres-
    ponds to the transmission regime ($|r| < 1$) and

2) surface wave (SW), $\xi = 0$, which corresponds to TIR
    ($|r| = 1$).

Both of them are nonlinear analogs of corresponding linear waves,
but differ now from those by dependence on the incidence light intensity. Using the expressions (4-7) with u and ξ constant for
the transition regime under conditions (1), it is easy to obtain
"Snell's nonlinear formula" for transmission angle ξ,[1,2]

$$\left(1 + \frac{\xi}{\psi}\right)^2 \left(\xi^2 - \psi^2 - \frac{\Delta\varepsilon_L}{\varepsilon_o}\right) = 4\,\frac{\varepsilon_2}{\varepsilon_o}\,|E_{in}^2| \tag{13}$$

or "Fresnel's nonlinear formula" for the reflection coefficient r,

$$4\,r\,\psi^2 + \frac{\Delta\varepsilon_L}{\varepsilon_o}\,(1+r)^2 + \frac{\varepsilon_2}{\varepsilon_o}\,|E_{in}^2|\,(1+r)^4 = 0\,, \tag{14}$$

(see Fig. 2).

If linear mismatch is negative ($\Delta\varepsilon_L < 0$) and $\psi < \psi_{cr}$, where
$\psi_{cr} = (|\Delta\varepsilon_L|/\varepsilon_o)^{\frac{1}{2}}$ is the critical angle of linear TIR, then nonlinear TIR can be excited. The profile of wave intensity now is
not exponential; it can be obtained from (11) with $u_\infty = 0$, that in
the nonlinear medium the surface wave has a "self-channel" shape
(which is well known in the theory of self-focusing),

$$u = \left(\frac{2\varepsilon_o}{\varepsilon_2}\right)^{\frac{1}{2}} \frac{\gamma}{\mathrm{ch}(k_o\gamma z + c)}\,; \quad \gamma = \left(\psi_{cr}^2 - \psi^2\right)^{\frac{1}{2}} \tag{15}$$

where c is determined from boundary conditions (5) and can have
from one to four values (for different ψ and E_{in}) that provides
multistability and the possibility of hysteresis jumps (see
Fig. 2).

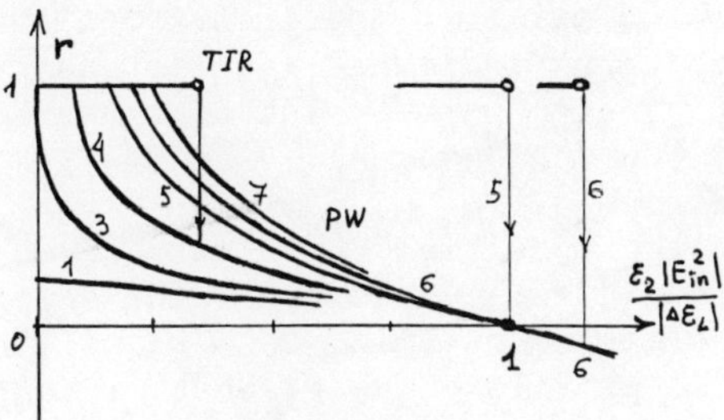

Fig. 2. Dependence of the reflectivity r on incident light inten-
         sity $|E_{in}^2|$ at different glancing angles ψ in the case of
         negative linear mismatch of susceptibilities ($\Delta\varepsilon_L < 0$) and
         positive nonlinearity ($\varepsilon_2 > 0$). Curves 1) $\psi > \psi_{cr}$,
         2) $\psi = \psi_{cr}$, 4–1) $\psi/\psi_{cr} > \tfrac{1}{2}\sqrt{2}$, 5) $\psi = \psi_{cr}/2\sqrt{2}$ (nonlinear
         bleaching), 6,7) $\psi < \psi_{cr}/2\sqrt{2}$.[1,2]

The jumps of reflection coefficient occur only when going from TIR
to transmission; in the opposite direction there is only a jump in
the phase of reflection. At negative mismatch there exists also
an incident light intensity,

$$E_{in}^2 = E_{b\ell}^2 \equiv |\Delta\varepsilon_L/\varepsilon_2|, \quad i.e., \quad \Delta\varepsilon_L = -\Delta\varepsilon_{NL} \tag{16}$$

at which there is no reflection at all (r = 0) for any glancing
angle ψ (see Fig. 2). This phenomenon is not the nonlinear analog
of the Brewster bleaching (since it does not depend on either the
angle or polarization of incident light), but is just due to the
fact that the field equalizes the susceptibilities of the two media
and makes the boundary completely transparent[1,2]. In this case,
the jump from total reflection to the total transmission takes
place at $\psi = \psi_{cr}/2\sqrt{2}$ (see curve 5 in Fig. 2). For CS_2, where
$n_2 \sim 10^{-11}$ cgs esu, we have $E_{b\ell} \sim 1.8 \times 10^6$ v/cm at $\Delta\varepsilon_L \sim -10^3$, in
which case $\psi \simeq 0.54°$.

III. EFFECTS AT "NEGATIVE" NONLINEARITY. LONGITUDINALLY
      INHOMOGENEOUS TRAVELLING WAVES (LITW)

In a number of physical situations, the nonlinearity can be
"negative" (i.e. $\varepsilon_2 < 0$ in (2)), for instance, due to nonlinear
resonant interactions[4,22]. In such cases, the nonlinear analogs
of linear waves (i.e., PW and SW) can exist as well. For plane
waves Snell's[13] and Fresnel's[14] formulas remain valid just by
taking into consideration that now the sign of ε_2 has already been
changed. Under this regime it is also possible to vary the value
of transition angle ξ and reflection r by changing the incident
light intensity (Fig. 3). In analogy to the "positive"

nonlinearity there also exists a nonlinear total bleaching of the interface; the "bleaching" incident light intensity is given by the same expression (16), but now the value of linear mismatch should be positive (Fig. 3, curves 3 and 4).

In contrast to the "positive" nonlinear case, the surface waves at the TIR regime should exist exactly in the same range of linear mismatches and glancing angles (i.e. $\Delta\varepsilon_L < 0$ and $\psi < \psi_{cr} = (\Delta\varepsilon_L/\varepsilon_o)^{\frac{1}{2}}$ but now the amplitude profile of SW given by integration of (11) is as follows:[2]

$$u = \sqrt{2\,\frac{\varepsilon_o}{|\varepsilon_2|}}\,\frac{\gamma}{\sinh(k_o\gamma z + c)}\;;\quad \gamma = \left(\psi_{cr}^2 - \psi^2\right)^{\frac{1}{2}} \tag{17}$$

where the constant c is determined by boundary conditions. The peculiarity of TIR in this case is that the depth of penetration of the field into the medium, which is defined by c, can be much smaller than the depth of linear penetration[2].

But the main physical peculiarity in the case of negative nonlinearity is the possibility of excitation (under special conditions) of a new kind of wave: longitudinally inhomogeneous travelling waves (LITW)[4],[7-9]. Being inhomogeneous and non-planar near the boundary, these waves reduce to plane waves sufficiently far from the boundary. The concept of longitudinal inhomogeneity was introduced by us[7] to distinguish this phenomenon from the self-actions due to transversely inhomogeneous waves, such as self-focusing[24] and self-bending[25] of bounded light beams.

The consideration of (11) shows that under conditions,

$$\frac{2}{3}Q < \frac{|\varepsilon_2|}{\varepsilon_o}\,u_\infty^2 < Q;\quad Q \equiv \frac{\Delta\varepsilon_L}{\varepsilon_o} + \sin^2\psi \tag{18}$$

the excitation of LITW is possible. Integration of (11) yields two possible intensity profiles of LITW in space:[7]

$$u^2 = u_\infty^2 \pm 2B^2 \left/ \begin{array}{c}\sinh^2\\ \cosh^2\end{array}\right.(Bk_o z + c)\;;\quad B \equiv \left(\frac{3}{2}\frac{\varepsilon_2}{\varepsilon_o}u_\infty^2 - Q\right)^{\frac{1}{2}}. \tag{19}$$

For the minimum possible energy density at infinity,

$$u_\infty^2 = u_m^2 = \frac{2}{3}\left|\frac{\varepsilon_2}{\varepsilon_o}\right|Q \tag{20}$$

the only one (limiting) type of LITW remains,

$$u^2_{lim} = 2\,\frac{|\varepsilon_2|}{\varepsilon_o}\left[\frac{Q}{3} + \frac{1}{(k_o z + c)^2}\right]\,. \tag{21}$$

The unique "continuum" problem arises which is connected with new solutions: we have only two conditions (which are equivalent to a single complex boundary condition (5)) to determine three unknown constants (v_∞, c and ϕ in (7)); the radiation conditions (8) have already been used for construction of LITW, i.e., (11) and (19). Therefore, there is a continuum of solutions even when boundary and radiation conditions are completely specified.

This situation differs radically from the situation in a linear medium and a "positive" nonlinear medium. The energetic criterion for selecting a unique kind of physically realized LITW was proposed by us[3,8] which is based on the minimization of wave energy density (at the same time it provides a maximum of wave energy flux). The only realized kind of LITW appears to be a limiting type (21), when v^2_∞ reduces to the minimum possible value for LITW, equal to v^2_n, (20). This choice was confirmed lately in numerical experiments[5b,10] by introducing weak absorption, and proved analytically[8,9] by asymptotic methods, implying a limiting transition from the absorbing medium to a transparent one (a small parameter of this transition appears to be of the order $(k_o L_{ab})^{-2/5}$, where L_{ab} is a characteristic length of wave absorption).

The LITW excitation provides a new physical effect. First of all, like a "positive" nonlinear case, it leads to the hysteresis behavior of a system under definite conditions, but now the hysteresis jumps are arising between states of two different travelling waves, PW and LITW (in contrast to "positive" nonlinearity, where they should arise between PW and TIR). Hysteresis takes place only under the following conditions:[3,8,9]

$$|\varepsilon_2\,E^2_{in}| > \Delta\varepsilon_L > 2\varepsilon_o \sin^2\psi\,. \tag{22}$$

The behavior of reflectivity for different parameters of a system is shown in Fig. 3 (in particular, one can see hysteresis, curve 5, and self-bleaching, curves 3 and 4). In this figure E^2_{cr} is the critical intensity of incident light required for excitation of LITW in every different case and $D \equiv \Delta\varepsilon_L/\varepsilon_o$.

Another effect which is of interest for applications consists of self-limitation of energy flux of LITW penetrated into a nonlinear medium[8,10]. It is a direct consequence of the selection of only a limiting wave (21) which has the only possible value of $u_\infty = u_n$; behavior of energy flux I is shown in Fig. 4 for hysteresis and non-hysteresis situations.

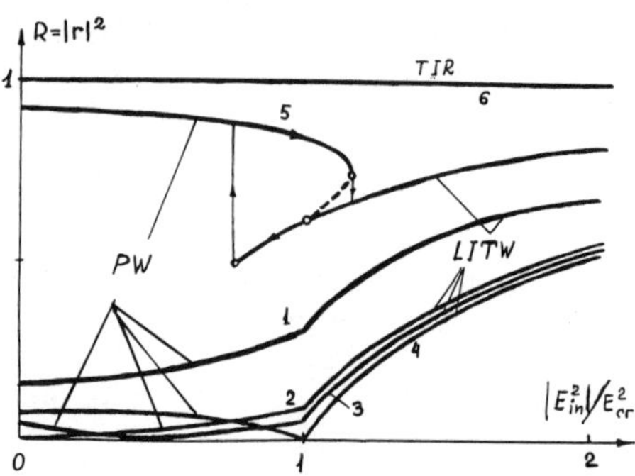

Fig. 3.   Reflectivity $R = |r|^2$ as a function of the incident light intensity $|E^2_{in}|$.   Curves: 1) $D < 0$, $\psi^2 > |D|$; 2) $D = 0$; 3) $\psi^2 > D/2 > 0$; 4) $\psi^2 = D/2$; 5) $\psi^2 < D/2$; 6) $D < 0$, $\psi^2 < |D|$.[9]

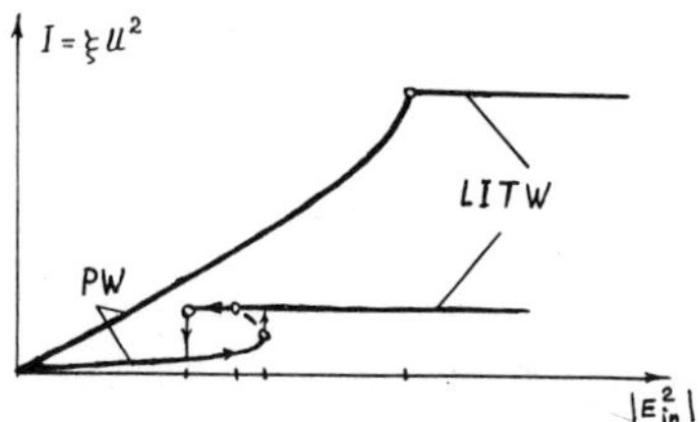

Fig. 4.   Energy flux I of transmitted waves as a function of the incident light intensity $|E^2_{in}|$ for hysteresis and non-hysteresis situations[9].

Finally, one of the interesting physics effects, caused by excitation of LITW, is self-parallax[8,9] i.e. displacement of the refracted rays along the interface. This effect is not the nonlinear analog of the well known Goos-Hänchen displacement of bounded beams under TIR, because in contrast to G.-H., self-parallax is valid for unbounded waves, and for the transmission regime, rather than for TIR. It corresponds to the initial "sticking" of the rays (Fig. 1, curve 2) because of conservation of energy flux (9) (the angle ξ should increase as the intensity u^2 decreases), and it does not depend on the existence of hysteresis. The ray trace in the space is given by the formula[9],

$$ x = z(3/Q)^{\frac{1}{2}} + \ell(1 + z_\ell/z)^{-1} \tag{23} $$

where ℓ is the parallax for $z \to \infty$ and z_ℓ is the characteristic depth for half of the parallax; ℓ and z_ℓ are determined by E_{in}, ψ and $\Delta\varepsilon_L$. Let $k_o = 10^5$ cm^{-1}; $\Delta\varepsilon_L/\varepsilon_o = -10^{-4}$; $\psi \sim 0.7°$ and $|\varepsilon_2/\varepsilon| = 10^{-10}$ cgs esu; then the critical value of E_{in} for excitation of LITW is $E_{cr} = 1.15 \times .10^5$ v/cm, and if $E_{in} = 2E_{cr}$, one should obtain $\ell = 1.2$ cm and $z_\ell = 0.01$ mm, so parallax is very strong and arises at a very small depth.

IV.   THEOREMS OF LITW EXISTENCE FOR ARBITRARY KINDS OF NONLINEARITY

The existence of LITW in the system with negative nonlinearity and its absence in a "positive" nonlinear system pose a number of questions comparable to arbitrary kinds of nonlinearity, such as: is it possible to predict what kinds of nonlinearities and system characteristics can produce the possibility for the existence of LITW; and, what kinds of nonlinearities do not allow LITW? What are the parameters of possible LITW if allowed?

All these questions have been answered by the theorem formulated and proved by us in Ref. 7. This theorem related the existence of different kinds of travelling waves with the behavior of a "characteristic" nonlinear function $F(u)$ (12). The main conclusion of the theorem is that the existence of LITW in a transparent medium is possible if, and only if: 1) there exists a range $V \ni u$ where $F(u) > 0$; 2) $F(u)$ falls at least somewhere in V (i.e., there is at least one interval $W \subset V$ where $F(u)$ falls monotonically); and 3) if these conditions are satisfied, the LITW, if excited, can have a value of u_∞ which must belong only to this falling interval, i.e. always $u_\infty \in W$. The proof of this theorem is based on investigation of the behavior of the integral in the right-hand part of Eq. (11).

One of the main consequences of this theorem is that LITW can be excited only in such a nonlinear medium whose nonlinearity $\varepsilon_{NL}(u^2)$ has at least one interval of fall. On the other hand, LITW can never be excited in the medium with increasing function $\varepsilon_{NL}(u^2)$. This is the reason why excitation of LITW is impossible for "positive" nonlinearity.

The "continuum problem" arises for all kinds of LITW. Therefore, the special direction of the theory of one-dimensional nonlinear waves should be devoted to the theory of "realization" of nonlinear transmitted waves in those cases of arbitrary nonlinearity. Several theorems concerned with this problem were formulated by us[9]. In particular, the complete class of functions $F(u)$ were found which allowed the realization of LITW in general, and it was proved that the principle of minimization of wave energy is valid throughout this class.

V.   BISTABLE REFLECTION BY "ARTIFICIAL" NONLINEAR INTERFACE

It was mentioned above that plane-wave theory (or precisely speaking, one-dimensional theory) can be applied to some real experimental situations rather than being just a simple physical model of phenomena. One such situation is the reflection of light from the interface between a linear and an "artificially" nonlinear medium proposed in Ref. 15. This can be achieved by using an electro-optic element as a nonlinear medium, which is driven by the signal from a detector which receives the reflected (or refracted) light. Such a light-feedback method is analogous to the one used in hybrid devices[26]. However, now it changes the refraction angle of the light behind the interface rather than the phase shift of wave in the resonator. The electro-optic element can change susceptibility of the same value practically in the entire working volume of the medium; this is the reason why the plane-wave theory can give a satisfactory description of this situation.

It can be proven that hysteresis arises only under conditions where the "output" beam is the reflected one and the "nonlinear" medium is the one of incidence. It means that the detector received the reflected light and its signal drives the input medium (i.e., incidence of light occurs from the electro-optical element) which has a susceptibility that is assumed to have the form $\varepsilon + \varepsilon_o + \Delta\varepsilon_L + kI_r$, where $\Delta\varepsilon_L > 0$, I_r is the intensity of reflected light and k is a constant which depends upon the detector and electro-optic modulator characteristics. By analogy with the theory in Refs. 1 and 2, the "nonlinear Fresnel formula" for amplitude reflectivity of the interface r can be obtained in the case of transmission regime[15]

$$\frac{kI_{in}}{\Delta\varepsilon_L} = \left[4\left(\frac{\psi}{\psi_{cr}}\right)^2 \frac{1}{(1+r)^2} - \frac{1}{r} \right] \cdot \frac{1}{r} \tag{24}$$

where ψ is again a glancing angle of incidence, $\psi_{cr} = \sqrt{\Delta\varepsilon_L/\varepsilon_o}$ is the critical angle of linear TIR and I_{in} is the intensity of incident light. Hysteresis jumps between the "transmission" and TIR states can only be observed if

$$\psi > \psi_{cr} \quad \text{and} \quad I_{in} > \frac{\Delta\varepsilon_L}{k} \left(\frac{\psi^2}{\psi_{cr}^2} - 1 \right) .$$

In contrast to "real" nonlinearities, very low power sources of light such as He-Ne laser can be used; mismatching $\Delta\varepsilon_L$ more than 10^{-2} and glancing angle ψ up to $5°-10°$ can be achieved. It provides a simple method for bistable nonresonant operation and optical signal processing such as an optical memory, switch and logic operations, pulse shortening, and so on. The hysteresis behavior of reflectivity of such devices for difference ratio ψ/ψ_{cr} is shown in Fig. 5.

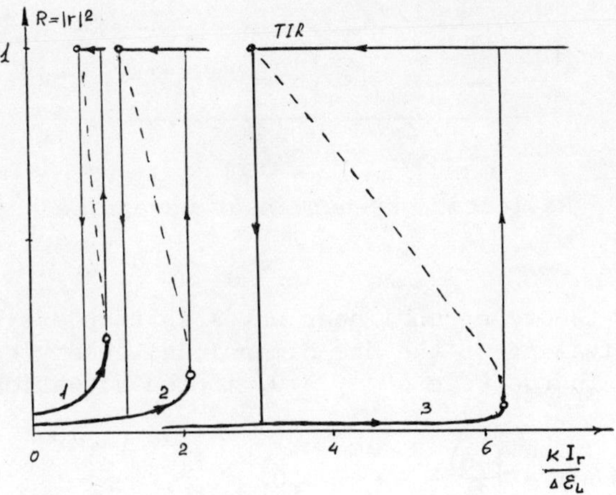

Fig. 5.   The reflectivity $R = |r|^2$ as a function of incident light intensity I_{in} for different ratios ψ/ψ_{cr}. Curves: 1) $\psi/\psi_{cr} = 1.3$;   2) $\psi/\psi_{cr} = 1.5$;   3) $\psi/\psi_{cr} = 2$.

VI. BISTABLE REFLECTION FROM THE BUTT–END OF A NONLINEAR WAVEGUIDE WITH METALLIC WALLS

Another case where the one-dimensional theory is valid is the reflection of light by the interface between linear and nonlinear single-mode waveguides with metallic walls[27], (Fig. 6).

It is not difficult to show that under conditions where waveguide size, a, is close to the "cutoff" value $\lambda_o/2 = \pi c/\omega\sqrt{\varepsilon_o}$, the wave equation for the nonlinear slab waveguide whose susceptibility can be represented in the form (2), is as follows,

$$\frac{d^2E}{dz^2} + k_o^2 E \left[1 + \frac{\Delta\varepsilon_L}{\varepsilon} + \frac{3}{4} \frac{\varepsilon_2|E|^2}{\varepsilon_o} - \left(\frac{\lambda_o}{2a}\right)^2 \right] = 0 \tag{25}$$

where λ_o is the wavelength in the linear waveguide, E is the field in the middle of the slab, and coefficient 3/4 appears due to non-uniformity of the field across the waveguide cross-section. It is equal to Eq. (4) if we introduce the "effective" nonlinearity, as well as "effective" glancing angle by,

$$\Delta\varepsilon_{NLef} = \frac{3}{4} \varepsilon_2 |E|^2 ; \quad \cos\psi_{ef} = \frac{\lambda_o}{2a} . \tag{26}$$

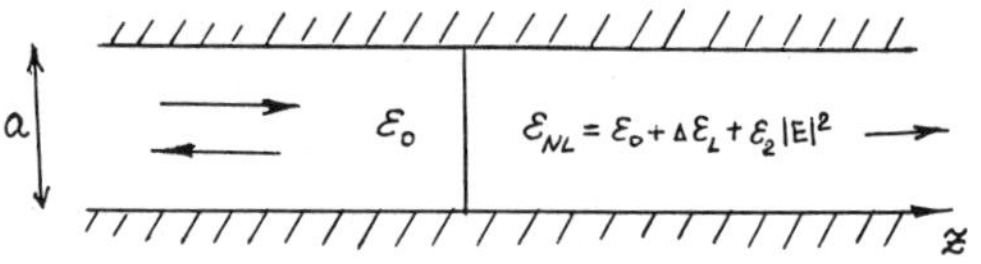

a

ε_0

$\varepsilon_{NL} = \varepsilon_0 + \Delta\varepsilon_L + \varepsilon_2 |E|^2$

z

Fig. 6.   Reflection by nonlinear waveguide butt-end.

Therefore, the theory of nonlinear waves in such a system should be completely equivalent to the one-dimensional theory considered above. In particular, in the case of $\varepsilon_2 > 0$, if the situation,

$$\Delta\varepsilon_L < 0 \text{ and } 1 - \left(\frac{\lambda_0}{2a}\right)^2 + \frac{\Delta\varepsilon_L}{\varepsilon_0} < 0 \qquad \left(a < \frac{\lambda_0}{2}\right)$$

is chosen (i.e. butt-end of the nonlinear waveguide is totally reflecting for a weak field), we can expect a nonlinear bleaching at the interface by strong fields as well as hysteresis jumping from the TIR state to the transmission regime. In such cases, the strong field "pushes through" the cutoff waveguide. All results mentioned above are applicable to those devices just by taking into account relationship (26). The only problem is that the requirement for the reflectivity of metallic walls should be very rigid; preliminary estimation shows this reflection has to be about 99% or better, which is not easy to obtain in the optical range.

VII.   CONCLUSION

It will be useful to point out several possible applications of phenomena considered above, proposed for the first time in Ref. (1,2) besides bistability:

1.   These phenomena can be used as a method of investigation of nonlinear properties of matter, for instance to measure non-linear coefficients ε_2;
2.   It is possible to develop a new kind of nonlinear surface spectroscopy, which would be analogous to linear spectroscopy of internal reflection;
3.   "Surface switching" devices can be used as a very fast and broad-band shutters in lasers for generation of giant pulses;
4.   Nonlinear refraction can be used for angular switching and scanning of the refracted beam (i.e. space switching and scanning), which can provide a declination angle of up to several angle degrees for a period of time less than a pico-second.

All these applications can be realized in hysteresis as well as in

nonhysteresis regimes. The discussion presented above points out
several possible directions for further work which is of great physical interest by itself, as well as of importance for possible applications. Not all of the phenomena discussed can be easily observed
under real physical conditions due to such factors as diffraction of
real bounded beams, possible instabilities of different kinds in
time and space, absorption, scattering and so on. But I would like
to believe that the main part of these effects should survive and
find their development and applications, especially in bistability,
angle and reflectivity scanning and switching, and new possibilities
to be investigated in nonlinear properties of matter.

ACKNOWLEDGMENTS

I would like to express my gratitude to P. W. Smith and W. J.
Tomlinson for discussions of some aspects of the problem, and to
P. L. Kelley for substantial support and assistance.

REFERENCES

1. A. E. Kaplan, JETP Lett. $\underline{24}$, 114 (1976).

2. A. E. Kaplan, Sov. Phys. JETP $\underline{45}$, 896 (1977).

3. a) P. W. Smith, J.-P. Hermann, W. J. Tomlinson and P. J.
    Maloney, Appl. Phys. Lett. $\underline{35}$, 846 (1979); b) P. W. Smith,
    W. J. Tomlinson, this volume.

4. V. S. Butylkin, A. E. Kaplan, Yu. G. Khronopulo and E. I.
    Yakubovich, Resonant Interaction of Light with Matter,
    Nauka, Moscow (1977), p. 331 (to be translated into English
    by Springer).

5. N. N. Rosanov, Sov. Tech. Phys. Lett. a) $\underline{3}$,   (1977);
    b) $\underline{4}$, 30 (1978).

6. B. B. Boiko, I. Z. Dzhilavdari and N. S. Petrov, J. Appl.
    Spectr. $\underline{23}$, 1511 (1975).

7. A. E. Kaplan, Sov. J. of Quantum Electronics $\underline{8}$, 95 (1978).

8. A. E. Kaplan, Proceedings of the 9th National Conference on
    Coherent and Nonlinear Optics, pp. 238 and 241, 1978,
    Leningrad-Moscow.

9. A. E. Kaplan, Radiophysics and Quantum Electronics $\underline{22}$, 229
    (1979).

10. V. A. Permyakov and O. V. Bagdasaryan, Radiophysics and
    Quantum Electronics $\underline{21}$, 92 (1978).

11. N. N. Rosanov, Optics and Spectroscopy $\underline{47}$, 335 (1979).

12. A. A. Kolokolov and A. I. Sukov, Radiophysics and Quantum
    Electronics $\underline{21}$, 1013 (1978).

13. D. Marcuse, Appl. Opt. (to be published).

14. W. J. Tomlinson, Opt. Lett. $\underline{5}$, 323 (1980).

15. A. E. Kaplan, XI IQEC, Boston, MA., U.S.A. (1980) paper T-10;

A. E. Kaplan, Appl. Phys. Lett., to be published.

16. H. Seidel, U. S. Patent No. 3,610,731 (1969).

17. A. Szöke, V. Daneu, T. Goldhar and N. A. Kurnit, Appl. Phys. Lett. $\underline{15}$, 376 (1969).

18. H. M. Gibbs, S. L. McCall and T. N. C. Venkatesan, Phys. Rev. Lett. $\underline{36}$, 1135 (1976).

19. H. M. Gibbs, S. L. McCall and T. N. C. Venkatesan, Optics News/ Summer, (1979).

20. F. S. Felber and J. H. Marburger, Appl. Phys. Lett. $\underline{28}$, 731 (1976).

21. C. M. Bowden and C. C. Sung, Phys. Rev. A$\underline{19}$, 2392 (1979); C. M. Bowden, XI IQEC, Boston, MA., U.S.A. (1980) and this volume.

22. N. Bloembergen and D. S. Pershan, Phys. Rev. $\underline{128}$, 606 (1962); N. Bloembergen and J. Ducuing, Phys. Lett. $\underline{6}$, 5 (1963).

23. V. S. Butylkin, A. E. Kaplan and Yu. G. Khronopulo, Sov. Phys. JETP $\underline{32}$, 501 (1971) and $\underline{34}$, 276 (1972).

24. P. L. Kelley, Phys. Rev. Lett. $\underline{15}$, 1004 (1965); R. Y. Chiao, E. Garmire and C. U. Townes, Phys. Rev. Lett. $\underline{13}$, 479 (1964); G. A. Askaryan, Sov. Phys. JETP $\underline{15}$, 1088 (1962).

25. A. E. Kaplan, JETP. Lett. $\underline{9}$, 33 (1969).

26. P. W. Smith and E. H. Turner, Appl. Phys. Lett. $\underline{30}$, 280 (1977).

27. A. E. Kaplan, to be published.

OPTICAL PROPERTIES OF NONLINEAR INTERFACES

P. W. Smith and W. J. Tomlinson

Bell Laboratories, Inc.
Holmdel, New Jersey 07733

Abstract: Under suitable conditions, the boundary between a
linear and a nonlinear (Kerr effect) medium has a reflection coefficient which exhibits complex nonlinear behavior as a function
of optical intensity. We present a detailed experimental study of
this effect. Numerical results are presented to show the effect of
using a Gaussian input beam. We conclude with a discussion of some
waveguide configurations of nonlinear interface devices. They
should provide fast, compact and relatively low-energy optical
signal processing elements.

I. INTRODUCTION

An interface between two dielectric materials, one of which
has an intensity-dependent refractive index (an optical Kerr effect), is a deceptively simple system, which is capable of exhibiting a wide range of complex and potentially useful optical phenomena.
The initial theoretical studies of such interfaces predicted that
under suitable conditions the reflectivity of the interface would
not only be intensity dependent, but would exhibit optical hysteresis
and bistability[1,2]. Subsequent theoretical studies have arrived at
conflicting conclusions,[3,4] and at present it is not possible to
specify exactly the conditions necessary to obtain bistability.

Previously published experimental results[5], and new results
reported in this paper, show clear evidence of optical hysteresis.
Further studies are required, however, to investigate the stability
of the two states involved and the switching time between them.

While it is currently unclear which particular configurations will exhibit bistability, it is clear that there are a number of configurations that will exhibit a highly nonlinear behavior that is of interest for various optical switching and signal processing applications.

A particular feature of devices based on nonlinear interfaces is that, because no optical resonator is involved, they should be capable of extremely fast response times. The present level of theoretical understanding of such devices does not permit us to compute an ultimate limit, but it seems clear that subpicosecond response times can be achieved.

In this paper we present and discuss experimental results on one particular type of nonlinear interface. We also present some ideas for waveguide devices using nonlinear interfaces, and the preliminary results of an experimental study of two such devices.

II. BASIC PHENOMENA AT A NONLINEAR INTERFACE

The nonlinear interface configuration that has received the most theoretical and experimental attention is illustrated in Fig. 1. In the negative-x half space we have an ordinary dielectric material with an index of refraction n_0. In the positive-x half space we have a nonlinear material. We assume that the nonlinear material has a zero-intensity refractive index that is less than than of the linear material by a small amount Δ, and that it has a positive optical Kerr coefficient n_2.

For a low-intensity incident beam in the linear medium, if the angle of incidence ψ is less than the critical angle $\psi_c \equiv (2\Delta/n_0)^{\frac{1}{2}}$, the beam will undergo total internal reflection (TIR) at the interface. There is then no transmitted beam, but there is, of course, an evanescent field in the nonlinear medium. Because of the positive Kerr constant, the evanescent field reduces the index difference across the interface, thus reducing the effective critical angle, and affecting the phase shift between the incident and reflected beams. Moving the effective critical angle closer to the incident angle generally results in an increase in the evanescent field. This produces a positive feedback effect in which an increase of the incident intensity increases the evanescent field, which reduces the effective critical angle, which further increases the evanescent field. Therefore, we expect that there will be some threshold input intensity at which there will be a sudden switch from TIR to a state in which part of the incident energy remains in the nonlinear medium.

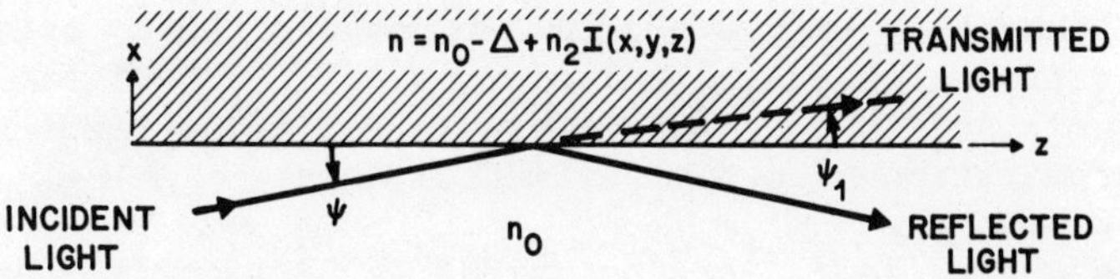

Fig. 1. Light incident on an interface between linear and non-
         linear media.

For an incident plane wave, theoretical studies[1] have shown
that the critical input intensity, I_0, for the switch from TIR is
given by:

$$\frac{n_2 I_o}{\Delta} = \begin{cases} \frac{1}{2}\left[1 - (\psi/\psi_c)^2\right] & 1/\sqrt{2} \lesssim \psi/\psi_c < 1 \\[2mm] & \qquad\qquad\qquad 0 < \psi/\psi_c \lesssim 1\sqrt{2} \\[2mm] \frac{1}{8}(\psi_c/\psi)^2 & \end{cases} \qquad (1)$$

The initial theoretical studies also predicted (for an incident plane wave) that once the switch from the TIR state has taken
place, the input intensity must be reduced by a finite amount before the interface will return to the TIR state, i.e., the reflectivity is bistable and will exhibit hysteresis as a function of incident intensity. Subsequent theoretical studies have presented
conflicting conclusions[3,4]. At present, the existence of bistable
behavior for this configuration with an incident plane wave must
be considered in doubt.

There are several other possible configurations of the nonlinear interface, and for some of these the plane-wave theories
also predict bistability. For the configurations with a negative
nonlinearity the bistability predictions would seem to be on somewhat firmer ground. This is an area that merits further theoretical and experimental study.

Experiments done with Gaussian laser beams do not provide a
direct test of the plane-wave theories. A complete theoretical
description, accounting for the spatial and angular characteristics
of a Gaussian beam, is a formidable problem, which has not yet been
attempted. In our initial experimental report[5] we compared our experimental pulse shapes to those calculated from the plane-wave
theory by neglecting the angular spread of the input beam, and by
treating each point on interface as independent, in what one might
call an "incoherent plane-wave" calculation. On the basis of subsequent analysis we no longer believe that this is an appropriate
comparison.

In collaboration with D. Marcuse we have recently obtained some results from a numerical calculation for a *two-dimensional* Gaussian beam[6], and in this paper we will use those results to aid in the interpretation of our experimental data.

For the waveguide devices described in Section IV, the combination of waveguide properties with nonlinear interface properties results in an extremely complicated theoretical problem. Our current analysis of the operation of these devices is based on a simple qualitative description of waveguide behavior which does not take into account the details of interactions at the nonlinear waveguide interfaces. Nevertheless, this analysis shows that these devices appear to have very interesting and useful optical characteristics, and these predictions are born out in the results of our preliminary experiments.

III.   EXPERIMENTS ON NONLINEAR INTERFACES

Experiments were performed with the apparatus shown in Fig. 2. The input pulse was generated by a mode-locked traveling-wave ruby laser. Internal Fabry-Perot mode selection was employed so that the laser output consisted of a train of $\approx$ 1-ns pulses. The absence of internal structure in these pulses was verified with a streak camera. A single pulse was selected from this train and directed onto the setup shown in Fig. 2. A portion of the beam was sampled and directed with an optical delay of $\approx$ 6 ns onto a fast photodiode. The remainder of the beam was focused into the glass cell containing CS_2. The glass was chosen to have an index of refraction close to that of CS_2 at room temperature. The entire cell was placed in a temperature-controlled holder and the temperature was adjusted so that for the 694.3 nm wavelength of the ruby laser, the index difference $\Delta \simeq 10^{-3}$. This corresponds to $\psi_c \simeq 2.0°$ in the glass cell.

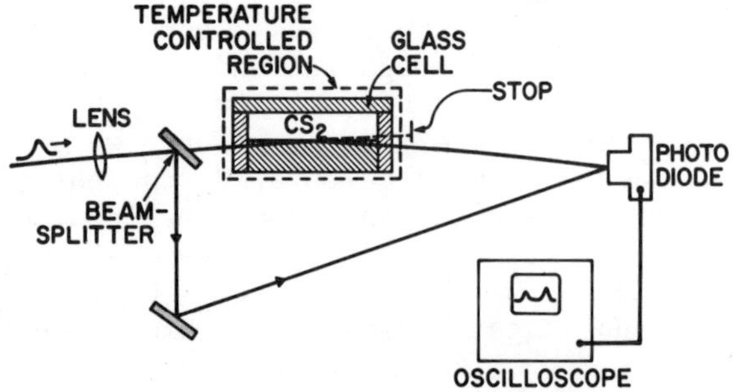

Fig. 2.   Experimental setup.

The reflected beam was monitored by the same photodiode used to monitor the incident pulse, and the diode output was displayed on a fast oscilloscope. The detector-oscilloscope combination had a measured response time of 320 ps.

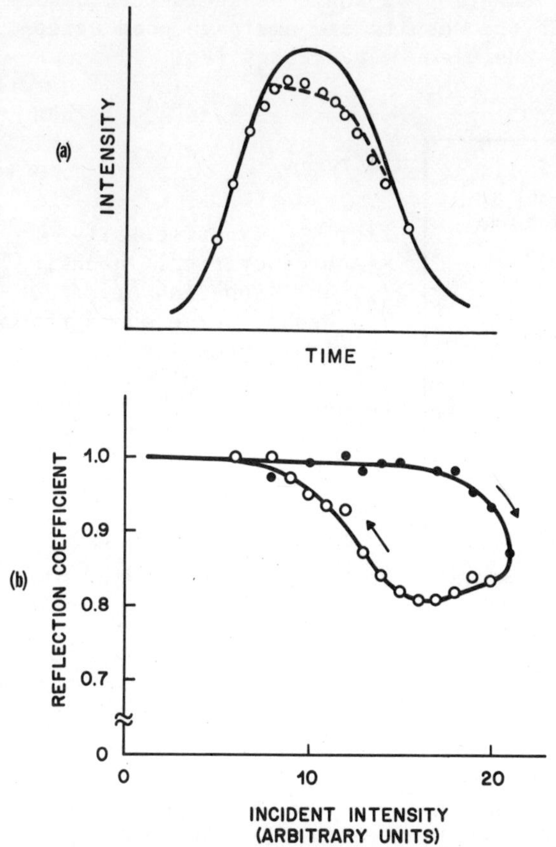

Fig. 3. Experimental measurements[5] of incident and reflected pulse shapes for $\psi/\psi_c = 0.707$. (a) The solid curve is the incident pulse, and the dots are the experimental measurements of the reflected pulse. The dashed line is the reflected pulse calculated from an incoherent plane-wave analysis for $I/I_0 = 1.125$ (see text).
(b) The experimental data in (a) plotted in a way that demonstrates the optical hysteresis observed.

Figure 3(a) shows the measured incident and reflected pulse shapes for an incident intensity slightly above the threshold for hysteresis[5]. The dashed curve is from the incoherent plane-wave calculations, and was fitted to the data by adjusting the ratio of the peak input intensity to the critical intensity, I_0. While the agreement between the shape of the experimental pulse and that obtained from the incoherent plane-wave calculation is probably fortuitous, we believe that the plane-wave theory should predict accurately the critical intensity [Eq. (1)] for the onset of nonlinear behavior. From the published value[7] $n_2 = 3 \times 10^{-8}$ $(MW/cm^2)^{-1}$ for CS_2, we compute a theoretical value of $I_0 = 8.1 \times 10^9 W/cm^2$. Experimentally, we measure $I_0 = 7.5 \times 10^9 W/cm^2$, in very good agreement with the theoretical value.

In Fig. 3(b) we show the same experimental data as in Fig. 3(a), but replotted in the form of reflectivity versus input intensity. This form emphasizes the optical hysteresis observed.

The measured values of critical intensity for the onset of reduced reflectivity, I_0, as a function of angle of incidence, ψ, are shown in Fig. 4. We see that the results are again in good agreement with the predictions of the plane wave theory [Eq. (1)].

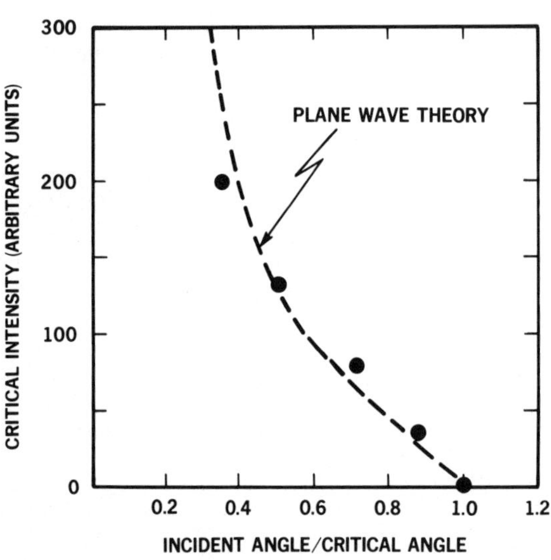

Fig. 4. Experimentally measured critical intensity, I_0, as a function of ψ/ψ_c. The dashed curve shows the critical intensity predicted by the plane-wave theory.

There are several areas in which the experimental results show substantial disagreement with incoherent plane-wave calculations. In Fig. 5, we show measured reflected pulse shapes for various values of the peak incident pulse intensity. The reflected pulse shapes deviate significantly from those obtained from incoherent plane-wave calculations. Figure 6 shows measurements of pulse-peak reflectivity as a function of peak incident intensity for two slightly different critical angles. It can be seen that the reflectivity falls off far more rapidly in one case than the other. Although a large background made absolute measurements imprecise. our data also indicated that the sum of the reflected and transmitted pulse energies was less than the incident pulse energy.

In order to investigate the form and stability of solutions of the problem of a Gaussian beam incident on a nonlinear interface, we have attempted a direct numerical solution of the wave equation[6]. Thus far it has only been possible to consider a two-dimensional Gaussian beam (i.e., we assume that the beam extends uniformly to infinity in the direction perpendicular to the plane

of incidence). Nevertheless, the results have already provided some
indications for the interpretation of our experimental results.

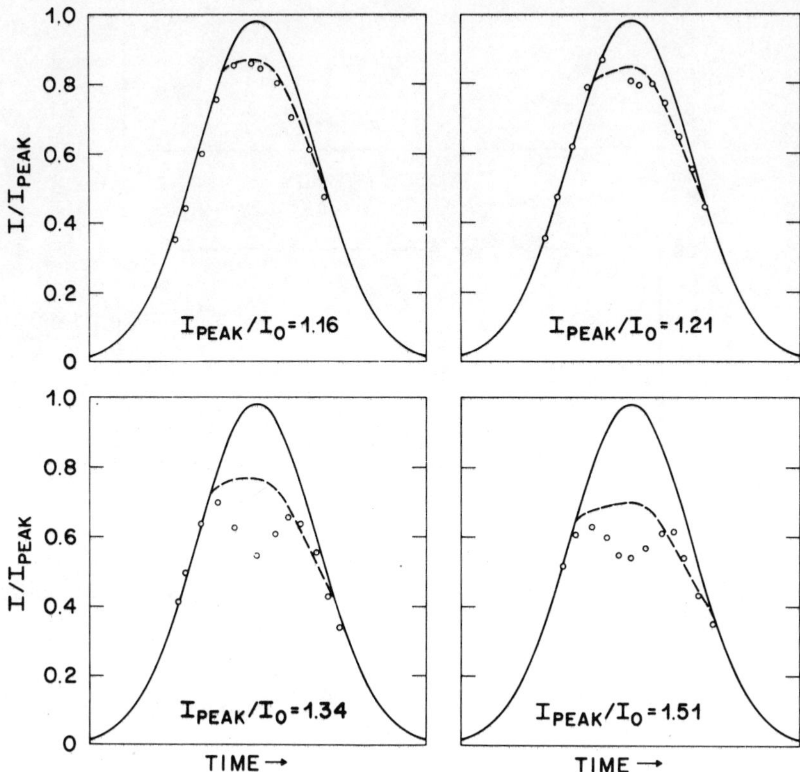

Fig. 5. Experimental measurements of incident and reflected pulse
shapes for $\psi/\psi_c = 0.707$. The solid curves are the incident
pulse and the dots are the experimental measurements of
the reflected pulses. The dashed lines are from the incoherent plane-wave analysis fitted to all four curves
with a single value of the adjustable parameter, I_0.

Figure 7 shows the results of these computations for three values
of the (normalized) incident intensity. In Fig. 7(a) the intensity is not sufficient to cause significant nonlinear behavior. The
output appears to be a Gaussian beam reflected from the interface.
Increasing the intensity by a factor of two - as shown in Fig. 7(b)-
causes dramatically different behavior. A significant portion of
the input energy is coupled into a surface wave[8] which propagates
along the interface. Figure 7(c) shows the case in which the input intensity is again increased - this time by a factor of 10.

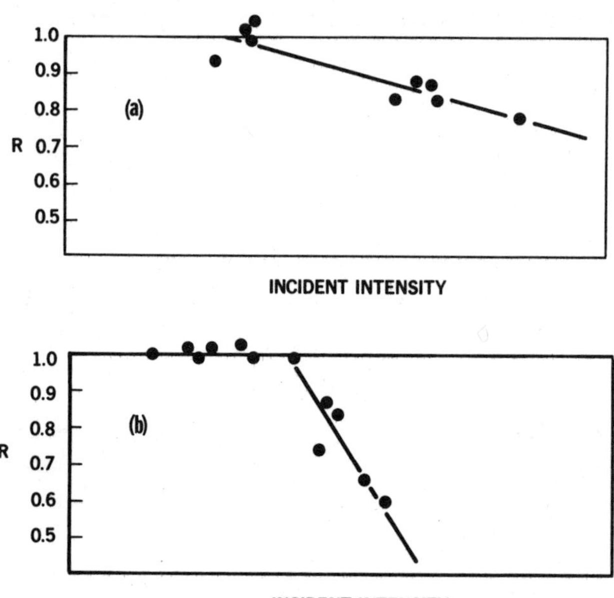

Fig. 6.   Experimental measurements of reflectivity at the peak of the input pulse as a function of peak incident intensity for $\psi/\psi_c = 0.707$.   (a) $\psi_c = 1.98°$; (b) $\psi_c = 2.39°$.

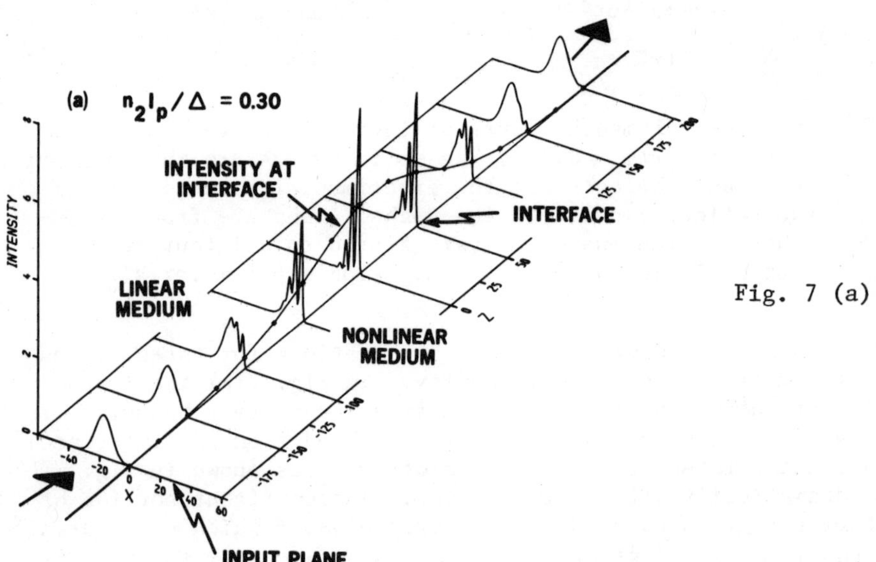

Fig. 7 (a)

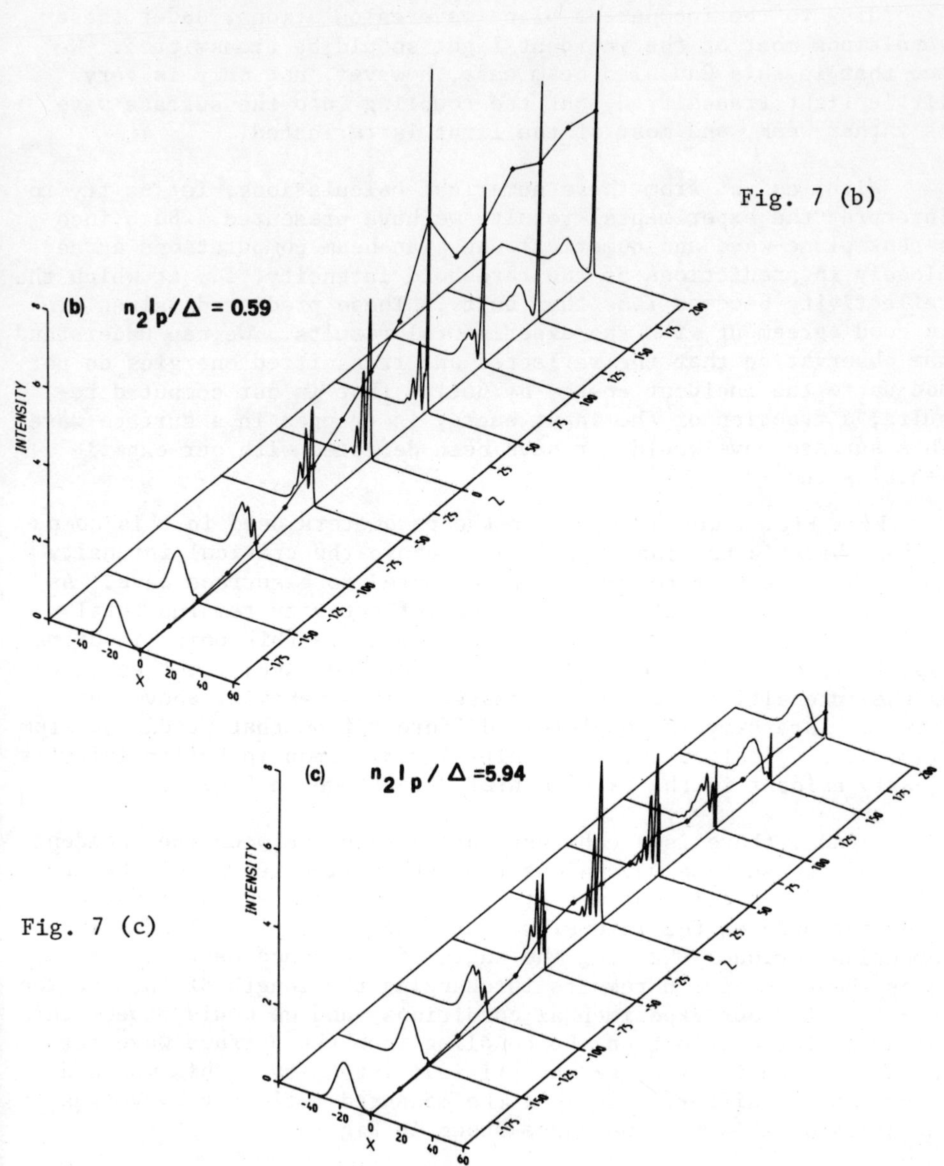

Fig. 7. Numerical computations of intensity profiles for a
Gaussian beam incident on the interface between a linear
and a nonlinear medium. The calculations are made for
the case $\psi/\psi_c = 0.53$, and for several values of the
normalized peak incident intensity $n_2 I_p/\Delta$ where I_p is the
peak intensity at the beam focus, $n_0 = 1.5$, $\Delta = 0.02$,
and $\psi = 5°$.   (a) $n_2 I_p/\Delta = 0.30$; (b) $n_2 I_p/\Delta = 0.59$;
(c) $n_2 I_p/\Delta = 5.94$.

According to the incoherent plane-wave calculations, under these conditions most of the incident light should be transmitted. We see that in this Gaussian beam case, however, not only is very little light transmitted, but the coupling into the surface wave is rather weak, and most of the light is reflected!

With insight from these numerical calculations, let us try to interpret the experimental results we have presented. Both incoherent plane-wave and numerical Gaussian-beam computations agree closely in predictions of the threshold intensity, I_0, at which the reflectivity becomes less than unity. These predicted values are in good agreement with the experimental results. We can understand the observation that the reflected and transmitted energies do not add up to the incident energy by noting that in our computed results, a fraction of the input energy is trapped in a surface wave. This surface wave would not have been detected with our experimental setup.

From Fig. 7 we see that for the parameters used in this computation there is no transmitted wave; above the critical intensity the light is either reflected or converted to a surface wave. As a function of incident intensity the reflectivity remains total until the critical intensity is reached; above this point it drops rapidly as energy is coupled into the surface wave, and then rises as the intensity is further increased. This behavior above the critical intensity is completely different from that predicted from plane-wave calculations. The initial rapid drop in reflectivity is clearly evident in the experimental data shown in Fig. 5.

Because there is a coherent interference between the incident wave and the surface wave along the interaction region of the interface, the fraction of energy remaining in the surface wave depends not only on the intensity, but also on the length of this interaction region. Changing the angle of incidence between the two cases shown in Fig. 6 results in changing the length of this region by $\approx$ 1 mm for our experimental conditions, and we would expect this to have a large effect on the coupling into the surface wave for incident intensities above the critical intensity. This would account for the different form of the measured reflectivity versus intensity data in the two cases shown in Fig. 6.

Additional experimental and theoretical studies will be required to fully elucidate the complex nonlinear behavior of these interfaces.

IV. WAVEGUIDE DEVICES

For the nonlinear interfaces described above, the input and output beams are freely propagating (and diffracting). Work in the field of integrated optics has shown numerous cases in which guided-wave devices can perform the same basic functions as bulk devices but more efficiently and in smaller volumes. Guided-wave devices are also attractive for situations in which one wants to inter-connect a number of devices to perform complicated signal process-ing functions. We have therefore investigated the possibility of incorporating nonlinear interfaces in optical waveguides. We pre-sent here a simple qualitative analysis of the types of behavior that might be achieved from various configurations, and report the results of some preliminary experiments on two such devices.

Some of the basic ideas are illustrated in Fig. 8. At the left of Fig. 8(a) is the input beam, which is focused on the input plane. If in the space between the input and output planes there is an optical waveguide, (illustrated schematically by shading) and the input beam is matched to the mode structure of the waveguide, all of the input light will be guided to the output detector. The output versus input characteristic of such a configuration is a straight line, with unity slope, as illustrated in Fig. 8(b). If, on the other hand, there is no waveguiding in the space between the input and output planes, the input beam will spread, as illustrated schematically by the dotted lines labeled "unguided beam." The output detector will then intercept only a small fraction of the input beam, and the output versus input characteristic will be a straight line with a slope of less than unity. The fraction of un-guided light intercepted by the detector, and thus this slope, will depend on the length of the waveguide. A length of a few confocal parameters of the incoming Gaussian beam is sufficient for strong discrimination between guided and unguided modes. For a waveguide radius of λ, this means a waveguide length of $\approx 10\lambda$ is sufficient for a useful device characteristic.

Using nonlinear interfaces we can make devices which will have little or no waveguiding for inputs below some critical power, and which will show strong waveguiding above that power, or vice versa.

At the bottom of Fig. 9 we show two waveguide configurations that we expect to give qualitatively similar characteristics. In the configuration on the left, the core of the waveguide is made from a material with a positive nonlinearity and the surrounding linear material is chosen to have a refractive index slightly higher than that of the core. In the configuration on the right the nonlinear materials has a negative nonlinearity, and is used to form the cladding for a linear core with an index slightly lower than that of the nonlinear material. Therefore, at low intensities

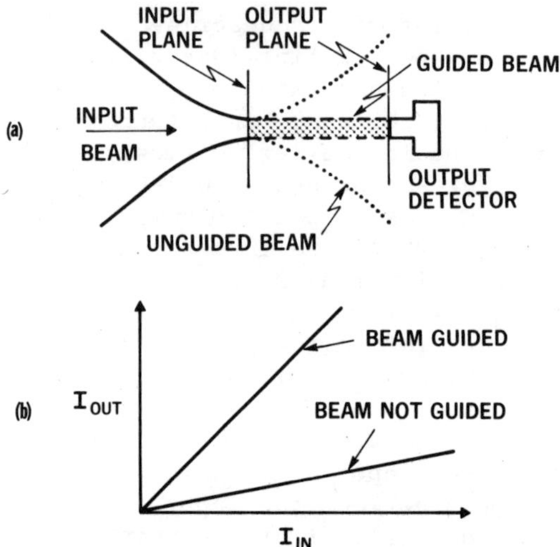

Fig. 8.   Schematic drawing illustrating some of the characteristics
of waveguide devices.

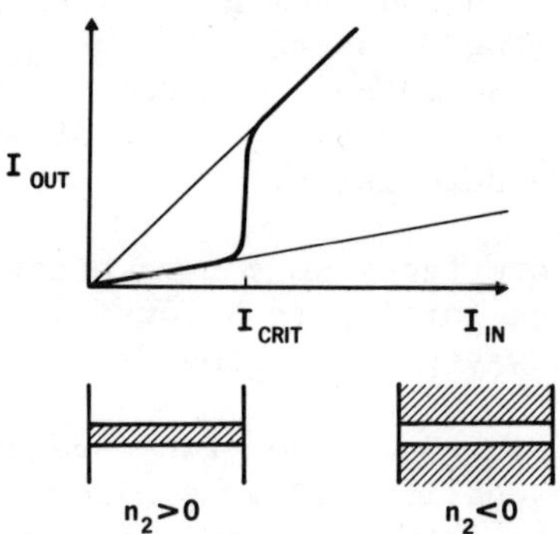

Fig. 9.   (Top) Qualitative output versus input characteristic for
optical triode configuration devices.   (Bottom) Two
waveguide configurations that should exhibit optical
triode behavior.   The shaded region is the nonlinear
optical material.

there will be little or no waveguiding. As the input intensity is
increased, the index difference between the core and cladding will
be reduced, and at some critical input intensity the core index will
exceed that of the cladding at the input plane. The input beam will
then start to be guided, and as it propagates down the guide it will
cause a similar index change all the way along the guide. Therefore, we expect the device to exhibit an output versus input characteristic of the general form illustrated at the top of Fig. 9.
Since a small change in the input level can cause a large change
in the output, we will refer to this as the "optical triode" configuration.

In Fig. 10 we show two configurations that give the opposite
behavior, that is they are guiding at low intensities, and stop
guiding when the input intensity exceeds a threshold value. However, in this case we do not expect a sudden change in output, because as the beam spreads, its intensity drops, and once it has
dropped below the critical intensity it will then be guided to the
output detector. Therefore, we expect an output versus input
characteristic of the type illustrated at the top of Fig. 10. For
obvious reasons, we will refer to this as the "optical limiter"
configuration.

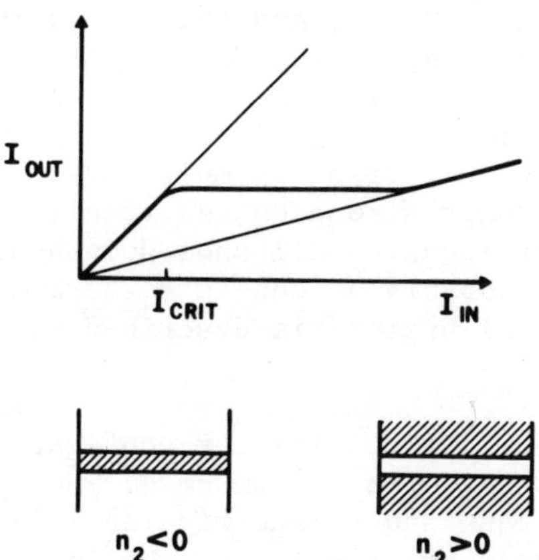

Fig. 10. (Top) Qualitative output versus input characteristic for
optical limiter configuration devices. (Bottom) Two
waveguide configurations that should exhibit optical
limiter behavior. The shaded region is the nonlinear
optical material.

In the above analysis we have not attempted to take account of the detailed behavior of nonlinear interfaces described in the preceding sections. The combination of a nonlinear interface with a waveguide having an intensity-dependent mode structure is a problem that will probably be solved only by brute force numerical techniques.

While in Fig. 8 we showed the input as a focused freely-propagating beam, and the output as going directly to a detector, we expect that for many applications of these devices the input source, and/or the output receptor, will be optical fibers.

For our initial experimental demonstrations of waveguide non-linear interface devices we chose to make use of a thermally-induced refractive index change, rather than a true optical Kerr effect. Although the thermal effect does not accurately model the Kerr effect in a number of significant ways, it has the feature that it is sensitive enough that we were able to do our experiments with moderate-power cw lasers.

For the optical triode configuration we used a 22-μm-diam. fused-silica fiber suspended in a solution of water and glycerol with a refractive index very close to that of fused silica. By varying the temperature of the water-glycerol solution we were able to fine tune its refractive index. Finely-divided carbon particles (India ink) were used to make the solution absorbing (O.D. 10 cm^{-1}). In Fig. 11 we show an experimental curve of output intensity versus input intensity. Note that there is negligible output until the input exceeds a threshold value, and that the output then increases very rapidly with increasing input, in agreement with our qualitative analysis (see Fig. 9). In the present experiments we do not see the reduction of slope at high intensities, but this is probably because our input excitation is multimode, and at high intensities the guiding becomes strong enough to allow additional guided modes. The experimental curves all show some hysteresis, but we believe that this is probably a long-term thermal effect, and is unrelated to the optical hysteresis described in the preceeding section.

To demonstrate the optical limiter configuration we used a hollow fused-silica fiber with a 5-μm bore. The fiber was filled with a mixture of acetone and xylene with an index close to that of the silica. As in the prior experiment, the index was fine tuned by adjusting the temperature. A small amount of dye was used to give the mixture a weak absorption. Figure 12 shows an experimental curve of output versus input, and clearly displays the expected limiter action. This particular curve only shows limiting over a 2:1 range of input intensities, but in other cases we have seen limiting over as much as a 20:1 range.

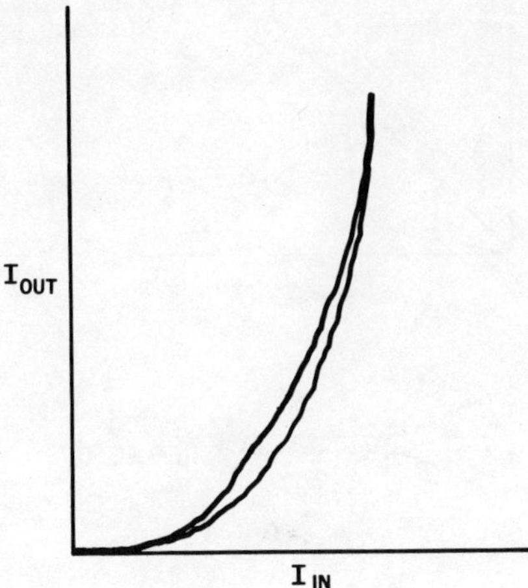

Fig. 11. Experimental results for an optical triode device using
a thermally-induced refractive index change.

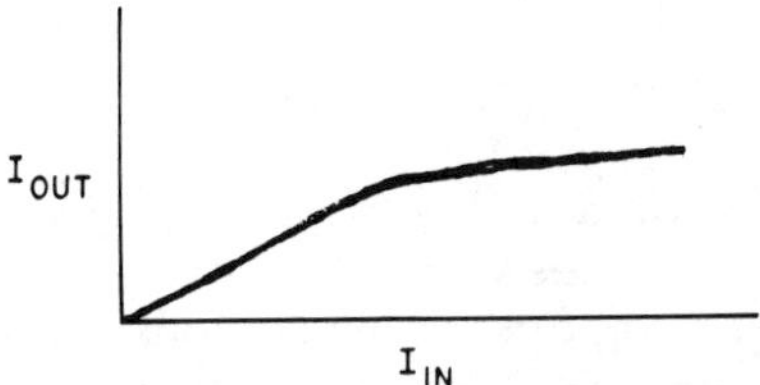

Fig. 12. Experimental results for an optical limiter device
using a thermally-induced refractive index change.

While we cannot be certain that either of the two waveguide
configurations we have just described will exhibit true optical
hysteresis, we have devised another configuration that is clearly
bistable and will exhibit hysteresis. As illustrated in Fig. 13,
we use an optical triode device, but with a retro-reflector at one
end. As the input intensity is increased, the reflector will have
negligible effect until the device switches to the guiding state.
The reflected beam will then be guided back down the guide, thus
increasing the average intensity in the guide. Therefore, it
should be possible to reduce the input intensity by almost a
factor of two before the device will switch back to the nonguiding
state, and a large region of bistability will exist.

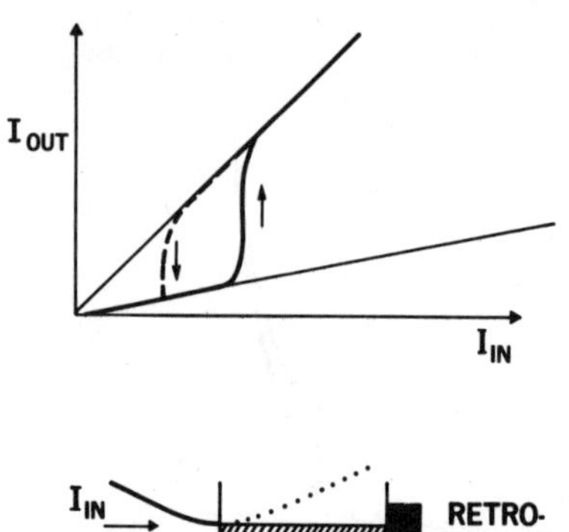

Fig. 13.   Schematic configuragion for a waveguide device that should exhibit optical hysteresis, and the expected characteristic curves. The shaded region is the nonlinear optical material.

V.   CONCLUSIONS

In this paper we have outlined the current state of experimental and theoretical progress towards the understanding of nonlinear interfaces. Many important questions remain to be investigated. Under what conditions does one expect to observe optical bistability? What are the limits on response time imposed by the transient field redistribution in the region of the interface? Much additional theoretical and experimental work will be necessary to fully answer these questions.

We can, however, make some general statements. Besides providing a considerable challenge for the theoretical analysis of their complex behavior, nonlinear interface devices may prove to be highly useful and versatile optical signal processing elements. Although the details remain to be studied, we can see in a general way that if the response time of the nonlinearity is sufficiently rapid, field redistribution effects will take place in a time less than the transit time through the device. For waveguide devices of the type described in Section IV, the transit time can be $<10^{-13}$ seconds. Nonlinear materials exist with extremely fast intrinsic response times. Recent measurements[9] indicate that the polydiacetylene P.T.S. (which has a response time estimated to be 10^{-14} seconds) has a nonresonant nonlinear coefficient of 150 times that of CS_2. Thus the potential exists for extremely fast and relatively sensitive nonlinear interface optical signal processing elements.

To conclude, we show in Table I how the projected response times and switching energies for nonlinear interface devices compare with other types of optical switching elements.

TABLE I

Device	Switching Power	Switching Time	Switching Energy
	Experimental Values		
Nonlinear Interface			
CS_2 - Glass[5]	2×10^5 W	2×10^{-12} s*	4×10^{-7} J*
Bistable Fabry-Perot Resonators			
GaAs[10]	2×10^{-1} W	4×10^{-8} s	8×10^{-9} J
InSb[11]	5×10^{-1} W	?	?
Hybrid $LiNbO_3$ Waveguide[12]	10^{-5} W	5×10^{-8} s	5×10^{-13} J
	Projected Values		
Waveguide Nonlinear Interface			
PTS - ?	70 W	10^{-13} s	7×10^{-12} J

*The switching time was not measured. The figure given is the response time of the CS_2 nonlinearity.

REFERENCES

1. A. E. Kaplan, JETP Lett. <u>24</u>, 114 (1976); Sov. Phys. JETP <u>45</u>, 896 (1977).
2. B. B. Boiko, I. Z. Dzhilavdari, and N. S. Petrov, J. Appl. Spect. <u>23</u>, 1511 (1975).
3. N. N. Rozanov, Sov. Tech. Phys. Lett. <u>4</u>, 30 (1978); Opt. Spectrosc. (USSR) <u>47</u>, 335 (1979).
4. A. A. Kolokolov and A. I. Sukov, Radio Phys. and Quantum Electron. <u>21</u>, 1013 (1978).

5.   P. W. Smith, J.-P. Hermann, W. J. Tomlinson and P. J. Maloney,
      Appl. Phys. Lett. $\underline{35}$, 846 (1979).
6.   D. Marcuse, P. W. Smith and W. J. Tomlinson, XI'th Int. Quantum
      Electron. Conf., Boston, paper AA-12 (1980); D. Marcuse,
      Appl. Opt. (to be published).
7.   M. J. Moran, C.-Y. She and R. L. Carmen, IEEE J. Quantum
      Electron. QE-11, 259 (1975).
8.   W. J. Tomlinson, Opt. Lett. $\underline{5}$, 323 (1980).
9.   J.-P. Hermann, and P. W. Smith, XI'th Int. Quantum Electron.
      Conf., Boston, paper T-6 (1980).
10.   H. M. Gibbs, S. L. McCall, T. N. C. Venkatesan, A. C. Gossard,
      A. Passner, and W. Weigmann, Appl. Phys. Lett. $\underline{35}$, 451 (1979).
11.   D. A. B. Miller and S. D. Smith, Opt. Comm. $\underline{31}$, 101 (1979).
12.   P. W. Smith, I. P. Kaminow, P. J. Maloney, and L. W. Stulz,
      Appl. Phys. Lett. $\underline{34}$, 62 (1979).

APPLICATIONS OF THE GENERALIZED P-REPRESENTATION TO OPTICAL

BISTABILITY

P. D. Drummond

Department of Physics and Astronomy
University of Rochester
Rochester, New York 14627

Abstract: The generalized (non-diagonal) coherent state
representation is defined for photons and atoms as a basis for
quantum theories of optical bistability.   Theorems on existence
properties and time-ordered operator products are given.   This
is used to derive Fokker-Planck equations that describe fluctuations
in various different optical bistability experiments. A comparison
of the physical applicability of these Fokker-Planck equations is
used to select the theory that best describes different types of
experiments.

I.   INTRODUCTION

In this paper, a theory of operator representations is developed
that is suitable for investigating the quantum theory of coherently
driven, nonlinear devices. The underlying problem is that of a
quantum mechanical open system, far from thermal equilibrium. Even
the steady-state distribution requires a careful analysis, and it
is not necessarily a canonical distribution. The theoretical methods
have similarities to those used in laser theory, but a significant
difference arises in the quantum statistical fluctuations. In many
cases these fluctuations are nonclassical in nature, so they cannot
be described using a distribution function with classical properties.
In other words, the Glauber-Sudarshan P-representation[1,2] which is
used in laser theory[6], develops singularities. This corresponds to
the prediction of photon antibunching[2,8,10] in the electromagnetic
field of the device output, even for a coherent input. While fluctuations of this type are normally negligible at high powers, they
could have significance in proposed bistable nonlinear optical devices operating at very low powers[3].

At the same time, it is worth emphasizing that optical bistability can occur (and has been investigated in experiment) in several
different systems[3]. These are not all described by the same types
of equation, and can have different transition and fluctuation properties. It is necessary, therefore, to define the microscopic origin
of the bistability and to know something of the relative relaxation
time-scales involved. Even including these factors, it is probably
true that the optical bistability theory developed here is oversimpli-
fied in its treatment of the nonlinear medium, for most experiments
to date. A treatment of realistic features like Gaussian modes and
inhomogeneous broadening is given, however. These prove to have a
very large quantitative effect on the critical point for absorptive
bistability.

II.   GENERALIZED P-REPRESENTATIONS

In this section the generalized P-representation is defined,
and some relevant theorems and examples are given. The representation is defined in terms of a characteristic function for the
quantum system, as follows[5]. Let $\hat{\chi}(\lambda)$ be an operator valued function, and $\chi(\lambda,\alpha)$ be a complex function, with an integration measure
$d\mu(\alpha)$. Then the generalized P-function $P(\alpha)$ is defined so that:

$$\bar{\chi}(\lambda) = \mathrm{Tr}[\hat{\rho}\hat{\chi}(\lambda)] = \int_D P(\alpha)\chi(\lambda,\alpha)d\mu(\alpha) \tag{2.1}$$

Usually, χ is obtained from $\hat{\chi}$ on simply replacing the n'th
operator by a variable complex number α_n. In most cases one would
specify;

$$\hat{\chi}(\lambda) = \exp(\lambda_1\hat{a}_1)\exp(\lambda_2\hat{a}_2)\ldots\ldots\exp(\lambda_n\hat{a}_n)$$

$$\chi(\lambda,\alpha) = \exp(\lambda\cdot\alpha) \ . \tag{2.2}$$

The following points are noteworthy:

(i) The operator ordering is given by $\hat{\chi}$, and in this paper
normal ordering (or generalized normal ordering[6]) is specified but
other orderings can also be used.

(ii) The c-number corresponding to a hermitian operator is
generally complex. Also, hermitian conjugate operators do not cor-
respond to complex conjugate c-numbers, i.e., $(\hat{a},\hat{a}^+) \neq (\alpha,\alpha*)$. However, for brevity the notation of "hermitian conjugate c-numbers"
is introduced, so $(\hat{a},\hat{a}^+) \rightarrow (\alpha,\alpha^+)$. This is for notational simplicity,
and has the same meaning as α where (α_1,α_2) are two variables in a
complex plane.

(iii) The integration measure can be chosen to have the form
$d^2\alpha d^2\alpha^+\delta^2(\alpha*-\alpha^+)$; this would then reproduce the usual Glauber-

Sudarshan P-representation.

(iv)   Other choices of integration measure are line integrals $d\alpha d\alpha^+$ (which define a complex P-representation) and area integrals $d^2\alpha d^2\alpha^+$ (which define a positive P-representation).

(v)   A normally ordered representation of harmonic oscillator operators can be written directly as:

$$\hat{\rho} = \int_D \left[\frac{|\alpha_1><\alpha_2^*|}{<\alpha_2^*|\alpha_1>} \right] P(\alpha_1,\alpha_2) d\mu(\underset{\sim}{\alpha}) \tag{2.3}$$

where $|\alpha>$ is a coherent state of the oscillator.   This corresponds to $\chi, \hat{\chi}$ in the form of Eq. (2.2), with normal operator orderings.

In the case of a harmonic oscillator, or an electromagnetic field mode, existence theorems for the measure $d\alpha d\alpha^+$ are as follows:

<u>Theorem 1</u>: At least one complex P-representation exists with line-integral measure for a density operator expanded in a finite set of number states.

If: $\hat{\rho} = \sum_{n,m} \rho_{nm} (a^+)^n |o><o| \hat{a}^n$

then:
$$P(\alpha_1,\alpha_2) = -e^{\alpha_1\alpha_2} \sum_{n,m} \rho_{nm} (n!m!) / (4\pi^2 \alpha_1^{n+1} \alpha_2^{m+1})$$

and the line integrals enclose the origin in α_1, α_2.

It can be recognized that this covers examples for which the complex P-function is well-behaved, even when the diagonal (Glauber-Sudarshan) P-function would have singularities. A more general theorem is:

<u>Theorem 2</u>: At least one complex P-representation exists for a density operator expanded in a bounded domain of coherent states.

If: $\hat{\rho} = \iint_D \rho(\alpha,\beta) |\alpha><\beta^*| d^2\alpha d^2\beta$

then:
$$P(\alpha_1,\alpha_2) = \left(\frac{-1}{4\pi^2}\right) \iint_D \rho(\alpha,\beta) <\beta^*|\alpha> \left(\frac{d^2\alpha}{\alpha_1-\alpha}\right) \left(\frac{d^2\beta}{\alpha_2-\beta}\right) \ .$$

For the measure $d^2\alpha d^2\alpha^+$, the generalized P-representation is related to Glauber's R-representation. The relationship is that the Glauber R-function, when multiplied by a variable complex normalization factor, is one example of a generalized P-function. However, this does not result in a positive P-function, which is obtained as follows:

$\underline{\text{Theorem 3}}$: At lease one positive P- function exists, for any $\hat{\rho}$ (with measure $d^2\alpha_1 d^2\alpha_2$) and is explicitly:

$$P(\underset{\sim}{\alpha}) = \left(\frac{1}{4\pi^2}\right) \exp\left(-\left|\frac{\alpha_1-\alpha_2^*}{2}\right|^2\right)\left\langle\frac{\alpha_1+\alpha_2^*}{2}\middle|\hat{\rho}\middle|\frac{\alpha_1+\alpha_2^*}{2}\right\rangle \tag{2.4}$$

A further theorem[5] which holds for any generalized P-representation (with measure $\Pi_n d^2\alpha_n$) provided $\chi(\lambda,\underset{\sim}{\alpha})$ is analytic in $\underset{\sim}{\alpha}$ is related to the time-development equation.

This theorem allows one to create equivalent positive-semi-definite Fokker-Planck equations in the complex plane, starting from a standard form which is analytic in α,α^*.

$\underline{\text{Theorem 4}}$: If any (second-order) Fokker-Planck equation exists, then an equivalent positive-semi-definite Fokker-Planck equation exists, and hence a stochastic differential equation. Specifically, for an (analytic) equation of form:

$$\frac{\partial P(\underset{\sim}{\alpha})}{\partial t} = \frac{\partial}{\partial\alpha_\mu}\left[A_\mu(\underset{\sim}{\alpha}) + \frac{1}{2}\frac{\partial}{\partial\alpha_\nu}\,d_{\mu\sigma}(\underset{\sim}{\alpha})d_{\nu\sigma}(\underset{\sim}{\alpha})\right]P(\underset{\sim}{\alpha}) \tag{2.5}$$

one can demonstrate the equivalence of the following stochastic equation:

$$\underset{\sim}{\dot{\alpha}} = -A(\underset{\sim}{\alpha}) + \underset{\approx}{d}(\underset{\sim}{\alpha})\cdot\xi(t)$$

$$\langle\xi_\nu(t)\xi_\mu(t')\rangle = \delta_{\mu\nu}\delta(t-t') \ . \tag{2.6}$$

This extends the usual stochastic theorem to include cases of non-positive definite diffusion by the technique of adding new dimensions.

Given equations for the time-development, it is necessary to calculate the correlation functions, which define the physical observables (i.e., the spectrum, intensity, and so on). In order to calculate correlation functions, one can prove the following theorem[5] (for $t > t'$):

$\underline{\text{Theorem 5}}$: Time-ordered correlation functions are given by: (for $t > t'$)

$$\langle\hat{A}(t)\hat{B}(t')\rangle = \int_D A(\alpha)P(\alpha',t')\mathcal{D}_B(\alpha')P(\alpha,t|\alpha',t')d\mu(\alpha)d\mu(\alpha')$$

where:

$$\langle\hat{A}\rangle = \int P(\underset{\sim}{\alpha},t)A(\underset{\sim}{\alpha})d\mu(\underset{\sim}{\alpha})$$

$$\langle\hat{\chi}\hat{B}\rangle = \int P(\underset{\sim}{\alpha},t)\mathcal{D}_B(\underset{\sim}{\alpha})\chi(\underset{\sim}{\lambda},\underset{\sim}{\alpha})d\mu(\underset{\sim}{\alpha}). \tag{2.7}$$

In the case of normally ordered correlations of harmonic-oscillator variables, one has:

$$<a^{+n}(t)a^{m}(t')> = \int_{D} (\alpha^{+})^{n}(\alpha')^{m} P(\alpha',t') P(\underset{\sim}{\alpha},t|\underset{\sim}{\alpha}'t') d\mu(\underset{\sim}{\alpha}) d\mu(\underset{\sim}{\alpha}') \; . \tag{2.8}$$

Similar results hold for $t'>t$. Although the derivation requires
new techniques, (as more than one equivalent P-function exists) the
theorem is similar to that of Louisell and other workers[6], obtained
for a "classical" phase-space.

An example of a linearized spectral calculation is for the
following Fokker-Planck equation, and its stochastic equivalent:

$$\frac{\partial}{\partial t} P(\underset{\sim}{\alpha}) = \frac{\partial}{\partial \alpha_{\mu}} [A_{\mu\nu}(\alpha_{\nu}-\bar{\alpha}_{\nu}) + \frac{1}{2} \frac{\partial}{\partial \alpha_{\nu}} d_{\mu\sigma} d_{\nu\sigma}]P(\underset{\sim}{\alpha})$$

$$\dot{\underset{\sim}{\alpha}} = -\underset{\approx}{A}\cdot(\underset{\sim}{\alpha}-\underset{\sim}{\bar{\alpha}}) + \underset{\approx}{d}\cdot\underset{\sim}{\xi}(t) \; . \tag{2.9}$$

Using theorems (4) and (5) one can derive a spectrum even when the
diffusion matrix dd^{T} is non-positive-definite. The result is
formally similar $\underset{\approx}{to}$ the well-known classical one

$$S_{\mu\nu}(\omega) = \lim_{t'\to\infty} \int <\alpha_{\mu}(t+t')\alpha_{\nu}(t')>e^{-i\omega t}dt/2\pi$$

$$= \bar{\alpha}_{\mu}\bar{\alpha}_{\nu}\delta(\omega) + \frac{1}{2\pi} \left[[\underset{\approx}{A}+i\omega]^{-1}\underset{\approx}{d}\;\underset{\approx}{d}^{T} [\underset{\approx}{A}^{T}-i\omega]^{-1} \right]_{\mu\nu} \; . \tag{2.10}$$

The new feature resulting is that while the observable spectrum
is positive, the Fourier transform of the linearized intensity correlation function can become negative as well as positive. This
cannot occur with classical or semiclassical theories, i.e., with
a Glauber-Sudarshan P-function that is positive semi-definite.
(However, intensity correlations of this type were observed by
Kimble, Dagenais and Mandel, as originally predicted by Carmichael
and Walls,[2] in the fluorescent radiation of one atom.)

If the linearized theory is applied to the simple case of an
electromagnetic mode with a coherent input field in an interferometer, the general Fokker-Planck equation has the tensor coefficients:

$$\underset{\approx}{A} = \begin{bmatrix} a & b \\ b^{*} & a^{*} \end{bmatrix} , \quad \underset{\approx}{D} = \underset{\approx}{d}\;\underset{\approx}{d}^{T} = \begin{bmatrix} -d & \Gamma \\ \Gamma & -d^{*} \end{bmatrix} \tag{2.11}$$

one can then easily derive,

486   P. D. DRUMMOND

$$I(\omega) = \int_{-\infty}^{\infty} e^{-i\omega t} \langle \hat{a}^+(t)\hat{a}(o) \rangle \frac{dt}{2\pi}$$

$$= |\bar{\alpha}|^2 \delta(\omega) + \frac{[\Gamma\omega^2 + 2\omega \mathrm{Im}(\Gamma a + db^*) + \Gamma(|b|^2 + |a|^2) + 2\mathrm{Re}(d^*ab)]}{2\pi\lambda(\omega)}$$

$$(2.12)$$

$$g^2(\omega) = \int_{\infty}^{\infty} e^{-i\omega t} \frac{\langle \hat{a}^+(o)\hat{a}^+(t)\hat{a}(t)\hat{a}(o) \rangle}{2\pi |\langle \hat{a}^+(o)\hat{a}(o) \rangle|^2} \, dt$$

$$= \delta(\omega) + [\Gamma\omega^2 + \Gamma(|b|^2 + |a|^2) + 2\mathrm{Re}(d^*ab)$$

$$-\mathrm{Re}(\frac{\bar{\alpha}^*}{\bar{\alpha}}[\omega^2 d + 2a^*b\Gamma + b^2 d^* + a^{*2}d]))]/[\pi|\bar{\alpha}|^2\lambda(\omega)]$$

where: (with eigenvalues $\lambda^{\pm} = \mathrm{Re}(a) \pm \sqrt{|b|^2 - (\mathrm{Im}(a))^2}$)

$$\lambda(\omega) = \omega^4 + \omega^2(a^2 + a^{*2} + 2|b|^2) + (|a|^2 - |b|^2)^2 \; .$$

The above results give, in principle, the linearized spectrum
and the intensity correlationspectrum for any single-mode linearized Fokker-Planck equation, and the technique of calculating coefficients is given later. However, it is worth remembering that
in practical experiments one has to reduce the effects of laser
fluctuations and mode-mode coupling, in order to utilize a simple
one-mode model with coherent input. These additional effects are
treated elsewhere in this volume.

III. ANHARMONIC OSCILLATOR MEDIUM

The simplest case of optical bistability[9],[10] involves a single-
mode interferometer with a nonlinear polarizability (as discussed
in the lecture of Walls et al). However, it is useful to have a
microscopic understanding of the nonlinear polarizability, to verify the macroscopic model, and to investigate the significance of
the medium response time. A well-known microscopic theory is that
of quantum anharmonic oscillator (which has been used to model SF_6,[9]
for example). With N quantum anharmonic oscillators, the Hamiltonian
for a single interferometer mode is:

$$H = \hbar\omega_o \hat{a}^+\hat{a} + \sum_{j=1}^{N} \hbar[\omega_j \hat{b}_j^+\hat{b}_j + ig_j(\hat{b}_j^+\hat{a} - \hat{b}_j\hat{a}^+) + \varepsilon\hat{b}_j^{+2}\hat{b}_j^2]$$

$$(3.1)$$

$$+ i\hbar[E(t)\hat{a}^+e^{-i\omega t} - E^*(t)\hat{a}e^{i\omega t}] + \{\mathrm{RESERVOIRS}\}.$$

Here the reservoir operators describe radiation damping of each
oscillator (with half-width $\gamma_\perp$) and damping of the interferometer
mode (half-width k'). E represents a coherent input field of power

P_I. For an interferometer round-trip time Δt, principal mirror reflectivities R_1, R_2 and round-trip losses of $(1-T)$, (where $(1-R_1)$, $(1-R_2)$, $(1-T) \ll 1$) one obtains for the input and transmitted powers, and the interferometer half-width,

$$P_I = |E|^2 \hbar\omega / [(1-R_1)\Delta T]$$
$$P_T = (1-R_2)\hbar\omega <\hat{a}^+\hat{a}>/\Delta T \qquad (3.2)$$
$$k' = (1-R_1 R_2 T)/(2\Delta T),$$

(these relations also hold in section 4). The frequencies ω_j represent an inhomogeneous broadening, ε is an anharmonicity, and the coupling g_j is given by:

$$g_j = \left| [\underset{\sim}{p}\cdot\underset{\sim}{e}][\frac{\omega}{2\hbar e_o}]u(\underset{\sim}{r}_j) \right| . \qquad (3.3)$$

Also, p is the oscillator transition dipole-moment, e the mode polarizatio͠n vector, e_o the permittivity, and $u(\underset{\sim}{r})$ the (no͠rmalized) mode function. The corresponding Fokker-Planck equation is (for $\hat{a} \to \alpha$, $\hat{b}_j \to \beta_j$),

$$\frac{\partial}{\partial T} P(\underset{\sim}{\alpha},\underset{\sim}{\beta}) = \left\{ \frac{\partial}{\partial\alpha} (k\alpha + \sum g_j\beta_j - E) + \frac{\partial}{\partial\alpha^+}(k^*\alpha^+ + \sum g_j\beta_j^+ - E^*) \right.$$

$$+ \sum_j \left[\frac{\partial}{\partial\beta_j}(\gamma_j\beta_j + 2i\varepsilon\beta_j^+\beta_j^2 - g_j\alpha) + \frac{\partial}{\partial\beta_j^+}(\gamma_j^*\beta_j^+ - 2i\varepsilon\beta_j\beta_j^{+2} - g_j\alpha^+) \right.$$

$$\left. - i\varepsilon \frac{\partial^2}{\partial\beta_j^2} \beta_j^2 + i\varepsilon \frac{\partial^2}{\partial\beta_j^{+2}} \beta_j^{+2} \right] + 2k'n_{th} \frac{\partial^2}{\partial\alpha\partial\alpha^+} \left. \right\} P(\underset{\sim}{\alpha},\underset{\sim}{\beta}) . \qquad (3.4)$$

Here n_{th} is the thermal photon number of the mode.

This has a corresponding stochastic equation with Gaussian random functions $\Gamma_\mu(t)$, $\xi_\mu^j(t)$:

$$\dot{\alpha} = E - k\alpha - \sum g_j\beta_j + \Gamma_1(t)$$

$$\dot{\alpha}^+ = E^* - k^*\alpha^+ - \sum g_j\beta_j^+ + \Gamma_2(t) \qquad (3.5)$$

$$\dot{\beta}_j = g_j\alpha - \gamma_j\beta_j - 2i\varepsilon\beta_j^+\beta_j^2 + \sqrt{-2i\varepsilon}\cdot\beta_j\xi_1^j(t)$$

$$\dot{\beta}_j^+ = g_j\alpha^+ - \gamma_j^*\beta_j^+ + 2i\varepsilon\beta_j\beta_j^{+2} + \sqrt{2i\varepsilon}\cdot\beta_j^+\xi_2^j(t)$$

where:

$$<\Gamma_1(t)\Gamma_1(t')> = <\Gamma_2(t)\Gamma_2(t')> = 0$$

$$<\Gamma_1(t)\Gamma_2(t')> = \delta(t-t')\ 2k'n_{th}$$

$$\langle \xi_\mu^i(t)\xi_\nu^j(t')\rangle = \delta_{ij}\delta_{\mu\nu}\delta(t-t') \qquad [i,j=1,\ldots N; \mu,\nu=1,2]$$

$$k = k' + i(\omega_o-\omega) = k' + i\Delta\omega_o$$

$$\gamma_j = \gamma_\perp + i(\omega_j-\omega) = \gamma_\perp + i\Delta\omega_j \quad .$$

In general, Eq. (3.4) is an intractable problem as it is a $2(N+1)$-dimensional complex nonlinear Fokker-Planck equation, although numerical simulation of Eq. (3.5) is possible. Adiabatic elimination, in either the limits of $|\gamma_j|>>|k|$ or $|\delta_j|<<|k|$ simplifies the problem (these are the "mode-dominated" or "polarization-dominated" regions respectively). These limits will be analyzed individually.

A. Adiabatic Elimination of Polarization Variables

The mode-dominated region has the interesting feature that the polarization equations alone have already bistable character (and, therefore, no unique value). For large detunings from resonance, $\Delta\omega_j>>\gamma_\perp$, they can be solved iteratively to allow adiabatic elimination (this requires detunings larger than the inhomogeneous line width). To lowest nontrivial order one has (neglecting higner orders in $[\gamma_\perp/\Delta\omega_j]$),

$$\beta_j \approx \left[\frac{g_j\alpha}{\gamma_j}\right] + 2i\varepsilon\left[\frac{g_j^3\alpha^2\alpha^+}{|\gamma_j|^4}\right] - \sqrt{-2i\varepsilon}\left[\frac{g_j\alpha}{|\gamma_j|^2}\right]\xi_1^j(t) \quad . \tag{3.6}$$

Hence the equation for the mode variable is,

$$\frac{\partial}{\partial t}\alpha = \varepsilon(t) - \bar{k}\alpha - 2i\bar{\varepsilon}\alpha^2\alpha^+ + \bar{\Gamma}(t)$$

where,

$$\bar{k} = k + \sum_j \frac{g_j^2}{\gamma_j}$$

$$\bar{\varepsilon} = \varepsilon \sum_j g_j^4/|\gamma_j|^4 \tag{3.7}$$

$$\bar{\Gamma}(t) = \Gamma(t) + \sqrt{-2i\varepsilon}\left[\frac{g_j^2\alpha}{|\gamma_j|^2}\right]\xi_1^j(t) \quad .$$

The fluctuation term has the correlations,

$$\langle \bar{\Gamma}(t)\bar{\Gamma}^{+}(t')\rangle = 2k'n_{th}\delta(t-t')$$

$$\langle \bar{\Gamma}(t)\,\bar{\Gamma}(t')\rangle = -2i\bar{\epsilon}\alpha^{2}\delta(t-t') \;.$$

$$(3.8)$$

This, therefore, corresponds exactly to the nonlinear polarizability theory mentioned before[10]. The bistable region is well known and the linearized spectrum can be calculated from Eq. (2.12). An interesting limit occurs for $n_{th} \to 0$ (no thermal fluctuations), when a potential-type solution of the quantum distribution function is obtained[10]. This seems to be the only exact solution of a Fokker-Planck equation for quantum fluctuations known in the mode-dominated region. For completeness, the linearized equation is given here (for comparison with Eq. (2.9)), for $\alpha = \bar{\alpha} + \delta\alpha$, $\alpha^{+} = \bar{\alpha}* + \delta\alpha^{+}$,

$$\frac{\partial}{\partial t}\begin{bmatrix}\delta\alpha \\ \delta\alpha^{+}\end{bmatrix} = -\begin{bmatrix}\bar{k}+4i\bar{\epsilon}|\bar{\alpha}|^{2}, & 2i\bar{\epsilon}\bar{\alpha}^{2} \\ -2i\bar{\epsilon}\bar{\alpha}*^{2}, & \bar{k}*-4i\bar{\epsilon}|\bar{\alpha}|^{2}\end{bmatrix}\begin{bmatrix}\delta\alpha \\ \delta\alpha^{+}\end{bmatrix} + \begin{bmatrix}-2i\bar{\epsilon}\bar{\alpha}^{2}, & 2k'n_{th} \\ 2k'n_{th}, & 2i\bar{\epsilon}\bar{\alpha}*^{2}\end{bmatrix}^{\frac{1}{2}}\begin{bmatrix}\xi_{1}(t) \\ \xi_{2}(t)\end{bmatrix} \;.$$

$$(3.9)$$

B. Adiabatic Elimination of the Mode

While the previous results hold for adiabatic elimination of the polarization variables, in some cases the inequality $|k|<<|\gamma|$ does not hold. In general the stability properties are difficult to obtain in this case. Simple results occur when the oscillators are equivalent, i.e., $\gamma_{j}=\gamma$, $g_{j}=g$ (this excludes standing-waves). In this case a "macroscopic" oscillator variable is defined with a new state-equation identical to that of nonlinear polarizability theory[10],

$$\beta = \sum_{j=1}^{N} \beta_{j} = N\beta_{j} \quad \text{(for any j)}$$

$$[Ng\epsilon/k] = \beta\left[\gamma + Ng^{2}/k + \left[\frac{2i\epsilon}{N^{2}}\right]\beta^{+}\beta\right]$$

$$(3.10)$$

$$\alpha = \frac{1}{k}[E - g\beta] \;.$$

This state-equation guarantees the existence of a steady-state for the deterministic equations. As it requires that $\beta_{j}=\beta/N$, it is not necessarily the only solution. Thus an effective anharmonicity and decay rate exists for the polarization variable (not the interferometer mode variable) so that,

$$\bar{\gamma} = \gamma + Ng^2/k$$
$$\bar{\epsilon} = \epsilon/N^2 \qquad\qquad (3.11)$$
$$\bar{E} = NgE/k \quad .$$

This state-equation does not require ϵ as a small parameter, and therefore (apart from the rotating wave approximation) is not a perturbative or asymptotic approximation. The linearized stochastic equation is,

$$\dot{\delta\alpha} = -k\delta\alpha - g\delta\beta + \Gamma_{th}(t)$$
$$\dot{\delta\beta} = Ng\delta\alpha - \bar{\gamma}\delta\beta - 4i\bar{\epsilon}|\bar{\beta}|^2\delta\beta - 2i\bar{\epsilon}\bar{\beta}^2\delta\beta* + \sqrt{-2i\bar{\epsilon}}\,\bar{\beta}\xi(t) \quad . \qquad (3.12)$$

In the limit $|k|>>|\gamma|$, the field variable can be adiabatically eliminated. This leaves a linearized equation in which the polarization variable $\beta=\bar{\beta}+\delta\beta$ replaces the mode variable α in Eq. (3.9). The new coefficients $\bar{\gamma}$, $\bar{\epsilon}$ are as in Eq. (3.11). In this case, therefore, a simple linearized theory is available just as in the mode dominated region. However, since β is a collective variable it does not completely describe large fluctuations from the steady state; for example, with some oscillators in the upper, some in the lower branch. Thus, in the polarization-dominated region, an exact steady-state distribution is not presently available. It is even possible that the distribution in $(\beta_1...\beta_N)$ is multi-modal, with more than two stable deterministic branches. Clearly, optical bistability for an anharmonic medium in the polarization dominated region has a different structure at the microscopic level than that for two-level atoms, and seems to merit a closer study.

IV.   TWO-LEVEL ATOMS

An alternative way to obtain switching behavior from a purely optical device, is via a nonlinear absorber with a transition that is resonant with the interferometer frequency[4,8]. A problem that occurs is that in practice, atomic spectral transitions usually involve several frequencies (for example, in the hyperfine structure of the sodium D1 transition). To simplify matters, in this section a single-frequency transition is treated. Although only atomic-beam experiments are expected to really have one frequency, this is a useful first step toward more general cases. Absorptive bistability has a great advantage over dispersive bistability, in that it does not display anomalous thresholds and other phase-dependent dynamical behavior during switching[12], and this could be of help in device applications. The chief problem in observing absorptive optical bistability, appears to be the practical limitations in real experiments, of non-plane-wave mode structures and inhomogeneously broadened absorption lines. In order to have a quantitative theory that relates to current experiments, these effects will be included in the state equations.

For simplicity, the theoretical treatment is mainly for a circularly symmetric Gaussian traveling-wave mode with negligible variation in the beam radius over the absorber length. The broadening is Lorentzian. A Gaussian inhomogeneous broadening can also be analyzed, giving rise to a more complicated state equation. Some factors that can occur in experiments, which are not included are:

(i) High absorption, low reflectivity, or changes in the mode structure over the interferometer length, i.e., self-focusing[9].

(ii) Multilevel atoms (although the state equation also holds for $J = \frac{1}{2} \rightarrow J = \frac{1}{2}$ transitions without hyperfine structure, with a redefinition of the coupling parameter).

(iii) Input laser amplitude fluctuations, or fluctuations in the interferometer medium and optical elements.

(iv) "Lamb dip" effects occurring with Fabry-Perot interferometers having large Doppler broadening.

(v) Coupling between different transverse or longitudinal modes in the interferometer (i.e., large mode spacing is necessary).

The Hamiltonian analogous to Eq. (3.1) (and with similar definitions) is, in terms of the Pauli atomic operators $\hat{\underset{\sim}{\sigma}}_j$,

$$
\begin{aligned}
\hat{H} = {}&\hbar\omega_0 \hat{a}^+\hat{a} + \sum_j \hbar[\omega_j\hat{\sigma}_j^z/2 - ig_j(\hat{\sigma}_j^+\hat{a} - \hat{\sigma}_j^-\hat{a}^+)] \\
&+ i\hbar[E(t)\hat{a}^+ e^{-i\omega t} - E*(t)\hat{a}\, e^{i\omega t}] + \{\text{RESERVOIRS}\}\,.
\end{aligned}
\tag{4.1}
$$

Here, the reservoirs describe radiative and collisional damping (leading to relaxation rates $\gamma_\perp, \gamma_\parallel$ for the polarization and inversion respectively). The other definitions are as in Eq. (3.1).

The coupling parameter in terms of γ is,

$$
g_j^2 = \left[\frac{3\pi\gamma_\parallel c^3}{2\omega^2}\right]|u(r_j)|^2 = g^2|u(r_j)|^2\,.
\tag{4.2}
$$

Using the rotating wave and Markov approximations, one derives a master-equation and hence a Fokker-Planck equation in a normally ordered, generalized P-representation, (for $\hat{a} \rightarrow \alpha, \hat{\underset{\sim}{\sigma}}_j \rightarrow \underset{\sim}{\sigma}$),

$$
\begin{aligned}
\frac{\partial}{\partial t} P(\underset{\sim}{\alpha},\underset{\sim}{\sigma}) = {}&\left\{\frac{\partial}{\partial\alpha}(k\alpha - \sum_j g_j\sigma_j^- - E) + \frac{\partial}{\partial\alpha^+}(k*\alpha^+ - \sum_j g_j\sigma_j^+ - E*)\right. \\
&+ \sum_j\left[\frac{\partial}{\partial\sigma_j^-}\gamma_j\sigma_j^- - g_j\alpha\sigma_j^z) + \frac{\partial}{\partial\sigma_j^+}(\gamma_j^*\sigma_j^+ - g_j\alpha^+\sigma_j^z)\right.
\end{aligned}
$$

$$+ \frac{\partial}{\partial\sigma_j} \left[\gamma_\parallel (\sigma_j^z + 1) + 2g_j (\sigma_j^+ \alpha + \sigma_j^- \alpha^+) \right]$$

$$+ \frac{\partial^2}{\partial\sigma_j^- \partial\sigma_j^+} (\sigma_j^z + 1)\gamma_P + \frac{1}{2}\left(\frac{\partial^2}{\partial\sigma_j^{-2}} [2g_j \alpha\sigma_j^-] + \frac{\partial^2}{\partial\sigma_j^{+2}} [2g_j \alpha^+\sigma_j^+] \right.$$

$$\left. + \frac{1}{2}\frac{\partial^2}{\partial\sigma_j^2}\left[2\gamma_\parallel (\sigma_j^z + 1) - 4g_j (\sigma_j^+ \alpha + \sigma_j^- \alpha^+) \right] \right] + 2k'n_{th} \frac{\partial^2}{\partial\alpha\partial\alpha^+} \Bigg\} P(\underset{\sim}{\alpha},\underset{\approx}{\sigma})$$

$$(4.3)$$

where

$$k = k' + i(\omega_o - \omega) = k'(1 + i\phi)$$

$$\gamma_j = \frac{1}{2}\gamma_\parallel + \gamma_P + i(\omega_j - \omega) = \gamma_\perp(1 + i\delta_j) = \gamma_\perp + i\Delta\omega_j$$

$$(\gamma_\parallel)/(2\gamma_\perp) = f.$$

Just as before, the general Fokker-Planck equation is non-trivial and now is equivalent to (3N+2) nonlinearly coupled stochastic equations. As there are three relaxation rates, there are three different adiabatic regions: mode-dominated ($|k| \ll \gamma_\parallel, |\gamma|$); polarization dominated ($\gamma_\parallel, |\gamma| \ll |k|$); and population dominated ($\gamma_\parallel \ll |\gamma|, |k|$). In the polarization dominated limit it is necessary to include 3N variables because of the relation $\gamma_\parallel \leqslant 2|\gamma|$. For equivalent atoms, macroscopic variables or collective atomic variables can be used to reduce this to three equations[8]. This simplification would not occur in Eq. (3.4), due to nonlinearities. Also, in the case of mode structure or inhomogeneous broadening, the simplification does not occur, and only the mode dominated (high-Q interferometer) limit is reasonably tractable. For this reason, the mode-dominated limit is the limit that will be investigated here. One can adiabatically eliminate the atomic variables (on setting $\underset{\sim}{\dot{\sigma}}^\pm, \underset{\sim}{\dot{\sigma}}^z = 0$), to obtain the following stochastic equation, which is also equivalent to a Fokker-Planck equation,

$$\dot{\alpha}_\mu = -A_\mu(\underset{\sim}{\alpha}) + [D_{\mu\nu}(\underset{\sim}{\alpha})]^{\frac{1}{2}}\xi_\nu(t) \qquad [\mu,\nu = 1,2]$$

where,

$$A_1(\underset{\sim}{\alpha}) = - E + \alpha\left[k + \sum_j \frac{g_j^2(1 - i\delta_j)}{\gamma_\perp \Pi_j(\alpha\alpha^+)} \right] \qquad \text{(and h.c.)}$$

$$\Pi_j(\alpha\alpha^+) = [1 + \delta_j^2 + (4g_j^2\alpha\alpha^+)/(\gamma_\parallel\gamma_\perp)] = [1 + \delta_j^2 + X_j]$$

$$D_{11}(\alpha\alpha^+) = -\left[\frac{1}{\gamma_\perp^2\gamma_\parallel}\right]\sum_j\left[\frac{4g_j^4\alpha^2}{\Pi_j^3(\alpha\alpha^+)}\right]\left[(1-i\delta_j)^3 f + i\delta_j X_j(1-f)(1-\delta_j) + \frac{X_j^2}{2}\right]$$

$$\text{(and h.c.)}$$

$$D_{12} = D_{21} = \left[\frac{1}{\gamma_\perp}\right]\sum_j\left[\frac{g_j^2 X_j}{\Pi_j^3(\alpha\alpha^+)}\right]\left[(1+\delta_j^2)(1-f)+X_j(2+\delta_j^2(1-f))+ \frac{X_j^2}{2}\right]+2k'n_{th}.$$

$$(4.4)$$

Thus, it is straight forward to obtain a general Fokker-Planck equation in this case. The linearized spectrum near a mean value $\bar\alpha$ of a positive slope branch of the state equation is obtained from Eq. (2.12), where $\bar\alpha$, a, b, d, Γ are given by,

$$A_\mu(\bar\alpha,\bar\alpha^*) = 0$$

$$a = k + \sum_j (1+\delta_j^2)(1 - i\delta_j)/[\gamma_\perp\Pi_j^2(|\bar\alpha|^2)]$$

$$b = -\sum_j 4g_j^4\bar\alpha^2(1 - i\delta_j)/[\gamma_\perp^2\gamma_\parallel\Pi_j^2(|\bar\alpha|^2)]$$

$$d = -D_{11}(\bar\alpha,\bar\alpha^*)$$

$$\Gamma = D_{12}(\bar\alpha,\bar\alpha^*) .$$

$$(4.5)$$

At this point, an explicit proof of stability of positive slope branches is necessary, to ensure that no additional instabilities exist in the single-mode problem. The Hurwitz criterion for stable eigenvalues is readily obtained and is found to be,

(A) $2\text{Re}[f(n) + nf'(n)] > 0$

(B) $|f(n)|^2 + n[f^*(n)f'(n) + f(n)f'^*(n)] > 0$ (4.6)

where,

$$f(n) = k + \sum_{j=1}^N \frac{g_j^2(1-i\delta_j)}{\gamma_\perp\Pi_j(n)} .$$

The first criterion for stability corresponds to no oscillatory instabilities (i.e., $\text{Re}(a) > 0$), and this is trivially satisfied in the high-Q limit, provided the spacing of the other mode frequencies, which are neglected, is large enough. The second criterion is exactly equivalent to a positive slope requirement. The unstable regions are, therefore, determined by the turning points of the state

equation, Eq. (4.5), which comes from the deterministic part of
Eq. (4.4).

It is useful to notice that for large detunings and radiative
relaxation, it is not necessary to use the full structure of Eq.
(4.4). Instead, provided atomic inversions are small, the equation
reduces approximately to the nonlinear polarizability model (Eq.
(3.7)) whose properties are well known, with,

$$\bar{k} = k + \sum_j \frac{|g_j|^2}{\gamma_j}$$

$$\bar{\epsilon} = \sum_j \left[\frac{g_j^4 \Delta\omega_j}{|\gamma_j|^4 f} \right] . \tag{4.7}$$

However, the diagonal correlations $\langle\Gamma_A(t)\Gamma_A(t')\rangle$ in Eq. (3.8) must
be multiplied by a factor of f when collisional relaxation is in-
cluded. This simplification is especially useful near the critical
point at large detunings as the state-equation is just a cubic equation, for general mode-structures.

A. Plane-Wave Mode, Homogeneous Broadening

This is the simplest case, although not a realistic one; as in
principle a plane-wave mode requires an infinite input power. The
results, which are well known[10,11] are summarized here. The Fokker-
Planck equation results from Eq. (4.4), on replacing the summations
by N, the number of atoms. Stability properties are easily derived[11],
leading to restrictions on N, δ,ϕ before bistability exists. These
restrictions are given on analyzing the turning points of Eq. (4.5)
for given C,δ,ϕ,

(A) $(27/4)C[1 + \delta^2][1 + \phi^2] \lesssim [\delta\phi + C - 1]^3$

(B) $2C \gtrsim \delta\phi - 1$ (4.8)

where

$$C = [g^2 N/2k'\gamma_\perp V] .$$

Here the first restriction is the main one. Condition (B) must
occur on the boundary of the region defined by (A) in order to have
a physical (i.e.,positive-intensity) bistable region, and is obtained trivially on examining the sign of the roots of the state
equation.

The situation with $\delta = \phi = 0$ corresponds to absorptive bistability, which occurs for $C \gtrsim 4$. This has been treated by Lugiato[8]
using a Fokker-Planck treatment equivalent to the present one, but

with a different operator ordering. His result includes atomic correlations, and gives exactly the same spectral results as linearizing Eq. (4.4) with detunings equal to zero. By contrast, some earlier
Fokker-Planck equations[4], which neglect atomic correlations, give different spectral results in general. Thus, neglecting atomic correlations is not a good approximation in optical bistability. New results[10] are obtained by allowing δ,ϕ to be non-zero. In particular,
a new type of transition occurs in the transmitted spectrum, which
is a transition from a single-peaked spectrum (on the lower branch)
to a twin-peaked non-symmetric spectrum (on the upper branch). This
is shown in Figs. 1 and 2.

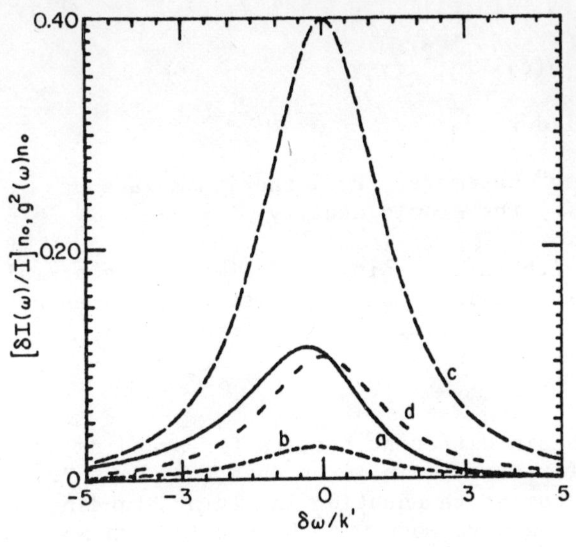

Fig. 1 Graph of $[I(\omega)n_0/I]$ [a,b] and $[g^2(\omega)]n_0$ [c,d]: vs. the relative output frequency $[\omega-\omega_I]k'$ for the plane wave mode below threshold $[C=40, \delta=5, \phi=5]$. The relative transmitted intensity (Eq. (4.10)) is X=.2. Curves [a,c] are for phase damping, [b,d] for radiative damping.

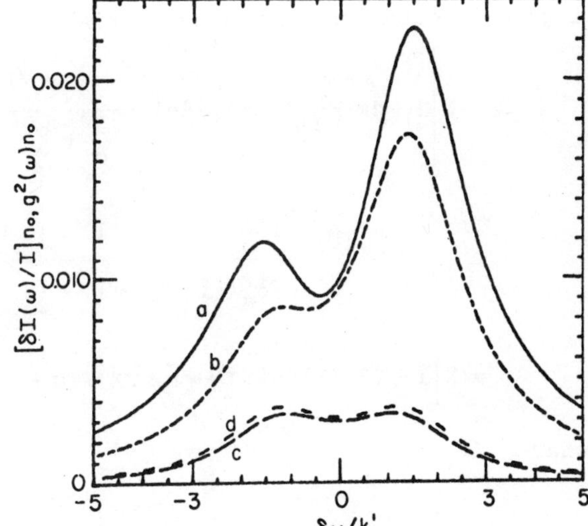

Fig. 2. As Figure 1, but above threshold [X=40].

B. Radially-Varying Mode, Lorentzian Inhomogeneous Broadening

In order to treat the more realistic problem of a radially vary-ing mode with inhomogeneous broadening, it is necessary to sum over different atomic positions and detunings in Eq. (4.4). The new state-equation is then,

$$E = k'<\alpha> \left[1 + i\phi + 2C \iint \frac{P(\delta)N(r)|u(r)|^2(1-i\delta)d^3r d\delta}{N_o \Pi(\delta,\underset{\sim}{r})} \right] \quad ,$$

where,

$$C \equiv [g^2/(2k'\gamma_\perp)] \int N(\underset{\sim}{r})|u(r)|^2 d^3\underset{\sim}{r} = g^2 N_o/(2k'\gamma_\perp) \qquad (4.9)$$

$$\Pi(\delta,\underset{\sim}{r}) = 1 + \delta^2 + 4g^2|u(r)|^2|\alpha|^2/(\gamma_\perp\gamma_\parallel)$$

$$N_o = \int N(\underset{\sim}{r})|u(\underset{\sim}{r})|^2 d^3\underset{\sim}{r} \quad .$$

Here C is the "cooperativity" parameter, $P(\delta)$ the inhomogeneous broadening distribution, $N(r)$ the atomic density.

For a Gaussian mode, with an absorbing region occupying a small fraction ℓ of the optical length in which the waist radius does not vary appreciably, one finds $N_o \approx N\ell$, i.e.,

$$C = \ell g^2 N/2k'\gamma_\perp.$$

In general, one can obtain different results for standing or traveling waves, as shown for example in laser theory by Haken[6], where a typical two-level atom state equation involving standing waves is given. The traveling wave mode with Lorentzian inhomo-geneous broadening is the simplest example, leading to the follow-ing state-equation in reduced variables,

$$Y = X\left| 1 + i\phi + \frac{C}{X}\left\{ (1-i\delta_o)\ln\left[\frac{\delta_o^2 + (\sigma + \sqrt{1 + 2X})^2}{\delta_o^2 + (\sigma + 1)^2} \right] \right. \right.$$

$$\left. \left. + \sigma\ln\left[\frac{\delta_o^2 + i\delta_o(\sqrt{1 + 2X} - 1) + (\sigma + 1)(\sigma + \sqrt{1 + 2X})}{\delta_o^2 - i\delta_o(\sqrt{1 + 2X} - 1) + (\sigma + 1)(\sigma + \sqrt{1 + 2X})} \right] \right\} \right|^2$$

$$= X\left| 1 + (2C/X)\ln[(\sigma + \sqrt{1 + 2X})/(\sigma + 1)] \right|^2 \quad (\delta_o = \phi = 0) \qquad (4.10)$$

where,

$$\left|u(r)\right| = \left[\frac{2}{\pi LW^2}\right]^{1/2} \exp\left[\frac{-r^2}{W^2}\right] \quad , \qquad X = \frac{|\alpha|^2}{n_o}$$

$$P(\delta) = \left[\frac{\sigma}{\pi}\right]/[\sigma^2 + (\delta-\delta_o)^2] \quad , \qquad Y = |E|^2/[k'^2 n_o]$$

$$n_o \equiv \frac{\gamma_\perp \gamma_\parallel}{4g^2} \int N(r)\left|u(r)\right|^4 \frac{d^3 r}{N_o} = \left[\frac{\pi LW^2}{4g^2}\gamma_\perp \gamma_\parallel\right] .$$

Here X, Y are proportional to the transmitted and incident power respectively (see Eq. (3.2)), while n_0 is the threshold photon number for an absorbing region occupying a small fraction ℓ of the total round-trip length L, near the mode waist of radius W.

The effect of this new state-equation is very significant for the observation of absorptive optical bistability. Firstly, even for $\sigma = 0$, the threshold value of C is approximately doubled. For larger inhomogeneous broadenings the value of C at the absorptive critical point is increased still further, growing approximately as σ^2, with a corresponding increase in threshold value of Y. This is illustrated in Figs. 3 and 4, which show the dependence of C_{crit}, X_{crit}, Y_{crit} on σ.

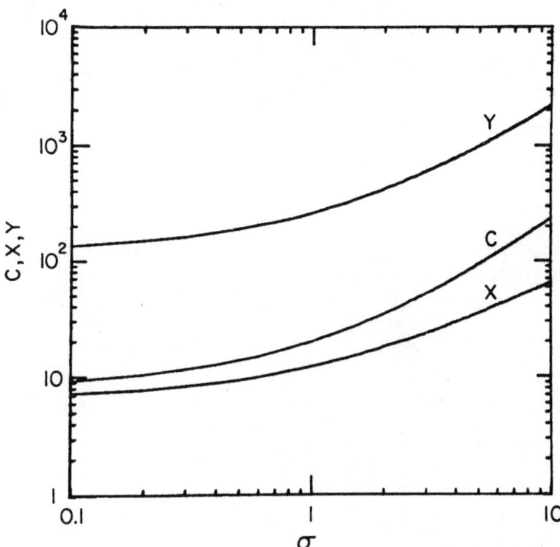

Fig. 3   Dependence of the absorptive critical cooperativity (C), transmitted intensity (X_{crit}) and input intensity (Y_{crit}), on the relative Lorentzian inhomogeneous broadening (σ) for σ=.1 to σ=10. In these calculations, $\delta=\phi=0$, and the mode is Gaussian.

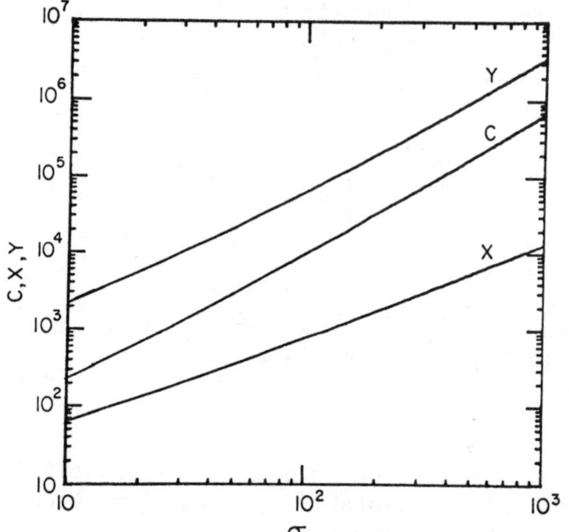

Fig. 4 As in Fig. 3, but for $\sigma = 10$ to $\sigma = 1000$.

However, even for large inhomogeneous broadening, there are still interesting results to be found. Near the threshold or critical point, there is a region of very large differential gain. This occurs due to power broadening, which increases the number of effectively interacting atoms as the local field increases (an alternative viewpoint is the view that power broadening reduces the absorptivity of resonant atoms relative to non-resonant atoms). This effect which causes a huge variation in the transmitted field of several orders of magnitude, for a very small increase in the input field, is shown in Fig. 5.

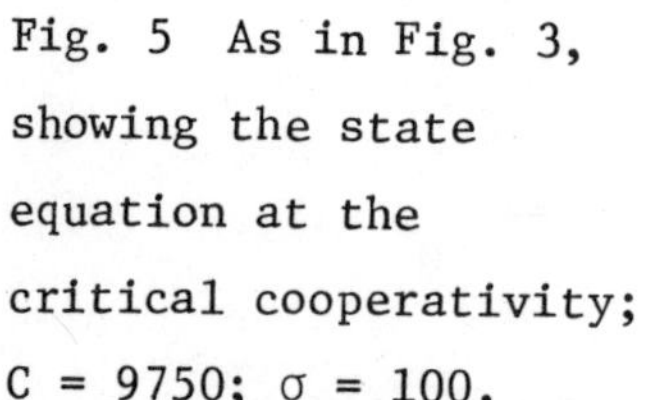

Fig. 5 As in Fig. 3, showing the state equation at the critical cooperativity; $C = 9750$; $\sigma = 100$.

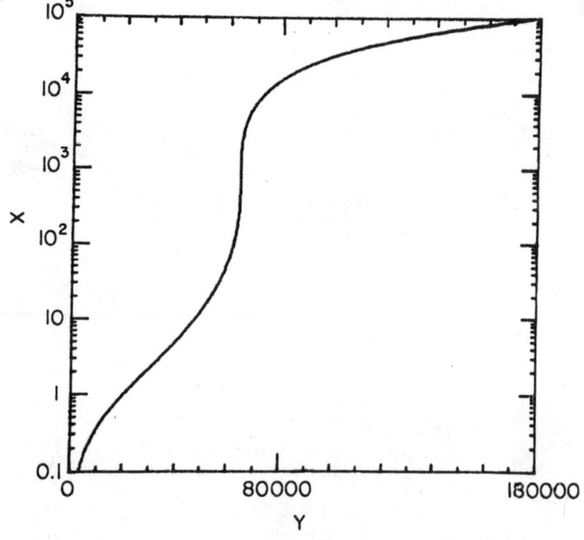

Numerical estimates of typical parameters in experiment are possible, and these are given in Table 1, for an experiment on a $\Delta m = 1$ transition with a dipole-coupling equal to that of the sodium D_1 transition (without inhomogeneous broadening), in a lossless spherical ring interferometer of finesse 250, round-trip length .24m, waist radius 100 μm. These parameters are in principle, obtainable in atomic beam experiments.

TABLE 1

Parameter	Numerical Value
$\gamma_{\parallel}$	$6 \times 10^7 \text{ s}^{-1}$
k'	$1.4 \times 10^7 \text{ s}^{-1}$
Δt	$.8 \times 10^{-9} \text{ s}$
ℓ	$.01$
g^2	$750 \quad \text{m}^3\text{s}^2$
$\gamma_{\perp}$	$3 \times 10^7 \text{ s}^{-1}$
n_o	4.5×10^3
C_{crit}, X_{crit}, Y_{crit}	$8,\ 7,\ 120$
$P_{IN(crit)}$	$2.7 \times 10^{-6} \text{ watt}$
$N_{(crit)}$	$9 \times 10^{14} \text{ m}^{-3}$

V.   SUMMARY

The nonlinear polarizability theory provides a straight-forward example of optical bistability, which has an exactly soluble Fokker-Planck equation for a high-Q single-mode interferometer in the steady-state. This theory is most readily treated including quantum fluctuations, in a generalized P-representation. More realistic problems are also tractable, and these often reduce to the non-linear polarizability theory in the limit of large detuning. The nonclassical behavior of the quantum fluctuations is especially significant in these cases, and would lead to difficulties if traditional representations (like the Glauber-Sudarshan representation) were utilized. By comparison the generalized P-representation is straight-forward to apply to this problem.

The treatment of problems involving nonlinear absorption is more complicated. In this situation (i.e., in absorptive optical bistability), the effect of inhomogeneous broadening and Gaussian mode structures makes a quantitatively large difference, especially in the critical cooperativity. It is, therefore, necessary to make a careful study of the state-equation for bistability, before calculating the effect of quantum fluctuations on the transmitted radiation. However, the linearized spectrum and correlation function due to quantum fluctuations can always be calculated, for a high-Q single-mode interometer. This requires a knowledge of the specific mode-function and inhomogeneous broadening in Eq. (4.5), which can then be utilized with Eq. (2.12) to calculate the relevant observable.

ACKNOWLEDGMENTS

I would like especially to thank D. F. Walls and C. W. Gardiner, who played a large part in the research outlined here, and also S. Chaturvedi, S. S. Hassan and W. Sandle. This work was partially supported by the United States Department of Energy and by the U.S. Navy.

REFERENCES

1.   R. J. Glauber, Phys. Rev. $\underline{131}$, 2766 (1963); E. C. G. Sudarshan, Phys. Rev. Lett. $\underline{10}$, 277 (1963).

2.   H. J. Carmichael and D. F. Walls, J. Phys. $\underline{B9}$, 1199; L43 (1976); H. J. Kimble, M. Dagenais, L. Mandel, Phys. Rev. Lett. $\underline{39}$, 691 (1977).

3.   H. M. Gibbs, S. L. McCall, and T. N. C. Venkatesan, Phys. Rev. Lett. $\underline{36}$, 1135 (1976); T. N. C. Venkatesan and S. L. McCall, Appl. Phys. Lett. $\underline{30}$, 282 (1977); D. Grischkowsky, J. Opt. Soc. Am. $\underline{68}$, 641 (1978); T. Bischofberger and Y. R. Shen, J. Opt. Soc. Am. $\underline{68}$, 642 (1978); H. M. Gibbs, S. L. McCall, T. N. C. Venkatesan, A. C. Gossard, A. Passner, W. Wiegmann, Appl. Phys. Lett. $\underline{35}$, 451 (1979); D. A. B. Miller, S. D. Smith, A. Johnston, Appl. Phys. Lett. $\underline{35}$, 658 (1979).

4.   A. Szoke, V. Daneu, S. Goldhar and N. A. Kurnit, Appl. Phys. Lett. $\underline{15}$, 376 (1969); S. L. McCall, Phys. Rev. $\underline{A9}$, 1515 (1974); R. Bonifacio and L. A. Lugiato, Phys. Rev. $\underline{A18}$, 1129 (1978); R. Bonifacio, M. Gronchi and L. A. Lugiato, Phys. Rev. $\underline{A18}$, 2266 (1978); C. R. Willis, Opt. Comm. $\underline{26}$, 62 (1978); G. S. Agarwal, L. M. Narducci, R. Gilmore and D. H. Feng, Phys. Rev. $\underline{A18}$, 620 (1978); J. Hermann, Optica Acta $\underline{27}$, 159 (1980).

5.   P. D. Drummond, D. Phil. Thesis (University of Waikato, 1979); P. D. Drummond and C. W. Gardiner, J. Phys. $\underline{A13}$, 2353 (1980).

6. H. Haken, "Handbuch der Physik", Vol. XXV/2c (Springer, Berlin, 1970); W. H. Louisell, "Quantum Statistical Properties of Radiation" (Wiley, New York, 1973).

7. L. Arnold, "Stochastic Differential Equations" (Wiley, New York, 1974).

8. L. A. Lugiato, Nuovo Cimento $\underline{59B}$, 89 (1979).

9. J. H. Marburger and F. S. Felber, Phys. Rev. $A\underline{17}$, 335 (1978); N. Bloembergen, Opt. Comm. $\underline{15}$, 416 (1975).

10. P. D. Drummond and D. F. Walls, J. Phys. $A\underline{13}$, 725 (1980), P. D. Drummond and D. F. Walls (to appear).

11. R. Bonifacio and L. A. Lugiato, Lett. Nuovo Cimento $\underline{21}$, 517 (1978); S. S. Hassan, P. D. Drummond and D. F. Walls, Opt. Commun. $\underline{27}$, 480 (1978); G. P. Agarwal and H. J. Carmichael, Phys. Rev. $A\underline{19}$, 2074 (1979); R. Bonifacio, M. Gronchi and L. A. Lubiato, Nuovo Cimento $\underline{53B}$, 311 (1979); R. Roy and M. S. Zubairy, Phys. Rev. $A\underline{21}$, 274 (1980).

12. F. A. Hopf, P. Meystre, P. D. Drummond, and D. F. Walls, Opt. Comm. $\underline{31}$, 245 (1979).

EFFECTS OF PROPAGATION, TRANSVERSE MODE COUPLING AND DIFFRACTION ON NONLINEAR LIGHT PULSE EVOLUTION

F.P. Mattar†

Aerodynamics Laboratory, Polytechnic Institute of New York, Farmingdale, New York 11735

Abstract: The effective computational methods developed to efficiently tackle transverse and longitudinal reshaping associated with single-stream and two-way propagation effects in cooperative light-matter interactions, using the semi-classical model are described. The mathematical methods are justified on physical grounds. Typical illustrative results of propagation in resonant absorbers, amplifiers and superfluorescence systems are presented.

I. INTRODUCTION

This paper reviews the unified mathematical methods developed for three-dimensional simulation of several physical phenomena previously studied independently. The same basic algorithm with some alterations will simulate both superfluorescence[1,2] and optical bistability[3,4]. With extra modifications, it can also analyze four-wave mixing[5] and phase conjugation[6] systems. Further applications include two-way Self-Induced Transparency[7] and Soliton Collision[8] studies.

The proposed model evolved as a result of close collaboration with the experimentalists, H.M. Gibbs[9-13], S.L. McCall[11-13] and recently, M.S. Feld[13], enhancing the rate of progress in the re-

†Work jointly sponsored by the Research Corporation, the International Division of Mobil Corporation, the University of Montreal, the U.S. Army Research Office, DAAG29-79-C-0148 and the Office of Naval Research, N000-14-80-C-0174.

search and leading to a better understanding of basic cooperative effects in light-matter interactions. Quantitative analyses in superfluorescence were obtained and are being developed in optical bistability.

The model encompasses propagation that includes rigorous diffraction[16,15], time-dependent phase variation, off-resonance[16] as well as nonuniform excitation[19] and transverse and longitudinal boundary conditions[18]. (An additional control probe-beam is being developed[21].)

The adoption of proven computational techniques, developed by Moretti[22-24] in aerodynamics, to solve problems in the laser field, is justified by the analogy between fluid and wave propagation problems described. The laser beam evolution can be interpreted in terms of an equivalent flowing fluid[25] whose density is proportional to the laser field intensity, and whose velocity is proportional to the gradient of the field phase. This description allows for the treatment of more slowly varying dependent variables and yields to governing equations of motion, which are a generalization of the Navier-Stokes equations[26]. In the fluid formulation, the equivalent fluid is compressible and is subjected to an internal potential, depending solely and nonlinearly upon the fluid density and its derivatives; this is called the "quantum mechanical potential." Furthermore, the field scalar wave equation mathematically corresponds to a complex heat diffusion equation with a non-uniform functional source; while the Bloch equations, in a rotating frame, are structurally similar to the torque equation[27]. For two-way problems, the simultaneous set of quasi-optic field equations (one for each traveling wave) play the same preponderant role as Euler equations in shock calculations for fluid dynamics problems.

Quite different effects, i.e., self-lensing[28], self-phase modulation[29], self-spectral broadening[30] and self-steepening[31], previously studied separately, combine here to modify the pulse behavior diversely at different positions and times. For example, the interplay of diffraction coupling through the Laplacian term and the inertial response of the non-uniform pre-excited medium will inevitably redistribute the beam energy spatially and temporally[32]. This transient one- or multi-beam transverse reshaping will profoundly affect the performance of any device that relies upon it. Specifically, this pragmatic, three-dimensional analysis helps in the interpretation of recent experimental results in superradiance, superfluorescence, optical bistability and active-mirror amplifiers for laser-fusion. It also accounts for deviations and departures between recent experimental observations and predictions of planar wave theory (see Fig. (1)).

To circumvent excessive memory requirements while insuring adequate numerical resolution, one must resort to nonuniform meshes. In this large computational problem, the calculational efficiency of the algorithm chosen is of crucial importance. A brute force, finite difference treatment of the governing equations is not feasible. Instead, by using the details of the physical processes to determine where to concentrate the computational effort, accuracy and economy are achieved. For example, if for self-focused beams, a fixed transverse mesh is used, a lack of resolution (see Fig. (2)) may result. A non-negligible loss of computational effort in the wings of the beam will also occur.

Coherent Pulse Propagation

I. Usual Theory

   1 Dim. $\xi=\xi(\rho)$

   'Uniform Plane Wave'

II. Usual Experiment

$$\xi=\xi(\rho)\, e^{-\rho^2}$$

   'Gaussian'

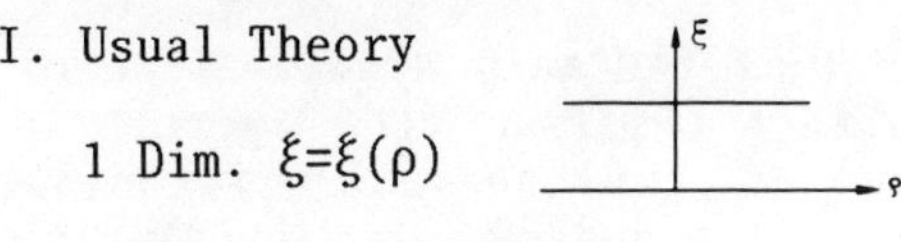

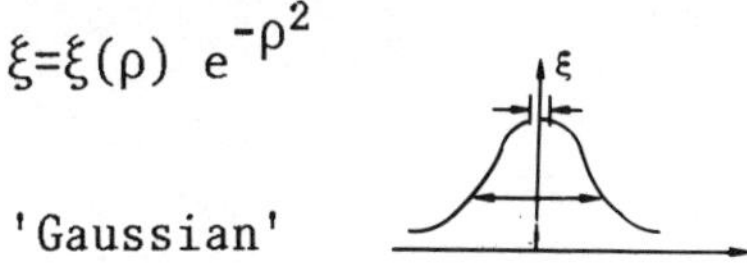

Fig. 1. The state of the art in coherent pulse propagation is displayed. The theoretical effort was restricted to a uniform plane wave prior to the work of Newstein et al; whereas the usual experiment was carried out using a Gaussian beam. To simulate a uniform plane wave, the smallest possible detector diameter was selected as compared to the Gaussian beam diameter (i.e., (i.e., $d_{detector} \ll d_{beam}$)

In particular, evenly-spaced computational grid points are related to variable grids in a physical space by adaptive stretching (Fig. (3)) and rezoning (Fig. (4)) techniques. This mapping consists either of an a priori coordinate transformation or an adaptive transformation (Fig. (5)) based on the actual physical solution. Both stretching transformation in time and rezoning techniques in space are used to alleviate the computational effort. The propagation problem is thus reformulated in terms of appropriate coordinates that will automatically accommodate any change in the beam profile[34-40].

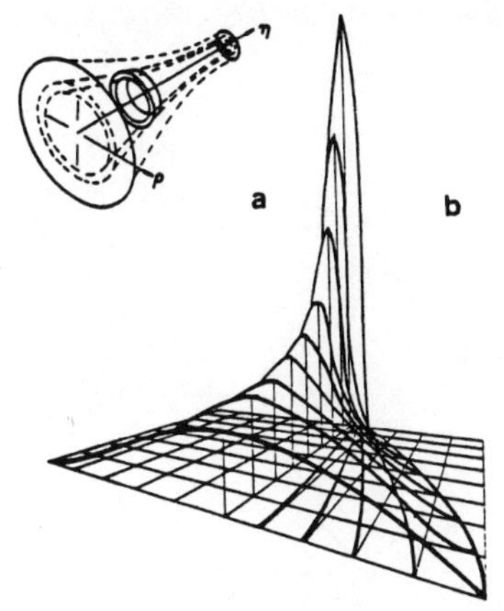

Fig. 2 (a) Isometric representation of the beam cross-section as it experiences self-focusing: The cross-section decreases as a function of the propagation distance; (b) An isometric display of the time integrated field energy as a function of ρ and η to illustrate the resolution limitation associated with uniform mesh.

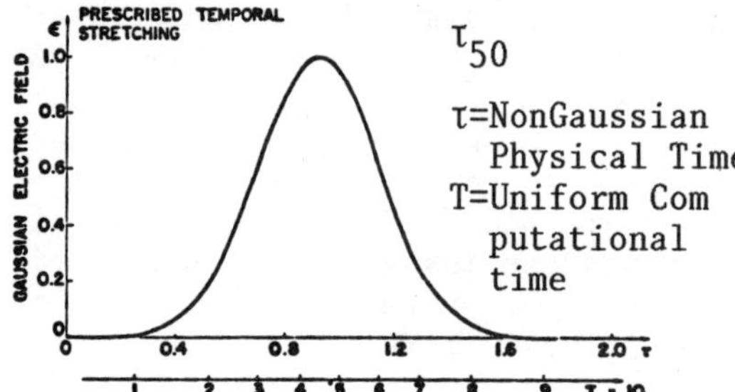

Fig. 3.   Non-uniform prescribed temporal stretching.

The resultant dynamic grid removes the main disadvantage of insufficient resolution, where uniform Eulerian codes generally suffer. Furthermore, the advantages of grid sensitivity can be obtained by either using adequate rezoning and mapping in Eulerian coordinates or by simply using traditional Lagrangian methods[41,42]. Thus, the convenience of moderate memory requirements can be combined with the desirable numerical resolution should one rezone the grids. The techniques due to Moretti[33] will economically generate precise results. Although this appears surprising because of the mesh coarseness, his technique succeeds because it discriminates intelligently between the different domains of the critical physical parameters.

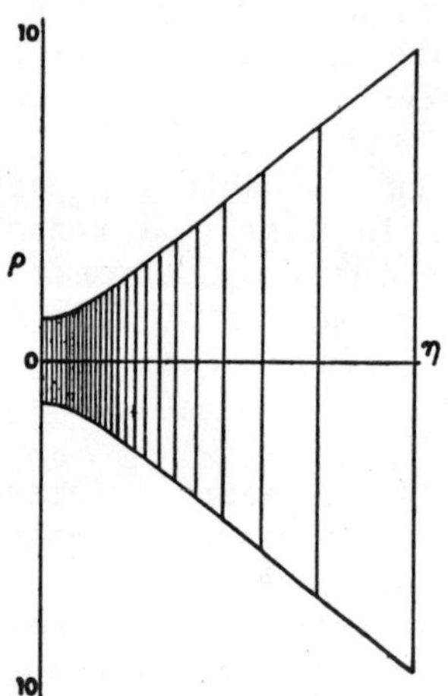

Figure 4. Two-dimensional prescribed rezoning for ρ and η. As the beam narrows the density of transverse points and the transmission planes increase simultaneously.

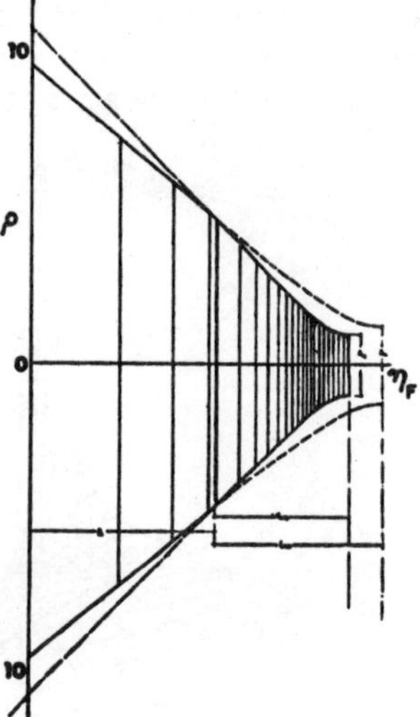

Fig. 5.   Self-adjusted two-dimensional rezoning for ρ and η to follow more closely the actual beam characteristics. The (normalizing) Gaussian reference beam is redefined during the calculation.

For the two-beam analysis, our approach relies on one-way nonsymmetric discretizations of the longitudinal and transverse derivatives as well as nonuniform grids. Numerical instrumentation is unavoidable. The role of characteristics as information carriers is emphasized and therefore the law of forbidden signals cannot be violated[43]. The physical subtleties of the nonlinear problem can be adequately implemented.

Interactive graphic software was developed to simplify the physics of extraction from these complex codes. Structural modular programming techniques are used, making the program easier to read, maintain and transport as well as for further extensions and generalizations of the planar wave theory. The resultant code is deceptively simple and easy to follow. This mathematical modeling, motivated by Gibbs' and McCall's experimental work, is engineering physics in its purest sense: its main goal is to obtain a numerical solution to and insight into a real physical problem, instead of reaching a neat analytical solution to an idealized problem of limited applications.

II. SIT/SUPERFLUORESCENCE EQUATIONS OF MOTION

In the slowly varying envelope approximation, the SIT dimensionless, semi-classical field-matter equations[15] (which describe a system in a cylindrical geometry with azimuthal symmetry), are:

$$-iF\, \nabla_T^2 e + \frac{\partial e}{\partial \eta} = \mathcal{P} \tag{1}$$

$$\partial \mathcal{P}/\partial \tau = eW - (i\Delta\Omega + 1/\tau_2)\mathcal{P} \tag{2}$$

and

$$\partial W/\partial \tau = -1/2(e^*\mathcal{P} + e\mathcal{P}^*) - (W-W^e)/\tau_1 \tag{3}$$

where

$$e = (2\mu/h)\tau_p e', \quad \text{and} \quad \mathcal{P} = (2/\mu)\,\mathcal{P}', \tag{4}$$

$$E = \text{Re}[e'\exp\{i(\kappa/c)z-\omega t)\}]; \tag{5}$$

with W^e the equilibrium value of W, subjected to the initial and boundary conditions.

1.   for $\tau \geq 0$: $e = 0$, $W = W_i$, $\mathcal{P} = \mathcal{P}_i$ known function to take into account the pumping effects or the initial tipping angle.

2.   for $\eta = 0$: e is given as a known function of τ and ρ;

3.   for all η and τ: $[\partial e/\partial \rho]_{\rho=0}$ and $[\partial e/\partial \rho]_{\rho=\rho_{max}}$ vanishes (with ρ_{max} defining the extent of the region over which the numerical solution is to be determined).

with $k/c = \omega$ (6)

and $\nabla_T^2 e = [\ \dfrac{1}{\rho} \dfrac{\partial}{\partial \rho}(\rho \dfrac{\partial e}{\partial \rho})]$;   (7)

after applying l'Hopital's rule, the on-axis Laplacian reads:

$$\nabla_T^2 = 2\ \frac{\partial^2 e}{\partial \rho^2}$$ (8)

$$P = i\ Re[\mathcal{P}'\ exp\{i(\kappa/c)z - kt\}].$$ (9)

The complex field amplitude e, the complex polarization density $\mathcal{P}$, and the energy stored per atom W, are normalized functions of the transverse coordinate $\rho = t/t_p$, the longitudinal coordinate $\eta = z \times \alpha_{eff}$, and the retarded time $\tau = (t-zn/c)\tau_p$ (see Fig. (6)). The time scale is normalized to the full width half maximum (FWHM) input pulse length, τ_p and the transverse dimension scales to the input beam spatial width r_p. The longitudinal distance is normalized to the effective absorption length,[44] $(\alpha_{eff})^{-1}$ where

$$\alpha_{eff} = [\frac{\omega \mu^2 N}{n \hbar c}]\tau_p \rightarrow [\alpha' \tau_p]$$ (10)

Here, ω is the angular carrier frequency of the optical pulse, μ is the dipole moment of the resonant transition, N is the number density of resonant molecules, and n is the index of refraction of the background material. The dimensionless quantities $\Delta\omega = (\omega - \omega_o)\tau_p$, $\tau_1 = T_1/\tau_p$, and $\tau_2 = T_2/\tau_p$ measure the offset of the optical carrier frequency ω from the central frequency of the molecular resonance ω_o, the thermal relaxation time T_1, and the polarization dephasing time T_2, respectively.

Even in their dimensionless forms, the various quantities have a direct physical significance. Thus ρ is a measure of the component of the transverse oscillating dipole moment (ρ has the proper phase for energy exchange with the radiation field). In a two-level system, in the absence of relaxation phenomena, a resonant field cause each atom to oscillate between the two states,

W=-1 and W=+1, at a Rabi frequency $f_R = e/2\tau_p = (\mu/\hbar)e'$. Thus e measures how far this state-exchanging process proceeds in rp.

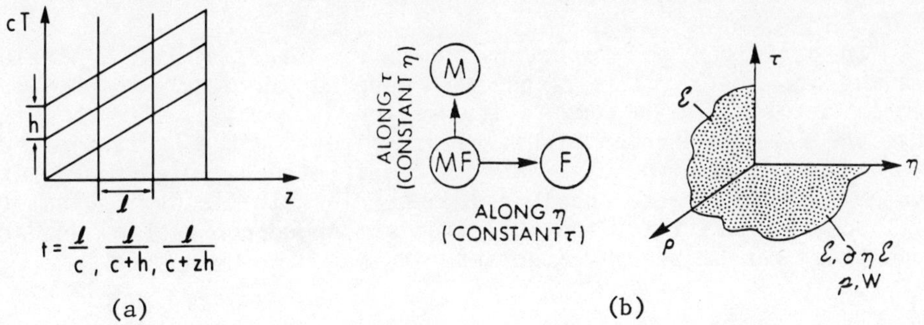

Fig. 6.   Graph (a) displays the retarded time concept. Graph (b) outlines the numerical approach: a marching problem along η for the field simultaneously with a temporal upgrading of the material variables along τ.

The dimensionless parameter, F, is given by $F=\lambda(\alpha_{eff})^{-1}/(4\pi r_p^2)$. The reciprocal of F is the Fresnel number associated with an aperture radius r_p and a propagation distance $(\alpha_{eff})^{-1}$. The magnitude of F determines whether or not one can divide the transverse dependence of the field into "pencils" (one per radius ρ), to be treated in the plane-wave approximation.

As outlined by Haus et al[45], the acceptance of equations (1-3) implies certain approximations: eq. (3) shows that the product 'e𝒫' of the electric field e and the polarization 𝒫 causes a time rate of change in the population difference leading to saturation effects. Inertial effects <u>are</u> considered.

III.   IMPORTANCE OF BOUNDARY CONDITIONS

When the laser beam travels through an amplifier, the transverse boundary has an increasingly <u>crucial</u> effect compared to the absorber situation. The laser field which resonates with the pre-excited transition, experiences gain; the laser which encounters a transition initially at ground state, experiences resonant absorption and losses. A greater portion of the pulse energy is diffracted outwardly in the amplifier than in the absorber[46]. Consequently, these boundary reflection conditions play a substantial role in the amplifier calculations and obscure the emergence of any new physical effects. Acceptable results are achieved <u>only</u>

by carefully coupling the internal points analyzed with the boundary points[47]. Special care is required to reduce the boundary effect to a minimum such as using non-uniform grids and confining the active medium by an absorbing shell.

In practice, the transverse boundary is simulated by implementing an absorbing surface and mapping an infinite physical domain onto a finite computation region (see Fig. (7)). In Fig. (8), the first and second radial derivatives and the Laplacian term are drawn. Figure (9) contrasts in the stretched radial coordinate system, the transverse coupling and the electric field. The numerical domain sensitivity and the physical dependence on the boundary conditions can be readily assessed.

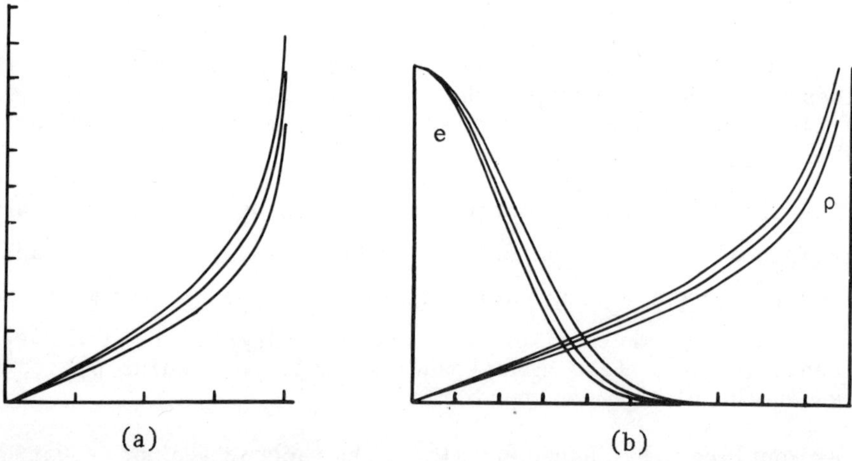

Fig. 7. Graph (a) shows non-uniform stretching of the transverse coordinate. Graph (b) contrasts the Gaussian beam e dependence with the nonuniform physical radius ρ. Both graphs are plotted versus the uniform mathematical radius R.

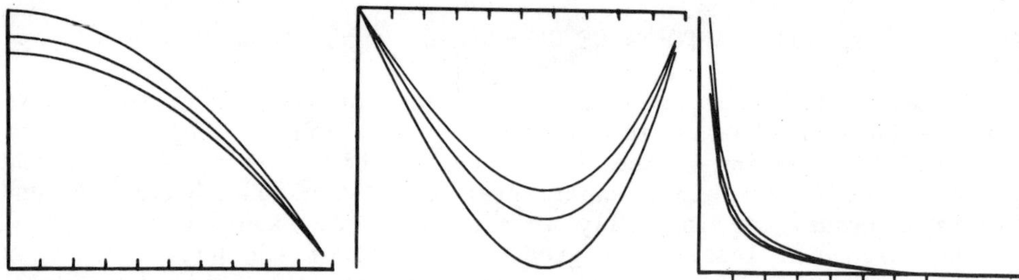

Fig. 8. This graph illustrates the dependence of the radial mapping and the derivatives on the different parameters versus the uniform mathematical radius: First weighting stretching factor $\partial R/\partial \rho$; 2nd weighting stretching factor, $\partial^2 R/\partial \rho^2$; weighted diffraction term, $\nabla^2_{T\rho} R$.

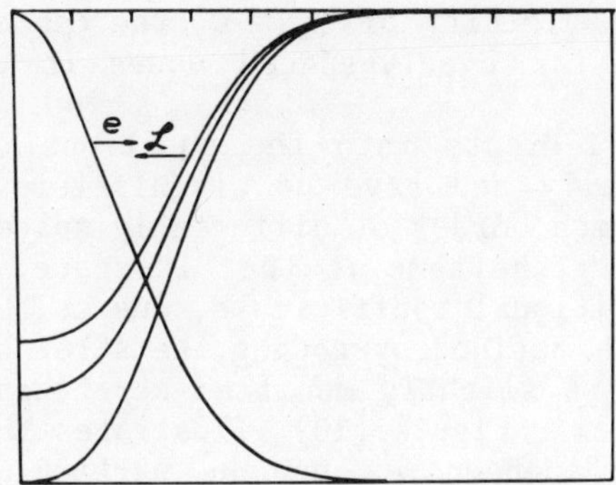

Fig. 9.   This figure contrasts the Laplacian dependence 'ℒ' for a given Gaussian profile 'e' for various non-uniform radial point densities.

IV.   PRESCRIBED STRETCHING

The numerical grid is defined by widely-spaced computational nodes in the area most distant from the plane of interest and by densely clustered nodes in the critical region of rapid change; the latter being in the neighborhood of maxima and minima, or for multi-dimensional problems, in the vicinity of saddle points. Resolution is sought only where it is needed. The costs involving computer time and memory size dictate the maximum number of points that can be economically employed. In planning such a variable mesh size, the following must be kept in mind:

(A) The stretching of the mesh should be defined analytically so that all additional weight coefficients appearing in the equations of motion in the computational space, and their derivatives, can be evaluated exactly at each node. This avoids the introduction of additional truncation errors in the computation.

(B) To assure a maximum value of ΔT, the mathematical grid step, the minimum value of $\Delta\tau$, the physical time increment, should be chosen at each step according to necessity. This means that the minimum value of $\Delta\tau$ must be a function of the pulse function steepness.

(C) The minimum value of $\Delta\tau$ should occur inside the region of the highest gradient which occurs near the pulse peak.

For example, following Moretti's approach,[32] if

$$T=\tanh(\alpha\tau) \tag{11}$$

and α the stretching factor must be larger than 1, the entire semi-axis τ greater than zero can be mapped on the interval $0 < T < 1$

with a clustering of points in the vicinity of $\tau_c = 0$, the center of gravity of the transformation for evenly-spaced nodes in t.

   This mapping brings new coefficients into the equations of motion which are defined analytically and have no singularities. It avoids interpolation at the common border of differently spaced meshes. The computation is formally the same in the "T" space as it was in the "τ" space. Some additional coefficients, due to the stretching function, appear and are defined by coding the stretching function in the main program. A slightly modified stretching function is used in the laser problem. Figure (10) illustrates the transformation and its different dependencies on the particular choice of its parameters.

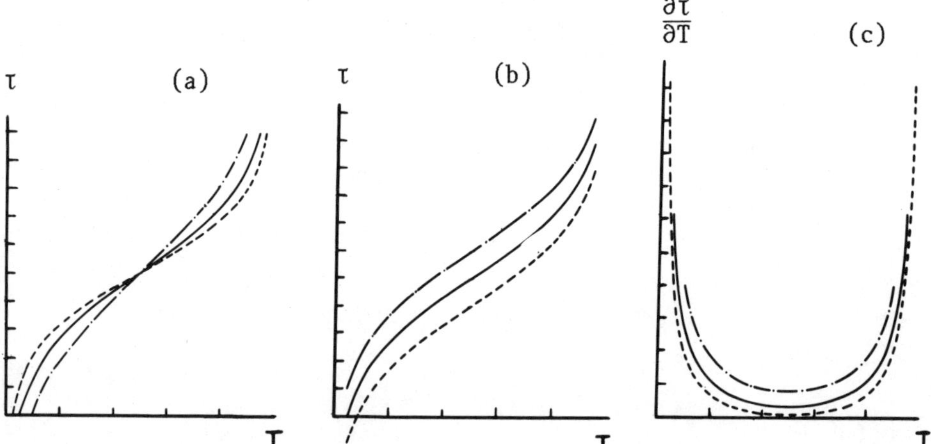

Fig. 10   Dependence of prescribed stretching τ and its derivatives $\partial\tau/\partial T$ on the point densities and the center of transformation versus the uniform computational T.

   The derivative of the mapping function produced by the gradual variation along the "T" axis is also defined analytically. In response, the computational grid remains unchanged while the physical grid (and the associated weighting factors) can change a lot.

   Should one need to study the laser field buildup due to initial random noise polarization (for superfluorescence), or an initial tipping angle (for superradiance), one must use a different stretching[32]. This stretching is like the one defined for treating radial boundary conditions. The mesh points are clustered near the beginning (small τ); their density decreases as τ increases.

V.   ADAPTIVE STRETCHING IN TIME

As the energy continues to shift back and forth between the field and the medium, the pulse velocity is modified disproportionately across the beam cross-section. This retardation/advance phenomenon in absorber/amplifier can cause energy to fall outside the temporal window. Also, due to nonlinear dispersion, various portions of a pulse can propagate with different velocities, causing pulse compression. This temporal narrowing can lead to the formation of optical shock waves. To maintain computational accuracy, a more sophisticated stretching is needed. The accumulation center of the nonlinear transformation is made to vary along the direction of propagation. This adaptive stretching will insure that the redistribution of mesh points properly matches the shifted pulse, Figure (11).

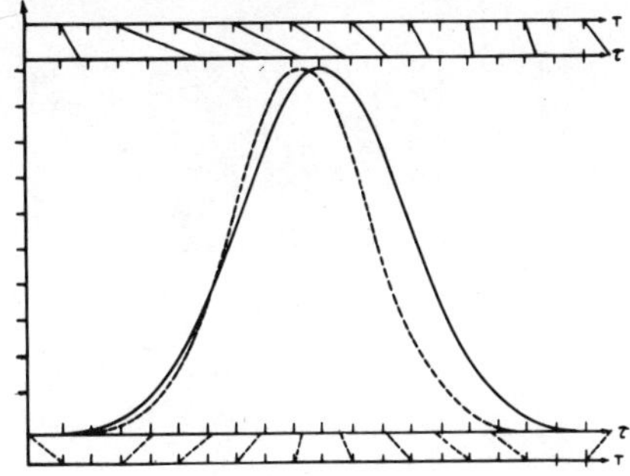

Fig. 11   Adaptive stretching with different centers of transformation.

Here, the transformation from τ to T is applied about a center τ_c which is a function of η. The stretching factor α could also be a function of η.

The field equation is similar to those of Section II, but contains an extra term:

$$-iF\ \nabla^2_{T\rho}e + \partial_\eta e + \frac{\partial e}{\partial T}\,[-\frac{\partial T}{\partial \tau}]\tau_c\ \frac{d\tau_c}{d\eta} = \wp \tag{12}$$

The role played by the time coordinate is different: an explicitly time-dependent term is now included.

VI.   REZONING

The main difficulty in modeling laser propagation through inhomogeneous and nonlinear media stems from the difficulty of pre-assessing the mutual influence of the field on the atomic dynamics and vice versa. Strong beam distortions should occur based on a perturbational treatment of initial trends. One must normalize out the critical oscillations to overcome the economical burden of an extremely fine mesh size. To insure accuracy and speed in the computation, a judicious choice of coordinate systems and appropriate changes in the dependent variables, which can either be chosen a priori or automatically redefined during the computation, must be considered (Figure (12))[33-40].

This coordinate transformation alters the dependent variables and causes them to take a different functional form. The new dependent variables are numerically identical to the original physical amplitudes at equivalent points in space and time.

The requirements of spatial rezoning will be satisfied by simultaneously selecting a coordinate transformation (from the

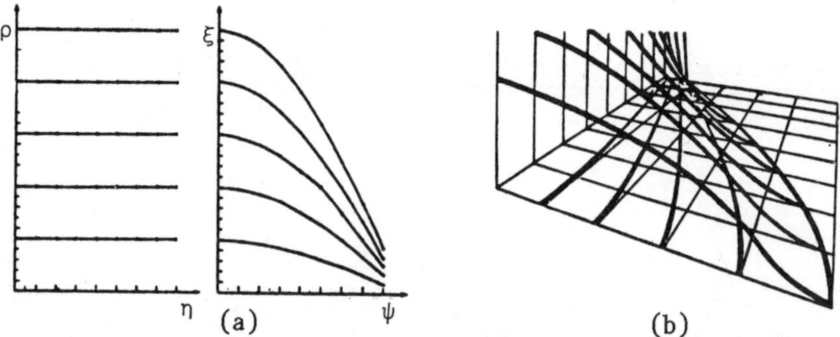

Fig. 12. The concepts of prescribed rezoning are shown in Graph (a); Graph (b) is a close-up of the nonuniform mapped grid of Fig. 2(b).

original coordinates ρ and η to new coordinates ξ and z) and an appropriate phase and amplitude transformation. The chosen function transformation will share the analytical properties of an ideal Gaussian beam propagating in a vacuum.

Since the parameter a, the measure of the transverse scale, shrinks or expands as the beam converges or diverges, it is logical to require the transverse mesh to vary as "a" varies. However, to assure stability and convergence, the ratio $[\Delta\eta/(\Delta\rho)^2]$ must be defined according to the chosen Fresnel number and it must be kept constant throughout the calculation. Accordingly, a new axial variable, z, must be introduced to keep this parameter constant as ρ varies. This should increase the density of η planes around the focus of the laser field where the irradiance sharply increases in magnitude causing a more extensive and severe field-material interaction to occur.

If the quadratic phase and amplitude variation are removed from the field and polarization envelopes, the new field equation varies more slowly than its predecessor; thus, the numerical procedure allows one to march the solution forward more economically by using larger meshes.

VII. ADAPTIVE REZONING

The foregoing concepts may be generalized by repeating the simple coordinate and analytical function transformations along the direction of propagation at each integration step. Figure (5) and graphs (13a) and (13b) illustrate this self-adjusted mapping in planar and isometric graphs.

The feasibility of such automatic rezoning was demonstrated by Moretti in his conformal mapping of supersonic flow calculations[34], and by Hermann and Bradley in their CW analysis of thermal bloom-

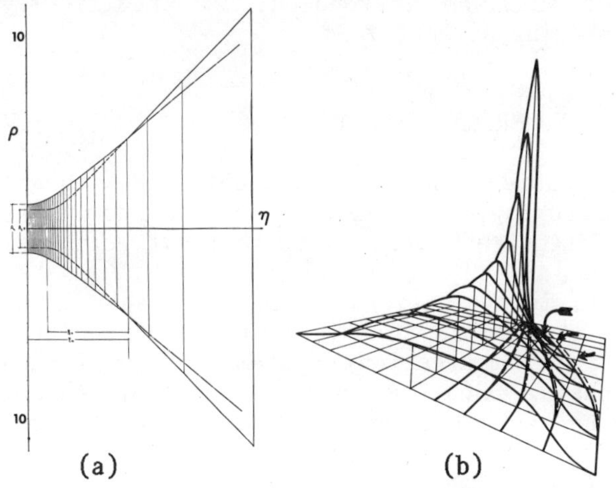

Fig. 13. Graph (a) illustrates the self-adjusted rezoned grid; Graph (b) shows the usefulness of adaptive two-dimensional mapping through isometric representation of the field fluency.

ing[35]. In particular, the change of reference wavefront technique consists of tracking the actual beam features and then readjusting the coordinate system. The new axial coordinate z is defined as before. Previously, the center of the transformation where the radial mesh points were most tightly bunched was at the focus ($z = \eta = 0$). Now the transformation is defined in terms of an auxiliary axial variable z_ξ as a function of z, which is calculated adaptively, in a way that reflects and compensates the changing physical situation.

In this adaptive rezoning scheme, the physical solution near the current z plane is described better by a Gaussian beam of neck radius $a_{\xi 0}$ whose point is a distance z_ξ away than by an initially assumed Gaussian beam with parameters a_0 and z. In addition, to

remove the unwanted oscillations, new dependent variables are
introduced without quadratic and quartic radial dependence in the
phases of the pulse and polarization envelopes. By minimizing the
local field phase gradient the relationship between the auxiliary
z_ξ and z is obtained. Thus the rezoning parameters are determined
appropriately from the local field variable at the preceding plane,
so the new variable at this present point has no curvature. Note
that the new equation varies less in its functional values than the
original. The numerical computation is significantly improved.
Notably, the instantaneous local rezoning parameters of the quadratic wavefront are determined by fitting the calculated phase of
the local field to a quartic in the nonuniform radius. More specifically, the intensity-weighted square of the phase gradient integrated over the aperture is minimized. Consequently, the curvature
at the highest intensity portion of the beam contributes the most.
Various moment integrals of the local field variable and the local
transverse energy current will be introduced, to specifically
evaluate the adjustable rezoning parameters.

VIII.   NUMERICAL RESULTS

This section outlines basic results in SIT, obtained with and
without rezoning and stretching, and illustrates why the more
sophisticated techniques required less computational efforts.

The first part of this investigation led to the discovery of
new physical phenomena which promise to have significant applications for proposed optical communications systems. It had been
shown that spontaneous focusing can occur in the absence of lenses,
and that the focusing can be controlled by varying the medium parameters. The second part of this analysis dealt with amplifiers.

The dependence of the propagation characteristics on the Fresnel number F^{-1} associated with an effective medium length, on the
on-axis input pulse "area," on the relaxation times and on the
off-line center frequency shift, has been studied. Furthermore,
particular care was exercised to ensure a perfectly smooth Gaussian
beam (see Figure (10)) thereby eliminating any possibility of
small-scale, self-focusing buildup[48].

The time-integrated pulse "energy" per unit area,
$\int_0^\tau |e(\rho,\eta,\tau')|^2 d\tau$, the fluency, is plotted for various values of the
transverse coordinate, as a function of the propagation distance
(see Fig. 14).

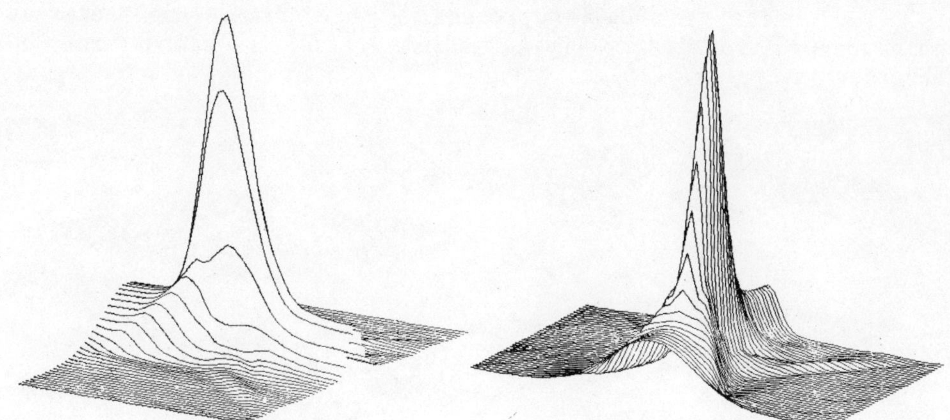

Fig. 14. The longitudinal orientation shown in the left-hand figure illustrates the gradual boosting mechanism that field energy experiences as it flows radially towards the beam axis (while η increases). The second orientation displays the severe beam distortion in its cross-section as a function of η.

The three-dimensional numerical calculations substantiate the physical picture based on a perturbational study of the phase evolution[10,15]. It could be visualized using selected frames from a computer movie simulation of the numerical model output data. In the left-hand curves of Figure (15) the transverse energy current is isometrically plotted against the retarded time for various transverse coordinates at four specific regions of the propagation process: (a) the reshaping region where the perturbation treatment holds; (b) the buildup regions; (c) the focal region; and (d) the post-focal region. The field energy is displayed for the specific regions in the right-most curves of Fig. (15). A rotation of the isometric plots is displayed in Figure (16), to emphasize the radially dependent delay resulting from the coherent interaction. Positive values of the transverse energy current correspond to outward flow, and negative values to inward flow. The results of the reshaping and buildup regions in Figures (15) and (16) agree with the physical picture related to the analytic perturbation discussed elsewhere.

The burn pattern, iso-irradiance level contours (against τ and ρ) for different propagation distances are shown in Figure (17). Severe changes in the beam cross-section are taking place as a function of the propagating distance. At the launching front, the beam is smooth and symmetrical; as the beam propagates into the nonlinear resonant medium, the effect of the nonlinear inertia takes place.

The general format for presenting three-dimensional coherent pulse propagation in amplifying medium will be the same as for the absorber (see Figs. (18) to (21)).

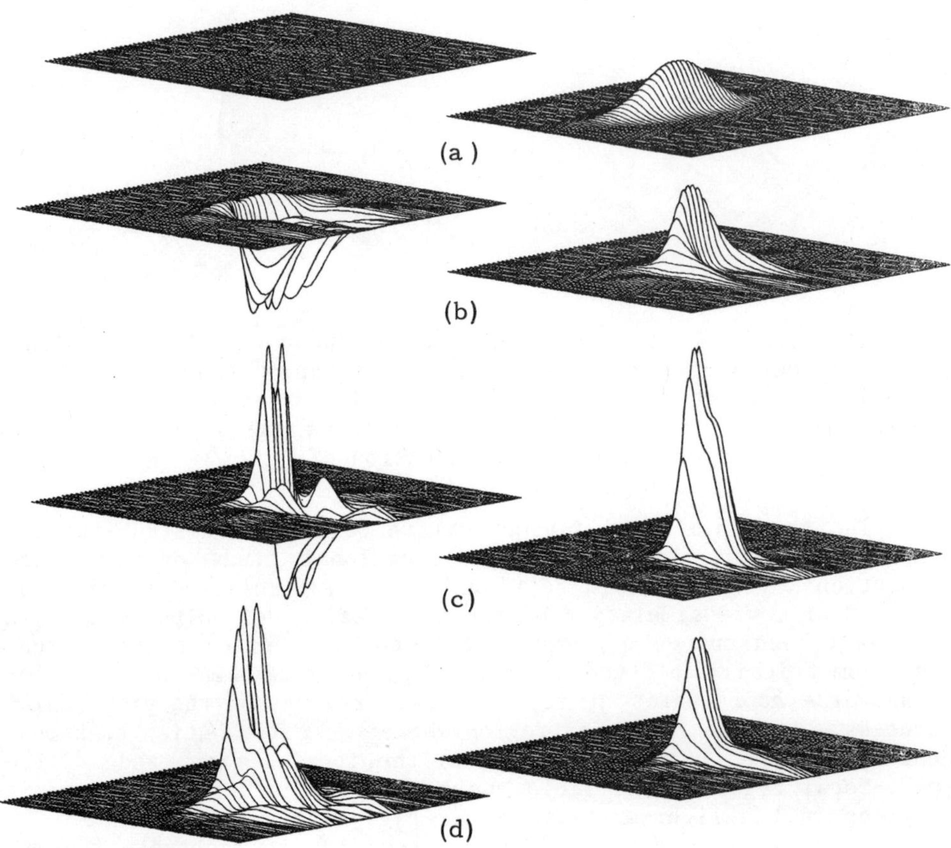

Fig. 15. Isometric plots of the absorber field energy and transverse energy flow, against the retarded time for various transverse coordinates at the four regions of interest.

IX.   TRANSVERSE EFFECTS IN SUPERFLUORESCENCE

With the help of Gibbs, the outstanding question dealing with the strong reduction (and elimination) of ringing observed in the low-density Cs [2] experiment from the amount predicted in the one-dimensional calculations [1(b)] was resolved. This was accomplished by developing a rigorous two-dimensional theory of Burnham-Chiao ringing [1b] and superradiance and superfluorescence (SF) in a pre-excited thick medium using a semi-classical formulation [1e] which includes one-way propagation effects as in SIT. The initiation of the SF emission process is characterized by a tipping angle θ_R. When the small signal field gain $\alpha_{eff}L/2$ (or equivalently, the characteristic radiation damping time τ_R of the collective atomic system) is sufficiently large, θ_R, the ratio of the length L to the coherence length L_c, and the Fresnel number $\mathcal{F}$ (equal to area/λL) completely characterize the system behavior. However, L/L_c is <u>not</u> a critical parameter as predicted by the mean field theory.

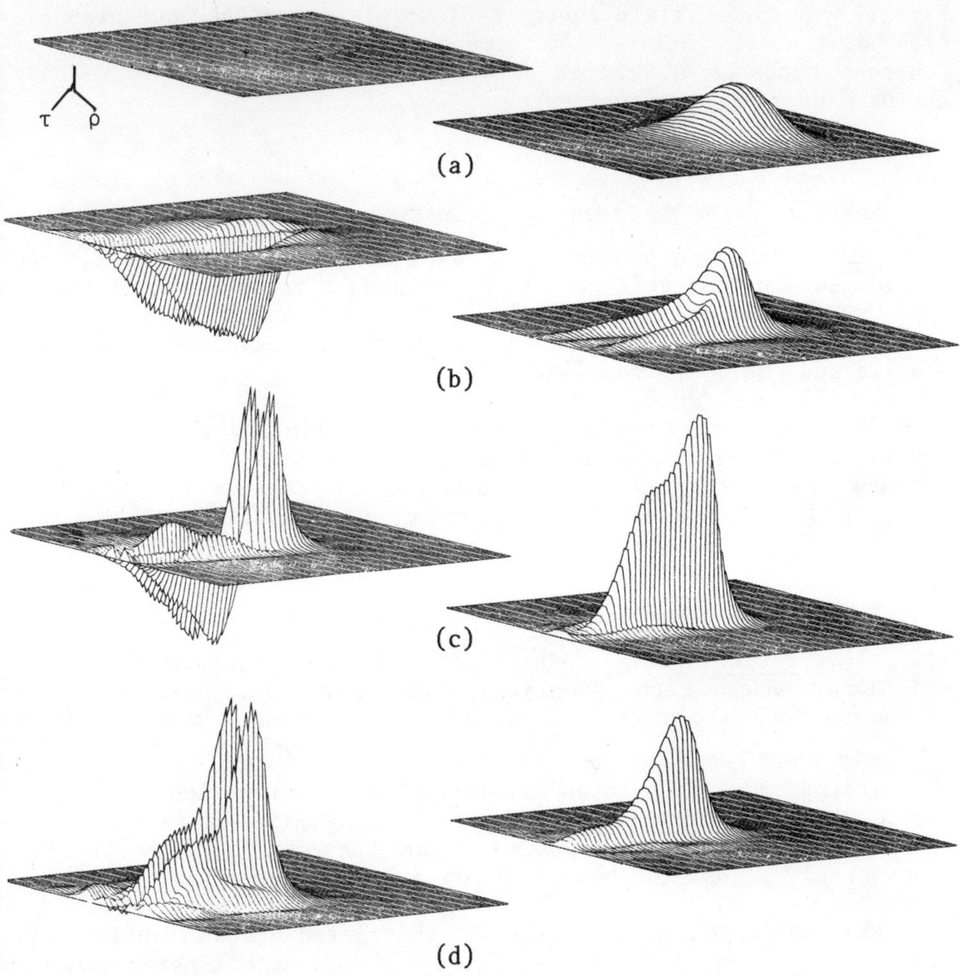

Fig. 16   Isometric plots of the absorber field energy and transverse energy flow profile for various time slices at the four regions of interest.

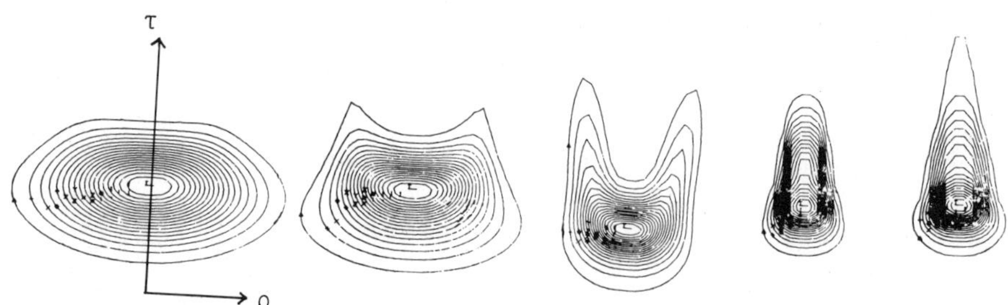

Fig. 17. Absorber field energy contour plots for the four propagation distances. Notice the temporal delay associated with the
coherent exchange of energy between light and matter, as well as
the beam cross-section narrowing.

Neither the mean-field approximations[1d], nor the substitution
of a loss term to account for diffraction coupling[1c,20d], are
considered; instead self-consistent methods similar to those developed for SIT studies are adopted[39,46]. The numerical simulation
takes fully into account both propagation and transverse (spatial
profile and Laplacian coupling) effects.

The previously reported pronounced SF ringing for plane-wave
simulation is reproduced for uniform input profile. The reduction
of ringing is studied for various radial profiles for the gain
$g_R = \alpha_{eff}[c\tau_R]$ (equivalently, the population inversion) and the small
input pulse area θ_R[11-13].

The ringing reduction can be explained by two physical mechanisms: (a) <u>a shell (ring) model</u>[32(d)]: spatial averaging of uncoupled planar modes, each associated with a particular shell and subjected to both a distinct θ_R and a radiation time. Radial averaging by a Gaussian gain profile of very large $\mathcal{J}$ eliminates most of
the ringing, resulting in an asymmetric pulse with a long tail; and
(b) <u>a rigorous diffraction coupling</u>: through the Laplacian term,
the adjacent shells interact, causing the field energy to flow
transversely across the beam from one region to another.

When diffraction coupling is considered concomitantly with
radial <u>variations</u> of θ_R and g_R (i.e., of τ_R), the ringing is more
subdued (see Fig. (23)). In other words, reducing $\mathcal{J}$ of a Gaussian
profile does reduce the asymmetry (in better agreement with the Cs
data) since the outer beam portions are stimulated to emit earlier

(a)

(b)

(c)

(d)

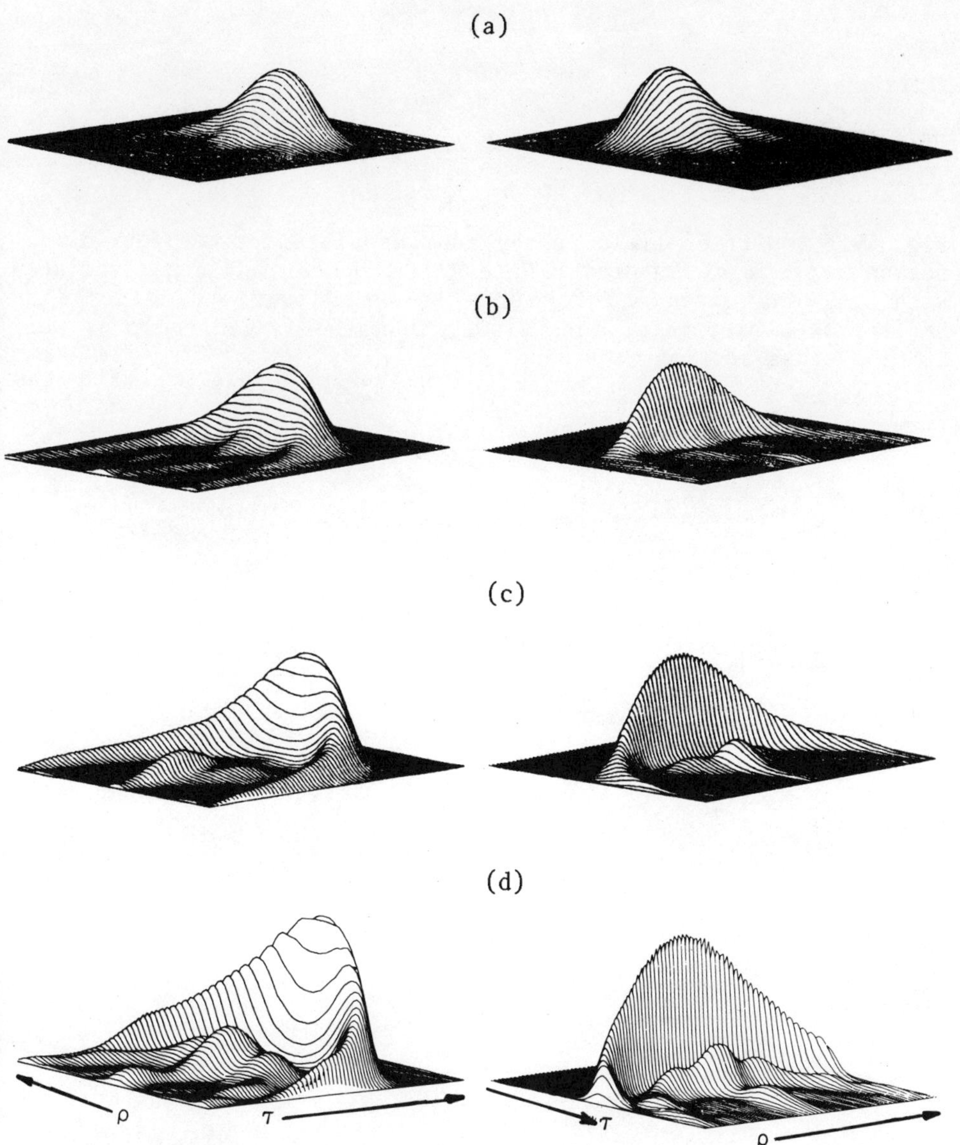

Fig. 18. Isometric plots of the amplifier field energy as a function of τ and ρ for two orientations $\pi/2$ apart at four locations along the propagation direction.

by diffraction from the inner portion. Thus, the effect of the Laplacian coupling is small for large $\mathcal{J}$ but becomes progressively greater at about $\mathcal{J} \leq 1$.

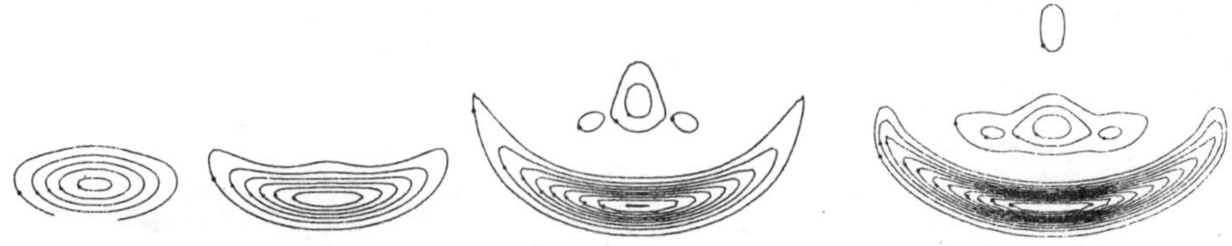

Fig. 19. Amplifier field energy contour plots for the four propagation regions of interest. Note the temporal advance associated with coherent exchange of energy between light and matter (the smaller area propagates more slowly than the larger one), as well as beam cross-section expansion.

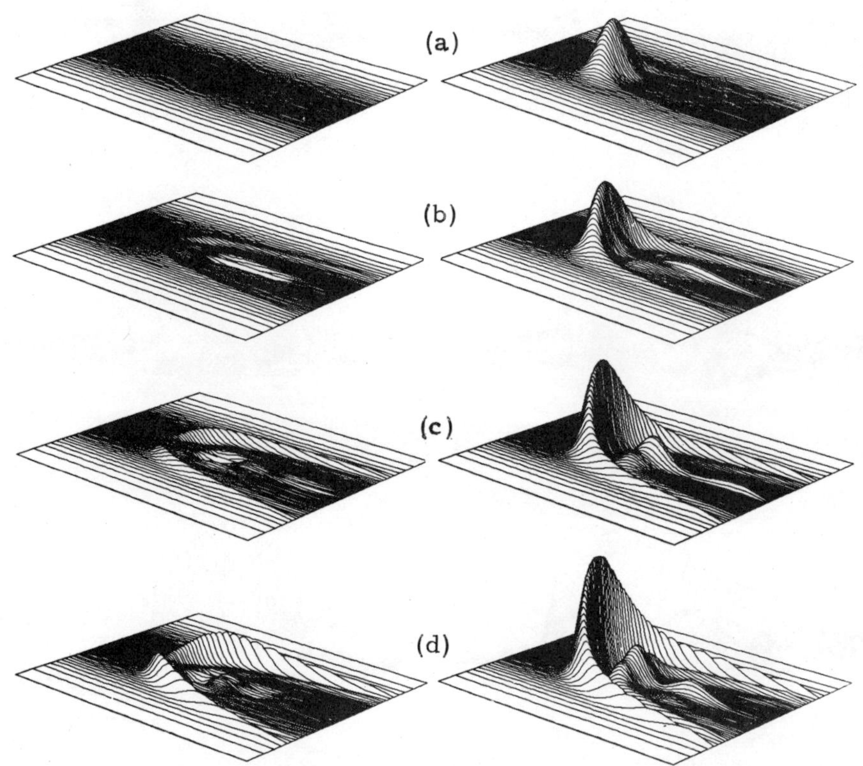

Fig. 20. Isometric plots of amplifier field energy and transverse energy flow against retarded time for various transverse coordinates at four propagation regions studied for absorbers. Stretched radial coordinate was adopted for proper accounting of transverse boundary condition. When these results are compared with those for an absorber, it is evident that a focusing phase is not restricted to the absorber, but develops also for the secondary pulses in amplifying media.

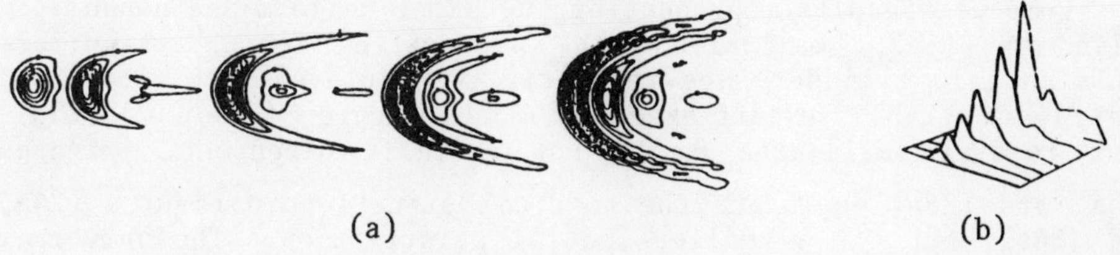

(a)   (b)

Fig. 21. Amplifier field energy contour plots for four propagation regions of interest with stretched radial coordinates. No severe reflection or abrupt variation in the field energy, at the wall boundary, is observed. The enhancement of diffraction by pre-excited two-level medium is clearly evident.

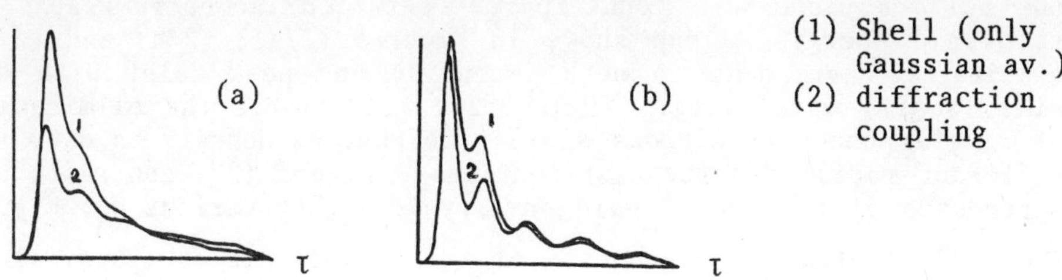

Fig. 22 Contrast the time dependence of the energy after integrating over ρ for the shell model (where θ_R and τ_R are both radially dependent) and the diffraction model (where the Laplacian coupling is rigorously present) for two population inversions: (a) Gaussian $g = g_0 \exp[-\rho^2]$, and (b) saturable inversion $g = g_0$ for $\rho < \rho_b$; $g = g_0 \exp[-\rho^2]$ for $\rho_b < \rho < \rho_{max}$.

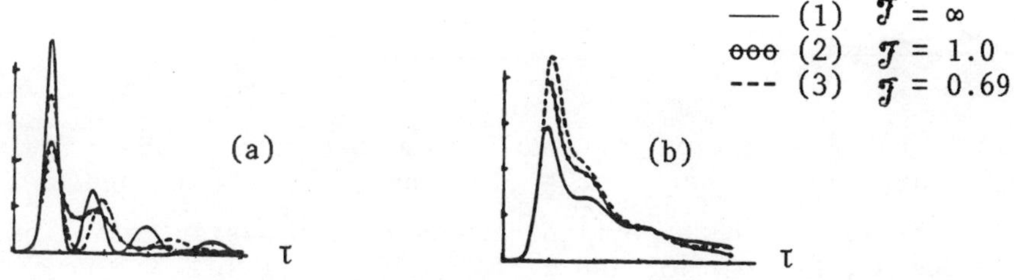

Fig. 23. Total energy per atom as a function of time with $\mathcal{J}$ as the labeling parameter. $\tau_R = 0.046$ ns and $L/L_c = 1.95$. $\theta_R = 3 \times 10^{-3}$ for all radii. (a) Superfluorescence of uniform cylinder or small-area pulse propagation through uniform gain cylinder; (b) Uniform small-area pulse propagation through Gaussian gain medium.

Computer results representing the SF of uniform and nonuniform cylinders (i.e., small-area pulse propagating through a uniform Gaussian gain cylinder) are respectively displayed in Figure (24a) and Figure (24b) for different $\mathcal{J}$. In Figures (25a) and (25b), this initial small-area θ_R is now radially dependent. Figures (26a) and (26b) duplicate the physical situation in Figures (24a) and (24b), but for a smaller initial polarization. The universal superfluorescence scaling law is seen not to hold; the calculated pulse length is much more sensitive to the magnitude of θ_R in the transverse case than it is in the planar case.

The ringing predicted by this two-spatial-dimensional theory agrees more with experimental observations than that predicted by the uniform plane-wave counterpart. Detailed isometric graphs of the field energy buildup show, in Figures (27a), (27b) and (27c) qualitative agreement in peak intensity and peak delay with the ring (shell) model [1c]. Figure (28) illustrates the elimination of ringing under conditions similar to the low-density Cs data for different radial density distributions. Figure (29) contrasts the dependence of the radial gain on a typical $\mathcal{J}$ by various θ_R; Figure (30) illustrates the dependence of the radial gain on a typical θ_R by different $\mathcal{J}$. Figure (31) shows the effect of varying τ_R on this output intensity. Various small-scale ripples were introduced in the gain profile (see Fig. 31).

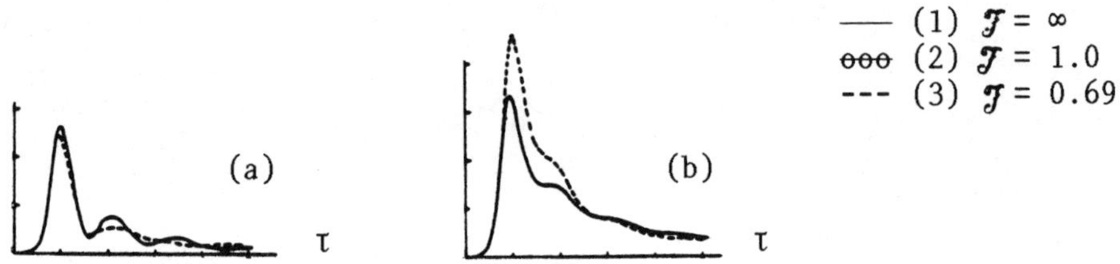

Fig. 24. (a) Propagation of small-area Gaussian profile pulse through uniform cylinders (τ_R = 0.046 ns, L/L_c = 1.35 and θ_R = 3×10^{-3} on-axis). (b) Superfluorescence with Gaussian radial gain (τ_R = 0.046 ns, L/L_c = 1.35 and θ_R = 3×10^{-3} on-axis).

Ringing is largely removed by a gain medium of $\mathcal{J}$ = 1, resulting in an asymmetric output pulse with a long tail. It now seems that a larger θ_R, see Fig. (33a) (unlikely, according to measurement of feedback effects and estimates of Raman effects during the excitation pulse[2d]), or smaller $\mathcal{J}$ (perhaps 0.4 consistent with the range $0.35 < \mathcal{J} < 1.39$ of ref. 1(a) which used a 1/e rather than a half width half maximum (HWHM) definition of r_p), see Fig. (33b),

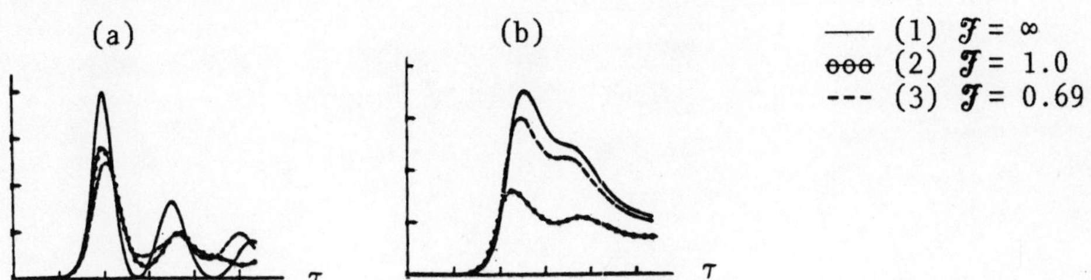

Fig. 25. Same parameters as in Fig. 23 but with a smaller $\theta_R = 10^{-4}$: (a) Small area propagation in a uniformly inverted cylinder. (b) Small-area propagation in a Gaussian inversion cylinder.

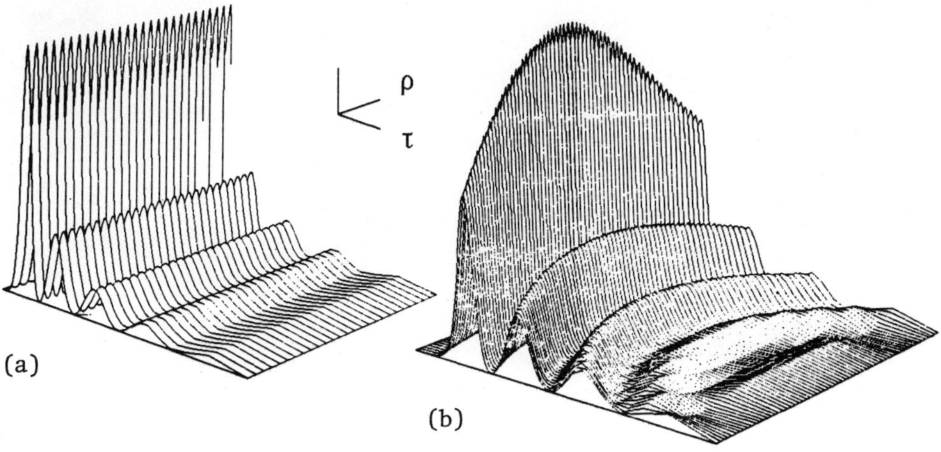

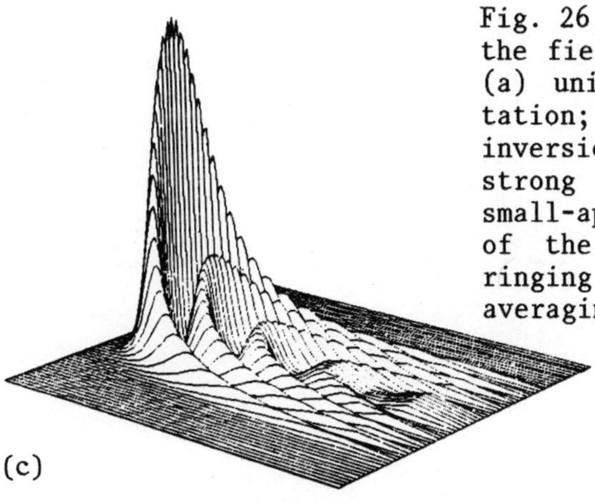

Fig. 26. Isometric representation of the field energy versus ρ and τ, for (a) uniform inversion and pre-excitation; (b) radial θ; (c) Gaussian inversion profile. Notice that strong ringing would be seen by a small-aperture detector in the center of the beam although very little ringing is in evidence after radial averaging.

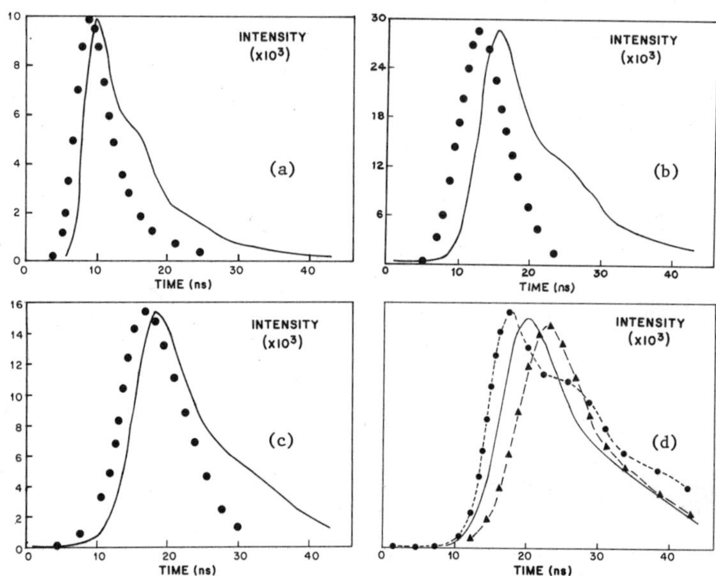

Fig. 27.   Comparison of pulse shapes for situations where L/L_c is similar to the low density Cs. Relaxation terms were not included in this analysis. Note the asymmetry associated with an atomic beam of $\mathcal{J} = 1$.   (a) $n = 1.9 \times 10^{11}$ cm^{-3}; $\theta_o = 2.64 \beta 10^{-4}$; (b) $n = 18.24 \beta 10^{10}$ cm^{-3}; $\theta_o = 1.37 \times 10^{-4}$; (c) $n = 11.9 \times 10^{10}$; $\theta_o = 1.69 \times 10^{-4}$; (d) $n = 8.75 \times 10^{-4}$; $\theta_o = 1.96 \times 10^{-4}$. Time is measured in nsec.

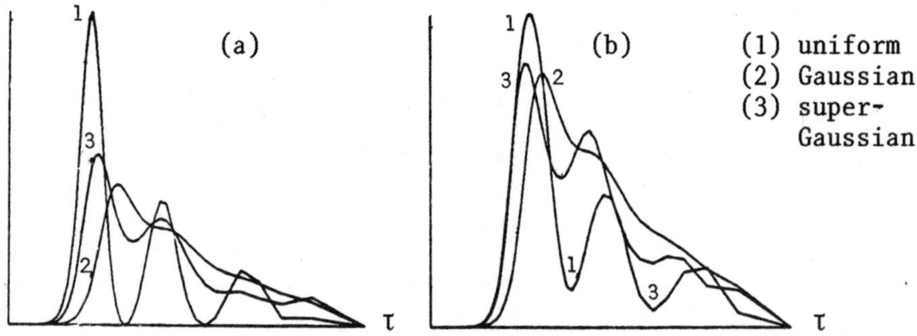

Fig. 28.   Contrast of the total energy per unit atom (versus time) for different radiation damping time τ_R for a chosen $\mathcal{J} = 0.7$ and a uniform $\theta_R = 3 \times 10^{-3}$ (for different inversion profiles.

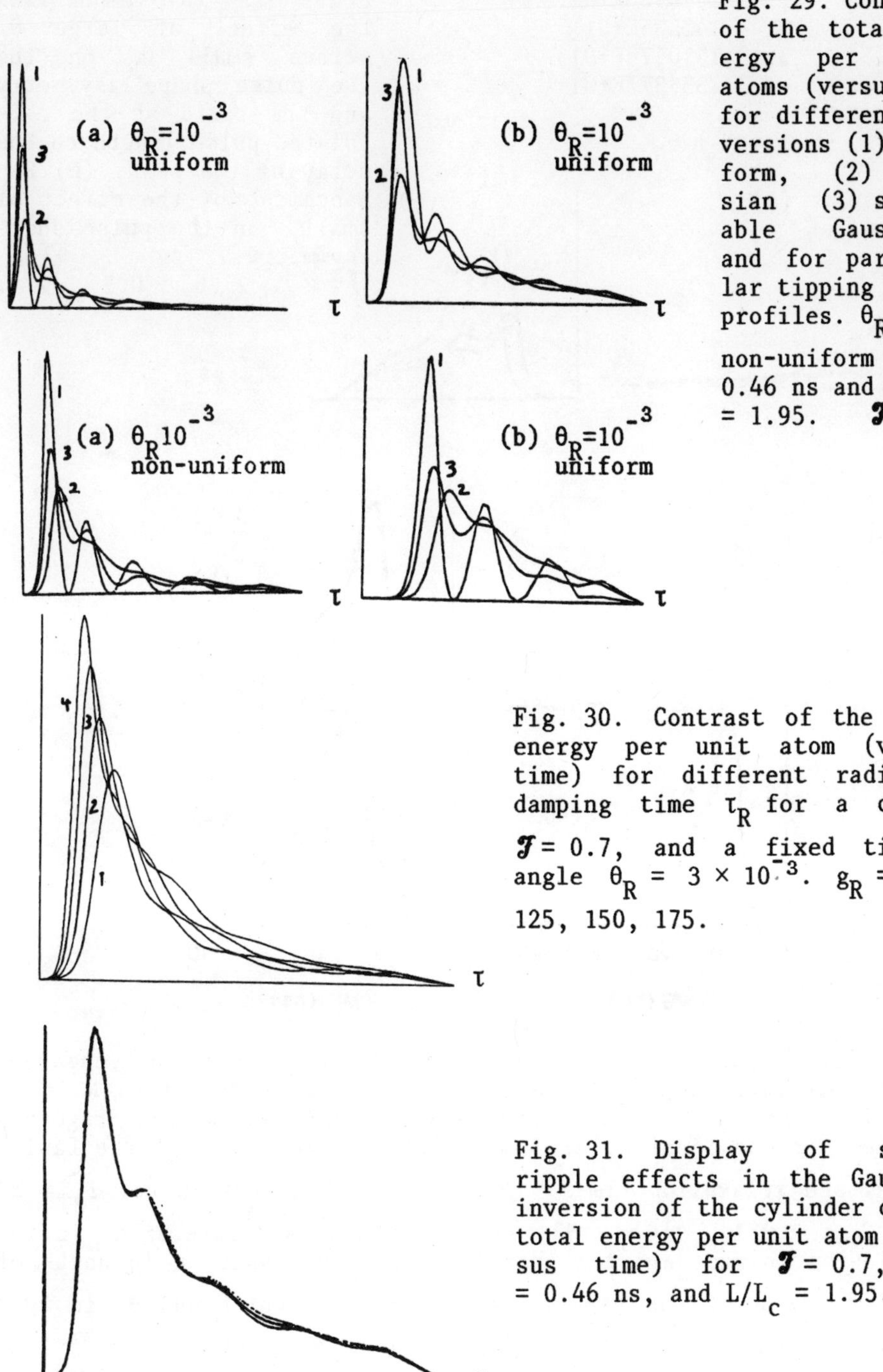

Fig. 29. Contrast of the total energy per unit atoms (versus time) for different inversions (1) uniform, (2) Gaussian (3) saturable Gaussian, and for particular tipping angle profiles. $\theta_R=10^{-3}$ non-uniform $\tau_R = 0.46$ ns and $L/L_c = 1.95$. $\mathcal{J} = \infty$.

Fig. 30. Contrast of the total energy per unit atom (versus time) for different radiation damping time τ_R for a chosen $\mathcal{J} = 0.7$, and a fixed tipping angle $\theta_R = 3 \times 10^{-3}$. $g_R = 100$, 125, 150, 175.

Fig. 31. Display of small-ripple effects in the Gaussian inversion of the cylinder on the total energy per unit atom (versus time) for $\mathcal{J} = 0.7$, $\tau_R = 0.46$ ns, and $L/L_c = 1.95$.

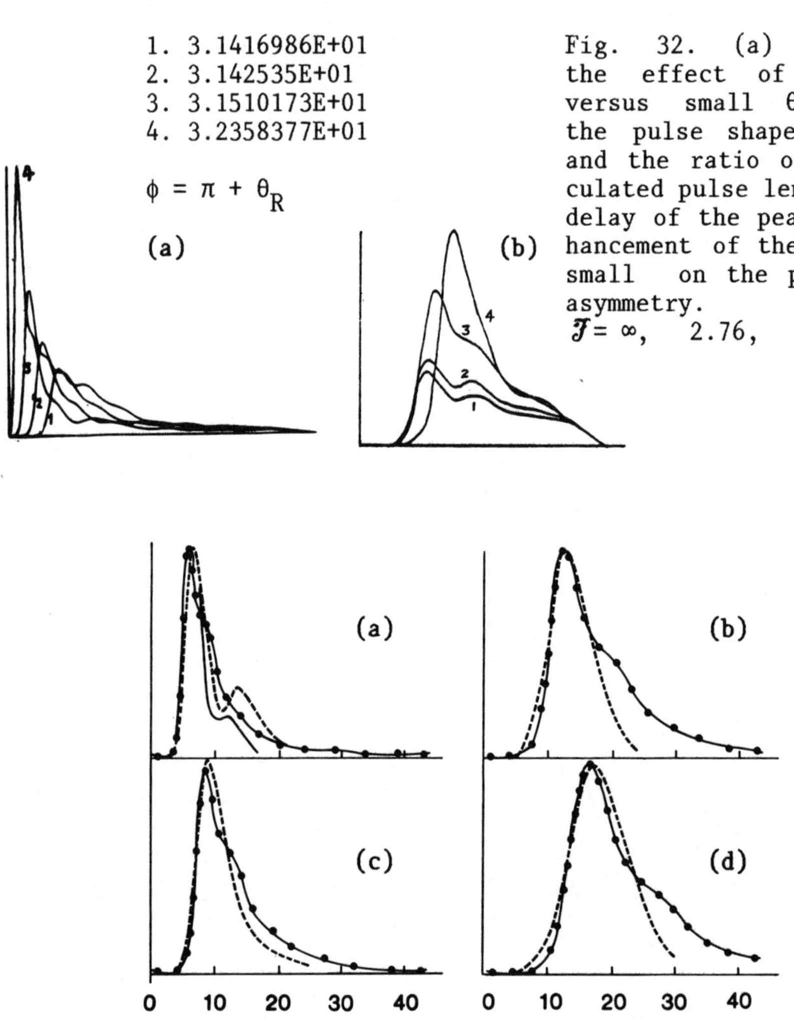

Fig. 32. (a) Emphasizes the effect of large θ_R versus small θ_R on the the pulse shape asymmetry and the ratio of the calculated pulse length to the delay of the peak. (b) Enhancement of the effect of small on the pulse shape asymmetry. $\mathcal{J} = \infty$, 2.76, 0.7, 0.4.

Fig. 33. Comparison of experimental and three-dimensional theoretical superfluorescence pulse shape for several densities N in an atomic beam of 2.0 cm length. The model encompasses rigorous radial dependence of N, τ_R and θ_R, diffraction (through the Laplacian) and relaxation times. $\mathcal{J} = 1$, L = 2 cm, $T_1 = 70$ ns, $T_2 = 80$ ns, $\lambda = 2.931\mu$, $\tau(0) = 551$ nsec, Gaussian and inversion; in the following columns are the on-axis inversion density n in units of 10^{11} cm^{-3}, n of the experiment in the same units and θ_o in 10^{-4} radians: (a) 3.1, 1.9, 1.07; (b) 3.1, 7.6, 1.37; (c) 1.2, 3.8, 1.69; (d) 0.885, 3.1, 1.96.

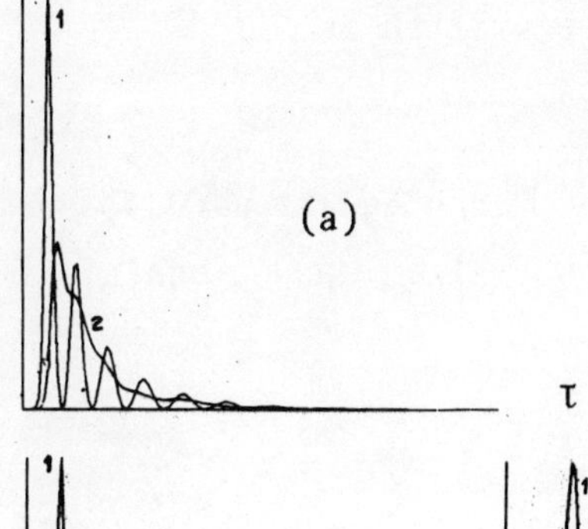

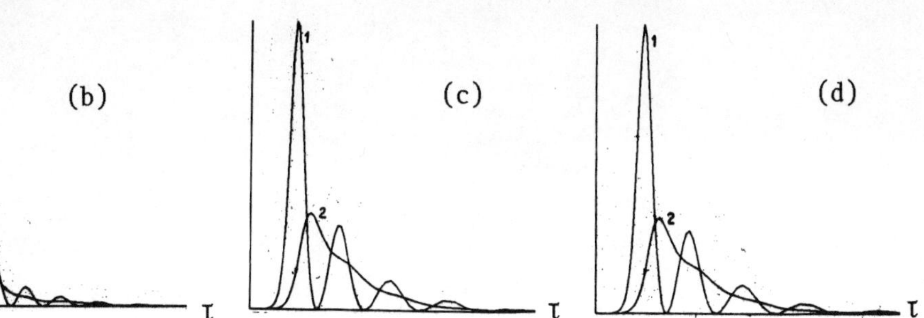

Fig. 34. Comparison of planar waves (curve 1) with three-dimensional calculations (curve 2) of the superfluorescence for the Cs experimental data. Note the lack of agreement between the two theories with respect to the ringing while much consistency occurs between diffraction calculations and experimental observations.

is needed to reduce the asymmetry and pulse width. But when relaxation terms are also included in the analysis and the densities are adjusted within quoted experimental uncertainties, a rather good agreement, (see Fig. (34)) is obtained between theory and experiments for a unity $\mathcal{J}$. These radial effects explain why the observed ringing in superfluorescence is less than that predicted by plane-wave simulations (see Fig. 34). Extensions of the present simulations to two-way propagation and random fluctuation of the tipping angle are planned. The agreement with experimental observations should be improved. [Recently, Bonifacio et al[1d] also reported the suppression of the ringing by using coupled-mode mean-field theory. However, their model does <u>not</u> encompass the propagational effects substantiated by both experimental observation and rigorous three-dimensional Maxwell-Bloch analysis.]

X.   FLUID DESCRIPTION

Consider the polar representation of the field

$$e = A \exp (+i\phi) \tag{13}$$

with A and ϕ real amplitude and phase. Also let the nonlinear polarization of the RHS of equation (1) be written as

$$P^{NL} = (\chi_R + i \chi_I)e + \chi_{NL}e, \tag{14}$$

where χ_R and χ_I are real functions of A. Using equation (13), one gets from equation (1) the transport and the eikonal equations $(n_o = k_o c/\omega_o)$

$$K_o \frac{\partial}{\partial \eta} A^2 + \nabla_T \cdot [A^2 \nabla_T \phi] = - \frac{4\pi\omega_o^2}{c^2} \chi_I A^2, \tag{15}$$

$$2k_o \frac{\partial}{\partial \eta} \phi + (\nabla_T \phi)^2 - \left[\frac{A \cdot \nabla_T^2 A}{A^2}\right] = \frac{4\pi\omega_o^2}{c^2} \chi_R \tag{16}$$

The transport equation (15) expresses conservation of beam energy over the transverse plane. When $\chi_I = 0$, total power is conserved along the direction of propagation. The eikonal equation (16) describes the evolution of the surface of constant phase. It has the form of the Hamilton-Jacobi equation for the two-dimensional motion of particles having unit mass and moving under the influence of a potential[49] given by

$$V = - \frac{1}{2k_o^2} \cdot (\nabla_T^2 A) A^{-1} - \frac{2\pi}{n_o^2} \chi_R$$

if $k_{oz}z$ is regarded as time coordinate and $k_{ox}x$, $k_{oy}y$ as spatial coordinates. Furthermore, if one adopts A^2 and $\nabla_T\phi$ as new dependent variables, the equations of motion become similar to the continuity and momentum transport equations of ordinary hydrodynamics[25,26]. By defining

$$\underset{\sim}{v} = k_o^{-1} \nabla_T \phi, \quad \text{and} \tag{17}$$

$$\rho = A^2 \tag{18}$$

and supposing $\chi_I = 0$, equations (15) and (16) can be written as

$$\frac{\partial \underset{\sim}{v}}{\partial \eta} + (\underset{\sim}{v} \cdot \nabla_T)\underset{\sim}{v} = \frac{1}{2k_o} \nabla_T[\rho^{-1/2}(\nabla_T^2 \sqrt{\rho})] + \frac{\gamma_2}{k_o} (\nabla_T\rho) \tag{19}$$

$$\frac{\partial \rho}{\partial \eta} + \nabla_T \cdot (\rho\underset{\sim}{v}) = 0. \tag{20}$$

These equations are the momentum and continuity transport equations of a fluid with a pressure

$$P = (\nabla_T^2 \sqrt{\rho})/\sqrt{\rho})). \tag{21}$$

It should be emphasized that this pressure depends here solely on the "fluid density" and not on the "velocity". Equation (19) and (20) can be rearranged into

$$\frac{\partial}{\partial \eta}(\rho \underset{\sim}{v}) + \nabla_T \cdot (\rho \underset{\sim}{v}\underset{\sim}{v}) = \frac{1}{2k_o^2} [\, \tfrac{1}{2}(\nabla_T^2 \rho) \underset{\sim}{I} -$$

$$- \frac{1}{2\rho}(\nabla_T\rho)(\nabla_T\rho)] + \frac{\gamma_2}{k_o} \rho(\nabla_T\rho), \tag{22}$$

where $\underset{\sim}{I}$ is the unit tensor.

XI. EQUATIONS OF MOTION FOR OPTICAL BISTABILITY

In the slowly varying envelope approximation, the dimensionless field-matter equations* are

$$-iF\nabla_T^2 e^+ + \frac{\partial e^+}{\partial \tau} + \frac{\partial e^+}{\partial z} = +g^+ < P^* \exp(ikz)> \tag{23}$$

$$-iF\nabla_T^2 e^- + \frac{\partial e^-}{\partial t} - \frac{\partial e^-}{\partial z} = +g^- <P \exp(+ikz)> \tag{24}$$

with g^+, g^- as the nonlinear form of the gain experienced by the forward (e^+) and backward (e^-) traveling waves associated with the pump. The quantities in the R.H.S. undergo rapid spatial variations; $<\cdots>$ spatial average of these quantities with a period of half a wavelength

$$\frac{\partial P}{\partial t} + (-i\Delta\Omega) + \tau_2^{-1})P = + \{W(e^+ + e^-)\} \tag{25}$$

$$\frac{\partial W}{\partial t} + \tau_1^{-1}(W^e - W) = -\tfrac{1}{2}(P^+ + P^-)(e^+ + e^-) \tag{26}$$

Equivalently,

$$\frac{\partial P}{\partial t} + (-i(\Delta\Omega)+\tau_2^{-1})P = W[e^+\exp(-ikz)+e^-\exp(+ikz)] \tag{27}$$

---

*As an aside, the nonlinear interface bistability effect[4(e)], though potentially important, is not considered.

$$\frac{\partial W}{\partial t} + \tau_1^{-1}(W^e - W) = \frac{1}{2}(Pe^{+*}\exp(ikz) + Pe^{-*}\exp(-ikz) + \underline{c.c.}) \qquad (28)$$

with

$$e^{\pm} = (2\mu\tau_p/\hbar)e^{\pm} \qquad (29)$$

$$P = (p'/2\mu), \qquad (30)$$

$$E^{\pm} = \mathrm{Re}\{e^{\pm}\exp[i(\omega t \mp kz)]\} \qquad (31)$$

and

$$P = \mathrm{Re}\{i\ p'\ \exp(i\omega t)\} \qquad (32)$$

The complex field amplitude $e^{\pm}$, the complex polarization density p and the energy stored per atom W are functions of the transverse coordinate

$$\rho = r/r_p, \qquad (33)$$

the longitudinal coordinate

$$z = \alpha_{eff}z' \qquad (34)$$

and the physical time

$$\tau = t/\tau_p. \qquad (35)$$

In the standing-wave problem, the two waves are integrated simultaneously along the physical time, as contrasted to S.I.T. retarded time.[50] Otherwise the physical parameters and variables have the same meaning.

The presence of opposing waves leads to a quasi-standing wave pattern in the field intensity over a half-wave length. To effectively deal with this numerical difficulty one decouples the material variables using Fourier series[18,19] namely,

$$P = \exp(-ikz)\sum_{p=0}^{\infty} P_{(2p+1)}^{+}\exp(-i2pkz) + \exp(+ikz)\sum_{p=0}^{\infty} P_{(2p+1)}^{-}\exp(+i2pkz) \qquad (36)$$

$$W = W_0 + \sum_{p=1}^{\infty} [W_{2p}\exp(-i2pkz) + c.c.] \qquad (37)$$

with W_o a real number. Substituting in the traveling equation of motion, one obtains

$$\partial_\tau P_1^+ + P_1^+/\tau_2 = W_o e^+ + W_2 e^- ; \tag{38}$$

$$\partial_\tau P_3^+ + P_3^+/\tau_2 = W_2 e^+ + W_4 e^- ; \tag{39}$$

$$\cdots \qquad \cdots \qquad \cdots \qquad \cdots$$

$$\partial_\tau P_{(2p+1)}^+ + P_{(2p+1)}^+/\tau_2 = W_{2p} e^+ + W_{2(p+1)} e^- ; \text{ and} \tag{40}$$

$$\partial_\tau P_1^- + P_1^-/\tau_2 = W_o e^- + W_2^* e^+ \tag{41}$$

$$\partial_\tau P_3^- + P_3^-/\tau_2 = W_2 e^- + W_4^* e^+ \tag{42}$$

$$\cdots \qquad \cdots \qquad \cdots \qquad \cdots$$

$$\partial_\tau P_{(2p+1)}^- + P_{(2p+1)}^-/\tau_2 = W_{2p}^* e^- + W_{2(p+1)}^* e^+ \tag{43}$$

$$\partial_\tau W_o + (W_o - W_o^e)/\tau_1 = -\frac{1}{2}(e^{-*} P_1^- + e^{+*} P_1^+ + c.c.) \tag{44}$$

$$\partial_\tau W_2 + W_2/\tau_1 = -\frac{1}{2}(e^{-*}P_1^+ + e^{+*}P_3^+ + e^+ P_1^{-*} + e^- P_3^{-*}) \tag{45}$$

$$\partial_\tau W_{2p} + W_{2p}/\tau_1 = -\frac{1}{2}(e^{-*}P_1^+ + e^+ P_{2p+1}^+ + e^+ P_{2p+1}^{-*} + e^{-*}P_{2p+1}^{-*}) \tag{46}$$

The field propagation and atomic dynamic equation are subjected to the following initial and boundary conditions:

1. INITIAL:

$$\text{for} \qquad t \geq 0$$

$$e^\pm = 0 \tag{47}$$

$$W_o = W_o^e \quad , \tag{48}$$

where W_o^e is a known function to take into account the pumping effects. For S.I.T. or soliton collision

$$P_{(2p+1)}^\pm = 0, \qquad \text{for all } p \tag{49}$$

while for the superfluorescence problem

$$P^{\pm}_{(2p+1)} \tag{50}$$

is defined in terms of an initial tipping angle θ_R.

2. LONGITUDINAL

For $z=0$ and $z=L$: e^+ and e^- are given in terms of a known incident function

$$e_{I0} \tag{51}$$

and

$$e_{IL} \tag{52}$$

of τ and ρ.

If enclosing mirrors delineating the cavity are used in the analysis, one must observe the longitudinal boundary equations

$$e^+ = \sqrt{(1-R_1)}\; e_{I0} + \sqrt{R_1}\; e^- \qquad \text{at } z = 0 \tag{53}$$

$$e^- = \sqrt{(1-R_2)}\; e_{IL} + \sqrt{R_2}\; e^+ \qquad \text{at } z = L \tag{54}$$

where R_1, R_2, $(1-R_1)$ and $(1-R_2)$ are the respective reflectivity and transmitting factor associated with each left and right mirror.

3. TRANSVERSE

For all z and τ $[\partial e^{\pm}/\partial \rho]_{\rho=0}$ and $[\partial e^{\pm}/\partial \rho]_{\rho=\rho_{max}}$ vanish. The previously described transverse boundary conditions (Section II) apply here for each of the fields.

It is noteworthy that the presence of the longitudinal mirrors will enhance the mutual influence of the two beams. Variations in polarization and population over wave-length distances are treated by means of expansions in spatial Fourier series, which are truncated after the third or fifth harmonic. The number of terms needed is influenced by the relative strength of the two crossing beams and by the importance of pumping and relaxation processes in restoring depleted population differences.

XII. CONCEPT OF TWO-WAY CHARACTERISTICS

An easy way to visualize the mutual influence of the two counter-propagating beams is to imagine their respective information carriers in the traveling wave description.

For a light velocity normalized to unity ($c/n = 1$), by introducing

$$\xi = \frac{1}{2}(t-z) \qquad \text{and} \qquad \eta = \frac{1}{2}(t+z) \tag{55}$$

or equivalently

$$t = \eta + \xi \qquad \text{and} \qquad z = \eta - \xi \ , \tag{56}$$

one obtains the new derivative as

$$\frac{\partial}{\partial t} = \frac{1}{2}\left(\frac{\partial}{\partial \eta} + \frac{\partial}{\partial \xi}\right) \qquad \text{and} \qquad \frac{\partial}{\partial z} = \frac{1}{2}\left(\frac{\partial}{\partial \eta} - \frac{\partial}{\partial \xi}\right) \cdot \tag{57}$$

Consequently

$$\frac{\partial}{\partial t} + \frac{\partial}{\partial z} = \frac{\partial}{\partial \eta} \ , \qquad\qquad \frac{\partial}{\partial t} - \frac{\partial}{\partial z} = \frac{\partial}{\partial \xi} \cdot \tag{58a}$$

The field equation reduces to

$$\frac{\partial e^-}{\partial \xi} = i\nabla_T^2 e^- + P^- \ ; \qquad\qquad \frac{\partial e^+}{\partial \eta} = i\nabla_T^2 e^+ + P^+ \ . \tag{58b}$$

This means that the field is integrated along its directional characteristic path. With the polarization having a dynamic functional dependence on the total field the full Bloch equations are required. Furthermore the two oppositely traveling waves must be integrated simultaneously.

$$P^{\pm} = P^{\pm}(P_1^{\pm},\ldots,P_n^{\pm},e^+,e^-) \tag{59}$$

An example of one of the material (Bloch) equations is

$$\frac{\partial P_k^{\pm}}{\partial \xi} + \frac{\partial P_k^{\pm}}{\partial \eta} + \gamma_k P_k = S_k(P_1^{\pm},\ldots,P_{k1}^{\pm},P_{k+1}^{\pm}\ldots P_n^{\pm},e^+,e^-) \tag{60}$$

By identifying as outlined in Courant and Hilbert [50], the characteristics variable, namely

$$\xi = \xi(s) \qquad \text{and} \qquad \eta = \eta(s) \ , \tag{61}$$

or equivalently

$$\xi = \xi_o + s \qquad \text{and} \qquad \eta = \eta_o - s \ . \tag{62}$$

one obtains

$$\frac{\partial \xi}{\partial s} = + 1 \qquad \text{and} \qquad \frac{\partial \eta}{\partial s} = -1 \tag{63}$$

which simplifies the Bloch equations as follows:

$$\frac{\partial P_k}{\partial s} + \gamma_k P_k = S_k \tag{64}$$

which can be rigorously[54,55] integrated to give

$$P_k(s+\Delta s) = P_k(s)\exp(-\Delta s/\gamma s) + \int_s^{s+\Delta s} \{\exp[-(s-s')\gamma]S_k(s')ds'\} \ . \tag{65}$$

Illustrating the method of solution (see Fig. (35), arrows indicate integration paths for reducing differential equations to finite difference equations. Paths AB are used for Field Equations, and while Paths CB are used for Material Equations.

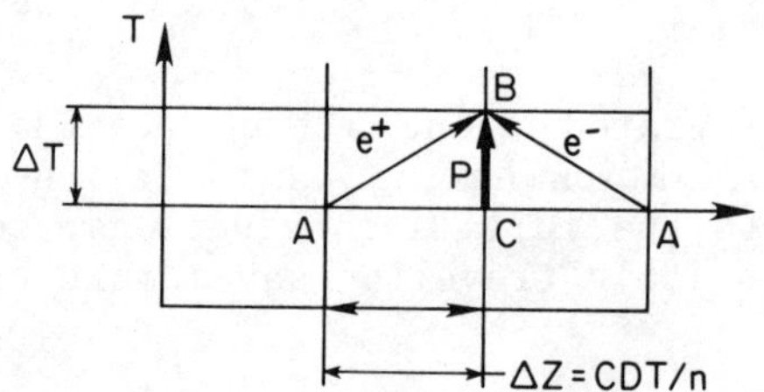

Fig. 35. Illustrates the two-way characteristic and the basis of the computational algorithm.

XIII. THE LAW OF FORBIDDEN SIGNALS

The effect of the physical law of forbidden signals on two-stream flow discretization problems was applied by Moretti to the integration of Euler equations[24,43].

For causality reasons, only directional resolution for spatial derivatives of each stream (forward and backward field) must be sought. This is achieved by using one-sided discretization techniques. The spatial derivative of the forward field is discretized using points lying to the left as all preceding forward waves have propagated in the same left-right direction; while the backward field is approximated by points positioned to the right. As a result, each characteristic (information carrier) is related to its respective directive history. Thus, violation of the law of forbidden signals is prevented.

In any wave propagation problem, the equations describe the physical fact that any point at a given time is affected by signals sent to it by other points at previous times. Such signals travel along lines known as the "characteristics" of the equations. For example a point such as A in Figure (36) is affected by signals emanating from B (forward wave) and from C (backward wave), while point A' will receive signals launched from A and D. Similar wave trajectories appear in the present problem, but the slopes of the lines can change in space and time.

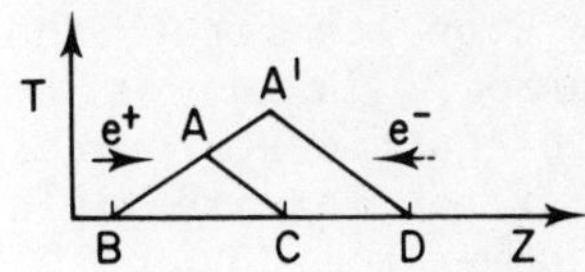

Fig. 36 Displays the role of characteristics as information carriers.

The slopes of the two characteristics carrying necessary information to define the forward and backward propagating variables at every point, are of different sign and are numerically equal to $\pm c/n$. For such a point A, Figure (37), the domain of dependence is defined by point B and C, the two characteristics being defined by AC and AB, to a first degree of accuracy. When discretizing the partial differential equations, point A must be made dependent on points distributed on a segment which brackets BC; e.g., on points D, E and F in Figure (38). This condition is necessary for stability but must be loosely interpreted. Suppose that one uses a scheme where a point A is made dependent on D, E and F, indiscriminately (this is what happens in most schemes currently used, including the MacCormack method). Suppose now, that the physical domain of dependence of A is the segment BC of Figure (38). The information carried to A from F is not only unnecessary;

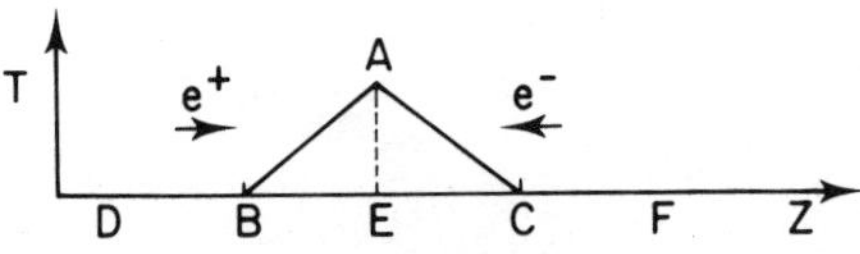

Fig. 37. Illustrates the concept of the law of forbidden signal for two-stream with characteristics of different sign.

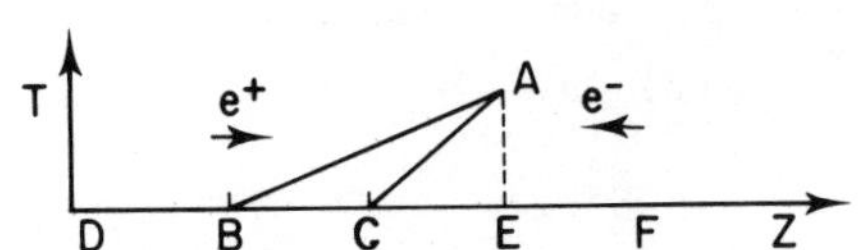

Fig. 38. Illustrates the concept of the causality for two-stream flow with characteristics of same (identical) sign.

it is also undue. Consequently, the numerical scheme, while not violating the Courant-Friedrick-Levy[54] (CFL) stability rule, would violate the law of forbidden signals. Physically, it is much better to use only information from D and E to define A, even if this implies lowering the nominal degree of accuracy of the scheme.

The sensitivity of results to the numerical domain of dependence as related to the physical domain of dependence explains why computations using integration schemes, like MacCormack's[52], show a progressive deterioration as the AC line of Figure (38) becomes parallel to the T-axis ($\lambda_1 \to 0$), even if λ_1 is still negative. The information from F actually does not reach A; in a coarse mesh, such information may be quite different from the actual values (from C) which affect A. On the other hand, since the CFL rules must be satisfied and F is the nearest point to C on its right, the weight of such information should be minimized. Moretti's λ-scheme, relying simultaneously on the two field equations provides such a possibility. Every spatial derivative of the forward field is approximated by using points lying on the same side of E as C, and every derivation of the backward-scattered field is approximated by using points which lie on the same side of E as B. By doing so, each characteristic relates with information found on the same side of A from which the characteristic proceeds also such information is appropriately weighted with factors dependent on the characteristic's slopes, so the contribution of points located too far outside the physical domain of dependence is minimized.

A one-level scheme which defines

$$\frac{\partial e^+}{\partial z} = (e_E^+ - e_D^+)/\Delta z \qquad \text{(forward wave)} \qquad (66)$$

$$\frac{\partial e^-}{\partial z} = (e_F^- - e_E^-)/\Delta z \qquad \text{(backward wave)} \qquad (67)$$

is Gordon's scheme [53], accurate to the first order. To obtain a scheme with second-order accuracy, Moretti considered two levels, in a manner very similar to MacCormack's. More points, as in Fig. (39) must be introduced. At the predictor level following Moretti's scheme one defines

$$\frac{\partial \tilde{e}^+}{\partial z} = (2e_E^+ - 3e_D^+ + e_G^+)/\Delta z \qquad \text{(forward wave)} \qquad (68)$$

$$\frac{\partial \tilde{e}^+}{\partial z} = (e_F^- - e_E^-)/\Delta z \qquad \text{(backward wave)} \qquad (69)$$

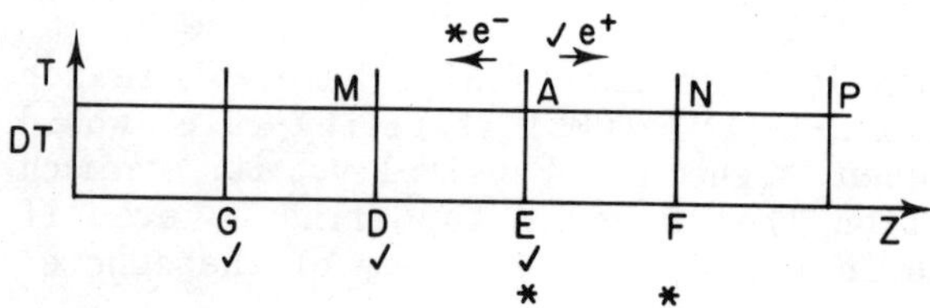

Fig. 39. Displays the computational grid for the λ-scheme.

At the corrector level, one defines

$$\frac{\partial \hat{e}^+}{\partial z} = (\tilde{e}_A^+ - \tilde{e}_M^+)/\Delta z \qquad \text{(forward wave)} \qquad (70)$$

and

$$\frac{\partial \hat{e}^-}{\partial z} = (-2\tilde{e}_A^- + 3\tilde{e}_N^- + \tilde{e}_P^-)/\Delta z \qquad (71)$$

It is easy to see that, if any function f is updated as

$$\tilde{f} = f + f_t \, \Delta t \qquad (72)$$

at the predictor level, with the t-derivatives defined as in (23) and (24) and the z-derivatives defined as in (68) and (69) and as

$$f(t+\Delta t) = \frac{1}{2} \, (f+\tilde{f}+f_t\Delta t) \qquad (73)$$

at the corrector level, with the t-derivatives defined again as in (23) and (24), and the z-derivatives defined as in (70) and (71), the value of f at 't+Δt' is obtained with second order accuracy. The updating rule (72) and (73) is the same as in the MacCormack scheme.

At the risk of increasing the domain of dependence, but with the goal of modularizing the algorithm, three- and four-point estimators were used for each first and second derivative respectively. Moretti's algorithm was also extended to non-uniform mesh to handle the longitudinal refractive left and right mirrors: the same one-sided differencing is used for both predictor and corrector steps. Nevertheless, the weights derived, using the theory of estimation, (presented by Hamming[53]), have improved the order of accuracy of the spatial derivative estimator at both predictor and corrector levels. In particular, the derivative estimators are of second order instead of first order as in Moretti's λ-scheme. Specifically, these weights are derived using a development in terms as a sum of Lagrangian polynomials at a set of points. As a result, the overall accuracy of Moretti's predictor/corrector scheme was increased[56] from second to third order. Either forward or backward longitudinal derivatives at both predictor and corrector stages are given for the point x_1, x_2 and x_3 as:

$$D_1 = \left(\frac{2x_1-x_2-x_3}{\pi_1(x_1)} \quad , \quad \frac{x_1-x_3}{\pi_2(x_2)} \quad , \quad \frac{x_1-x_2}{\pi_3(x_3)} \right) \qquad (74)$$

$$D_2 = \left(\frac{x_2-x_3}{\pi_1(x_1)} \quad , \quad \frac{2x_2-x_1-x_3}{\pi_2(x_2)} \quad , \quad \frac{x_2-x_1}{\pi_3(x_3)} \right) \qquad (75)$$

$$D_3 = \left(\frac{x_3 - x_2}{\pi_1(x_1)} \;\;,\;\; \frac{x_3 - x_1}{\pi_2(x_2)} \;\;,\;\; \frac{2x_3 - x_1 - x_2}{\pi_3(x_3)} \right) \tag{76}$$

$$\text{with } \pi_j(x) = \prod_{i \neq j = 1}^{3} (x - x_i) \tag{77}$$

Here D_1, D_2 and D_3 represents forward, central and backward differencing estimators for the (first-order longitudinal spatial) derivative.

XIV. TREATMENT OF LONGITUDINAL BOUNDARY

When treating any point within the cavity or at either longitudinal boundary (where a partially reflecting mirror is situated), there is no problem. For example, at $z = 0$, e^+ is determined by equation (53) and not through previous predictor/corrector formulas (68-71), as only e^- is calculated at $z = 0$ in that predictor/corrector manner (68-71). However, for a point one increment ($\delta = \Delta z$) from the left mirror, one encounters difficulties calculating the forward wave. The second needed point, which is vital to the formulas, would fall outside the cavity. An identical difficulty arises from the counterpart backward wave with respect to the right hand mirror. The field traveling from the right is defined at $z = L$ by equation (54).

To deal with this situation one has to modify the predictor/corrector schemes so the increment "δ^2" is used instead of δ. The loss of that second point reduces the accuracy of the derivative estimator. To maintain the same order of accuracy near the mirror, one must compensate for this loss by reducing the mesh size.

XV. NUMERICAL PROCEDURE FOR SHORT OPTICAL CAVITY

An alternate procedure to carry out the computation is to integrate the field along the longitudinal propagational distance. This approach is particularly attractive for a short cavity. It was developed with the help of McCall[57] as an attempt to relax the restrictive relation between the temporal t and spatial meshes z and r. It is presently being implemented and will be outlined here.

The reflecting effect of the partially refracting mirror can be built into the determining equations. Forward and backward field and polarization terms will appear explicitly as driving sources in each traveling field equation (see Fig. 40). One can readily contrast the two physical situations of long and short cavity. To illustrate the methodology the diffraction is neglected. For no reflection, the fields are described by

$$e^+(t+\Delta t,z) = e^+(t,z-c\Delta t) + \int_{z-c\Delta t}^{z} dz' \; P^+(t+\Delta t - \frac{z-z'}{c} \, , \, z') \tag{78}$$

which applies if $z > c\Delta t$. Also

$$e^-(t+\Delta t,z) = e^-(t,z+c\Delta t) + \int_{z}^{z+c\Delta t} dz' \; P^-(t+\Delta t + \frac{z-z'}{c} \, , \, z') \tag{79}$$

applies if $L-z > c\Delta t$. For one reflection, the fields are obtained by

$$e^+(t+\Delta t,z) = \sqrt{T} \; e_{I0}(t+\Delta t - z/c) + \int_{0}^{z} dz' \; P^+(t + \Delta t - \frac{z-z'}{c} \, , \, z')$$

$$+ \sqrt{R} e^-(t,c\Delta t-z) + \sqrt{R} \int_{0}^{c\Delta t-z} dz' \; P^-(t+\Delta t- \frac{z+z'}{c}, \, z') \tag{80}$$

whenever $z < c\Delta t$, and if $L-z < c\Delta t$, then one reflection

$$e^-(t+\Delta t,z) = \sqrt{T} \; e_{IL}(t+\Delta t - \frac{L-z}{c}) + \sqrt{R} \; e^{i\beta} \; e^+(t,2L-z-c\Delta t)$$

$$+ \int_{z}^{L} dz' \; P^-(t+\Delta t + \frac{z-z'}{c} \, , \, z')$$

$$+ \sqrt{R} \; e^{i\beta} \int_{2L-z-c\Delta t}^{L} dz' \; P^+(t+\Delta t - \frac{2L-z-z'}{c} \, , \, z') \tag{81}$$

In all of the above it is assumed that $c\Delta t < L$ (so that two reflections cannot occur in time Δt). To correctly include the influence of diffraction, appropriate weighting coefficients must be used as summarized below:

(1) For no reflection-correct by $\frac{1}{2} \nabla_T^2(e^+ c\Delta t)$, $\frac{1}{2} \nabla_T^2(e^- c\Delta t)$

(2) For one reflection-

    (a) Term $\sqrt{T} \; e_{I0}$ only propagates z ($c\Delta t > z$) so correct only by $z \, \nabla_T^2$

(b) Term $\int_0^z dz' \, P^+$ goes a distance of an average of $(\frac{1}{2})z$; correct

      by $\frac{z}{2} \, \nabla_T^2$

(c) Term $e^-(t, c\Delta t - z)$ goes a distance of $c\Delta t$; full correction by $c\Delta t - z$

(d) Term $\sqrt{R} \int_0^{c\Delta t - z} dz' \, P^-$ goes $\frac{c\Delta t - z}{c} + z$; correct bya distance of $\frac{c\Delta t + z}{2} \, \nabla_T^2$

(e) Term $\sqrt{T}$ goes e_{IL} goes a distance of $(L-z)$; correct by

      $(1/2)(L-z)\nabla_T^2$

(f) Term $\sqrt{R} \, e^{i\beta} \, e^+$ goes full distance; correct $\frac{1}{2} \, c\Delta t \, \nabla_T^2$

(g) Term $\int_z^L dz' \, P^-$ goes a distance of $\frac{L-z}{2}$; correct by

      $\frac{1}{2} \, (L-z) \, \nabla_T^2$

(h) Term $\sqrt{R} \, e^{i\beta} \int_{2L-z-c\Delta t}^L dz' \, P^+$ goes a distance of $\frac{(L-z)+c\Delta t}{2}$ on the

      average; correct $\frac{L-z+c\Delta t}{2} \, \nabla_T^2$

and similarly for any time correction.

Instead of the usual predictor/corrector weighting of 1/2 for each of predicted and corrected values, a more complicated procedure must be used.

XVI. TWO-LASER THREE-LEVEL ATOM

An extension of the SF calculations presented in Section IX should include such pump dynamics and its depletion on a three-level system similar to the model suggested by the Bowden et al[59]. The simulation of the dynamic interactions of two intense, ultrashort laser pulses propagating simultaneously through a gas of three-energy level atoms was considered[60]. The rigorous diffraction and cross-modulation interplay of the two laser beams with the inertial response of the doubly resonant medium is studied using an extension of the numerical algorithm developed for SIT analysis. It is expected that by altering the pump characteristics, one encodes information in the pulse that evolves in the nonlinear media resulting in a light by light control. An intermediate study will be Double Coherent Transients[61,62]. Another benefit of this study would be an analysis of Wall's[63] scheme for optical bistability in a coherently-driven three-level atomic system. However, some material equation modifications must be made as the novel mechanism relies on the nonlinear absorption resonances associated with a population trapping, coherent superposition of the ground sublevel. When one defines dimensionless variables in a parallel manner to SIT, the physical problems are described by the following equations: τ_{pa} and τ_{pb} are the pulse τ_p of laser a and laser b respectively. Q is the quadrupole slowly varying envelope.

$$-iF \, \nabla_T^2 \, e_{a,b} + \partial_\eta \, e_{a,b} = g_{a,b} \, P_{a,b} \tag{82}$$

with

$$g_{a,b} = (\mu_a/\mu_b) \, (\tau_{pa}/\tau_{pb})^{1/2} \tag{83}$$

$$\partial_\tau P_a = e_a \, W_a - i(\Delta\Omega_a)P_a - P_a/\tau_{2a} + \frac{i}{2} \, e_b^* \, Q \tag{84}$$

$$\partial_\tau P_b = e_b \, W_b - i(\Delta\Omega_b) \, P_b - P_b/\tau_{2b} - \frac{i}{2} \, e_a^* \, Q \tag{85}$$

$$\partial_\tau Q = -i[(\Delta\Omega_a + \Delta\Omega_b)]Q + \frac{i}{2} \, (e_a \, P_b - e_b \, P_a) - Q/\tau_{2ab} \tag{86}$$

$$\partial_\tau W_a = -\frac{i}{2}(e_a^* \, P_a + e_a \, P_a^*) - (W_a - W_a^e)/\tau_{1a} + \frac{1}{4}(e_b^* \, P_b + e_b \, P_b^*) \tag{87}$$

$$\partial_\tau W_b = -\frac{1}{2}(e_b^* \, P_b + e_b \, P_b^*) - (W_b - W_b^e)/\tau_{1b} + \frac{1}{4}(e_a^* \, P_a + e_a \, P_a^*) \tag{88}$$

If one uses the identity

$$W_a + W_b = W_{ab} \tag{89}$$

a further equation (not absolutely necessary) is introduced:

$$\partial_\tau W_{ab} = + 1/4[(e_a^* P_a + e_a P_a) + (e_b^* P_b + e_b P_b^*)] - (w_{ab} - W_{ab}^e)/\tau_{ab} \tag{90}$$

when $W_{a,b}^e$ and W_{ab}^e are the equilibrium values of $W_{a,b}$ and W_{ab}, subjected for infinite relaxation times to a conservation of probability

$$\partial_\tau \{|P_a|^2 + |P_b|^2 + |Q|^2 + (W_a^2 + W_b^2 + W_{ab}^2)\} = \text{zero}. \tag{91}$$

Equivalently:

$$|P_a|^2 + |P_b|^2 + |Q|^2 + 2/3(W_a^2 + W_b^2 + W_{ab}^2)$$
$$= |P_{a,i}|^2 + |P_{b,i}|^2 + |Q_i|^2 + 2/3(W_{a,i}^2 + W_{b,i}^2 + W_{ab,i}^2). \tag{92}$$

Figure (40) illustrates W_a, W_b and W_{ab} as a function of time for a particular radius in the reshaping region.

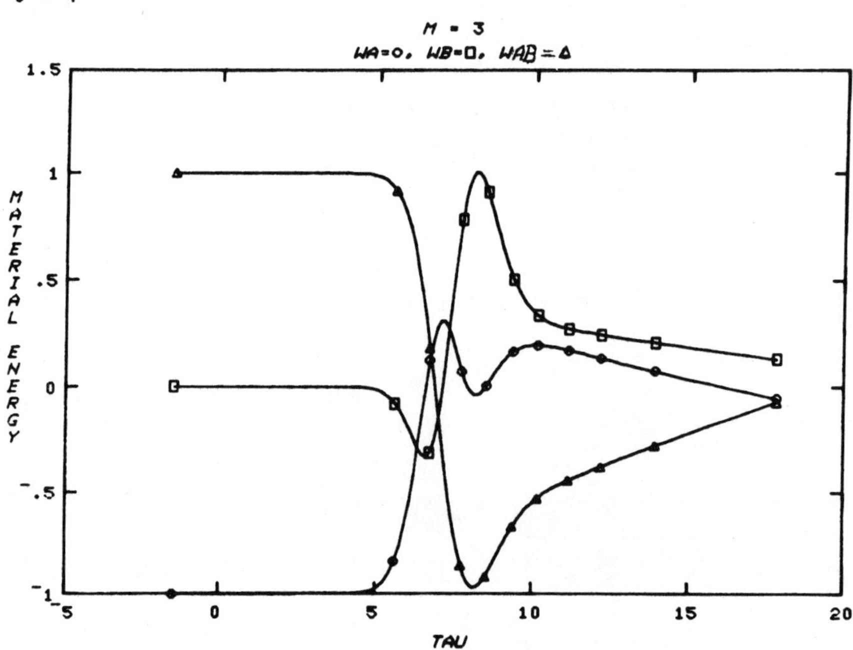

Fig. 40. Contrast of the material energy for a double self-induced transparency calculation.

Numerical Refinements

If the two laser beams which propagate concomitantly are severely disparate from each other, the normal stretching technique must be generalized into a double stretching transformation[60c] to ensure that the nonuniform temporal grids simultaneously match the two different pulses. No spatial rezoning is as yet designed.

Prescribed Double Stretching

Due to the essential nonlinear nature of the cooperative effects associated with a coherent light-matter interaction, dif-

ferent speeds are associated with pulses of different strengths.
So particular attention must be given to deal effectively with two
concomitant longitudinal speeds (one for each laser). Mathematically this is

$$T = a\tau + b \sin \omega_s \tau$$

$$\frac{\partial T}{\partial \tau} = a + b \omega_s \cos \omega_s \tau$$

and is shown in Fig. 41. Evenly spaced grid points in T are clearly related to non-uniform variable grid points in the physical time
τ.

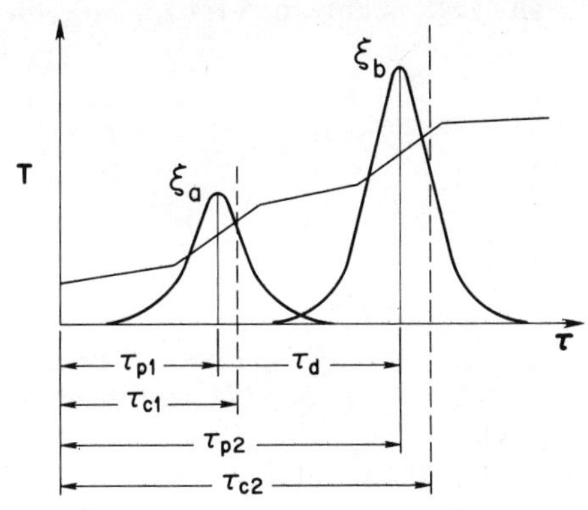

Fig. 41. Displays the prescribed   double
stretching.

ω_s	0	$\pi/2$	π	$3\pi/2$	2π
$\cos \omega_s \tau$	1	0	-1	0	1
$\partial T/\partial \upsilon$	$a + b\omega_s$	a	$a - b\omega_s$	a	$a + b\omega_s$

For $\omega_s \tau = \pi$, $\partial T/\partial \tau$ is minimum.

Several noteworthy facts must not be overlooked, i.e., (i) ω_s
is related to the frequency of oscillations; and (ii) the steepness
of the slopes must depend on the concentration points.

The various stretching parameters are given by

$$a = 1/2 \left[\frac{\partial T}{\partial \tau}\Big|_{max} + \frac{\partial T}{\partial \tau}\Big|_{min} \right]$$

$$b = \{1/2\ w_s\} \left[\frac{\partial T}{\partial \tau}\Big|_{max} - \frac{\partial T}{\partial \tau}\Big|_{min} \right]$$

$$w_s(\tau_{c2} - \tau_{c1}) = 2\pi \Rightarrow w_s\ \tau_d = 2\pi$$

If τ_d increases, w_s decreases - a smaller frequency yields to a larger b, if τ_d decreases, w_s increases - a larger frequency yields to a smaller b parameter.

To ensure monotonicity of the function T in τ (so that multi-valued possibilities are excluded), an important condition which must never be violated (see Fig. 42), is

$$g = \frac{\partial T}{\partial \upsilon}\Big|_{min} \equiv (a - bw) > 0 \ \cdot$$

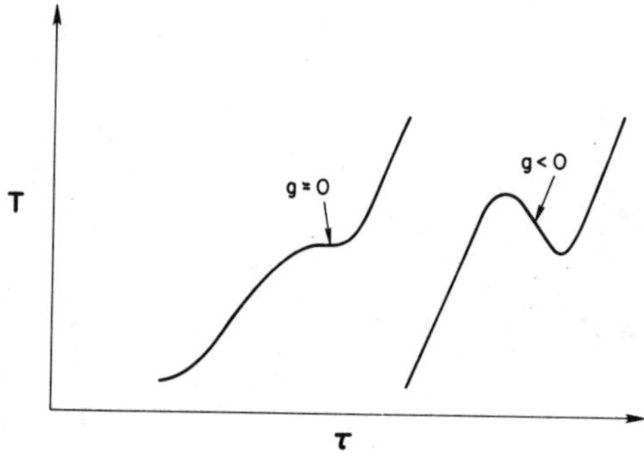

Fig. 42. Displays the limitations on the parameter choice to the double stretching transformation.

Adaptive Double Stretching

Following the spirit of adjusted stretching for a single pulse, described in Section V, the sampling frequency w_s can vary along the direction of propagation η.

Prescribed Triple Stretching

For a correct treatment of the pulses propagating concomitantly while one of the two lasers may have broken up into two small pulses, successive double stretchings are applied

Step 1 $\zeta = A\ x^2 + Bx$

from
$$x = x_2, \quad \zeta = \zeta_0 = Ax_2^2 + Bx_2$$
$$x = x_3, \quad \zeta = 2\zeta_0 = Ax_3^2 + Bx_3$$

and
$$x = 0, \quad \zeta = 0.$$

$$\zeta_1 - \zeta_2 = \zeta_2 - \zeta_3 = 0 - \zeta_1$$

one gets,
$$A = \frac{\zeta_0(x_3 - 2x_2)}{x_3 x_2 (x_2 - x_3)} \quad \text{and } B = \frac{\zeta_0(2x_2^2 - x_3^2)}{x_2 x_3 (x_2 - x_3)} \quad ;$$

Step 2
$$Y = a\,\zeta + b\,\sin\,\omega_s\,\zeta$$

Cumulative step
$$Y = a(Ax^2 + Bx) + b\,\sin\,\omega_s\,(Ax^2 + Bx)$$

$$Y_x = a(2x\,A + B) + b\omega_s\,(2Ax + B)\,\cos\,(Ax^2 + B)$$

$$= (2Ax + B)\,(a + b\omega_s\,\cos\,(Ax^2 + B))\,.$$

The coefficients are readily found (see Fig. 43).

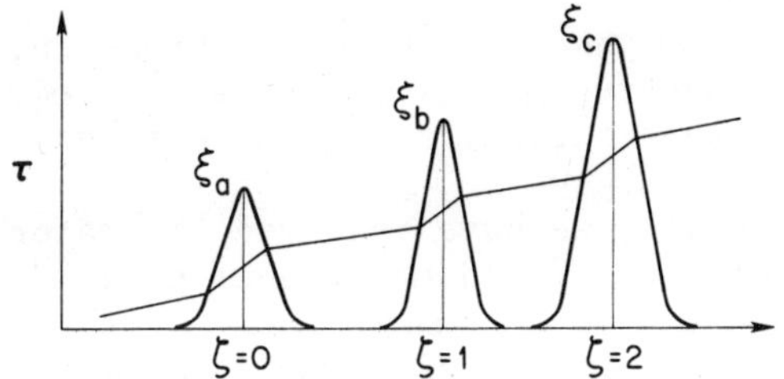

Fig. 43. Illustrates a pre-scribed triple stretching.

XVII. CONCLUDING REMARKS

Most of the features of the numerical model used to study temporal and transverse reshaping effects of single and multiple short optical pulses propagating concomitantly in active, non-linear, resonant media have been presented. The calculations strive to achieve a rigorous analysis of this nonlinear interaction with maximum accuracy and minimum computational effort. The applicability of computational methods developed in gas and fluid dynamics to the detailed evolution of optical beams in nonlinear media have been demonstrated.

By introducing adaptive stretching and rezoning transformations wherever possible, the calculations improved considerably.

In particular, self-adjusted rezoning and stretching techniques consisting of repeated applications of the same basic formulae were reviewed as a convenient device for generating computational grids for complex nonlinear interactions. The techniques are well-suited for each programming because the mapping functions and all related derivatives are defined analytically as much as possible. Enhancement of speed and accuracy was realized by improving the integration technique/algorithm which was general and simple in its application compared with its analogue, the two-dimensional Lagrangian approach[42].

This method was applied to a number of SIT situations with and without homogeneity in the resonant properties of the atomic medium. Note that the theoretical predictions defined with the single stream SIT code, when applied to absorbing media, were quantitatively found[64] by independent experimental observations[65], and recent independent perturbational[66] and computational analysis[67]. The design of the first of these experiments dealing with sodium vapor, was based on qualitative ideas, quantitative analysis and numerical results obtained with the code described in this paper. More recently, King et al also reported[68] the experimental observation in iodine atomic vapor of the coherent on-resonance self-focusing. This is a novel manifestation of the phenomenon as it deals with a magnetic dipole instead of an electric dipole moment.

Also, the severe beam distortion and on-axis pulse break-up, when the problem of transverse boundary is rigorously addressed, was observed in high power lasers used in Laser Fusion experiments.

With the help of Gibbs and McCall, we have resolved the major discrepancies between planar calculations (as done by Hopf et al[69]) and the Cs experimental observations. The main sources of these discrepancies[60] were the occurrence of transverse effects in the experiments and the uncertainty in the tipping angle values.

Optical bistability shares with the previous SIT and SF the same basic physical features; however, the initial and boundary conditions are different and complicate the problem. Nevertheless, the similarities predominate; therefore, a unified numerical description with some modifications can apply to all these problems. This new computational approach, based on the concept of absolute consistency of the numerics with the physics, should be successful.

ADDENDUM

An alternate solution to eliminate rapid oscillations from the two-mode Bloch equation without recourse to harmonic expansion could be to adopt Moore and Scully[71] multiple-scaling perturbation expansion. They have applied the techniques of multiple-scaling perturbation theory, described in hydrodynamics textbooks, to the free-electron laser problem and the pico-second transient phenomena.

ACKNOWLEDGMENTS

F.P. Mattar thanks his thesis advisor Professor M.C. Newstein, his mentors Professors H.A. Haus and Gino Moretti, for their guidance in the physics and numerics of the work. He has benefitted from discussions with Drs. H.M. Gibbs, S.L. McCall, J.H. Marburger, D.C. Brown, P.E. Toschek, and M.S. Feld in the physics; and J. Hermann, B.R. Suydam and J. Fleck, on the numerics. He is indebted to Dr. Gibbs for his faith in the work that led to the first experimental verification of the coherent on-resonance self-focusing. The hospitality of Dr. J. Teichman at the Univ. of Montreal and Dr. T.C. Cattrall at Mobil, (which made the computations possible), is gratefully acknowledged. F.P. Mattar particularly thanks Dr. C. Hazzi and D.J. Steele for their patience, support and encouragement during his convalescence. Furthermore, the editing efforts of D.J. Steele the artwork of Kerop Studio in Cairo, W. Roberts and D.J. Steele in New York; and the skillful and laborious word processing efforts of E. Cummings are joyfully appreciated.

REFERENCES

1. (a) R.H. Dicke, Phys. Rev. <u>93</u>, 99 (1954) and in <u>Proc. Third Int. Conf. on Quant. Elec.</u>, Paris, 1963, ed. by P. Grivet and N. Bloembergen (Columbia University Press, N.Y., 1964); (b) D.C. Burnham and R.Y. Chiao, Phys. Rev. <u>188</u>, 667 (1979) and S.L. McCall, Ph.D. thesis, Univ. of California, Berkeley (1968); (c) J.C. MacGillivray and M.S. Feld, Phys. Rev. <u>A14</u>, 1169 (1976); (d) R. Bonifacio and L.A. Lugiato, Phys. Rev. A11, 1507 and <u>12</u> 587 (1975).

2. (a) <u>Cooperative Effects in Matter and Radiation</u>, ed. by C.M. Bowden, D.W. Howgate and H.R. Robl (Plenum Press, N.Y., 1977), and (ii) panel discussion compiled by M. Konopnicki and A.T. Rosenberg, p. 360, (ii) H.M. Gibbs, p. 61 and (iii) Q.H.F. Vrehen, p. 79; (b) lecture on Superfluorescence experiments, <u>1977 NATO/ASI on Coherence in Spectroscopy and Modern Physics</u> (Plenum Press, N.Y., 1977); (c) Q.H.F. Vrehen, <u>Coherence and Quantum Optics IV</u>, ed. by L. Mandel and E. Wolf (Plenum Press, N.Y., 1978), pp. 78 and <u>Laser Spectroscopy IV</u>, ed. H. Walther and I.W. Rothe; (d) Q.H.F. Vrehen, H.M.J. Hikspoors and H.M. Gibbs, Phys. Rev. Lett. <u>38</u>, 764 (1977), Phys. Rev. Lett. <u>39</u>, 547 (1977) and <u>Laser Spectroscopy III</u>, ed. by J.I. Hall and J.I. Carlsten, Springer-Verlag (1977);

(e) Q.H.F. Vrehen and M.F.H. Schuurmans, Phys. Rev. Lett. <u>42</u>, 224 (1979).

3. (a) H. Seidel, <u>Bistable optical circuits using saturable absorber within a resonant cavity</u>, U.S. Patent 3 610 731 (1969); (b) H.M. Gibbs, S.L. McCall and T.N.C. Venkatesan, Optics News <u>5</u>, 6 (1979), J. Opt. Soc. Am. <u>65</u>, 1184 (1975), Phys. Rev. Lett. <u>36</u>, 1135 (1976) (U.S. patents #012,699 [1975] and #121,167 [1976], lecture on Optical Bistability and Differential Gains at <u>1977 NATO/ASI on Coherence in Spectroscopy and Modern Physics</u> (Plenum Press, 1977); (c) A. Szöke, V. Daneu, J. Goldher, and N.A. Kurint, Appl. Phys. Lett. <u>15</u>, 376 (1969).

4. (a) T.N.C. Venkatesan and S.L. McCall, Appl. Phys. Lett. <u>30</u>, 282 (1977), H.M. Gibbs, S.L. McCall, T.N.C. Venkatesan, A.C. Gossard, A. Passner and W. Wiegmann, CLEA/1978 (IEEE J. Quantum Electronics <u>15</u>, 108D [1979]) and Appl. Phys. Lett. <u>35</u>, 451 (1979); S.L. McCall and H.M. Gibbs (to be published in Opt. Comm. 1980) and H.M. Gibbs, S.L. McCall and T.N.C. Venkatesan (to be published in special issue on <u>Optical Feedback in Opt. Eng.</u>); (b) T. Bischofberger and Y.R. Shen, Appl. Phys. Lett. <u>32</u>, 156 (1978), Opt. Lett. <u>4</u>, 40 & 175 (1979) and Phys. Rev. <u>A19</u>, 1169 (1979); (c) P.W. Smith, J.P. Hermann, W.J. Tomlinson and P.J. Maloney, Appl. Phys. Lett. <u>35</u>, 846 (1979); (d) <u>Proc. of the Int'l Optical Bistability Conf.</u>, Asheville, North Carolina (June 1980), ed. C.M. Bowden, M. Ciftan and M.R. Robl, to be published (to be published) by Plenum Press.

5. R. W. Hellwarth, J. Opt. Soc. Am. <u>67</u>, 1 (1977) and IEEE J. Quantum Electron., <u>15</u>, 101 (1979).

6. A. Yariv, IEEE J. Quantum Electron., <u>14</u>, 650 (1978) and <u>Compensation for Optical Propagation Distortion through Phase Adaptation</u> (preprint Caltech 1979).

7. (a) S.L. McCall, Ph.D. thesis, Univ. of California, Berkeley (1968); (b) J.-C. Diels, Ph.D. thesis, Univ. of Calif. Berkeley (1973); (c) A. Icsevgi and W.E. Lamb, Jr., Phys. Rev. <u>185</u>, 517 (1969).

8. D.W. Dolfi and E.L. Hahn (to appear in Phys. Rev. a, 1980).

9. H.M. Gibbs, B. Bölger, F.P. Mattar, M.C. Newstein, G. Forster and P.E. Toschek, Phys. Rev. Lett. <u>37</u>, 1743 (1976).

10. F.P. Mattar, M.C. Newstein, P. Serafim, H.M. Gibbs, B. Bölger, G. Forster and P. Toschek, <u>Fourth Rochester Conf. of Coherence and Quantum Optics</u>, Rochester, NY (June 1977), Plenum Press, ed., L. Mandel and E. Wolf, p. 143-164 (1978).

11. F.P. Mattar, H.M. Gibbs, <u>Laser 1980</u>, New Orleans(1980).

12. S.L. McCall, F.P. Mattar, and H.M. Gibbs, <u>Optics in Four-Dimensions</u>, Encenade, Mexico (1980).

13. F.P. Mattar, H.M. Gibbs, S.L. McCall and M.S. Feld submitted to Phys. Rev. Lett. and Phys. Rev. A.

14. N. Wright and M.C. Newstein, Opt. Commun., 9, 8 (1973) and IEEE
    J. Quantum Electron., 10, 743 (1974).
15. M.C. Newstein and F.P. Mattar, Proc. 7th Conf. Numerical Simu-
    lation of Plasmas (Courant Institute, NYU, June 1975, p. 223;
    and Proc. of 10th Congress of the Int'l Commission of Optics,
    Prague, Czechoslovakia (1975), Recent Advances in Optical
    Physics, p. 199, B. Havelka and J.Blabla, distributed by the
    Soc. of Czechoslovak Math. and Phys. (1976); IX Int'l Conf.
    of Quantum Electronics, Amsterdam (1976), see Opt. Comm. 18,
    70 (1976), and IEEE J. Quantum Electron, 13, 507 (1977) and
    see ref. 2(a) 139.
16. J.C. Diels and E.L. Hahn, Phys. Rev. A8, 1084 (1973) and Phys.
    Rev. A10, 2501 (1974).
17. (a) S.L. McCall and E.L. Hahn, Phys. Rev. Lett. 28, 308 (1967),
    Phys. Rev. 183, 487 (1969) and Phys. Rev. A2 (1970); (b)   A.
    Icsevgi and W.E. Lamb, Jr., Phys. Rev. 185, 517 (1969).
18. S.L. McCall, Phys. Rev. A9, 1515 (1974).
19. (a) J.A. Fleck, Jr., Appl. Phys. Lett. 13, 365 (1968); (b) J.A.
    Fleck, Jr., Phys. Rev. B1, 84 (1970); (c) See Ref. 1(c) and
    ref. (11)
20. (a) S.L. McCall and H.M. Gibbs (to be published in Optics Comm.
    1980); (b) H.J. Carmichael (to be published in Optica Acta
    (1980) and H.J. Carmichael and G.P. Agrawal, Inhomogeneous
    Broadening and Mean Field Approximation for Optical Bistabil-
    ity in a Fabry-Perot (preprint 1979); Steady State Formula-
    tion of Optical Bistability for a Doppler Broadened Medi-
    um in a Fabry-Perot (preprint 1980); (c) J.H. Marburger and
    F.S. Felber, Appl. Phys. Lett. 28, 831 (1976) and Phys. Rev.
    A17, 335 (1978); (d) R. Sanders and R.K. Bullough in Cooper-
    ative Effects in Matter and Radiation, ed. by C.M. Bowden,
    D.W. Howgate and H.R. Robl, 209, Plenum Press (1977); and R.
    Saunders, S.S. Hassan and R.K. Bullough, J. Phys. A9, 1725
    (1976); (e) J.H. Eberly, K.G. Whitney and M. Konopnicki,
    unpublished Final Research Report to ONR (Fall 1977) Roches-
    ter, NY; (f) P. Meystre, Opt. Comm. 26, 277 (1978) and R.
    Bonifacio and P. Meystre, Opt. Comm. 27, 147 (1978); and F.A.
    Hopf and P. Meystre (to be published in Opt. Comm.).
21. (a) C.M. Bowden and C.C. Sung, Phys. Rev. A18, 1558 (1978) and
    Phys. Rev. A20, 2033 (1979);   (b) R. Brewer and E.L. Hahn,
    Phys. Rev. A11, 1641 (1975); (c) M. Sargent III and P. Hor-
    witz, Phys. Rev. A13, 1962 (1976); (d) N.S. Feld, Frontiers
    in Laser Spectroscopy, 2, 203 (Les Houches Lectures 1976),
    Ed. R. Balian, S. Haroche and S. Liberman, North Holland
    (1977); (e) F.P. Mattar and J.H. Eberly, Proc. of the Phys-
    ics and Chemistry of Laser- Induced Process in Molecules,
    Edinburgh (1978) ed. by K.L. Kompa and S.C. Smith, 61, Sprin-
    ger-Verlag (1979); (f) M. Konopnicki and J.H. Eberly, Proc.
    10th Pittsburgh Annual Simulation and Modeling Conf., Pitts-
    burgh (1979) ed. Vogt and Publ. Instrument Soc. of America
    (ISA), Pittsburgh, PA.

22. G. Moretti, Polytechnic Institute of New York Rept 69-25, 1969;
    Polytechnic Institute of New York Rept 69-26, 1969; Proc. of
    the 1974 Heat Transfer and Fluid Mechanics Inst., Stanford
    Univ. Press, 1974, Polytechnic Institute of New York-AE/AM
    Rept 74-9; (POLY-AE/AM Rept 74-23); Polytechnic Institute of
    New York Rept 73-18, 1973; Polytechnic Institute of New York
    Rept 69-26, 1969; Polytechnic Institute of New York Rept
    70-48, 1970; Polytechnic Institute of New York Rept 71-25,
    1971; POLY-AE/AM Rept 74-15, 1974; Proc. Symposium Transsoni-
    cum II, Gottingen, Germany, September 8-13, 1975, Springer-
    Verlag, Berlin, 439 (2976); (Polytechnic Institute of New
    York-AE/AM Rept 76-06).
23.   G.   Moretti, The Chemical Kinetics Problem in the Numerical
    Analyses of Nonequilibrium Flows, Proc. of the IBM Scientific
    Computing Symposium on Large Scale Problems in Physics, Dec.
    1963, IBM Research Ctr, Yorktown Heights, NY); and Polytech-
    nic Institute of New York Rept 68-15 (1968). Polytechnic
    Institute of New York-AE.AM Rept 74-15, 1974.
24. G. Moretti, AIAA 14, 834 (1976), and Computers and Fluids 7,
    191 (1979).
25. (a) E. Madelung, Z. Physik 40, 322 (1966). (b) H.E. Wilhelm,
    Phys. Rev. D1, 2278 (1970). (c) M. Jammer, The Philosophy of
    Quantum Mechanics 21, J. Wiley (1974).
26. (a) F.P. Mattar, Proc. of the Ninth Conf. of Numerical Methods
    in Plasma, Monterey, California (June 1978). Lawrence Liver-
    more Lab. (LLL) Tech. Rept 78-004. (b) F.P. Mattar, J.
    Teichman, L. Bissonnette and R.W. MacCormack, Proc. of the
    Second Int'l Symp. on Gas Flow and Chemical Lasers, ed. J.
    Wendt, Western Hemisphere Pub. (1979).
27. R.P. Feyman, F.L. Vernon, Jr. and R.W. Hellwarth, J. Appl.
    Phys. 28, 43 (1957); and E.L. Hahn, Heritage of the Bloch
    Equations in Quantum Optics, to appear in The Felix Bloch
    Festschrist 75 Birthday.
28. (a) G. Askary'yan, Sov. Phys. JET P42, 1568 (1962) Moscow; and
    Usp. Fiz. Nauk., 111, 249, October 1973. (b) S.A. Akhamanov,
    A-P. Sukhorukov, and P.V. Kokhlov, Laser Handbook, F. Ar-
    rechi, Ed., Amsterdam, The Netherlands: North Holland, 1972,
    1151. (c) B.R. Suydam, Self-Focusing in passive media I, Los
    Alamos Scientific Lab, Tech. Rep. L4-5002-MS, March 1973, and
    IEEE J. Quant. Elec. 10, 837 (1974), and IEEE J. Quant. Elec.
    11, 225 (1975).
29. (a) J.H. Marburger, Progress in Quantum Electronics, 4, 35 ed.
    J.M. Sanders and Stenholm, Pergamon Press (1975), and Theory
    of Self-focusing with Counter-propagating Beams,   preprint
    (1979). (b) O. Svelto, Progress in Optics XII, E. Wolf, ed.
    Amsterdam, The Netherlands: North Holland, 1974, 1.
30. Ibid Ref. 16.
31. T. Gustafson, J.-P. Taran, H.A. Haus, J. Lifsitz and P. Kelley,
    Phys. Rev. 177, 60 (1969).

32. (a) F.P. Mattar, Kvantovaya Electronika 4, 2520 (1977). (b)
    S.L. McCall and H.M. Gibbs (ibid ref. 20(a) and ref. 4(d)).
    (c) D.A.B. Miller and S.D. Smith, Opt. Comm. 31, 101 (1979)
    and D.A.B. Miller, S.D. Smith and A. Johnston, Appl. Phys.
    Lett. 35, 658 (1979). (d) M.S. Feld and J.C. MacGillivray in
    Advances in Coherent Nonlinear Optics, ed. M.S. Feld and V.S.
    Letokhov (to be published by Springer-Verlag, 1980). (e) R.
    Bonifacio, J.D. Farina and L.M. Narducci, Opt. Commun. 31,
    377 (1979).
33. G. Moretti, Polytechnic Institute of New York Rept 69-25, 1960;
    Polytechnic Institute of New York Rept 69-26, 1969.
34. G. Moretti, Proc. of the ASME Symp. on Numerical/Laboratory
    Computer Methods in Fluid Mechanics, ASME, Dec. 1976. Poly-
    tecnhic Institute of New York -M/AE Rept 76-06.
35. L. Bradley and J. Hermann, MIT-Lincoln Lab Tech. Rept LTP-10
    (July 1971) and Internal note on Change of reference wave-
    front in the MIT CW Nonlinear Optics Propagation Code (Fall
    1974) private communication.
36. P.B. Ulrich, NRL Rept 7706 (May 1974).
37. H.J. Breaux, Ballistic Research Labs, Aberdeen Proving Ground,
    Maryland; BRL Rept 1723 (1974).
38. J.A. Fleck, Jr., J.R. Morris and M.D. Feit, Appl. Phys. 10, 129
    (1976) and 14, 99 (1977).
39. F.P. Mattar, Appl. Phys. 17, 53 (1968) Springer-Verlag.
40. K.G. Whiteney, G.L. Nader, and P.B. Ulrich, Naval Research Lab.
    Washington, DC, NRL Rept 8074 (1977).
41. W.A. Newcomb, Nuclear Fusion, Suppl. 2, 451 (1962).
42. F.P. Mattar and J. Teichmann, IEEE Conf. Plasma Science, Mon-
    treal (1979) and submitted to Comp. Phys. Comm.
43. (a) G. Moretti, AIAA 14, 894 (1976) and Poly-M/AE Tech. Rep. 78
    (1980), PINY; (b) B. Gabutti, La stabilita de una schema
    alle differenze finite per le equazioni della fluide dinamica
    (preprint, Polytechnico di Milano).
44. H.M. Gibbs and R. E. Slusher, Appl. Phys. Lett. 18, 505 (1971),
    Phys. Rev. A5, 1634 (1972) and Phys. Rev. A6, 2326 (1972).
45. H.A. Haus and T.K. Gustafson, IEEE J. Quant. Elec. 4, 519
    (1968).
46. F.P. Mattar and M.C. Newstein, PINY Rept. ADL-M/AE 79-63 (1979)
    (to be published in Comp. Phys. Comm. 1980).
47. G. Moretti, Polytechnic Institute of New York Rept 68-15
    (1968). Polytechnic Institute of New York-AE. AM Rept 74-15,
    1974.
48. (a) V.I. Bespalov and V.I. Talanov, Soviet Phys., JETP letters,
    3, 471 (1966), Eng. Trans. 3, 307. (b) V.I. Talanov, Soviet
    Phys. JETP Letters 2,218 (1965) Eng. Trans. 2, 138 and Zh.
    Eksp. Teor. Fiz. Pisma. Red. 11, 303, Eng. Trans. 133, JETP
    letters, 1970. (c) B.R. Suydam, Laser Induced Damage in
    Optical Material, 1973 NBS special publication 387, 42 and
    IEEE J. Quant. Elec. 10, 837 (1973) and IEEE J. Quant. Elec.

$\underline{11}$, 225 (1975). (d) J.B. Trenholme, <u>(LLL) Laser Fusion program 2nd 1973 semi-annual program report</u>, 47 (UCRL-60021-73-2). (e) N.B. Baranova, N.E. Bykovskii, B. Ya. Zel'dovich and Yu. V. Senatskii, Kvant. Elektron. $\underline{1}$, 2435 (1974). (f) D.C. Brown, <u>The Physics of High Peak Power Nd-Glass Lasers</u>, Springer-Verlag (to be published in the Spring 1980). (g) L.A. Gol'shov, V.V. Likhanskii and A.O. Napartovich, Zh. Eksp. Teor. Fiz. $\underline{72}$, 769 (197˜) Moscow. (h) M.J. Ablowitz and Y. Kodama, Phys. Lett. (1979). (i) L.A. Bol'shov, T.K. Kirichenko, A.P. Favolsky, U.S.S.R. Acad. of Sci., Math. Div. preprint, Fall 1978), Inst. of Appl. Math. 1978, 53 (3A) in Russian.

49. W.G. Wagner, H.A. Haus and J.H. Marburger, J. Phys. Rev. (1970).

50. R. Courant and K.O. Friedrichs, <u>Supersonic Flow and Shock Waves</u>, NY Interscience (1948).

51. F.A. Hopf and M.O. Scully, Phys. Rev. $\underline{179}$, 399 (1969).

52. R.W. MacCormack, <u>AIAA Hypervelocity Impact Conf.</u>, 1969, paper 69-554 and <u>Lecture Notes in Physics</u>, Springer-Verlag, 151 (1971).

53. P. Gordon, General Electric Final Rep. NOL Contrace #N60921-7164 (1968).

54. R.W. Hamming, <u>Numerical Methods for Scientists and Engineers</u>, McGraw-Hill Book Co. (1962).

55. (a) B.R. Suydam, <u>A Laser Propagation Code</u>, Los Alamos Sci. Lab., LA-5607-MS Tech. Informal Rept (1974). (b) F.P. Mattar (with the help of Dr. B.R. Suydam, LASL) <u>Eighth Conf. on Numerical Simulation of Plasmas</u>, Monterey, Calif. (June 1978). Distributed by the Univ. of Calif./LLL, Livermore, Calif. (rept Conf-780612).

56. F.P. Mattar and R.E. Francoeur, <u>Transient Counter-Beam Propagation in a Nonlinear Fabry Perot County (Long Sample)</u>, PINY, ADL-M/AE Tech. Rept. 80-12 (1980) and <u>Proc. of the Ninth Conf. of Numerical Simulation of Plasma</u> (ed. G. Knorr and J. Denavit) Northwestern Univ. Evanston, IL, 1980.

57. F.P. Mattar and S.L. McCall, <u>Transient Counter-Beam Propagation In a Nonlinear Fabry Perot County II. Short Sample</u>, PINY, ADL-M/AE Tech. Rept. 80-15 (1980).

58. (a) Ibid ref. 1(c); (b) Ibid ref. 2

59. Ibid ref. 21(a).

60. (a) Ibid ref. 21(d); (b) Ibid ref. 21(e); (c) F.P. Mattar, <u>Proc. 10th Pittsburgh Annual Simulation and Modeling Conf.</u>, Pittsburgh (1979), ed. W. Vogt and M. Mickle, Publ. ISA, Pittsburgh, PA.

61. R.G. Brewer, (1975) in <u>Frontiers in Laser Spectroscopy</u>, ed. R. Balian, S. Haroche and S. Liberman 338 (North Holland 1977).

62. V.S. Letkhov and B.D. Sov. J. Quant. Elec. $\underline{4}$, 11 (1975).

63. D. Walls (ref. 4(d)).

64. (a) H.M. Gibbs, B. Bölger, F.P. Mattar, M.C. Newstein, G. Forster and P.E. Toschek, Phys. Rev. Lett. 37, 1743 (1976) (prep. by H.M. Gibbs); (b) Ref. (10); (c) F.P. Mattar, G. Forster and P.E. Toschek, Spring meeting of the German Phys. Soc., Mainz, F.R. Germany (Feb. 1977). Kvantovaya Electronika 5, 1819 (1978).

65. (a) H.M. Gibbs, B. Bölger and L. Baade, IX Int'l Conf.of Quant. Elec. (1976), Opt. Comm. 18, 199 (1976). (b) G. Forster and P.E. Toschek, Quantum Optics Session of the Spring Annual Meeting of the German Phys. Soc., Hanover, F.R. Germany (Feb. 1976). (c) W. Krieger, G. Gaida and P.E. Toschek, Z. Physik B25, 297 (1976).

66. (a) L.A. Bol'shov, V.V. Likhanskii and A.O. Napartovich, Zh. Eksp. Teor. Fiz. 72, 1769 (1977) Moscow; (b) M.J. Ablowitz and Y. Kodama, Phys. Lett. A70, 83 (1979); (c) L.A. Bolshov and V.V. Likhansky, Zhurnal Eksperimental'noi i Teoreti cheskii Fiziki, 75, 2947 (1978) in Russian (Eng. trans.: Sov. Phys. Jrnl of Exp. and Theo. Phys. 48, 1030 (1979)).

67. L.A. Bol'shov, T.K. Kirichenko, A.P. Favolsky, (USSR Acad. of Sci., Math. Div. preprint, Fall 1978), Inst. of Appl. Math. 1978, 53 (3A) in Russian.

68. J.J. Bannister, H.J. Baker, T.A. King and W. G. McNaught, Phys. Rev. Lett. 44, 1062 (1980) and XI Int'l Conf. of Quant. Elec., Boston (1980).

69. F.A. Hopf (private communication).

70. (See ibid ref. 11-13.)

71. G.T. Moore and M.O. Scully, Coherent Dynamics of a Free-Electron Laser with Arbitrary Magnet Geometry. I-General Formulation, Opt. Sci. Ctr, Univ. of Arizona, preprint (brought to the author's attention by Professor F.A. Hopf).

PANEL DISCUSSION ON THE PHYSICS OF OPTICAL BISTABILITY

INTERNATIONAL CONFERENCE ON OPTICAL BISTABILITY
ASHEVILLE, NORTH CAROLINA, U.S.A.
JUNE 5, 1980

CHAIRPERSON:

Professor Elsa Garmire

University of Southern California, Center for Laser
Studies, Los Angeles, California 90007

TRANSCRIBED BY:

Edward Seibert and Lambros Johnson
Physics Department, Drexel University
Philadelphia, Pennsylvania 19104

TOPICS FOR DISCUSSION

1. Minimum size, energy and time projections for use of optical
    bistability.

2. Best materials and techniques.

3. Limitations for use of optical bistability in systems.

4. Cavity effects; transverse effects.

5. What are sources of noise and their effects?

6. What experiments would please the theorists?

7. What theories would please the experimentalists?

8. What are general properties of optical bistability systems?

GARMIRE: Let's begin with estimating the minimum size, energy
            and time response for bistable devices and the projected
            uses given those estimates.

BOWDEN: For item number one, I nominate Hyatt Gibbs.

GIBBS: In terms of size we are now talking about devices which
            are of order of 5 to 10 times the wavelength in vacuum.
            One may think of going down to dimensions which are of
            the order of the wavelength in the material λ/n. This
            will obviously involve guided wave techniques and may be
            in planar geometries or in fiber configurations. There
            is a lot of work that must be done before one can real-
            ize actual systems along these lines. But in principle,
            I think, the limitation does not stop until you approach
            that λ/n; and there are materials such as semiconductors,
            which allow one, in those kinds of volumes, to get down
            to numbers of atoms of the order of 1000 to 10,000.
            This is probably about as low as one wants to go from a
            statistical consideration anyway. Those energies are,
            I think, at least at the GaAs wavelength, 0.25 fJ for
            1000 such photons. So you are talking about a few fJ
            which does not come out badly in relation to high speed
            electronics. So I think initially one would talk about
            systems which would be of specialized applications where
            you may want "all optics" or you may want it to be immune
            to electromagnetic interference or whatever. You might
            see these things begin to come first. The ultimate
            speeds; again, there is reason to believe that you can
            switch devices on in picoseconds even if it takes con-
            siderably longer time to cut them off. We are beginning
            to learn from Scientific American articles that some
            Josephson-Junction devices work the same way. They flip
            on rather rapidly, in tens of picoseconds and can't cut
            off rapidly. So they cut the whole thing off every nano-
            second to return to the other state. So the point is
            that you can, if you can flip something one way quickly,
            do a chain of computations very rapidly. You can make
            100 such computations in a series configuration before
            you reset and go again.

S.D.SMITH: Just to follow up Hyatt's comments. I'm looking at semi-
            conductors. What I would like to forcast is that we'll
            discover another 5 or 6 different systems which produce
            nonlinearities for dispersive bistability in the next two
            years. Bell Labs have found one good system resonant
            with an exciton; we at Heriot-Watt have found another
            good system resonant with a band gap. And there are
            numerous other systems: an important point is that some
            of them definitely have picosecond relaxation times
            (i.e.p-Ge). They won't necessarily have the other

parameters optimum, but I'm sure it would be possible to switch off in picoseconds. The other point on the semiconductor side undoubtedly is that we are already in agreement and already have shown that the cavity size should be small for fast build-up times, and I go along also with the prediction of the dimensions, anything between 1 μm^3 and 10 μm^3; this is possible. Also, with energies: between 1 pJ and 1 fJ (but it might be that for certain purposes rather larger energies are necessary for very fast switching).

MILLER: I would like to emphasize something else which I said in my talk. I feel there is a basic physical tradeoff between switching speeds and switching powers, and I feel that perhaps with semiconductor systems, where we know a lot about things going on in picoseconds, it will be possible to play about with the systems a bit for that tradeoff. I think that's a basic tradeoff; you can't have both, you either have one or the other.

GIBBS: You can have a small switching energy that takes you one way.

MILLER: Yes, but if you want to come back you've got to trade off and use more power overall in the system.

GERLACH: In this last discussion one thing that I'm wondering about is this: If we are going to pack an optimum number of such devices, say one separated from the other by roughly 5 or 10 times the wavelength of light, if we're going to start firing these various devices, is the power necessary to drive them not going to melt the whole system?

GARMIRE: It's important to stay away from absorptive materials.

P.SMITH: We may ask where it is likely that bistable devices could compete favorably with other technologies which are already available, say Josephson junctions or semiconductor electronics. It seems to me there is one area that we can now identify where we have some clear advantages and that's in the area of high speed response. If we can start talking about picoseconds of response, we are already out of the range of Josephson devices, out of the range of semiconductor electronics. We have the field to ourselves and there are a number of devices which we can at least project down to these picosecond times. Maybe I can point out that one of the reasons we are interested in pursuing these nonlinear interfaces has to do with the fact that there is no resonator involved. Using a nonlinearity with a fast response we expect to make very fast responding devices.

S.D.SMITH: Can I follow the last two remarks by commenting that
               there seems to be no problem in making a two-dimensional
               bistable device. There are crystals of InSb that are
               1 cm across; so, in principle, you can put a lot of de-
               vices on that just by putting beams on it. So a com-
               pletely new data processing procedure by using the bi-
               stable switching, storing and amplifying, could be
               thought of here wihch has never been considered. It
               means that you might not have to switch so fast, and at
               least from the packing point of view it's promising.

GARMIRE: There are plenty of people who might have thought about
               that, perhaps the people from Ohio University should
               comment.

GERLACH: What we're using is a liquid crystal where we have been
               able to put quite a number of independent flip-flops,
               for example, on a 1 inch crystal. The thing here is to
               think in terms of systems and how to actually use them.

S.D.SMITH: That's what I meant ...

GERLACH: We've got the basic flip-flop; it can be switched so to
               speak, but the point is a matter of data processing -
               data routing is the key concept really. That's a big
               area - some new ideas have to be introduced here. The
               purpose being the utilization of the parallel processing
               capability of such crystal devices.

GARMIRE: I point out that there is an entire field of image pro-
               cessing in the holography parallel processing field
               that has been looking for nonlinear optical processing
               devices for a long time. There are bistable liquid
               crystal light valves in similar devices that do that.
               These devices are of great interest because you can send
               an image through the bistable device and process it and
               enhance the image.

P.SMITH: Maybe we should emphasize this point. If you make a
               Fabry-Perot resonator which has a lot of degenerate
               modes then you could essentially resonate your energy
               with some resolution depending on how many modes have
               low losses, and because you are using an intrinsic non-
               linearity with spatial resolution you can do image pro-
               cessing in a simple direct way with a slab of material
               in a suitable resonator. This intrinsic nonlinearity
               allows you to do that, very simple.

GERLACH: The problem would be, of course, to have this bistability,
               say, over one mode, work independently of the bistability
               over another mode.

P.SMITH: You need a highly degenerate resonator to do that, or a
               large plane parallel resonator.

GARMIRE: In a simple way, each little element must be such that
               diffraction would not occur over the length of the reso-
               nator. This gives a rough rule of thumb of what you'd
               expect the independent element size to be.

SANDLE: The thing to do would be, I think, to explicitly move
               into the self trapping regime and you could develop a
               stable element.

P. SMITH: That's a bit tricky, experimentally.

GARMIRE: We'll get to the transverse effects in a while. Let's
               stay for now on the first question. So we've estimated
               speed and projections for use, we identified the optical
               computer where we're talking about very small sizes and
               volumes, and we've identified the signal processing or
               a parallel computer.

TOMLINSON: We're talking about an optical computer, I think. The
               sort of array processors we've been discussing is: doing
               a very simple operation with a large number of elements.
               I think it's very difficult to build complex processing-
               functions. This sort of thing will be useful if you
               want to do simple operations very fast, possibly with a
               very large number of inputs.

GARMIRE: With parallel operations.

TOMLINSON: Yes, of course.

MILLER: I think where we've talked about the idea of using the
               optical systems to do something that we can already do
               with electronics; we must be very careful, because the
               laser is much more expensive than an electrical cell.
               So I think we should have something in mind which the
               optical system can do, which cannot be achieved in other
               ways, as Peter has said, which is into the high speed
               area. I would see that as being a very likely first
               application.

McCALL: I have a couple of things: In the semiconductor business
               there are elements, for example, charge coupled devices,
               which remember essentially without any holding power.
               They just stay that way without dissipation of power
               while they're on; certainly we can think of bistable
               devices that require no energy to stay in one state,
               namely, just burn a hole in a piece of paper. The
               trouble is they don't turn off ever again. Can someone
               think of a way to, say, make a bistable device which
               does not need holding power? If you could, there will
               probably be another conference, where we would show
               each other's companies...

GARMIRE: Use light to put atoms in an excited state and let them
              stay there for a long time.

McCALL: A long time, yes, right, and when you want to turn it
              off, connect with another level and then they decay back
              down. The other thing, from the theoretical work: Hyatt
              and I spent a lot of time saying, "well, what material
              shall we use next," and learned a lot about materials
              and so forth. So we spent a lot of time trying to fig-
              ure out which material we should use in our experiments.
              We weren't originally trained in materials science;
              there are other people who know more about materials.
              If any of the theorists here know about materials please
              make suggestions.

GARMIRE: Well, you went smoothly into the second question. I
              think, a very relevant one. What are the best materials?

S.D.SMITH: I'm not going to answer, InSb. Just a sideline from
              what Sam said about energies in the last remark. It
              might be of interest to notice the minimum holding power
              we got in our last experiment is almost within the range
              of semiconductor diode lasers which is an important point
              from the point of view Sam's making, the choice of the
              best material. I'm going to forecast there will be a
              paper by Gibbs, McCall, Miller and Smith, In Ga Sb As.
              That would have an energy gap of 1.8 µm. It will have
              an absolutely infinitely sharp absorption edge with a
              little (excitonic) effect to keep Sam happy. It'll
              have a χ^3 of approximately 10 esu. Then we can all go
              home quite happy!

TOMLINSON: Another response to Sam's question about a device without
              holding power. There was a recent paper by Chang and
              Boyd at Bell Labs, a bistable liquid crystal device.
              There are two stable configurations of the liquid crys-
              tal which have different polarizations. It's still in
              an early stage, they don't have good ways of switching
              it. However, it is something that is stable.

GERLACH: What's the response time on that type of liquid crystal?

TOMLINSON: Slow.

GERLACH: Milliseconds?

TOMLINSON: Lots of milliseconds.

GIBBS: I would like to add, under the second item, the search
              for materials. I think that there are various ways to
              study things and one is to decide to study a particular
              area, of, say, solid state physics and then learn all
              you can about it. I think that Sam would agree that we
              found it equally fascinating to direct the research

from the point of view of bistability. You start looking at materials and see what kinds of things give interesting effects; and the physics we ran into in this
process is just as interesting as the physics encountered
if one is just trying to study materials. I don't know
if the agencies are listening, but it seems to me that
there are various ways to approach the study of solid
state systems, and having another motive in terms of the
direction of the kinds of things you look at can be very
helpful. For example, specifically looking for nonlinearities which are very large. In so many experiments,
people use such high power lasers that anything that we
are interested in appears down at the origin and would
not be considered as a nonlinearity at all, since they
are looking at nonlinearities that occur for MW or GW/
cm^2. So that kind of a guide through the subject of
studying materials, I think, is equally as good as con-
centrating on a particular class of materials.

GARMIRE: That's important because obviously there is a trade-off
              here between nonlinearity and absorption. For example,
              in your materials search, one must include the size of
              your material and the intensity of the beam required
              and there are many requirements you can put into the
              search; nonlinear index is just one of them.

MILLER: I would like to go on record by saying again that the
              real figure of merit for a refractive bistable system
              is the ratio of the nonlinearity to the absorption co-
              efficient. In terms of practical materials like GaAs
              and InSb, in both cases you can work even with quite
              low absorption coefficients, from a semiconductor view-
              point, and even with these you do not have long cavities.
              So you are not talking about switching times limited by
              cavity effects. Once you get down below that level where
              your cavities are so thin anyway you are not bothered
              about the cavity buildup; then this ratio is the figure
              of merit you are looking for.

GARMIRE: I would like to have some of the people that were work-
              ing on the two-photon type of effects speak as to
              whether or not those are really viable competition to
              the one-photon effects. Is there anybody willing to
              address that?

S.D.SMITH: Well, you can say something straight away. If you can
              take what Christos Flytzanis was implying this morning,
              could one do something like that in InSb with a 10 μ
              laser and 5 μ gaps? The power densities that you have
              in order to have any appreciably two-photon effect is
              up by 10^6 or more, so that the holding powers and the

energies for that process really will be up by that sort of amount. I don't see any alternative to that; the reason for that is that you don't get such a multiply resonant situation in your nonlinearities.

GARMIRE: Any other comments on that? There seems to be a universal agreement that the multiphoton situations are not very practical.

DeTEMPLE: It seems to me that if you go to a three-level system, just forgetting about the two-photon case, you might be able to use the third level as a metastable level, and do what Sam was talking about. You need to deal with a state that lives there for a reasonable amount of time, if we are going to talk about a long-term storage. You turn it on and you take your radiation away; maybe in that case we are really talking about a two-photon process, one strong wave and one weak beam.

GARMIRE: That's a good suggestion. Any other comments on materials or techniques for fabricating bistable optical devices? How about limitations?

P.SMITH: Let me make one short comment in response to Dave's remark about the ratio of the nonlinearity to the absorption. I agree in the case of the Fabry-Perot device; however, there are other types of devices such as this nonlinear interface where the considerations are different and that would not be an appropriate index.

MILLER: Yes, I would like to pin you down on what you think the figure of merit really is because there you are talking about a very small penetration into the nonlinear material and basically linear absorption is not going to matter until it gets down to a wavelength.

P.SMITH: The linear absorption is likely to create a problem because of thermal effects way before you get to this point of absorption in the wavelength. It's likely to be a kind of a dirty factor that is doing your limitation rather than the fundamental limitation of destroying the Q of the resonator.

GARMIRE: Would you like to comment on how important you think the interfaces would be compared to the small Fabry-Perots?

P.SMITH: The main point I can see for the interface is that we have no resonator, we are simply pulsing the light off the surface and because the interaction region can be made very small we are going to have a very fast response providing, of course, we are using a fast responding nonlinearity. Now there are several questions that we cannot answer precisely because we don't have a

good enough handle on the theory. One of them is how
tightly can we focus our beam - that's obviously important in terms of using the minimum power and having
the minimum interaction length. The other is what is
the limitation on the switching time. It seems obvious
from physical reasons that it must be shorter than the
transit time of the light across this interaction region,
which for the particular experiments I described yester-
day, is 25 picoseconds for a particular set-up. It's
obvious that it's faster than that. We believe that the
physical picture has to do with the differential time
between the propagation of the surface wave along the
interface and the propagation of the light that is incident on this surface and that the actual time response
has to do with the different propagation constants of
these two light beams; but we don't have a good theoretical picture for this, so I can't give you an exact answer. It's going to be very fast.

GARMIRE: How is your intensity compared with the holding intensities for the other Fabry-Perots?

P.SMITH: You will have to compare one of the same kind, a Fabry-
Perot full of Cs_2.

GARMIRE: Is it not going to be a lot higher?

P.SMITH: No. The powers are of the same order.

S.D.SMITH: If one uses the semiconductor materials and the nonlinearities we discovered and combine them with Peter's
method for interfaces, then there are immense possibilities. And the holding powers will be lower.

GARMIRE: If one were to use a semiconductor, what will be the advantages of the nonlinear interface over the Fabry-Perot?

MILLER: I think, Peter has just said that the interface gets
around the difficulty of making a Fabry-Perot with de-
cent finesse, but the analogous question you've got to
ask for the interface is, how good the interface has to
be, how smooth does it have to be, how good does it have
to be to compare with the wavelength and that's an engineering point that has to come up.

GARMIRE: Making a Fabry-Perot is just an engineering problem,also.

MILLER: Yes, I mean, I think that we are nowhere near knowing
how good the interface has to be. By contrast, we can
sit here and do sums with finesses and talk about how
to polish, we know what finesse of a cavity is and how
it is easily made.

S.D.SMITH: There's a second advantage, you don't really have a deep
               absorption problem should you use a semiconductor above
               its absorption edge and frequency. It's quite possible
               this criterion about the absorption may be different.

TOMLINSON: However, in the nonlinear interface, the field does not
               penetrate very deeply, it does travel along the inter-
               face for a considerable distance. You don't want to
               have too much absorption over that length.

MEYSTRE: I just want to make a comment about the surface quality
               of the interface. I think you might be right that this
               could be a problem for bistability, but on the other hand,
               these surface waves might be useful for spectroscopic
               studies of surface defects. The surface wave phenomenon
               is something that must be investigated, not only in the
               context of bistability.

McCALL: In the context of these two devices we've been talking
               about, if heat dissipation is the limitation, then given
               two materials and requiring certain fractional absorp-
               tion in order to achieve a given change in refractive
               index, the criterion would be: which requires more change
               in refractive index times volume. I don't know the an-
               swer, but, off hand, it sounds like the Fabry-Perot re-
               quires less.

P.SMITH: It requires less by approximately the finesse of the
               Fabry-Perot.

McCALL: Yes.

P.SMITH: So, we're talking about factors of 5 or 10 in the kind
               of cases that have been done experimentally.

McCALL: Also, it will be interesting to put a mirror at each
               end.

GARMIRE: It might be easier to heat sink the nonlinear interface
               than a Fabry-Perot, which would require a transparent
               heat conductor, or something like that.

P.SMITH: Perhaps.

KAPLAN: Some additional comments. I would like to say that the
               problem of absorption remains important because although
               the depth of penetration can be small, the length of
               travel of the wave in the material remains relatively
               long. The length of propagation of the wave along the
               interface can be very comparable with the waves in the
               Fabry-Perot resonator. About the possibility of new
               spectroscopy, I like to remember that in my first papers
               there were suggested two possible applications of this
               phenomenon for nonlinear spectroscopy. You know, there

is a linear spectroscopy of the total internal reflection, and I believe that similar spectroscopy, but nonlinear spectroscopy, can be proposed.

GARMIRE: Any more comments to what we've talked about so far?

GERLACH: One wishes to take more seriously this idea of parallel processing as compared to the series processing and then talk about the figure of merit. One might here in essence talk about the nonlinearity per unit area of that parallel processing device. In other words, how many nonlinear devices, bistable or multistable, with the order of stability being proportional to the nonlinearity, can you fit into a unit area? That might be an additional figure of merit that has to be considered in that context.

GARMIRE: That leads very directly into the next subject, that of transverse effects. Obviously, there were a lot of theoretical papers discussing plane waves and some preliminary theories discussing transverse effects. Obviously, because of self-focusing and other phenomena, transverse effects have been demonstrated to be important in a lot of these nonlinear situations. I am opening up this subject for comments. What are these transverse effects going to do?

BOWDEN: I would like to pose the following question in regard to the limit cycle oscillations that have recently been predicted by Bonifacio and Lugiato in the upper bistable state of absorptive optical bistability in the presence of several of the cavity modes: If transverse effects are considered, what happens to these limit cycle oscillations?

McCALL: I was talking with Rodolfo about that last night. Of course, you have to take a suitable average across the beam to see if the excess gain due to the Rabi frequency still persists, since the gain shoots suddenly negative at that region. I think that these would be big effects. I tried to do the integrals and I did them several times and I didn't get the same answer twice. I don't know the answer. Rodolfo tells me he is going to do the integrals.

GARMIRE: So you're saying there will be a big effect and it will probably wash out limit cycle predictions?

McCALL: It would certainly reduce it. Also standing waves will reduce the effects.

MEYSTRE: Not in the ring cavity.

McCALL: Not in the ring cavity; of course not.

GARMIRE: There are no standing waves in the ring cavity.

GIBBS: I'd be surprised if it does anything important to the
                switching times involved just because the gain is
                Gaussian instead of plane wave. Basically if you are
                working in a diffraction limited region then whatever
                happens in the center rather quickly gets communicated
                to the rest. If you are diffraction-limited and some-
                thing breaks loose in the center, that diffracts and
                will affect the rest. In these calculations with Mattar,
                we've looked at the superfluorescence case. If you have
                a Gaussian gain profile and you don't allow for any
                radial communication, you see very long delays out in
                the wings as you would expect. But if you allow the
                communication, as soon as the center starts to super-
                fluoresce it spreads the message to the rest of the beam
                and it very quickly starts to superfluoresce as a whole.
                So, I think that if you have a transverse profile, the
                critical intensity is still at the center of your beam,
                as that switches up then soon the rest follows.

GARMIRE: I suppose an analogous situation would be to say that a
                laser mode has a given threshold. Even though the in-
                tensity is higher at the center of the mode than at the
                edge of the mode, the entire mode begins to lase when
                threshold is reached.

S.D.SMITH: I wonder if we can get on record a statement of the
                actual conditions used in the experiments where disper-
                sive bistability has been seen. I'd like to know, for
                example, exactly what part of the beam or how much of
                the beam Hyatt used in his experiment, and Dave can
                state what we had in ours.

MILLER: Well, the simple fact is we usually use something that
                is pretty much like a Gaussian beam and the whole thing
                switches on at once and you look at the whole curve.

GARMIRE: The dimension of your beam at the resonator is what?

MILLER: We have been working with beams of the order of 100 μ
                at the resonator.

GARMIRE: And your resonator is how long?

MILLER: The resonator is of the order of a few hundred microns –
                very much less than a diffraction length, a Rayleigh
                length. That's the situation we are in all the time.
                We also think, incidentally (just an order of magnitude
                argument) that there is very little defocusing taking
                place inside the crystal. But we always look at the
                whole beam. It doesn't matter at all if the beam on a
                particular day is not a very good beam. We still get
                bistability and the whole thing switches at one time.

GARMIRE: Have you at any time looked at any portion of your beam?

MILLER: No, it's not meaningful in our experiments because we
               always observe at the diffraction far field.   So there's
               no point in looking at a portion of the diffraction far
               field because it's giving you information from all of
               the near field.

GARMIRE: Do you see any expansion of your beam at the far field?

MILLER: Yes, there's an expansion of the beam in the far field
               as you go through a transition.

MATTAR: When Professor Smith was at Max Planck I understand that
               there were some problems... Now apparently they have
               vanished?

S.D.SMITH: I think that you misunderstood actually. We were in the
               early stages of observation; our nonlinearity is nega-
               tive. We do get self-defocusing and we thought in the
               beginning it was necessary to select part of our beam to
               observe bistability. In fact it isn't. We can observe
               it over the whole beam. That's what David stated here.

GARMIRE: And it looks the same over a part of the beam as it does
               over the whole?

MILLER: Well, in the far field, yes; that's not to be thought
               surprising at all. At the time we spoke to Mattar we
               hadn't solved the propagation problem; we didn't know
               what the propagation was inside the crystal. Since then
               we have put together a very simple model; all that hap-
               pens is that basically the phase alters and the shape
               of the beam stays the same inside the crystal.

GARMIRE: Let's have Hyatt's comments.

GIBBS: I think there are certainly situations where the self-
               focusing can be very important. Zelber and Marburger
               have shown that under certain conditions self-focusing
               can lower the threshold. I would agree with essentially
               what Dave said relative to the semiconductor devices
               where they are very short.

GARMIRE: What was the dimension of your beam?

GIBBS: Essentially 10μ diameter, but again the focal length was
               long compared with the sample.

GARMIRE: And essentially Gaussian?

GIBBS: Yes.

S.D.SMITH: Were you looking at a complete Gaussian profile? Or were
               you looking at a portion of it?

GIBBS: We were looking at the entire thing. I mentioned the
              other day that in the Na experiments we did have other
              situations where we had a beam which was not focused
              down tightly because it was going through a long cavity.
              It was a plane wave-like propagating beam, but with a
              Gaussian profile. Due to instabilities of a nature
              which we didn't track down, dust particles or whatever
              else, there were phase variations on the front. Having
              such a long distance and the proper detuning from line
              center, we saw the beam breaking up into filaments. You
              could clearly see that there was a different threshold
              for different filaments depending on how they happened
              to trap. We showed some of those crazy bistable curves
              that have those bistable hysteresis loops stuck all over
              the place, which I think brings up something that is quite
              interesting from the transverse point of view. That is,
              how close can you bring two beams next to each other.
              If you want separate devices you may have to really
              separate them by other materials, and there might be
              other reasons to do that in order to shorten the re-
              laxation time. But take the simple case, how close can
              you bring two beams and have them operate as independent
              devices?

GARMIRE: Very important. Something that theoreticians can work
              on for us which will be very useful.

GERLACH: We haven't been talking so far, until you mentioned this
              last point, about the integrity of the beam. How good
              is the integrity of the beam? We could take the alterna-
              tive viewpoint that you just raised - take the beam as
              consisting of different units and ask now about inter-
              action between different beams so to speak. And now ask
              and answer questions about cross talk between those
              beams. Now, cross talk is ordinarily an undesirable
              thing, but if it is controlled with the help of feedback
              or what have you, then we are getting into the central
              area. This area concerns itself with this idea of data
              routing. The data is to be transferred from one beam
              to the other and so forth.

SANDLE: Two separate points: One is that a Gaussian beam has a
              pronounced effect on absorptive bistability.

GARMIRE: These are experimental results?

SANDLE: Well, yes, theoretical as well. In a diffraction dom-
              inated mode, such as a single mode of a spherical sym-
              metric Fabry-Perot, it has an insignificant effect on
              dispersive bistability.

GARMIRE: Can you be quantitative? How much effect?

SANDLE: Yes, there's roughly doubling of the threshold for absorptive bistability.

P.SMITH: You mean Gaussian as opposed to plane wave?

SANDLE: As opposed to plane dominated. The reason for that is that several passes and some buildup is needed to trigger the absorptive case. Each successive pass involves diffraction spreading and so the beam doesn't take advantage of having been eaten through. Maybe it gets concentrated through the center, but on the next time through it's a very spread-out beam so that the advantage of multiple buildup is lost in the absorptive case. But in the dispersive case, as long as one can ignore self-focusing and defocusing effects, there's no change, and therefore there is really very little difference for the Gaussian beam situation compared to the plane wave with similar average intensities.

GARMIRE: Can you give some of your experimental results? Experimental dimensions?

SANDLE: Our Na length was 2 cm, beam diameter was 80 μm, the cavity finesse without the Na vapor was 55, cavity gain 4.5, the switching power for absorptive bistability was 135 mW.

GARMIRE: What is the cavity gain?

SANDLE: How much more intense the field is inside compared with what it is outside. Theoretically, that is related to the finesse. Every experimentalist knows it isn't related to the finesse very clearly because you have additional losses and they have different effects on the gain than they do on the finesse. You need to specify both of those quantities. The second point I wish to discuss is on self-focusing and defocusing. This is important to dispersive devices because one is operating at high intensities in order to get significant phase changes. However, if one is near the absorptive-homogeneously-broadened condition and is operating either in the absorptive region or across the absorptive region with αL being small then self-focusing and defocusing is negligible.

BOWDEN: I have a comment to inject in regard to the importance of intrinsic J^2 conservation in models for optical bistability. Models based upon the Dicke Hamiltonian in one mode are intrinsically J^2-conserving, and this condition has to be taken as an artifact in any discussion of steady-state properties. For instance, atomic cooperative effects in resonance fluorescence have been analyzed by several authors and the results are not

independent of the intrinsic J^2 conservation in the
models. The question now arises as to the connection
between predicted limit cycle oscillations in optical
bistability and intrinsic J^2 selection rules in the
models. If steady-state results are not independent of
J^2 conservation, it would appear that they cannot con-
nect with experiments, since it would be difficult to
design an experiment that selects and preserves J^2
selection rules in the true steady-state.

CARMICHAEL: The way I understand the limit cycle that Professors
Bonifacio and Lugiato expect to have, they exist on the
upper branch,which from my understanding,has quite a
different value of J^2 from the lower branch.

BOWDEN: It still has definite quantum numbers for the angular
momentum.

CARMICHAEL: It does?

BOWDEN: Sure.

CARMICHAEL: Throughout the cycle?

BOWDEN: I think so.

McCALL: The atoms are in phase with one another, somewhat. So,
you add to get the total Bloch vector and have a finite
expectation value.

BOWDEN: It certainly does if you describe the situation by a
Hamiltonian which is J^2 conserving; it has a definite
value for the angular momentum quantum number for any
given steady state.

CARMICHAEL: How are these limit cycles understood in terms of mode-
mode coupling - that's the question?

McCALL: The limit cycles will be understood in terms of gain at
another frequency and it happens to be the inverse of
the cavity length.

SENITZKY: Can you say something about limit cycles? I'm sorry I
don't understand what you mean by limit cycles.

BOWDEN: By limit cycles, I'm referring to the particular paper
given by Professor Lugiato, which predicts instabilities
associated with the upper state of bistability leading
to oscillations of the internal field in the asymptotic
sense.

GARMIRE: There are regions, in what we call the steady state
curve, which,he discovered, under certain conditions,
will oscillate.

BOWDEN: In a nutshell, what it means is that since in the upper
state you have Stark splitting, and if in addition to the

principal mode in the cavity you also have another mode which happens to overlap with the Stark splitting of the atoms, the modes get coupled by the atomic system.

CARMICHAEL: Limit cycles are quite different to conservative oscillations. They're dissipative things. Instead of having a dissipative relaxation to a fixed point, you have a relaxation to an oscillatory solution.

BOWDEN: It's really a steady state.

CARMICHAEL: It's a steady state, yes; but it's basically in a dissipative non-converving context.

KAPLAN: In terms of the phase space, what does this mean?

CARMICHAEL: If you plot trajectories in the phase space then in the limit they tend towards oscillations.

McCALL: But in the context of small devices, the oscillations that Bonifacio and Lugiato talked about won't occur because there aren't cavity modes close enough. However, the oscillations that I talked about in a publication several years ago could be a blessing or a problem, depending on the application. That is the case when there are two different kinds of n_2 and they oppose each other, for example.

GARMIRE: Are there any other comments on transverse effects in cavities?

MILLER: I'll just return to what we've been discussing before. The idea of deciding when you have two separate beams and when you have one beam operating at once, this is something that is of great importance, I think, experimentally for the idea of setting up two-dimensional types of devices. If anybody wants to solve it, please will they do the defocusing case first because when I was trying to solve this business of propagation in InSb I had to wade through an awful lot on this self-focusing question. Nobody did the defocusing and it's something like 100 times easier to do!

GARMIRE: That's in the thermal blooming literature.

MILLER: But thermal blooming is a different kind of defocusing. Thermal blooming is a diffusion type defocusing, whereas if you just have a χ^3 which is negative you get a different kind of defocusing. It's very easy to solve, but I couldn't find it anywhere in the literature. Please, will you do the defocusing case first?

DeTEMPLE: Does anybody know what effect level degeneracies have on bistability?

GARMIRE: What level degeneracy?

DeTEMPLE: So far most of the analysis is in regard to non-degenerate
              two-level systems. Most real systems that you consider
              may require this. For a gas you have the degeneracy
              built into it. What is that going to do?

McCALL: For absorptive bistability the polarization, which I re-
              fer to as P, so often will have different matrix elements
              in the case of degeneracy and this will reduce and will
              harm optical bistability because the maximum in the cur-
              rent will be reduced so that the value of Γ or C will be
              larger. For dispersive bistability well out of resonance,
              it will make no difference.

GARMIRE: If I'm understanding you right, what you are saying is
              that if you have more than two levels and you try to
              saturate the absorption, obviously it takes a lot more
              intensity to saturate a degenerate level than it does a
              single level.

McCALL: I think the important thing is that they have different
              saturation intensities, so that the nonlinearity be-
              comes less nonlinear.

DeTEMPLE: The degenerate system should be less susceptible to self-
              focusing than the nondegnerate system, because you have
              a spread in the Stark split level.

GIBBS: But it's not clear that that is the case for some of the
              semiconductors, because the mechanisms are quite dif-
              ferent. Whether you can talk about a two-level system
              or not you can produce carriers that screen out exciton
              features regardless of whether or not you're using a two-
              level description.

GARMIRE: It certainly seems though there is a place for someone
              to use an accurate model for the semiconductor and do
              the theoretical calculations. All the theoreticians
              laugh; you know, they want two-level atoms. Well, it
              looks like the experiments are going to be in semicon-
              ductors. So let's have some models of semiconductors;
              they exist in the literature, with densities of states
              and all the rest, and apply these to optical bistability.

S.D.SMITH: There's nothing wrong with using two-level models as a
              first model for semiconductors. Just put a lot of them
              together and change their frequencies.

SENITZKY: The basic principle is the resonance. And there must
              exist a resonanace in semiconductors also, although it's
              probably much wider.

MILLER: Yes

SENITZKY: The essence is the resonance and the nonlinearity. Those
               are the constituents that produce the phenomenon.

GIBBS: My point is that the absorption at one frequency which
               produces carriers helps to saturate not only that par-
               ticular transition but also has just as strong an effect
               on saturating excitonic transitions far off resonance
               as well.

SENITZKY: But through the resonance?

GIBBS: Yes, but it might saturate rather sharply; that's my
               point.

McCALL: I think the point is this: If you have degeneracy but
               there's rapid population equilibration among the sub-
               levels, population between the various sublevels, in
               the ground and in the excited state, then it's effectively
               a two-level system.

GARMIRE: Sure, but we're also talking about building these things
               to respond on the order of picoseconds; now we're getting
               close to the intraband relaxation times and clearly the
               next conference is going to have to have some semicon-
               ductor theorists present.

SENITZKY: As long as we are dealing with the subject of resonance,
               perhaps I can suggest a question that is of conceptual
               interest. What is the relationship between optical bi-
               stability and resonance fluorescence? Does anyone have
               any ideas on the subject?

MEYSTRE: Gigi Lugiato explained that to me one day. The big dif-
               ference is the cavity (a big crucial difference) and you
               have more than one atom, but the big difference is the
               boundary conditions.

CARMICHAEL: The boundary conditions are necessary to give feedback.
               If you put all the atoms in a small enough volume, as
               Bowden pointed out, they get feedback anyhow because
               they are so close to each other.

SENITZKY: My feeling is that optical bistability is associated
               with coherent resonance fluorescence and its collective
               aspects. It's the collective aspects of coherent
               resonance fluorescence that create the phenomenon of
               optical bistability.

KAPLAN: Only one possible kind, when you are using two-level
               systems. There's another optical bistable system which
               doesn't use a resonance of the medium.

SENITZKY: You're talking about nonresonant phenomena.

KAPLAN: Sure. There's a lot of approaches, such as J. Marburger's,
               a simple analytic analysis.

SENITZKY: Maybe other types, yes. Certainly the collection of resonant two-level systems and the resonant cavity produce the phenomenon of coherent resonance fluorescence and a collective effect.

GARMIRE: That's one form of bistability.

SANDLE: Can I come back to this question of changes with the Gaussian beam? I have quantitative data. The value of C is round trip absorption over 1 minus R, and for the plane wave ring cavity it is 4, and as everybody knows for the plane wave Fabry-Perot standing wave it is 4.97, whereas for the Gaussian ring cavity it is 8.31 and for the Gaussian Fabry-Perot cavity it is 10.04. So that's what, roughly doubled? The threshold power however is very different. In the absorptive regime Y^2, which is defined as the on axis average intensity, is just the intensity in a ring cavity, and it's the sum of the propagating and counterpropagating intensities for a plane wave civity divided by the saturation value. Y^2 as defined, is, of course, 27 for the plane wave ring cavity and it's 28 for the plane wave Fabry-Perot cavity. However, it rises to 249, an order of magnitude increase in the absorptive region. By comparison, in the dispersive region, the dispersion, of course, depends on the detuning, and on C. The power is proportional to (Δ^3/C) and the coefficient of proportionality depends on the model. It is 0.77 for the plane wave ring, 0.52 for the plane wave Fabry-Perot, 1.54 for the Gaussian ring and 1.02 for the Gaussian Fabry-Perot. There is very little difference for the dispersive regime, an order of magnitude increase in threshold in the absorptive.

GARMIRE: Those are basically all theoretical?

SANDLE: That's theoretical, they are supported by our data.

GARMIRE: Based on Bloch equations, mean-field theory?

SANDLE: That's the Bloch equations, two-state, mean-field theory, supported by the relative values as supported by our experiments in the standing wave cavity case. To follow along that line, in connection with the non-linear interface, it was implicit in the data that Peter showed. For the Gaussian beam that we considered there, the normalized intensity at which we begin to see the surface is .6 → .59. That's the normalized intensity at the peak, at the center of the Gaussian. In the same units, the plane wave intensity, for which the plane wave theory predicts a threshold is, .5. So that the threshold intensity at the peak of the Gaussian seemed to have to

be only a bit greather than 20%, greater than the value
predicted by plane wave theory.

GARMIRE: I'd like to go on to the next question concerning the
effects of noise in optical bistability because there's
been a lot of talk about noise here from the point of
view of the theorists, and because the experimentalists
are very interested in it. We want some real numbers.
I'd like to make an invitation to see if we can come up
with anything definitive. For one thing, I'll ask the
experimentalists what do they think the real sources of
noise are in their experiments, so they can tell the
theoreticians what they should be calculating. I would
like the theoreticians to tell us, for example, where
should we do our switching. Where should we sit with
our devices when they're in the bistable regime in order
that the signals have good integrity and not be going
into the other state via noise?

P. SMITH: Can I add an additional question that will throw in some
light? I would be interested to hear the theorists address the question of what are the noise mechanisms and
sources that are going to limit us in trying to go to
small, low power devices? What are going to be the
fundamental limits in terms of making devices smaller,
faster, and lower switching power?

GARMIRE: How few atoms can be used?

KAPLAN: It is so small, I think, that some instrumentation noise
must be more important.

FARINA: Sam gave the example of a thousand atoms; I took his
numbers literally. And, as far as quantum fluctuations
go, we said that C was equal to 40, Q was equal to .02.
I'm sorry, N_S was about a thousand, so Q came out to
be something like .02. Now, just by taking the data
from the numerical solutions that we got, we get something like 10^{300} cavity decay times for the effect of
the quantum fluctuations on that system. Now those are
rough estimates, you know; N_S may not be quite right,
or there may be other deviations. There could be systems in which quantum fluctuations are important and
when you get maybe a thousand atoms, where N_S is not
equal to a thousand, then if C is not equal to 40, then
you may run into the problem where Q is one. Now, I
have to weigh the quantum fluctuations with fluctuations
of the laser source, thermal fluctuations of the cavity
or whatever the situation is.

GARMIRE: In a very practical way, I don't understand the steady-
state operating point for the bistable device. Are you

sitting in the center of the hysteresis curve for that calculation?

FARINA: For example, not at the center of the hysteresis curve, but if you plot the expectation value of the transmitted field right in the transition region, right in the middle there, that's the indication that we have where the longest tunneling time exists.

GARMIRE: Yes, because many people who consider switching talk about reducing their switching energies by operating right at the edge of the hysteresis near the switch-up point. Obviously, that gets to be a much noisier situation and it would be nice to see some real engineering numbers. In order to design any kind of information processing system, there are considerations like "bit error rates" and there will be a tradeoff between the energy requirements and the "bit error rates." I would invite any of the theoreticians to do a little looking into some of this engineering, because that will be very important.

McCALL: I want to comment that the devices that we have worked with so far, the main source of noise, of course, is not quantum. However, as you make the devices smaller and smaller, the quantum noise will grow whereas the effects of other kinds of noise will roughly stay constant, so that quantum noise is a subject, valid today.

GARMIRE: It's a very valid point.

McCALL: Now switching up using small energies, I don't believe, is a very imporatant consideration because the "fan out" capabilities of the bistable device is, at least as well as we can envision it now, quite large. You have available large quantities of energy for switching another device. The amount of energy required to turn on a device is considerably smaller than the change in output intensity during a time for which one device can drive quite a few other devices.

GIBBS: Fan out?

McCALL: Fan out, yes. Fan out capability.

GIBBS: The bistability is larger than the widths you have.

McCALL: But that's just supposing that you switch it on by injecting light into the side of the cavity and not through the mirror so the fan out capability is roughly the reciprocal of the transmission of the mirrors.

GARMIRE: Yes, but there are reasons why you need to reduce the power consumption. That's not a power consumption argument.

McCALL: That's the switching energy required to switch the de-
               vice.

GARMIRE: You're saying one device can switch equivalent devices?

McCALL: Yes.

GARMIRE: The switching energy should be a minimum if you don't
               want to heat up your devices.

McCALL: I think you mean holding power, too much holding power.

P. SMITH: We should really distinguish between two kinds of energy
               here; energy which is dissipated in the form of heat is
               very important if you're considering making a large num-
               ber of devices in a small space. A small space is, of
               course, important if you want these things to communicate
               fast and so on. However, there is also the question of
               how much light intensity do you need to cause this thing
               to switch over, though the dissipation is very low, and
               this energy can be used somewhere else. These are really
               two different things.

GARMIRE: I disagree that they're two different things because if
               you're going to switch a lot of information, you're
               going to have a duty cycle which is 50 percent, then the
               two are comparable.

P. SMITH: No! Not true! One is going directly into heat and that
               causes a problem which in many cases is the limiting
               factor in the usefulness of devices. I refer you to
               Keyes paper of some years ago, at IBM. He did a study
               of fundamental limitations on devices. Heat is very
               important from a practical point of view. But another
               important consideration is simply how much power or how
               much energy do you need to put in to switch, even though
               most of this you may get back in the form of some re-
               flected wave or some transmitted beam or something, and
               these are separate. These affect the system in very
               different ways.

GERLACH: Is the reflected energy really going to be dissipated
               in some external heat sink, or is the reflected energy
               going to be dissipated by another little device that
               caused the switching in the first place?

P. SMITH: Now you have some flexibility here. Now maybe you can
               do clever designing and utilize the reflected energy to
               do some other processing function. The energy is avail-
               able there, whereas if it is dissipated in heat, it's
               not only not available, but it's caused you a serious
               problem which you have to worry about.

KAPLAN: I think that eventually all energy must be transferred
               to heat.

GARMIRE: Exactly, that's what I'm saying, in the ultimate device.

P. SMITH: No, no, wrong! You can have a device and you can go to
              10^3 switching operations and most of the light can come
              out the end.

FARINA: Using a Keyes type argument and a...

P. SMITH: I still don't see that. It just seems to me the light
              can jump out the ends of the system.

FARINA: Using a Keyes type argument, saying that the fundamental
              limits for a memory tape operation is somewhere around
              kT, using an estimate at room temperature for a 100
              picosecond switching time, over an area of square wave-
              length, say a micron, you get something like 300 Watts
              per square centimeter over a 100 picosecond time scale.
              That's a tremendous amount of power to get rid of. There
              will be instantaneous heating and this is a problem that
              people had with semiconductors for a long time. Thermal
              resistance measurements came out but that doesn't tell
              you what the instantaneous thermal resistance is. The
              junctions will be over 200°C when thermal resistance
              measurements tell them it's only supposed to rise 1
              degree C. What happened?

GARMIRE: Well, clearly, cooling is important, 300 Watts per square
              centimeter is by no means absurd. Those are standard
              numbers in electronics, to be cooled.

FARINA: No. When you're trying to make a memory, you want to
              make a disc or something and you find that when you cram
              those things together, all these things are sensitive to
              thermal effects.

GARMIRE: Yes.

FARINA: And there is a question of cross talk between devices.
              Now, if you want to make these things very dense, you
              have to design devices that have very little thermal
              effect. They don't feel anything thermally.

GARMIRE: Certainly. These Fabry-Perot's have to be stable
              thermally or you'll drift the frequency of the cavity,
              among other things. OK, let's go back to the noise
              problem, are there other sources of noise?

GERLACH: Well, it seems to me so far we discussed, at least men-
              tioned, quantum noise, and thermal noise.

GARMIRE: Consider incident laser intensity fluctuations. If you
              buy a laser it's typically rated at 3% stability. Now,
              what is 3% fluctuation in the laser intensity going to
              do to the hysteresis curves?

MEYSTRE: Well, I think that intensity fluctuations will not be as
              important as phase fluctuations. They are your basic
              fluctuations. You cannot avoid them, and I have a feel-
              ing that in a long cavity, this may be a problem. In a
              short cavity I don't see any difficulty because there
              the laser linewidth will be much smaller than the cavity
              bandwidth.

GARMIRE: Can you look at it in the sense that the laser has a
              coherence time and the coherence time of the laser has
              to be longer than the round trip time of the cavity
              times the finesse?

MEYSTRE: Yes, I made a few computer runs, and although it's much
              too early to know really what's going on, my preliminary
              results seem to indicate that when the two bandwidths,
              laser and cavity, become comparable, you start running
              into problems.

MILLER: Can I make a comment on the whole idea of phase switch-
              ing? I think this is going to be quite unimportant if
              we work with devices where the cavity buildup time is
              much shorter than the material response time. And none
              of these, phase switching, frequency switching, none of
              these things is going to be important. If the devices
              are in the short cavity limit, then very happily we can
              forget about that problem which is a fundamental problem
              with the laser. The laser goes along for perhaps a long
              time and then it does a very sudden "glitch" in the phase.
              It could be a real problem if we do not work in the short
              cavity limit.

SANDLE: I was just going to say that, getting back to the long
              cavity situation, we suspect laser frequency fluctua-
              tions as the most probable reason for our observed dis-
              agreement of a factor of 10 (in addition to the factor
              of 10 I spoke about before) between observed and ex-
              pected thresholds. In fact, we talked to Pierre about
              that.

GARMIRE: There is another advantage for using short cavities or
              nonlinear reflections.

S.D.SMITH: Coming back to this question of what the practical noise
              mechanism will be. It's very hard to make a laser with
              an amplitude stability of less than about a percent, but
              in the information processing role of bistable devices,
              there are two interesting points here. One is that one
              will use pulses to switch it and the second is that the
              power limiting properties of the bistable devices them-
              selves will, in fact, probably fit quite nicely into
              this. And one will have to have, so to speak, a pre-
              optical circuit with power limiters to condition your

pulses before you start to use them. The properties of
the power limiter are likely to be good enough, they
will be much the same as the device itself, to operate
the bistable switch. I think that's slightly helpful.

P. SMITH: Yes, good point.

GARMIRE: I wondered about this question of phase switching. I
would like to know what convenient, small, phase switch
one can buy that's loss less, since presumably a phase
switch is loss less. How can I come along with a 180
degree glitch in the phase to my light? I don't know
where to buy one. Have you looked at how long electro-
optic modulators are on the 4 micron scale that we're
talking about?

McCALL: Clearly, you take a length of material and you hit it
with a light pulse. It changes refractive index very
fast.

GARMIRE: How? Well, isn't that called a bistable device?

McCALL: No, a switching device.

GARMIRE: How small can you make it? That's my question. Unless
we can get devices down to this 4 micron size that we're
talking about, I think we're talking about devices that
won't ultimately be used.

McCALL: Well, there are big phase switches, big phase changes in
the output of dispersive devices when they change state.

GIBBS: To hit one of these bistable devices (semiconductor)
from the side and change its index, you're doing a very
similar thing, aren't you?

GARMIRE: Well, right. So, why bother with a phase switch?
You're doing an intensity switch.

GIBBS: If it's operating in a dispersive mode, I'd suspect
it's very similar to a phase switch.

P. SMITH: You're making it more sensitive because you're inside a
resonantor so you don't have to make a bigger refractive
index change.

GARMIRE: But, in principle, you're doing an intensity switch,
namely you have something that's a function of intensity that's doing the switch. We're getting down to
semantics now.

SENITZKY: To say something qualitative rather than quantitative,
the operation in what is called the cooperative branch,
seems to be very sensitive to the phase relationship
between the incoming beam and the oscillating dipole
moment of the device, while in the noncooperative branch

	there is very little dependence on phase relationship. Thus phase instability will be much more important in the cooperative branch than in the noncooperative branch.
GARMIRE:	Pierre, can you respond to that?
MEYSTRE:	I would like to hear you say the opposite, but I don't know how to justify it right now. I have to think about it, because we find the opposite to be true.
GIBBS:	It's more sensitive in the upper state?
GARMIRE:	So you find the upper state more phase sensitive. Well that would go along with the results of Lugiato that see oscillations in the upper state. I don't know quite how to translate what you said to compare it to theirs.
MILLER:	I would just like to make a comment about this whole idea of cooperation. I very much doubt whether in the semiconductor devices cooperation actually has anything to do with the operation at all. I think we're operating in really dispersive or refractive bistability (which ever you prefer to call it) regime, and I really think that cooperation just doesn't come into it at all. And it only becomes important, and I think I'm right in saying, when you get near a large absorption case or into the case where you are getting some contribution from both absorption and dispersion.
GERLACH:	One additional source of noise that will become important ultimately is cosmic rays. When we're going to make these gadgets smaller and smaller they're going to have switching energies which will be comparable to the amount of energy that a cosmic ray would deposit into this refractive medium via Cerenkov radiation or what have you.
GARMIRE:	That is one of the ultimate limits of VLSI and it seems to me that if we want to know what the ultimate limits are we should look at the technology that already exists in the understanding of VLSI which defines the minimum transistor size.
KAPLAN:	To exhaust this problem, maybe the systems can be good detectors for cosmic rays.
CIFTAN:	The technology of VLSI does not yet exist in our strict sense. We are trying to develop it, we are pushing toward it. I think you may contribute to it in these sort of discussions.
GARMIRE:	It is not clearly understood what the ultimate limit is in VLSI; there is a lot of theoretical work; there should be more.

CIFTAN: This problem is very fundamental.

GARMIRE: Yes, it's very fundamental.

MEYSTRE: The theorists around here don't seem to know what VLSI
              is.

GARMIRE: Very large scale integration. The basic question in
              very large scale integration is how many transistors
              can you put on one silicon chip? Well, you make them
              very small. How small can they get? When you get down
              to where you have only a few electrons or holes in your
              transistor region, you're going to start having statisti-
              cal deviation and a lot of noise. This has been observed
              experimentally by Texas Instruments, among other people,
              who have made these very small devices. The transistors
              become very noisy and they cease to work.

CIFTAN: Just to give you an idea, we are thinking of something
              like, 40-60 angstrom devices. Transport, Botlzmann
              transport equation cease to apply. You have to go into
              quantum effects. Energy levels go berserk. Surface
              effect is everything. At interfaces, just a little bit
              of roughness changes all.

GARMIRE: They also get to be very fast in which case they may
              put us all out of business.

FARINA: Electronic devices?

GARMIRE: Well, if the electron transit distance is 40 angstroms,
              that's very fast.

CIFTAN: This is speculative. That's what I want to say.

KAPLAN: I would like to repeat my suggestion. Maybe it is pos-
              sible to use bistable devices like Geiger devices.

FARINA: Even optically?

GARMIRE: It's certainly possible to make picosecond detectors
              with a bistable device.

KAPLAN: Sure. It is possible to make some special devices.

DeTEMPLE: Especially related to the cosmic rays, they found that
              just the alpha particle background from the encasement
              to these semiconductor devices is contributing to the
              "bit error rates" and, in fact, they distinguish be-
              tween two different modes of operation. A static memory
              type system is more sensitive to these errors as opposed
              to rapid dynamic memory. What that means as far as what
              we're talking about here is that the instabilities in
              the switching have to be taken within the context of
              how you're ultimately going to use the device. If you
              operate it over a very long period of time, noise

instabilities are going to be a much more severe problem than if you operate a very, very short cycle.

McCALL: I think it's relevant here. Computers when they work make mistakes all the time. They have algorithms built in to correct their own mistakes.

GIBBS: Biological systems do that too. There are ways to check, compute in parallel and check your answers. If you get a different answer, you just do it again.

GARMIRE: Well, the field of information theory deals with that.

FARINA: The system also gets bigger when errors are checked.

GIBBS: I know, but if you're saying that that's the fundamental limitation on how small you can get, then you can make it 10 times smaller and do two in parallel.

GARMIRE: I would like to solicit some of the theorists to justify why their approach to the noise and fluctuation problems are better than the other theorists' approach to the fluctuation problem. Being an experimentalist, I was interested in the different approaches and I just wondered if anybody is willing to summarize differences.

CARMICHAEL: I don't think that the theorists feel that there's that much difference in their approaches.

FARINA: As a matter of fact, I think a lot of what we've done seems as if we're all doing the same thing.

CARMICHAEL: It's irritating.

GARMIRE: Well, then, how come you don't have any hard numbers for us? OK, the next question on the list to discuss is what experiments would the theorists like to see the experimentalists do and what theories would the experimentalists like the theoreticians to do. We've obviously touched on a lot of this, but I would like to open it up for more.

HASSAN: The small volume situation described here, is it possible or not?

GARMIRE: You want us to come up with N atoms in lambda?

HASSAN: Something like that, within a very small volume.

GARMIRE: How many is that?

BOWDEN: I suggest looking at thin atomic beams as Professor Agarwal suggested at the end of my talk, or thin semiconductor films at low temperature, or something like that. The important thing would be to verify optical bistability without mirrors, and for that you would need a small volume. The atoms must be able to "talk" to each other before they can relax or become dephased. Therefore, the photon transit time in the material must be shorter than any other characteristic time. This does not mean that you need to work with a volume smaller than a wavelength, necessarily. There has to be a preferred mode, however, into which the atoms tend to radiate; this is determined by the polarization of the input field and also, probably, by the geometry of the material. I believe that defraction could play a very important role.

McCALL: Ammonia molecules might do the job for you. If you want to do that you really want to do r.f. experiments, don't you?

SENITZKY: I think an experiment in which a sudden phase switch in the incoming beam can be made would be very interesting and shed more light on the phenomenon.

GARMIRE: I think so.

DeTEMPLE: With regards to that, let me ask Hyatt – Do you have a feeling for what T_2 is in the semiconductors that you've been looking at? Is that truly in the picosecond regime?

GIBBS: I assume so.

S.D.SMITH: From the mobility values it's got to be picoseconds.

MILLER: It is definitely known that the processes taking place inside semiconductor bands are on a timescale of picoseconds. It's very easy to calculate. It doesn't matter what scattering mechanism you're looking at if it's phonons, if it's ionized impurities, anything, they all happen on that sort of time scale.

DeTEMPLE: Well, does that suggest that if you overlook phase effects, you should be in a gas phase then because of the longer T_2 time?

MILLER: I would think so.

MEYSTRE: I don't know because if you want to compare with our predictions you have to remember that we eliminate the atoms. That means that you want to have short T_1 and short T_2 in terms of cavity time.

McCALL: What's the cavity time?

MEYSTRE: The inverse cavity bandwidth. The system was cavity dominated.

McCALL: Long cavity?

MEYSTRE: Yes.

McCALL: Long cavity time.

BOWDEN: What about the other case when you're atom-dominated?

MEYSTRE: It's hard.

BOWDEN: Something I would like to see someone do is an experi-
            ment to observe the fluorescence spectrum. This has
            not been done to my knowledge.

MEYSTRE: Maybe I can just mention that there is a group in Munich
            setting up an experiment to do this kind of measurement
            with several atomic beams in parallel in an evacuated
            Fabry-Perot and stabilized lasers. The beams are work-
            ing.

GARMIRE: These are what? Sodium?

MEYSTRE: Yes, I think so.

HASSAN: Is that transmitted spectrum or scattered spectrum?

MEYSTRE: I think they will want to look at the side. I don't
            know; that's my guess.

McCALL: I would like to know what is the difference between
            looking at the spectrum from the side from a bistable
            device or just looking at the medium from the side with
            two laser beams going through the material. Is there
            a difference?

GARMIRE: That's the difference between bistability and super-
            fluorescence.

McCALL: Would the spectrum be different? That's essentially
            my question.

GARMIRE: Isn't that the whole point of the theoretical work
            they've been doing?

McCALL: Yes, I think the answer is yes; the spectra are the
            same.

KIMBLE: Well, there is a difference. In one point in the
            hysteresis cycle, as you come down the upper branch and
            before you switch to the lower branch, at that point
            the internal cavity fields are high enough to see an
            A.C. Stark effect, but in fact what's predicted is a
            very narrow spectrum. So it's different than having
            an atom in two counter-propagating fields of the same
            strength as the intracavity field.

GARMIRE: Isn't it true then that what is predicted is that at
            critical slowing down the line width goes to zero?

McCALL: I don't think so.

GIBBS: What's the difference between an atom inside this cavity
              undergoing optical bistability and the same atom sub-
              jected to those same fields? Why can't you say it's
              just resonance fluorescence? What is it that's so novel?

SENITZKY: Although optical bistability means you have two stable
              states for the same incoming field, it's not for the
              same outgoing field or for the same field in the cavity.
              In the cavity there's always a unique field for a given
              atomic behavior. In other words, if you look at the
              curves along one axis you can see several values but
              along the other axis there is only one value. So, in-
              side the cavity there is a unique field for a certain
              atomic behavior.

GIBBS: I certainly agree. Now why isn't the spectrum what you
              would expect for that field subjected to a single atom?

McCALL: How does the atom know it's part of a bistable device in
              that situation?

SENITZKY: The atom knows what field it sees. For a given field,
              it behaves in a certain way. It's the collective be-
              havior that's different.

McCALL: In that case if I take the mirrors away and shine the
              same field on the atom, then it doesn't know the dif-
              ference?

SENITZKY: That's right.

CARMICHAEL: Well, I don't know what really to say, perhaps I'm just
              restating what a number of people have said. I think
              what a number of questions are getting at is that people
              want to resolve what this word collective means. I
              would very much like to see an experiment done, with
              collaboration between experimentalists and theorists, to
              resolve this situation and tell us whether there is col-
              lectivity in the specific case of the 2-level medium in
              a Fabry-Perot, whether there is collectivity and if
              there is, exactly what it is.

DeTEMPLE: What are the signatures that you're looking for?

CARMICHAEL: What are the signatures? Well, that's part of the ques-
              tion, isn't it? The model I'm talking about is the
              Fabry-Perot or "ring cavity." It's not important as to
              the cavity geometry, but the model which closely cor-
              responds to the 2-level system, resonance fluorescence
              generalized by feedback.

DeTEMPLE: Yes, but some of these people are throwing in that atom-
              atom correlations and things like that are important.
              Doesn't that give you a different answer if you don't
              have these?

CARMICHAEL: What I'm asking about is in the laboratory; atom-atom
             correlation may be there or not. Well, what are the
             signatures then? What are the theoretical signatures?
             Between the problems, you put that in, you get an answer,
             or you take it out and get a different answer, a dif-
             ferent spectrum.

McCALL: Look,just take a lot of atoms and just shine a light
             beam on them; they adopt similar phases. You don't need
             the mirrors; you don't need bistability for that. I
             think that's just a way of describing the polarizability
             of the medium.

CARMICHAEL: But it is different phases that they adopt, implicit in
             the mean value of the polarization. The questions of
             collectivity should be referred to statistics and not to
             mean behavior which, indeed, of course, is just deter-
             ministic behavior, as you say. You force a certain
             oscillatory phase on the mean dipole moment. Collective
             behavior must be defined in terms of statistics and to
             get at the statistics the easiest way is to look at the
             fluorescence spectrum. So you need prediction for what
             that spectrum is, an identification of what the features
             are which distinguish it from a noncollective spectrum
             and then experiments to show which case you observe.

McCALL: The last point is that, what's the difference if you
             have mirrors or not? You still have the same behavior,
             collective or whatever you wish to call it.

CARMICHAEL: When you have mirrors, the atom at this end can talk to
             the atom at that end, and the atom at that end can talk
             to the atom at this end. Without your mirrors they're
             only talking in one direction.

McCALL: If you have two light beams going opposite to each other,
             the same thing applies, because one atom absorbs a
             little light or it radiates some more and it goes down
             and interacts with the other atoms.

CARMICHAEL: I don't think that's correct, because even in the case
             of the two modes in a ring laser, which go in opposite
             directions, there's fundamental differences in that be-
             havior and the forward and backward waves of a Fabry-
             Perot. The couplings are different.

McCALL: Well, I can't agree about an atom down stream in a ring
             laser being the same. Yes.

GARMIRE: Well, I think the question is well taken and we'll pose
             it to the theoreticians to go home and think about what
             is exactly the thing that experimentalists will see when
             they measure the spectrum in this case that's different
             from the case of resonance fluorescence.

BOWDEN: There is one point I wish to inject. I believe it
            should be emphasized that the key experiments will be
            with regard to observation of the fluorescent spectrum.
            For instance, the equation of state looks the same for
            absorptive optical bistability as it does for optical
            bistability caused entirely by collective effects and
            you can't really tell the difference, at least for the
            on resonance, steady state case. The difference will
            occur in observation of the fluorescent spectrum. That
            will be the key measurement. This gives details of the
            statistical behavior of the system.

GARMIRE: Yes, but the question is what is unique to bistability
            as opposed to resonance fluorescence?

WALLS: There is the spectrum of the fluorescent light. This
            has been calculated for absorptive bistability by Lugiato,
            and Agarwal, Narducci, Gilmore and Feng. The general
            fluorescent spectrum for both absorptive and dispersive
            bistability has been calculated by Carmichael, Drummond,
            Hassan and Walls.

GARMIRE: OK. What is different? Look I'm an experimentalist,
            what do I go and look for?

HASSAN: Before I say something about the spectrum, I would like
            to come back to the basic definitions of resonance
            fluorescence and optical bistability. Resonance fluores-
            cence essentially concerns the interaction of a single
            atom (or a small group of atoms) in free space, e.g. in
            an atomic beam, with a resonant cw laser field, which
            implicitly means that the processes of absorption, re-
            emission and hence scattering of radiation is taking
            place. Now obtical bistability describes the behavior
            of a relatively large group of atoms placed inside a
            cavity and driven by a cw laser field: in general there
            is both a cooperative interaction between the atoms and
            feedback through the mechanism of the cavity. The cavity
            is not essential, but the interatomic cooperation is
            (this is driven by the external field and disappears
            in the strong field limit where single atom chaotic be-
            havior dominates). In the weak field limit the group of
            atoms exhibits the usual dielectric behavior, that is
            it develops a refractive index and a behavior dependent
            on the geometry of the group of atoms. Thus, essentially,
            optical bistability arises through the cooperative in-
            teraction of resonantly fluorescing single atoms. Our
            work shows that in this many atom resonance fluorescence
            case, if the system has J^2 (total angular momentum) not
            a constant of the motion (due to e.g. independent atomic
            decay), the system is bistable and the spectrum on the

lower branch is a single broad Lorentzian; whilst on the
upper branch it is just the three peak Stark spectrum,
similar to the one atom case. On the other hand if J^2
is a constant (and for this I am talking about the small
volume situation in which only cooperative atomic decay
is allowed for) the system does not exhibit bistable behavior (as mentioned in my talk) and the spectrum is
again a broad Lorentzian for a weak laser field, whilst
for a high intensity laser, the spectrum is a three peak
Stark spectrum with additional side-bands which occur
at all harmonics.

WALLS: In bistability we predict a very broad spectrum on the
lower branch with a linewidth of the order of $N\gamma$ where
N is the cooperativity parameter and γ is the natural
linewidth. This is indicative of a cooperative decay.

KIMBLE: But that's not a cooperative effect. That is to say it
is no manifestation of cooperativity in bistability because if I take N independent absorbers, as long as
their absorption rate is fast compared to the cavity
decay rate, I'll see that same broad line. What you
said is simply a statement that if you take a high
finesse cavity and muck it up with an absorber, you'll
have a broad linewidth. This point is demonstrated by
a classical calculation for independent atoms. I think
that the manifestation of cooperativity is when you
approach the switching point and the line narrows.

WALLS: No, the manifestation of the cooperativity occurs on the
lower branch where the atoms decay collectively with
decay time $1/N\gamma$ giving rise to the broad spectrum. This
occurs far from the instability point of the lower
branch. The line narrowing which occurs as one approaches
the instability point indicates a "critical slowing down"
of the time scale for any fluctuations away from the
lower branch to return to the branch. This occurs in
both equilibrium and non-equilibrium systems near an
instability point. To continue with my discussion of
the spectrum of the fluorescent light for optical bistability, as one increases the driving field the spectrum develops a narrow peak which grows and narrows to
zero as the instability point is approached. The system then makes a discontinuous transition to the upper
branch where since the atoms evolve independently the
spectrum is the usual three peaked spectrum of single
atom resonance fluorescence. In single atom resonance
fluorescence the three peaked spectrum evolves continuously from a single peaked spectrum as the driving
field is increased, whereas in optical bistability the
three peaked spectrum with well separated Rabi sidebands

will appear discontinuously as one increase the driving field past the threshold for stability of the lower branch.

CARMICHAEL: But Sam McCall would say that you can understand that simply in terms of the fact that the intensity inside the cavity has now gone up and everybody expects that the intensity inside the cavity is now strong enough to produce sidebands, but it's not strong enough to give you sidebands on the lower branch in optical bistability.

GARMIRE: Several of us feel we understand how the device works and that we think we would understand the spectrum. Obviously it would be nice for someone to measure the spectrum to check out our understanding of it. I think those of us who feel we understand it, know what the experimentalists will get. Let's suggest the experimentalists measure it and find out.

MILLER: I think, getting back to your simple point, Sam, (nice and straightforward) what is the difference between putting in two beams in opposite directions and putting a cavity there with only one beam? Of course, there is no difference in terms of the internal field inside the Fabry-Perot from the laser in the first place; it looks exactly the same. The difference must be that you're feeding back the atomic polarization in a way that you don't get if you took the atoms sitting in free space. So I get some atomic polarization from a given atom coming back not just in one direction, but it comes back in the opposite direction as well. So it's a different physical situation. If it happens in free space, the atomic polarization of that atom only goes one way and never comes back on itself. It's different physics.

McCALL: Shine two beams at once.

MILLER: I'm talking about the atom talking to itself; it never talks back to itself unless it knows its own reflected field.

GARMIRE: It's a question of feedback.

McCALL: Look, I understand a little bit about this, and I know that in the beam that comes out of the bistable device in the forward direction, the statistics are different. It's the process to the side I'm asking about. To look at noise in the forward direction is essentially impossible right now, until we get to the very small devices. It's the fluorescence to the side that's the question.

GARMIRE: Allright. That's a very good point. I hope all of the theoreticians understand the difference. He's talking

about specifically the noise coming out the side of the
atoms and indicating that the fluorescence is going to
be different than the statistics of the light coming out
the front.

SENITZKY: Let me say something about the cavity. If you look at
an individual atom and you irradiate it, you will get
both coherent resonance fluorescence and incoherent
resonance fluorescence. Now what the cavity does is
concentrate the coherent resonance fluorescence which
in turn changes the field, the coherent field, that each
atom sees. What the cavity does is collect the coherent
resonance fluorescence of the atoms and, in that sense,
the cavity is important because it affects the total coherent field that the atoms see.

TOMLINSON: I would just like to suggest that perhaps some theorist
might like to consider the problems of the nonlinear
interface. Here you're very much intimately involved
with the transverse properties and not only that but you
really need to take care of both transverse dimensions
because they're quite different, unlike a Fabry-Perot
type device where you have a cylindrical symmetry. Here
the two transverse directions are very different and
with our brute force numerical calculations so far we've
only been able to take account of one of them, and even
there, only over a limited range of parameters. We
can't handle as big a beam as we would like to do, and
some new theoretical approaches, theoretical tricks or
tools that would give you at least some analytical
handle on at least one of these dimensions would be very
helpful because there's still a very fundamental question in my mind. What is the limiting time response of
these devices?

S.D.SMITH: As an experimenter, I would like to see some theory
relevant to the circumstances where observations have
been made. To be a little kinder, to say that: could
some of the lovely material that's been presented this
week for atomic states be reworked for the conditions
in semiconductors, please? That would be very interesting and it's bound to yield a lot of interesting
results, I think. And the second point is even more
practical. Please obtain some numbers out of transverse
effect calculations that again we could think of applying to an experimental situation. Neither of these
things has been done and there's a big field here.

CARMICHAEL: Could I ask what specific conditions in semiconductors
one must model? You mentioned previously that, essentially, one can take a particular model and just take
different detuning and frequencies. Is that asking

anything more sophisticated than this?

MILLER: It's not adequate, it's a start. It's something to get
               you off the ground, but there's no doubt that semicon-
               ductors are very complicated quantum mechanical systems.
               I mean, if you're talking about atoms which are separate
               in space you don't get transfer of excitation from one
               atom to another. Potentially, however, every state in
               a semiconductor is coupled to every other state and all
               extended states attempt an optical and nonoptical coupl-
               ing with other states in the system. Now we take the
               very simplest possible model because experimentalists
               essentially want something to see whether we're in the
               right ball park. But there's an awful lot of room if
               somebody wants to move in and really look at semicon-
               ductors. For example, is there a cooperative response?
               That's an extremely involved theoretical point.

SENITZKY: And without some resonance, one wouldn't get anywhere.

MILLER: With *some* resonance, yes. But I say we're in the regime
               where, with two-level atoms, equivalently you've got
               nonlinear refraction, but linear absorption. That is,
               as you move away from resonance, the absorption becomes
               linear faster than the refraction. That's a simple
               theoretical regime that we think we're in. You then
               start to sophisticate that as you move to take in more
               resonance terms, and there's lots of scope.

S.D.SMITH: I think the essential difference in the physical param-
               eters is time, the time concerned with things such as
               what the Rabi frequency is, and in relation to various
               sorts of scattering processes, and the relation between
               the buildup of the fields. These sort of times are im-
               portant considerations, just restating what David says,
               but breaking it down to what you would do to get from
               your atomic time theory into solid state transient theory
               of bistability. That would be a tremendous step, of
               course.

GARMIRE: One place you might look is in the theory of semicon-
               ductor lasers, in which there's obviously been a lot of
               models made of semiconductors and how they behave from
               band to band, assuming this is a band to band nonlinear-
               ity. Are there other experiments that the theorists want
               or theories that experimentalists want? I think we've
               already covered a lot of it.

MATTAR: Sam, Hyatt, and I, are building a theory which is
               modular enough so you can remove the Bloch equations as
               a black box and put in different nonlinearities that
               you want. There was a lot of effort being done to make

the program as self-consistent as possible in order to make the change that you want. It was tested, each part alone, modularly, and hopefully in the Fall we will have a sample computer comparison like what was shown this morning in optical fluorescence. But the main idea of going to this large computation was to collaborate with experiment. This is what we have tried to do for the last 4 or 5 years: work hand in hand, theory and experiment. We solicited things, detailed explicitly what we may want to see computed in the range of parameters and already, Bonifacio, and Lugiato suggested that we may ask some questions for the computer to compare between their mean field theory and our rigorous calculation.

MEYSTRE: I would like to see an experiment done where people would look at the steady state behavior in any system they want, I guess, as a function of laser line width. Maybe this has been done but I don't know.

McCALL: What is that again now?

MEYSTRE: What I would like to see is if the hysteresis cycle is reduced if you increase the laser line width.

McCALL: I might make one comment about that. In our sodium vapor experiment I believe it was half a megaHertz shift that completely wiped out a nice bistability curve.

SENITZKY: I think experiments that would evaluate the effects of the cavity, a good cavity, a poor cavity, or no cavity, would be interesting and would shed light on the understanding of the phenomenon of optical bistability.

GARMIRE: Certainly ring cavity experiments would be welcome, since the theoreticians find ring cavity theories more correct than standing waves. It would be nice to do ring cavity experiments as checks.

SENITZKY: A ring cavity is still a cavity even though you have a traveling wave.

MEYSTRE: Yes, but the point that she's trying to make is that you don't have standing waves in a ring cavity.

SENITZKY: Yes. You eliminate standing waves, but there are people here who think one could have the effect without a cavity.

CARMICHAEL: I don't think there's anyone here who thinks you could have an effect without a cavity. A lot of people have posed questions saying if you let your cavity transmission go down to zero, then nothing happens.

SENTIZKY: But didn't you just say a while ago that if you have a large number of atoms close together you can get optical bistability?

McCALL: You need feedback.

GARMIRE: We're now down to the last question which is what we're
               really already talking about. What are the requirements
               really necessary for bistable systems? What is a bi-
               stable system and what ones can be predicted for the
               future in physical systems and theoretical systems?

S.D.SMITH: Could I start by just summarizing the numbers that have
               been quoted here?

GARMIRE: Surely.

S.D.SMITH: The first point is that bistability has been observed with
               cavities as bad as a finesse of one and as good as a
               finesse of 55. Holding power densities as low as 10
               Watts per square centimeter and as high as megawatts per
               square centimeter; a tremendous range. Absorption co-
               efficients as low as 0.1 cm^{-1} to 10^4 cm^{-1}. Cavity
               lengths from 1 micron to a meter.

GARMIRE: That's a good set of numbers.

BOWDEN: Well, it seems to be generally regarded that there are
               really 3 categories of intrinsic optical bistability:
               absorptive, dispersive, and what I call cooperative
               optical bistability. It appears that the current experi-
               ments are more pertinent to the second category, that is
               dispersive optical bistability, whereas the theories,
               for the most part, have addressed absorptive optical
               bistability. So, now, where does that leave us?

CARMICHAEL: I think a lot of theories have addressed dispersive bi-
               stability as well. I don't think it's any longer the
               case of just absorptive bistability.

BOWDEN: How well do they explain the experiments?

CARMICHAEL: Certainly as well as the experimentalists.

GARMIRE: I think there's a long way to go before we have con-
               trolled experiments. The experiments are very hard to
               do and very hard to get good control of.

McCALL: I think the experiments that have been done have been
               explained by the people who've done the experiments.

GARMIRE: Well, within orders of magnitude.

McCALL: Within factors of 10 or something like that.

GARMIRE: Yes, but the theoreticians are presumably looking beyond
               factors of 10.

McCALL: But when we do experiments, we are confined by what is
             available and what we are willing to do. For example,
             we're not willing to work with "kilometer" long lasers.

GARMIRE: Gaussian beams are available.

McCALL: Gaussian beams are available. We have those plane waves,
             and so forth.

GARMIRE: Infinite plane waves are not available.

KAPLAN: I would like to remember the situation in nonoptical
             fields, this hysteresis and bistability. You know that
             there is a bistability of the kind of feromagnetic bi-
             stability. And what is the main feature? Well, there
             is a pumping; the feromagnet can change state and when
             you switch out, turn off the pumping, it can enter the
             new state. It is one kind of multistability. And
             there's another kind which is known, at least in Russia,
             as oscillating hysteresis, oscillating multistability in
             nonlinear circuits. And all kinds of optical multi-
             stability which we've discussed at this meeting are kinds
             of oscillation hysteresis and multistability.

P. SMITH: I think what you're trying to describe are flip-flops.
             We call these flip-flops.

KAPLAN: Maybe flip-flops. And one question arises; I would like
             to point out an additional optical problem, which was
             not discussed in our meeting. Maybe it is a very well
             known problem. We always are interested in shortening
             the time of single operations, switching and so on.
             But you know that every computer system requires a long
             term memory, as well as logic elements and a short term
             memory.   So the question is why are we not interested
             in the phenomena which give you the possibility
             of long term optical bistability and long term optical
             memory which must actually be equivalent to the fero-
             magnetic hysteresis? It must be something, some new
             gyromagnetic affair, but I'm sure there must be a lot of
             effects which can exist without pumping. All the first
             pulses, switch, turn on such devices. When you turn off
             your field, the light field, this phenomenon must give
             you the possibility of conserving the new value, the new
             state of the medium. Why are we not interested in such
             phenomena, which give you the possibility of providing
             a memory of the pumping? It is a question. And if such
             phenomena in such media are known now, well-known now,
             why are we not interested in shortening the relaxation
             time of such devices. I am sure that in some regions
             they can be used in short term logic operations.

GARMIRE: On those questions we will end. Thank you all for your
             participation.

BOWDEN: Just a few final announcements, a farewell, and then we
             adjourn. We had, I think, a very fruitful conference
             and I know it's been very stimulating to me and I want
             to thank you all for coming and making it possible that
             we should have such a conference. Before you go, how-
             ever, one final word, and since it has been such a good
             conference, it's on the other side of it to end with a
             very sad note. We just received word that Mario Gronchi
             died of cancer. He is a friend of many here and a
             colleague of some and we end on that sad note. The
             organizers would like to propose that we dedicate the
             Proceedings volume to Mario Gronchi, since he has con-
             tributed a great deal to this field and subject. So,
             with unanamous agreement the meeting is adjourned.

Genossar, J.	Technion, Haifa, Israel
Gerlach, U. H.	Ohio State University
Gibbs, H. M.	Bell Laboratories, Murray Hill
Goldstone, J. A.	University of Southern California
Gozzini, A.	Instituto di Fisica, dell'Universitá di Pisa, Italy
Gragg, R. F.	University of Texas at Austin
Graham, R.	Universität Essen, Federal Republic of Germany
Gronchi, M.	Instituto di Fisica, dell'Universitá, Milano, Italy
Hassan, S. S.	The University of Manchester, United Kingdom
Hermann, J. A.	Imperial College of Science and Technology, England
Ho, P. T.	University of Southern California
Hopf, F. A.	Optical Sciences Center, University of Arizona
Johnson, L. G.	Drexel University
Kaplan, A. E.	Massachusetts Institute of Technology
Kimble, H. J.	University of Texas at Austin
Lovitch, L.	Instituto di Fisica, dell'Universitá di Ferrara, Italy
Lugiato, L. A.	Instituto di Fisica, dell'Universitá, Milano, Italy
Mandel, L.	University of Rochester
Mattar, F. P.	Polytechnic Institute of New York
McCall, S. L.	Bell Telephone Laboratories, Murray Hill
McNeil, K. J.	University of Waikato, New Zealand
Meier, D.	Physik-Institut, Universität Zürich, Switzerland
Meystre, P.	Max-Planck-Gesellschaft zur Förderung der Wissenschaften e.v., Projektgruppe für Laserforschung, West Germany
Miller, D. A. B.	Heriot-Watt University, United Kingdom
Narducci, L. M.	Drexel University

Pistelli, E.	Instituto di Fisica, dell'Universitá di Pisa, Italy
Roy, R.	University of Rochester
Sandle, W. J.	University of Otago, New Zealand
Schenzle, A.	Universität Essen, Federal Republic of Germany
Schieve, W. C.	University of Texas at Austin
Seaton, C. T.	Heriot-Watt University, United Kingdom
Seibert, E. J.	Drexel University
Senitzky, I. R.	Technion, Haifa, Israel
Shakir, S. A.	Optical Sciences Center, University of Arizona
Shenoy, S. R.	University of Hyderabad, India
Singh, S.	University of Rochester
Smith, P. W.	Bell Telephone Laboratories, Holmdel
Smith, S. D.	Heriot-Watt University, United Kingdom
Thompson, B. V.	University of Manchester, United Kingdom
Tomlinson, W. J.	Bell Telephone Laboratories, Holmdel
Venkatesan, T. N. C.	Bell Telephone Laboratories, Murray Hill
Walls, D. F.	University of Waikato, New Zealand
Willis, C. R.	Boston University
Yuan, J. M.	Drexel University
Zurek, W.	University of Texas at Austin